AF412315

Advances in the Computer Simulations
of Liquid Crystals

NATO Science Series

A Series presenting the results of activities sponsored by the NATO Science Committee. The Series is published by IOS Press and Kluwer Academic Publishers, in conjunction with the NATO Scientific Affairs Division.

A.	**Life Sciences**	IOS Press
B.	**Physics**	Kluwer Academic Publishers
C.	**Mathematical and Physical Sciences**	Kluwer Academic Publishers
D.	**Behavioural and Social Sciences**	Kluwer Academic Publishers
E.	**Applied Sciences**	Kluwer Academic Publishers
F.	**Computer and Systems Sciences**	IOS Press
1.	**Disarmament Technologies**	Kluwer Academic Publishers
2.	**Environmental Security**	Kluwer Academic Publishers
3.	**High Technology**	Kluwer Academic Publishers
4.	**Science and Technology Policy**	IOS Press
5.	**Computer Networking**	IOS Press

NATO-PCO-DATA BASE

The NATO Science Series continues the series of books published formerly in the NATO ASI Series. An electronic index to the NATO ASI Series provides full bibliographical references (with keywords and/or abstracts) to more than 50000 contributions from international scientists published in all sections of the NATO ASI Series.
Access to the NATO-PCO-DATA BASE is possible via CD-ROM "NATO-PCO-DATA BASE" with user-friendly retrieval software in English, French and German (WTV GmbH and DATAWARE Technologies Inc. 1989).

The CD-ROM of the NATO ASI Series can be ordered from: PCO, Overijse, Belgium

Series C: Mathematical and Physical Sciences – Vol. 545

Advances in the Computer Simulatons of Liquid Crystals

edited by

Paolo Pasini

Istituto Nazionale di Fisica Nucleare,
Sezione di Bologna,
Bologna, Italy

and

Claudio Zannoni

Dipartimento di Chimica Fisica ed Inorganica,
Università di Bologna,
Bologna, Italy

Kluwer Academic Publishers

Dordrecht / Boston / London

Published in cooperation with NATO Scientific Affairs Division

Proceedings of the NATO Advanced Study Institute on
Advances in the Computer Simulations of Liquid Crystals
Erice, Italy
11-21 June 1998

A C.I.P. Catalogue record for this book is available from the Library of Congress.

ISBN 0-7923-6098-2

Published by Kluwer Academic Publishers,
P.O. Box 17, 3300 AA Dordrecht, The Netherlands.

Sold and distributed in North, Central and South America
by Kluwer Academic Publishers,
101 Philip Drive, Norwell, MA 02061, U.S.A.

In all other countries, sold and distributed
by Kluwer Academic Publishers,
P.O. Box 322, 3300 AH Dordrecht, The Netherlands.

Printed on acid-free paper

CONTENTS

5   Liquid crystal lattice models I. Bulk systems   99

P. Pasini, C. Chiccoli and C. Zannoni

6   Liquid crystal lattice models II. Confined systems   121

P. Pasini, C. Chiccoli and C. Zannoni

7   Computer simulation of lyotropic liquid crystals as models of biological membranes   139

O.G. Mouritsen

8 Flow properties and structure of anisotropic fluids studied by non-equilibrium molecular dynamics, and flow properties of other complex fluids: polymeric liquids, ferro-fluids and magneto-rheological fluids

189

S. Hess

9   Self atom-atom empirical potentials for the static and dynamic simulation of condensed phases   235

A. Gavezzotti and G. Filippini

13 Parallel molecular dynamics techniques for the simulation of anisotropic systems   389

M.R. Wilson

PREFACE

Computer Simulations provide an essential set of tools for the understanding of the macroscopic properties of liquid crystals and of their phase transitions in terms of molecular models. While simulations of liquid crystals are based on the same general Monte Carlo and Molecular Dynamics techniques used for other fluids, they present a number of specific problems and peculiarities connected with the intrinsic properties of these mesophases. An example is the need to develop suitable algorithms to calculate anisotropic static properties such as order parameters and tensorial observables as well as dynamic quantities like diffusion tensors, viscosities, susceptivities etc. Another series of problems is connected with the need of predicting the properties of liquid crystals from molecular models. This involves the determination of phase transitions and their characteristics, and requires on one hand the development of intermolecular potentials for modelling the essential molecular features of mesogens, and on the other performing large scale simulations, with a number of particles often an order of magnitude greater than those used for simple fluids. Such large numbers require exploiting state of the art resources in computing, and particularly parallel techniques with the development of appropriate algorithms. Further peculiarities of liquid crystals, compared to ordinary isotropic fluids, are their topological defects and it is now becoming feasible to simulate their core structure and optical textures. Similar methods can even be used to perform direct microscopic level simulations of simple devices and displays.

The field of computer simulations of anisotropic fluids is clearly interdisciplinary and evolving very rapidly and this has brought us to organize the NATO Advanced School Institute (ASI) on Advances in the Computer Simulations of Liquid Crystals, held at the Ettore Majorana Centre of Scientific culture (CCSEM) in Erice, Sicily from June 11th to June 21st, 1998. The School, that has tried to cover all the above issues considering various techniques and model systems (from lattice to hard particle and Gay-Berne up to atomistic) for thermotropics, lyotropics and some liquid crystals of biological interest, has had an excellent panel of international Lecturers and has attracted over 70 very keen students and young researchers from different disciplines and 16 countries. The successful organization of the NATO ASI was only possible thanks to the help of many people and we would like xiv

to mention particularly our collaborators R. Berardi, A. Porreca, S. Orlandi and the competent and helpful staff of CCSEM. A particular thought is due to the memory of Dr. Alberto Gabriele, in charge of organization at the Centre, whose efficiency together with kindness, courtesy and dedication to CCSEM we had the fortune to appreciate.

The present book contains a large part of the lectures given at the NATO ASI starting with some introductory papers followed by more applicative and technical chapters. We believe it provides a timely account of the techniques and problems in the field and we hope it will convey some of the stimulating atmosphere of the School.

We are particularly grateful to the Lecturers, the members of the Scientific Committee, Professors N.A. Clark, G.R. Luckhurst, D. Frenkel and to the CCSEM Director, Professor A. Zichichi. We wish to thank NATO and its Science Committee for their essential support which made the organization of the School and the production of this book possible. We also acknowledge financial contributions from Ente per le Nuove tecnologie, l'Energia e l'Ambiente (ENEA), Consorzio Interuniversitario per la Chimica dei Materiali (now INSTM), TECDIS S.p.A., Consiglio Nazionale delle Ricerche (CNR) and Quadrics Supercomputers World (QSW).

Paolo Pasini
INFN, Bologna

Claudio Zannoni
Università di Bologna

INTRODUCTION TO SIMULATIONS AND STATISTICAL MECHANICS

MICHAEL P. ALLEN
University of Bristol
H. H. Wills Physics Laboratory
Royal Fort
Tyndall Avenue
Bristol BS8 1TL
United Kingdom

Abstract. This chapter is intended as a very basic introduction to statistical mechanics and to computer simulation, in the context of understanding liquids and liquid crystals. Reference is made to the ideas of thermodynamic ensembles and averages, distribution functions, and ensemble conversion. Then space and time correlation functions are defined, emphasizing their relation to experimentally measurable quantities such as spectra and transport coefficients, through linear response theory. Lastly, some technical aspects of Monte Carlo simulation are mentioned, namely importance sampling and the Metropolis method, and, very briefly, the kinds of algorithms used in molecular dynamics.

1. Introduction

Computer simulation and statistical mechanics are intertwined. If it is desired to do any more with a simulation than present pictures of molecular configurations, then an understanding of statistical mechanics will be needed: this provides the link between (time-resolved) atomic coordinates and measurable properties. Also in order to develop a new, possibly more efficient, simulation method, in a correct way, a proper understanding of the underlying statistical mechanics will be essential. This link works the other way too. Increasingly, statistical mechanical theories rely on support from simulation for their underlying assumptions and their final predictions. In the area of liquid crystal research, both the statistical mechanics and the

1

P. Pasini and C. Zannoni (eds.), Advances in the Computer Simulations of Liquid Crystals, 1–16.
© 2000 *Kluwer Academic Publishers. Printed in the Netherlands.*

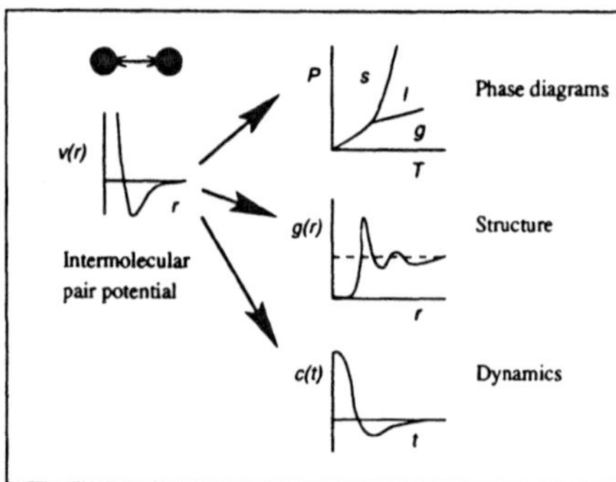

Figure 1. Simulations produce thermodynamic, structural, and dynamical information from molecular parameters.

simulation techniques tend to be challenging: many of the phenomena occur over long time and length scales, and the reduced symmetry of these phases makes the description of structure and dynamics quite complicated. Moreover, the constituent molecules themselves are expensive and difficult to model in a realistic way: frequently they have both flexible and rigid parts, both polar and polarizable groups, and of course they tend to be highly non-spherical. None of this should dissuade the reader from attempting to make a contribution in the area: progress has been rapid over the last few years and shows every indication of accelerating.

Computer simulation and statistical mechanics are both suitable subjects for lengthy postgraduate lecture courses, and in this short chapter it will not be possible to give any more than a superficial overview. Hopefully this will set the scene for later chapters, in which other authors will give much more detail. Also the following books [1–4] (amongst many other excellent ones) are recommended as starting points for further study.

1.1. MICROSCOPIC AND MACROSCOPIC

Computer simulations act as a bridge between microscopic length and time scales and the macroscopic world of the laboratory. We provide a guess at the interactions between molecules and obtain predictions of bulk properties which are 'exact' (subject to computer budget). Simulations then reveal the hidden detail behind bulk properties. Examples are the link between the diffusion coefficient and velocity autocorrelation function (the former easy to measure experimentally, the latter much harder) and the connection between equations of state, structural correlation functions, and phase transitions.

Simulations act as a bridge in another sense: between theory and experiment. One may test a theory using idealized models, conduct 'thought

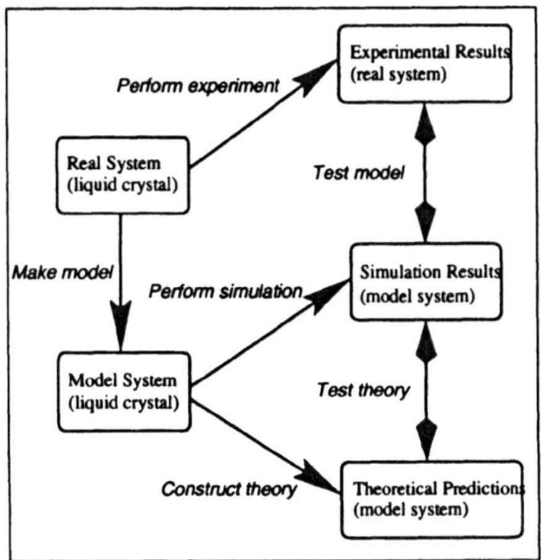

Figure 2. Theory, simulation and experiment

experiments' and perform 'ab initio' computer simulations. A well-focused simulation may help one understand what is measured in the laboratory, and test a postulated explanation at a fundamental level. Further clarification may result from carrying out simulations on the computer that are difficult or impossible in the laboratory (for example, working at extremes of temperature or pressure, or applying an unphysical perturbing force). Ultimately one may wish to make direct comparisons with particular experimental measurements on real materials, in which case a good model of molecular interactions is essential. The ultimate aim of so-called *ab initio* molecular dynamics is to reduce the amount of fitting and guesswork in this process to a minimum, while providing an accurate description of the molecular framework and electronic properties. In view of the role of liquid crystals in electronic devices, and the sensitivity of their properties to molecular details, this is an important aim. On the other hand, one may be interested in phenomena of a rather generic nature, or simply wish to discriminate between good and bad theories. When it comes to aims of this kind, it is not necessary to have a perfectly realistic molecular model; one that contains just the essential physics may be quite suitable.

1.2. MOLECULAR DYNAMICS AND MONTE CARLO

Molecular dynamics consists of the brute-force solution of Newton's equations of motion. It is necessary to encode in the program the potential energy and force law of interaction between molecules; the equations of motion are solved step-by-step. Advantages of the technique are that this corresponds closely to what happens in 'real life'; the method gives thermo-

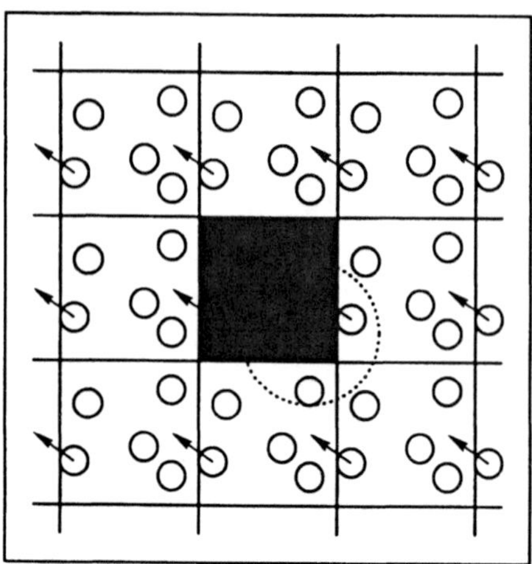

Figure 3. Periodic boundary conditions

dynamic, structural, and dynamical properties; it may allow efficient relaxation of 'collective modes'; and standard packaged routines are available. Monte Carlo can be thought of as a prescription for sampling configurations from a statistical ensemble. The interaction potential energy is coded into the program, and a procedure adopted to go from one state of the system to the next. Advantages of this technique are that it is easy to program; relatively simple to specify external conditions $(T, P$ etc.); and many tricks may be devised to improve efficiency.

Simulation time scales are typically short, $t \sim 10^3$–10^6 MD steps or MC sweeps, corresponding to perhaps a few nanoseconds of real time. It is always necessary to test whether equilibrium has been reached, and one needs to subject simulation averages to a statistical analysis, designed to make a realistic assessment of the error bars, just as in a real experiment. Simulation length scales are also typically small, with system sizes $N \sim 10^3$–10^6 particles. This corresponds to nanometre – submicron phenomena. It is often necessary to do a system-size analysis of simulation results, to quantify these effects, and to take special measures to avoid domination by surface properties, unless these are of intrinsic interest. Usually simulations employ periodic boundary conditions, illustrated in Fig. 3. The system is represented by a finite simulation cell, often cuboidal, surrounded by replicas of itself. In the course of the simulation, if an atom leaves the basic simulation box, attention can be switched to the incoming image. For short-ranged forces the minimum image convention is applied: each atom interacts with the nearest atom or image in the periodic array. For long-ranged forces, e.g. the Coulomb interaction, special measures such as the Ewald sum must be used.

2. Statistical mechanics and why we need it

The formal mechanism of statistical mechanics enables the calculation of bulk properties from microscopic details (exactly or approximately). For example, we use thermodynamic relations to investigate equations of state and order parameters; we describe liquid-state structure using spatial correlation functions which may be related to diffraction experiments; and we investigate dynamical properties through time correlation functions which may be compared with spectroscopy and non-equilibrium experiments.

Statistical mechanics also enables us to invent new simulation methods and ensure their correctness. This is especially important when we need more efficient sampling schemes, for example Monte Carlo methods to simulate polymers and colloids, to calculate free energies, to study phase transitions, and to investigate rare events and quantum effects.

Lastly, statistical mechanics enables us to construct theories of simple and complex liquids. It may be useful to refer to Refs [1, 5, 6], and examples of theoretical models of liquid crystals will appear in the other chapters of this volume.

3. Ensembles and simulation

The usual starting point for statistical mechanics is the canonical ensemble partition function

$$Q_{NVT} = \int d\Gamma \; e^{-\beta E(\Gamma)} = \int dE \; \Omega_{NVE} \; e^{-\beta E} . \qquad (1)$$

Here Γ stands for the microscopic coordinates and momenta of all the particles, $E(\Gamma)$ the energy of a particular configuration of atoms (kinetic and potential), and $\beta = 1/k_{\mathrm{B}}T$. This ensemble represents a system at constant volume V and number of particles N, in thermal equilibrium with a heat bath at temperature T. The second form of the equation is an integral over energies involving the density of states Ω_{NVE} at each energy

$$\Omega_{NVE} = \int d\Gamma \; \delta(E - E(\Gamma))$$

where E is the chosen energy, and $\delta(\ldots)$ is the Dirac delta function. To simplify notation, multiplying prefactors (e.g. involving Planck's constant and corrections for indistinguishability) are omitted. The most common form of Monte Carlo simulation samples states from the canonical probability distribution

$$\varrho_{NVT}(\Gamma) = Q_{NVT}^{-1} \; e^{-\beta E(\Gamma)} .$$

Molecular dynamics, typically, probes the microcanonical ensemble,

$$\varrho_{NVE}(\Gamma) = \Omega_{NVE}^{-1} \; \delta(E - E(\Gamma))$$

corresponding to an isolated system of constant energy E. Other ensembles (in both MC and MD) may be used to allow variations of V (at constant pressure) or N (at constant chemical potential).

3.1. THERMODYNAMIC QUANTITIES

The connection with thermodynamics is made by

$$F = -k_{\mathrm{B}} T \ln Q_{NVT}$$

where F is the Helmholtz free energy. It is also possible to write many quantities as 'ensemble averages', involving $\varrho_{NVT}(\Gamma)$, and these in turn become ratios of two sums or integrals over states:

$$\langle A \rangle_{NVT} = \int \mathrm{d}\Gamma \; \varrho_{NVT}(\Gamma) \; A(\Gamma) = \frac{\int \mathrm{d}\Gamma \; A(\Gamma) \mathrm{e}^{-\beta E(\Gamma)}}{\int \mathrm{d}\Gamma \; \mathrm{e}^{-\beta E(\Gamma)}} .$$

For classical systems, Q_{NVT} may be factorized into ideal and excess parts, the free energy written $F = F^{\mathrm{id}} + F^{\mathrm{ex}}$, and many other quantities decomposed in a similar way.

Simple manipulations give expressions for common thermodynamic functions. Differentiation with respect to β gives

$$\begin{aligned}
E &= \left(\frac{\partial \beta F}{\partial \beta} \right) = -\left(\frac{\partial \ln Q_{NVT}}{\partial \beta} \right) = -Q_{NVT}^{-1} \frac{\partial}{\partial \beta} \int \mathrm{d}\Gamma \; \mathrm{e}^{-\beta E(\Gamma)} \\
&= \int \mathrm{d}\Gamma \; \varrho_{NVT}(\Gamma) \; E(\Gamma) = \langle E \rangle \\
k_{\mathrm{B}} T^2 C_V &= \langle E^2 \rangle - \langle E \rangle^2 = \left\langle \delta E^2 \right\rangle .
\end{aligned}$$

The last relation, between heat capacity and energy fluctuations, also shows that the latter are small $\sqrt{\langle \delta E^2 \rangle} \sim \mathcal{O}(\sqrt{N})$ compared with extensive quantities such as the energy $E \sim \mathcal{O}(N)$. Differentiating with respect to volume gives

$$\begin{aligned}
\beta P &= -\left(\frac{\partial \beta F}{\partial V} \right) = \left(\frac{\partial \ln Q_{NVT}}{\partial V} \right) = Q_{NVT}^{-1} \frac{\partial}{\partial V} \int \mathrm{d}\Gamma \; \mathrm{e}^{-\beta E(\Gamma)} \\
\Rightarrow \quad PV &= N k_{\mathrm{B}} T - \frac{1}{3} \left\langle \sum_i \sum_{j \neq i} w_{ij} \right\rangle .
\end{aligned}$$

Here we assumed a pairwise additive potential energy $\mathcal{V} = \sum_i \sum_{j \neq i} v_{ij}$, defined the virial function $w(r) = r(\mathrm{d}v(r)/\mathrm{d}r)$, and calculated the ideal gas part exactly. The result is obtained by introducing scaled coordinates

$\mathbf{r}_i \equiv L\mathbf{s}_i$, $V = L^3$ into the integral. Differentiating the free energy with respect to number gives

$$\beta\mu = \left(\frac{\partial \beta F}{\partial N}\right) = -\left(\frac{\partial \ln Q_{NVT}}{\partial N}\right) \approx -\ln\left(\frac{Q_{N+1}}{Q_N}\right)$$

which yields the Widom test-particle formula for the excess chemical potential

$$\mu^{\mathrm{ex}} = -k_{\mathrm{B}}T \ln \int \mathrm{d}\mathbf{s}_{N+1} \left\langle \mathrm{e}^{-\beta\Delta\mathcal{V}_{N+1}} \right\rangle = -k_{\mathrm{B}}T \ln \left\langle\left\langle \mathrm{e}^{-\beta\Delta\mathcal{V}_{N+1}} \right\rangle\right\rangle .$$

Here the double angle brackets indicate both a simulation average and an unweighted average over the position of the extra particle.

3.2. PROBABILITY DISTRIBUTIONS

The probability distribution function for the energy may be written as the average:

$$\mathcal{P}_{NVT}(E) \;=\; \langle\delta(E(\Gamma) - E)\rangle_{NVT} = \frac{\Omega(E)\,\mathrm{e}^{-E/k_{\mathrm{B}}T}}{Q_{NVT}} .$$

We have abbreviated $\Omega(E) \equiv \Omega_{NVE}$. Since the density of states is related to the entropy by $S = k_{\mathrm{B}} \ln\Omega$, $\Omega(E)$ is a very rapidly increasing function, but the $\mathrm{e}^{-E/k_{\mathrm{B}}T}$ factor in $\mathcal{P}_{NVT}(E)$ cuts this off leaving an extremely sharp peak. This also applies to the integrand of the integration over energies appearing in eqn (1). In a first approximation, neglecting energy fluctuations and counting only the contribution of the peak, this equation gives $Q_{NVT} \approx \Omega_{NVE}\mathrm{e}^{-\beta E}$ from which the thermodynamic relation $F = -k_{\mathrm{B}}T \ln Q = E - TS$ follows; this illustrates the equivalence of ensembles. At the next level of approximation, energy fluctuations giving the width of $\mathcal{P}_{NVT}(E)$ (dictated by the heat capacity, as we have seen) are taken into account, leading to small correction factors in ensemble conversion formulae for free energies, ensemble averages and other quantities. Away from phase transitions, $\mathcal{P}_{NVT}(E)$ may be represented by a Gaussian fun ction; near phase transitions this breaks down. Typically, near a continuous phase transition, fluctuations produce a non-Gaussian form, while near a first-order phase transition the function may become doubly peaked.

3.3. TIME AND LENGTH SCALES

Suppose one is interested in a variable a, defined such that $\langle a \rangle = 0$. The time correlation function $\langle a(t_0)a(t_0 + t)\rangle$ relates values calculated at times

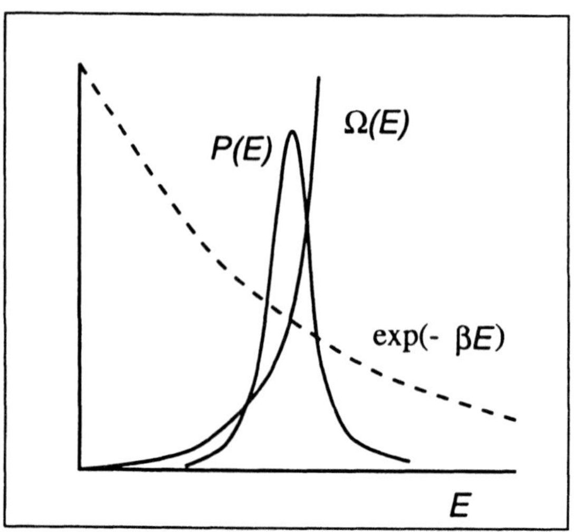

Figure 4. Density of states $\Omega(E)$ and Boltzmann factor combine to give a sharply peaked canonical energy distribution $\mathcal{P}_{NVT}(E)$.

t apart; assuming that the system is in equilibrium, this function is independent of the choice of time origin and may be written $\langle a(0)a(t)\rangle$. It will decay from an initial value $\langle a(0)a(0)\rangle \equiv \langle a^2\rangle$ to a long-time limiting value

$$\lim_{t\to\infty} \langle a(0)a(t)\rangle = \lim_{t\to\infty} \langle a(0)\rangle \langle a(t)\rangle = 0$$

as the variables $a(0)$ and $a(t)$ become uncorrelated; this decay occurs over a characteristic time τ_a. Formally we may define a correlation time

$$\tau_a = \int_0^\infty dt\, \langle a(0)a(t)\rangle / \langle a^2\rangle .$$

Alternatively, if time correlations decay exponentially at long time, τ_a may be identified approximately from the limiting form

$$\langle a(0)a(t)\rangle \propto \exp\{-t/\tau_a\} .$$

Similarly, define a spatial correlation function $\langle a(0)a(r)\rangle$ relating values computed at different points r apart. Spatial isotropy allows us to write this as a function of the distance between the points, r, rather than the vector $\mathbf{r}$: note that this symmetry is broken in a liquid crystal. Spatial homogeneity, which applies to simple liquids (but not to solids or smectic liquid crystals) allows us to omit any reference to an absolute origin of coordinates. This function decays from a short-range nonzero value to zero over a characteristic distance ξ_a, the correlation length.

It is almost essential for simulation box sizes L to be large compared with ξ_a, and for simulation run lengths τ to be large compared with τ_a, for all properties of interest a. Only then can we guarantee that reliably-sampled statistical properties are obtained. Roughly speaking, the statistical error in a property calculated as an average over a simulation run of length τ is proportional to $\sqrt{\tau_a/\tau}$: the time average is essentially a sum of $\sim \tau/\tau_a$ independent quantities, each an average over time τ_a. Within the time periods τ_a, values of a are highly correlated. A similar statement can be made about properties which are effectively spatial averages over the simulation box volume L^3: root-mean-square variations of such averages are proportional to $\sqrt{(\xi_a/L)^3}$. This means that collective, system-wide properties deviate by only a relatively small amount from their thermodynamic, large-system, limiting values; the deviation becomes smaller as the averaging volume increases, and is also determined by the correlation length.

Near critical points, special care must be taken, in that these inequalities will almost certainly not be satisfied, and indeed one may see the onset of non-exponential decay of the correlation functions. In these circumstances a quantitative investigation of finite size effects and correlation times, with some consideration of the appropriate scaling laws, must be undertaken. Liquid-crystal phase diagrams often include continuous phase transitions, or weakly first-order transitions exhibiting significant pre-transitional fluctuations.

3.4. CORRELATION FUNCTIONS AND EXPERIMENT

The importance of correlation functions lies in their relation to experimental measurements, and to the definition of transport coefficients. Neutrons and X-rays, for instance, scatter from density fluctuations. Defining Fourier components of the density

$$\hat{\rho}(\mathbf{k}) = \int d\mathbf{r}\ e^{-i\mathbf{k}\cdot\mathbf{r}}\ \rho(\mathbf{r}) = \sum_i e^{-i\mathbf{k}\cdot\mathbf{r}_i}$$

the relevant quantity for an isotropic, atomic, system is the structure factor

$$\begin{aligned}
S(\mathbf{k}) &= N^{-1}\langle \hat{\rho}(\mathbf{k})\hat{\rho}(-\mathbf{k})\rangle \\
S(k) &= 1 + 4\pi\rho \int d\mathbf{r}\ r^2 g(r)\frac{\sin kr}{kr}.
\end{aligned}$$

The structure factor, or the pair distribution function $g(r)$, may be calculated in a simulation. The above formulae apply to atomic liquids; for molecular liquids, there is an additional dependence on orientations [7] and for liquid crystals the single-particle distribution function is non-trivial, so the corresponding formulae become more complicated.

Spectroscopy and scattering experiments involve applying a perturbation which varies with time, typically in a periodic way, and measuring the system's response. For example, the electric vector $\mathbf{E}(t)$ of electromagnetic radiation couples to the dipole moment $\mathbf{M}$ through a term $\mathbf{E} \cdot \mathbf{M}$ in the hamiltonian. Applying a time-varying electric field $E_x(t) = E_0 \cos \omega t$, linear response theory may be used to relate the absorption of energy to the Fourier transform of $C(t) = \langle M_x(0) M_x(t) \rangle$, evaluated at frequency ω. More detailed aspects of linear response theory, sometimes called Green-Kubo theory, will be treated in the chapter of Zannoni. In the following section, the essential ideas and the relationship with transport processes will be summarized very briefly.

3.5. TRANSPORT COEFFICIENTS

A perturbation applied to the system is represented by an extra term in the hamiltonian, and a corresponding change in the distribution function ϱ:

$$\begin{aligned}
\mathcal{H}_A &= \mathcal{H} + \Delta\mathcal{H}_A = \mathcal{H} - A(\Gamma) F_A(t) \\
\varrho_A(\Gamma, t) &= \varrho(\Gamma) + \Delta\varrho_A(\Gamma, t) .
\end{aligned}$$

The perturbation couples through a dynamical variable $A(\Gamma)$, multiplied by a time-varying external field $F_A(t)$. In the classical canonical ensemble, the nonequilibrium linear response in a variable B may be written

$$\langle B(t) \rangle_A = \beta \int_0^t \mathrm{d}t' \, F_A(t') \, \langle B(t - t') \dot{A} \rangle$$

where we assume that the perturbation is switched on at $t = 0$. On the right we have an equilibrium time correlation function; $\dot{A}$ is the time-derivative of A. For the special case that the perturbation is a constant in time, and examining the response in the limit $t \to \infty$, the integration may be cast into the following form

$$\begin{aligned}
\langle B(\infty) \rangle_A &= \beta F_A \int_0^\infty \mathrm{d}t'' \left\langle B(t'') \dot{A} \right\rangle = -\beta F_A \int_0^\infty \mathrm{d}t'' \left\langle \dot{B}(t'') A \right\rangle \\
&= -\beta F_A \left\langle (B(\infty) - B(0)) A \right\rangle .
\end{aligned} \tag{2}$$

Making the assumption that $\langle B(\infty) A \rangle = 0$ this yields the well-known static result $\langle B \rangle_A = \beta F_A \langle AB \rangle$.

However, for suitably chosen perturbations and responses, things work out differently, and a transport property may be calculated. Consider a dilute solution of ions, and focus on one ion. Choose "A" in our linear response theory to be $Q r_{ix}$ where Q is the charge and r_{ix} is the x component of the position of the chosen particle i. Choose "F_A" to be E_x, an electric field applied in the x direction, applied at time $t = 0$ and constant thereafter. Then $\Delta \mathcal{H}_A = Q r_{ix} E_x$. Measure, as "$B$", the x component of the induced current $I(t) = Q \dot{r}_{ix}(t) = Q v_{ix}(t)$. So

$$\langle I(t) \rangle_E = \frac{Q^2 E}{k_B T} \int_0^t dt' \ \langle v_{ix} v_{ix}(t') \rangle$$

and at long times $t \to \infty$ the steady-state current is

$$\langle I(\infty) \rangle_E = \frac{Q^2 E}{k_B T} \int_0^\infty dt \ \langle v_{ix} v_{ix}(t) \rangle \ .$$

The essential point here is that the time integral must be treated differently from the one in Eqn. (2). Instead of a static quantity $\langle r_{ix} v_{ix} \rangle$ (which is meaningless, as r_{ix} is unbounded) it gives a finite, non-zero value dictated by the dynamics

$$\langle (r_{ix}(\infty) - r_{ix}(0)) v_{ix} \rangle = \langle \Delta r_{ix}(\infty) v_{ix} \rangle \ .$$

We can identify the conductivity σ

$$\text{Conductivity} = \frac{\text{current}}{\text{field}} \qquad \Longrightarrow \qquad \sigma = \frac{Q^2}{k_B T} \int_0^\infty dt \ \langle v_{ix} v_{ix}(t) \rangle \ .$$

However the charge and the field just provide a convenient physical 'handle' on the chosen particle. We have equally well calculated its mobility μ

$$\text{Mobility} = \frac{\text{drift velocity}}{\text{force}} \qquad \Longrightarrow \qquad \mu = \frac{1}{k_B T} \int_0^\infty dt \ \langle v_{ix} v_{ix}(t) \rangle \ .$$

In a real system of ions at high concentration the total current would involve all of them, the applied field would couple to all of them, and the correlation function would involve self- and cross-terms.

Mobility is related to diffusion $\mu = D/k_B T$. In one, a deterministic force is applied to produce a drift; in the other, random thermal forces operate. At long times, the mean-square displacement of an atom increases linearly in time, according to Einstein's law of diffusion.

$$\lim_{t \to \infty} \langle |\mathbf{r}_{ix}(t) - \mathbf{r}_{ix}(0)|^2 \rangle = 6Dt \qquad \text{or} \qquad \lim_{t \to \infty} \langle (r_{ix}(t) - r_{ix}(0))^2 \rangle = 2Dt \ .$$

Define the diffusion coefficient by

$$D = \frac{1}{2} \lim_{t \to \infty} \frac{d}{dt} \left\langle (r_{ix}(t) - r_{ix}(0))^2 \right\rangle = \frac{1}{2} \lim_{t \to \infty} \frac{d}{dt} \left\langle \Delta r_{ix}(t)^2 \right\rangle \ . \tag{3}$$

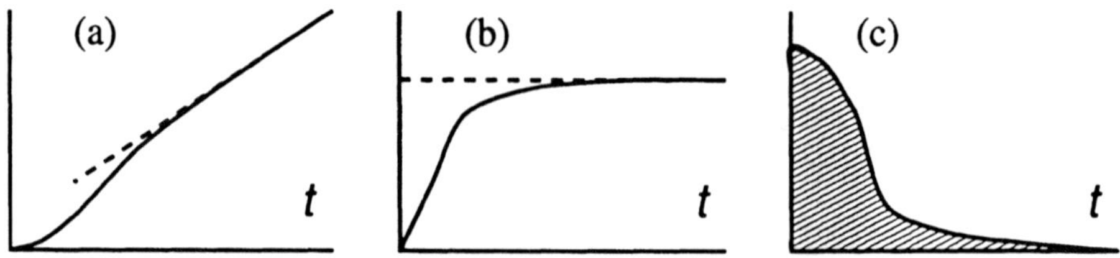

Figure 5. The diffusion coefficient may be evaluated by measuring: (a) the limiting long-time gradient of the mean-squared displacement, eqn. (3); (b) the limiting long-time value of the correlation between displacement and velocity, eqn. (5); (c) the limiting long-time integral of the velocity autocorrelation function, eqn. (4).

Using $r_{ix}(t) = r_{ix}(0) + \int_0^t dt' v_{ix}(t')$ it is easy to relate D to the velocity autocorrelation function $\langle v_{ix}(0)v_{ix}(t)\rangle$ and hence the mobility.

$$D = \int_0^\infty dt\ \langle v_{ix}(0)v_{ix}(t)\rangle = \mu k_{\mathrm{B}}T\,. \tag{4}$$

Performing the integral as indicated earlier gives another expression for D

$$D = \lim_{t\to\infty} \langle v_{ix}(0)\left(r_{ix}(t) - r_{ix}(0)\right)\rangle = \lim_{t\to\infty} \langle v_{ix}(0)\Delta r_{ix}(t)\rangle \tag{5}$$

Equations (3)–(5) are all mathematically equivalent, and any of the three may be used in a simulation. The different methods of arriving at D are illustrated in Fig. 5. The important point, in all cases, is to ensure that the calculation is extended to sufficiently long time t.

Similar relationships may be written down for shear and bulk viscosities and thermal conductivity. The same kind of expressions apply to liquid crystals, with additional complications arising from reduced symmetry, the possibly of coupling to director rotation, and orientational elasticity [8, 9]. More details of the underlying relationship between transport coefficients and conservation relations, which play a subtle role in determining whether or not integrals in expressions of this kind vanish, may be found elsewhere [1].

4. Sampling and summing over states

Summing over all the states in the partition function is impractical: there are too many, for a system of more than a few particles. This can easily be deduced from the relations $\Omega_{NVE} = e^{S/k_{\mathrm{B}}}$ and $Q_{NVT} = e^{-F/k_{\mathrm{B}}T}$ (recalling that S and F are extensive quantities) or by considering a simple system of N Ising-like spins (which will have 2^N microstates in all). Instead, in Monte Carlo simulation, states are sampled from the appropriate distribution (i.e. the higher-probability states occur more often). This allows the effort to be concentrated on the *important* states, the ones which make a significant contribution to the sum.

Suppose we devise a mechanism to select states in a biased way such that the probability of observing a state Γ_m is proportional to $\varrho_m = \varrho(\Gamma_m)$ (this is easier than it sounds, see below). Then a sequence of N_t 'observations' or 'steps' may be used to produce an average which, in the limit of a long run, will be equivalent to the ensemble average

$$\langle A \rangle_{NVT} = \lim_{N_t \to \infty} \frac{1}{N_t} \sum_{t=1}^{N_t} A_t \,.$$

The Boltzmann weight appears *implicitly* in the choice of states. This is formally like a time average as calculated in molecular dynamics. A limitation of this approach is that the partition function, and hence the free energy, cannot easily be calculated in this way.

The importance sampling method involves designing a stochastic algorithm for stepping from one state of the system to the next. Consider an *ensemble* of systems all evolving at once. Specify a matrix π whose elements $\pi_{m \to n}$ give the probability of going from state Γ_m to state Γ_n, for every m and n. π must satisfy $\sum_n \pi_{m \to n} = 1$ for all m since a state must go somewhere. At each step, jumps are implemented with this transition matrix. This generates a *Markov chain* of states. Feller's theorem [10, 11] then applies: subject to some reasonable conditions, there exists a limiting (equilibrium) distribution of states and the system will tend towards this limiting distribution.

Designing a matrix π, to give the limiting distribution

$$\varrho_m \propto \exp\{-\beta E(\Gamma_m)\}$$

is relatively straightforward. Consider an arbitrary initial ensemble, defined by a probability for each state m, $\varrho_m^{(0)}$. One application of the transition matrix gives a new distribution $\varrho_m^{(1)}$. Writing this in vector form, and iterating, gives

$$\varrho^{(1)} = \varrho^{(0)} \cdot \pi \,, \qquad \varrho^{(2)} = \varrho^{(1)} \cdot \pi \,, \qquad \cdots \qquad \varrho^{(t)} = \varrho^{(t-1)} \cdot \pi \,.$$

The limiting distribution as $t \to \infty$, $\varrho^{(\infty)} \equiv \varrho$, will satisfy

$$\varrho^{(\infty)} = \varrho^{(\infty)} \cdot \pi \qquad \text{or simply} \qquad \varrho = \varrho \cdot \pi \,,$$

an eigenvalue equation, *independent* of the starting distribution.

Writing the eigenvalue equation out in full,

$$(\varrho \cdot \pi)_m = \sum_n \varrho_n \pi_{n \to m} = \sum_n \varrho_m \pi_{m \to n} = \varrho_m$$

we see that the left side is the rate of arrival of systems at state m from everywhere, and the right side is the rate of departure from m to everywhere. The truth of the above equation may be guaranteed by imposing *microscopic reversibility*

$$\varrho_n \pi_{n \to m} = \varrho_m \pi_{m \to n}$$

so that each term in the sum on the left matches the corresponding term on the right. This choice is not essential, but it is a simple one. There are still many ways to choose $\boldsymbol{\pi}$. The Metropolis approach [12] is to write the elements of $\boldsymbol{\pi}$ as follows:

$$
\begin{aligned}
\pi_{m \to n} &= \alpha_{m \to n} & \varrho_n \geq \varrho_m & \qquad m \neq n \\
\pi_{m \to n} &= \alpha_{m \to n}(\varrho_n / \varrho_m) & \varrho_n < \varrho_m & \qquad m \neq n \\
\pi_{m \to m} &= 1 - \textstyle\sum_{n \neq m} \pi_{m \to n} \, .
\end{aligned}
$$

$\boldsymbol{\alpha}$ is the underlying matrix: essentially dictating the probability of *attempting* a move like $m \to n$. The other factors give the probability of *accepting* such a move. $\boldsymbol{\alpha}$ is symmetrical, $\alpha_{m \to n} = \alpha_{n \to m}$, i.e. the probability for attempting a move to state n (given that the system is currently in state m) should be the same as the probability for attempting a move to state m (if it is in state n).

It is important to be able to prove microscopic reversibility for this scheme (and related schemes). Suppose that $\varrho_n \geq \varrho_m$, without loss of generality. Then the move $m \to n$ is 'downhill': $\pi_{m \to n} = \alpha_{m \to n}$, selected with probability $\alpha_{m \to n}$ and accepted unconditionally. The reverse move $n \to m$ is 'uphill': $\pi_{n \to m} = \alpha_{n \to m}(\varrho_m / \varrho_n)$, selected with the same probability $\alpha_{n \to m} = \alpha_{m \to n}$, but only accepted with probability ϱ_m / ϱ_n. We see that $\varrho_n \pi_{n \to m} = \varrho_m \pi_{m \to n}$ as required.

This scheme only requires a knowledge of the ratio ϱ_n / ϱ_m. It does not require knowledge of the factor normalizing the ϱ's, i.e. the partition function. Standard canonical ensemble Monte Carlo samples *configurational* quantities, so the relevant quantity is the potential energy difference between old and new states, $\varrho_n / \varrho_m = \exp\{-\beta(\mathcal{V}_n - \mathcal{V}_m)\}$. The decision to accept or reject is taken with a random number generator. If the move is rejected, the current state is counted again in the calculation of averages.

A typical move selection for a constant-NVT Monte Carlo simulation of a system of atoms consists of picking an atom at random, and randomly displacing the atom uniformly within a small surrounding cube. The new state n differs slightly from m; a very large number of such states exists, so the value of $\alpha_{m \to n}$ is correspondingly small but we don't need to know it. All that is needed is that it be equal to $\alpha_{n \to m}$, the selection probability for the reverse move. A moment's consideration of the procedure that would be followed should the system be in state n shows that this is indeed the case, provided the same methods are employed for randomly choosing the atom and the displacement. This would not be the case if, for example, we decided to attempt moves more frequently for atoms in a certain region of space (for example, near a solid surface) than those in other regions. Such a choice of α is not symmetric, and hence introduces a bias.

As an example where things are not quite so simple, consider selecting moves for rigid linear molecules: here we make a small translation and rotation. Suppose that we represent molecular orientation with polar angles θ, ϕ. Angular configuration space has a weighting factor $d\Omega = \sin\theta d\theta d\phi$: the $\sin\theta$ factors must be included in $\alpha_{m\to n}$. Uniform sampling of θ and ϕ is wrong and will generate biased distributions; uniform sampling of $\cos\theta$ and ϕ would be one way of generating a correct distribution.

In any case, especially when some new sampling method is being proposed, it is essential to consider the forward and reverse move selection probabilities. For complex fluids it may be advantageous to deliberately introduce some bias into the selection of moves, but in this case it is essential to correct for the bias either by adjusting the move acceptance/rejection procedure, or by introducing a weighting in the simulation average to compensate. Examples appear in some of the other chapters in this volume.

5. Molecular dynamics algorithms

In the molecular dynamics method, the potential $\mathcal{V}(\mathbf{r})$ is specified, and this is analytically differentiated to obtain the forces $\mathbf{f} = -\nabla\mathcal{V}(\mathbf{r})$. These enter Newton's equations of motion $\dot{\mathbf{r}} = \mathbf{v}$, $\dot{\mathbf{v}} = \mathbf{f}/m$, which may be solved in a step-by-step way. The sequence of events for the popular velocity Verlet algorithm is

1. Advance the positions, and 'half-advance' the velocities, using the current values of the forces:

$$\begin{aligned} \mathbf{r}(t+\Delta t) &= \mathbf{r}(t) + \Delta t\, \mathbf{v}(t) + \tfrac{1}{2}\Delta t^2\, \mathbf{f}(t)/m \\ \mathbf{v}(t+\tfrac{1}{2}\Delta t) &= \mathbf{v}(t) + \tfrac{1}{2}\Delta t\, \mathbf{f}(t)/m \end{aligned}$$

2. Calculate the new forces $\mathbf{r}(t+\Delta t) \to \mathbf{f}(t+\Delta t)$.
3. Complete the advancement of the velocities

$$\mathbf{v}(t+\Delta t) = \mathbf{v}(t+\tfrac{1}{2}\Delta t) + \tfrac{1}{2}\Delta t\, \mathbf{f}(t+\Delta t)/m$$

Desirable properties of a simulation algorithm are that it should be reversible and 'area-preserving' (this term refers to the mapping in phase space which occurs during one time step); these properties lead to good

16

long-time energy conservation, allowing a long time step. The Verlet-like algorithms have been shown to satisfy these criteria, being derivable from a Trotter-like factorization of the classical dynamical propagator. This approach leads to many extensions of MD algorithms, including a robust multiple time-step approach; more details appear in the chapter of Procacci in this volume.

An algorithm should also require just one force evaluation per time step, as the force calculation is typically the most expensive part of the operation, involving up to $\frac{1}{2}N(N-1)$ distinct pairs, assuming pairwise additivity of the potential. Applying a potential cutoff reduces this number, but not all parts of the potential are short-ranged, and there are some standard tricks to optimize the efficiency: (i) make a neighbour list for each atom, updated at intervals; (ii) use a domain decomposition approach, dividing space into a lattice of cells; (iii) use multiple time step approaches, with longer timesteps for the longer-ranged, more slowly-varying forces.

Many other authors in this volume will give further details of molecular dynamics methods, especially concentrating on methods to handle long-range forces and parallel algorithms; additionally one may consult the references [1–4].

References

1. Hansen, J.-P. and McDonald, I.R. (1986) *Theory of Simple Liquids*. Academic Press, London, second edition.
2. Frenkel, D. and Smit, B. (1996) *Understanding Molecular Simulation : from Algorithms to Applications*. Academic Press, San Diego, ISBN 0–12–267370–0.
3. Allen, M.P. and Tildesley, D.J. (1989) *Computer Simulation of Liquids*. Clarendon Press, Oxford, paperback edition, ISBN 0–19–855645–4.
4. Binder, K. and Ciccotti, G. (eds.) (1996) *Monte Carlo and Molecular Dynamics of Condensed Matter Systems*, Italian Physical Society, volume 49, Bologna. Proceedings of Euroconference, Como, Italy, July 3–28, 1995, ISBN 88–7794–078–6.
5. Doi, M. and Edwards, S.F. (1986) *The Theory of Polymer Dynamics*. Clarendon Press, Oxford.
6. de Gennes, P.G. and Prost, J. (1995) *The Physics of Liquid Crystals*. Clarendon Press, Oxford, second, paperback edition.
7. Gray, C. and Gubbins, K.E. (1984) *Theory of Molecular Fluids*. Clarendon Press, Oxford.
8. Forster, D. (1974) *Ann. Phys.*, **85**, 505.
9. Forster, D. (1975) Hydrodynamic Fluctuations, Broken Symmetry and Correlation Functions, volume 47 of *Frontiers in Physics*. Benjamin, Reading.
10. Feller, W. (1957) *An Introduction to Probability Theory and its Applications*. Volume 1. Wiley, New York, second edition.
11. Feller, W. (1966) *An Introduction to Probability Theory and its Applications*. Volume 2, Wiley, New York, second edition.
12. Metropolis, N., Rosenbluth, A.W., Rosenbluth, M.N., Teller, A.H. and Teller, E. (1953) *J. Chem. Phys.*, **21**, 1087.

LIQUID CRYSTAL OBSERVABLES: STATIC AND DYNAMIC PROPERTIES

CLAUDIO ZANNONI

Dipartimento di Chimica Fisica ed Inorganica,
Università, Viale Risorgimento 4, 40136 Bologna, Italy

Abstract. In this Chapter we introduce the description of single and pair particle static properties of liquid crystals, and discuss their calculation from computer simulations. We also briefly describe the calculation of dynamic properties from molecular dynamics simulations using Linear Response theory.

1. Single particle properties

We consider a system of N molecules at certain specified thermodynamic conditions. Typically we shall consider that volume V and temperature T are fixed together with N (canonical conditions), but we shall also refer to the case where pressure P is fixed together with T (isobaric conditions). We assume the molecules to be classical, rigid particles with centre of mass at position $\mathbf{r}$ and orientation ω given for instance by a set of three Euler angles (α, β, γ), or only two angles (α, β) as illustrated in fig. 1 if we can assume that the molecules have cylindrical symmetry [1].

We shall discuss the calculation of observables in liquid crystals [2] in fairly general terms, but adopting a rather special point of view, that of computer simulations. As will be clear from the contributions in this book, computer simulations techniques [3] actually generate configurations of the system, i.e. sets of positions $\mathbf{r}_i = (x_i, y_i, z_i)$ and orientations ω_i of all the particles. In particular the Monte Carlo (MC) method generates equilibrium configurations, albeit non necessarily in the proper time order, while Molecular Dynamics (MD) actually generates configurations time step after time step in their natural time sequence. In this last case configurations consists not only of positions and orientations, but also of the full set of linear and angular velocities.

P. Pasini and C. Zannoni (eds.), Advances in the Computer Simulations of Liquid Crystals, 17–50.
© 2000 *Kluwer Academic Publishers. Printed in the Netherlands.*

18

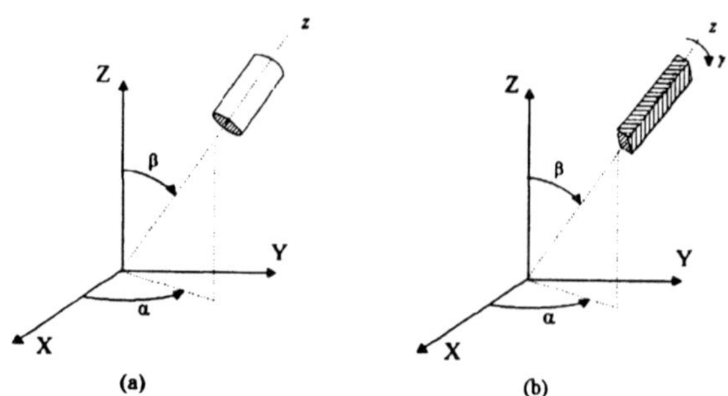

Figure 1. The two angles α, β defining the orientation of a cylindrically symmetric molecule (a) and the three Euler angles α, β, γ required for a generic rigid particle (b).

A complete static information about the system is represented by a sufficiently large set of M of its configurations. Indeed if we can calculate the value of a property dependent on molecular positions and orientations $A(\mathbf{r}_1, \omega_1, ..., \mathbf{r}_N, \omega_N)$ in each of these equilibrium configurations $(\mathcal{J})$, then the average value of A is

$$\langle A \rangle = \frac{1}{M} \sum_{\mathcal{J}}^{M} A(\mathbf{r}_1, \omega_1, ..., \mathbf{r}_N, \omega_N)^{(\mathcal{J})}. \tag{1}$$

The enormous number of positional and orientational coordinates specifying the various configurations is fortunately unnecessary if, as it is often the case, we are only interested in calculating average properties of single or pair molecule properties. In the next sections we shall discuss these two cases in turn.

1.1. THE SINGLET DISTRIBUTION AND ITS EXPANSION

Let us suppose that the probability density for a molecule to have a certain position $(\mathbf{r} + d\mathbf{r})$ and orientation $(\omega + d\omega)$ i.e. $P^{(1)}(\mathbf{r}, \omega)$ [4] is known. In this case the average of any property $A(\mathbf{r}, \omega)$ relating to a single molecule can be calculated as

$$\begin{aligned} \langle A \rangle &= \langle A(\mathbf{r}, \omega) \rangle_{\mathbf{r}, \omega} \\ &= \int d\mathbf{r} d\omega A(\mathbf{r}, \omega) P^{(1)}(\mathbf{r}, \omega)/N, \end{aligned} \tag{2}$$

where we use the angular brackets $\langle ... \rangle$ to indicate a statistical average over the relevant degrees of freedom (here on positions and orientations), that we indicate explicitly only when needed. The volume elements $d\mathbf{r}$, $d\omega$ are respectively $dxdydz$ and $d\alpha \sin \beta d\beta$ or $d\alpha \sin \beta d\beta d\gamma$ for a rigid molecule of arbitrary symmetry. Thus the integral over positions gives the volume

V of the sample and that on orientations gives the total angular measure Ω ($\Omega = 4\pi$ for cylindrical symmetry and $\Omega = 8\pi^2$ for the general case). The factor $1/N$ in eq. 2 comes from the normalization of the distribution $P^{(1)}(\mathbf{r}, \omega)$. The singlet distribution $P^{(1)}$ therefore contains all the microscopic information necessary to calculate one particle properties and in turn the structure and ordering of the system will be reflected by $P^{(1)}$. $P^{(1)}(\mathbf{r}, \omega)$ counts the average number of particles that are found in a small volume element centred at $(\mathbf{r}, \omega)$. In practice having all the configurations at hand we could scan $\mathbf{r}$ and ω space adding one to a suitable multidimensional histogram bucket when we find a molecule with that $(\mathbf{r}, \omega)$. A useful way of writing this definition for the singlet distribution is through the introduction of Dirac delta functions. Indeed, since, for example, $\delta(\mathbf{r}_1 - \mathbf{r}_1')$ is different from zero only when the position $\mathbf{r}_1$ of molecule 1 is at $\mathbf{r}_1'$ we can use a delta function as a device for counting the molecules at a certain position-orientation:

$$P^{(1)}(\mathbf{r}_1, \omega_1)/N = \langle \delta(\mathbf{r}_1 - \mathbf{r}_1')\delta(\omega_1 - \omega_1') \rangle_{\mathbf{r}_1', \omega_1'}, \tag{3}$$

which gives the average number of molecules with the desired position-orientation. The formula, very useful in extracting distributions from simulated configurations, can be easily checked using the definition of single particle average eq. 2. Thus,

$$\begin{aligned}
\langle A(\mathbf{r}_1, \omega_1) \rangle_{\mathbf{r}_1, \omega_1} &= \left\langle \int d\mathbf{r}_1' d\omega_1' \delta(\mathbf{r}_1 - \mathbf{r}_1')\delta(\omega_1 - \omega_1') A(\mathbf{r}_1', \omega_1') \right\rangle_{\mathbf{r}_1, \omega_1} \tag{4} \\
&= \int d\mathbf{r}_1' d\omega_1' A(\mathbf{r}_1, \omega_1) \langle \delta(\mathbf{r}_1 - \mathbf{r}_1')\delta(\omega_1 - \omega_1') \rangle_{\mathbf{r}_1, \omega_1} \\
&= \langle A(\mathbf{r}_1', \omega_1') \rangle_{\mathbf{r}_1', \omega_1'}, \tag{5}
\end{aligned}$$

giving eq. 3. We now concentrate on the description of long range orientational order. This is a central issue for liquid crystals, since this kind of order is common to all the various mesophases. In general we can obtain a purely orientational distribution $P(\omega)$ integrating out positions in eq. 3. For a uniform fluid, such as a nematic (but not a smectic), the singlet probability will be anyway independent of the position of molecules:

$$P^{(1)}(\mathbf{r}, \omega) = \rho P(\omega), \tag{6}$$

where $\rho \equiv N/V$ is the number density. In the limiting case of an ordinary isotropic liquid $P(\omega)$ will just be a constant.

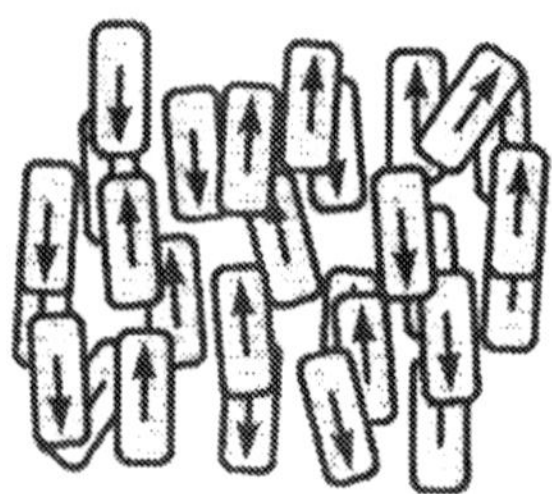

Figure 2. Schematic molecular organization for a system of polar molecules without overall polarization.

2. Orientational order of cylindrical molecules in uniaxial phases

2.1. EXPANSION OF THE ORIENTATIONAL DISTRIBUTION

We start by considering the simplest case of a molecule that can be considered of cylindrical symmetry, be it or rod-like or disc-like shape and that is embedded in uniaxial phase, $P(\omega) = P(\alpha, \beta)$. We do not distinguish by now if the molecule is a solute one or if it actually belongs to the mesophase.

If we take the laboratory Z axis parallel to the director and if the mesophase is uniaxial around the director then rotating the sample about Z should leave all observable properties unchanged. This means that the probability for a molecule to have an orientation (α, β) should be the same whatever the angle α [5]. More concisely

$$P(\alpha, \beta) = P(\beta)/2\pi. \tag{7}$$

Another experimental finding for nematics and some smectics is that nothing changes on turning the aligned sample upside down. Thus we should have

$$P(\beta) = P(\pi - \beta). \tag{8}$$

This is quite reasonable if we can think of the molecules of interest as spherocylinders or other cylindrically symmetric objects in which head and tail are not distinguishable. However, most mesogen molecules are not like this and for instance have dipole moments like *p-n* alkyl *p'*-cyano biphenyls (nCB). In practice the symmetry eq. 8, that is normally verified experimentally, means that the molecular arrangement will be such as to have on average no overall polarization (no *ferroelectricity*) as we show schematically in fig. 2. There is no fundamental argument that forbids uniaxial ferroelectric fluids and indeed these have been predicted by theory and simulations (see [6] and references therein), although not yet experimentally found. Notice that here we have used the same notation for $P(\beta)$ and $P(\cos \beta)$, that we assume to be renormalized to 1.

In a real experiment it will be extremely difficult to get this kind of complete information on the orientational distribution. A useful approach is, however, that of expanding $P(\beta)$ and approximating it in terms of a set of quantities that we can obtain from experiment. We need for this a set of functions that are orthogonal when integrated over $d\beta \sin\beta$. Such a set of functions is that of Legendre polynomials $P_L(\cos\beta)$, for which we have

$$\int_0^\pi d\beta \sin\beta \, P_L(\cos\beta)P_N(\cos\beta) = \frac{2}{2L+1}\delta_{LN}. \tag{9}$$

The explicit form of these Legendre polynomials is really very simple [7] and the first few terms are

$$P_0(\cos\beta) = 1 \tag{10}$$

$$P_1(\cos\beta) = \cos\beta \tag{11}$$

$$P_2(\cos\beta) = \frac{3}{2}\cos^2\beta - \frac{1}{2} \tag{12}$$

$$P_3(\cos\beta) = \frac{5}{2}\cos^3\beta - \frac{3}{2}\cos\beta \tag{13}$$

$$P_4(\cos\beta) = \frac{35}{8}\cos^4\beta - \frac{30}{8}\cos^2\beta + \frac{3}{8}. \tag{14}$$

Notice that $P_L(\cos\beta)$ is an even function of $\cos\beta$ if the rank L is even and an odd one if L is odd. Since $\cos(\pi-\beta) = -\cos\beta$ this means that in writing our even orientational distribution in terms of $P_L(\cos\beta)$ functions only even L terms need be retained. Clearly the odd terms will be present if $P(\beta)$ is not even, as for ferroelectric liquid crystal phases. Limiting ourselves to the more common even (see eq. 8) case we can write

$$P(\beta) = \sum_{L=0}^{\infty} \frac{2L+1}{2}\langle P_L\rangle P_L(\cos\beta); \quad L \quad \text{even}, \tag{15}$$

where the coefficients have been obtained exploiting the orthogonality of the basis set. The averages $\langle P_J\rangle$:

$$\langle P_J\rangle = \int_0^\pi d\beta \sin\beta P_J(\cos\beta)P(\beta) / \int_0^\pi d\beta \sin\beta P(\beta) \tag{16}$$

represent our set of orientational order parameters. The knowledge of the (infinite) set of $\langle P_J\rangle$ would completely define the distribution. From eq. 15 we can write

$$P(\beta) = \frac{1}{2} + \frac{5}{2}\langle P_2\rangle P_2(\cos\beta) + \frac{9}{2}\langle P_4\rangle P_4(\cos\beta) + \ldots \tag{17}$$

22

The first term contains the second rank order parameter

$$\langle P_2 \rangle = \frac{3}{2}\langle \cos^2 \beta \rangle - \frac{1}{2}. \tag{18}$$

It is easy to see that $\langle P_2 \rangle$ has the properties that we would intuitively expect an order parameter to possess and that can be identified with the empirical parameter introduced by Tsvetkov [8]. For a system of perfectly aligned molecules, where $\beta = 0$ for every molecule, $\langle P_2 \rangle = 1$. At the other extreme, for a completely disordered system such as an ordinary isotropic fluid we have $\langle \cos^2 \beta \rangle = 1/3$ and thus $\langle P_2 \rangle = 0$. In general

$$-\frac{1}{2} \le \langle P_2 \rangle \le 1 \tag{19}$$

because $0 \le \langle \cos^2 \beta \rangle < 1$. On going from an ordered to a disordered system the order parameter jumps discontinuously to zero if the transition is of the first order type, like the nematic–isotropic one. Notice that the same $\langle P_2 \rangle$ can be compatible with rather different molecular organizations and thus that it is important to try to determine $\langle P_4 \rangle$ or indeed as many as possible order parameters, to discriminate between alternative structural physical models [12]. The treatment has been generalized to molecules of arbitrary symmetry and to phases more complex than uniaxial by expanding $P(\alpha, \beta, \gamma)$ in a set of Wigner matrices $D^L_{mn}(\alpha, \beta, \gamma)$, orthogonal in (α, β, γ) space and systematically applying the symmetry properties of the molecule and of the phase [4].

2.2. EXPERIMENTAL DETERMINATION OF ORDER PARAMETERS. AN EXAMPLE

The order parameter $\langle P_2 \rangle$ is proportional to the anisotropy in experimentally measurable second rank properties. Just to make contact with real life measurements, we briefly consider as an example its determination from the *diamagnetic anisotropy* of liquid crystal materials.

When a diamagnetic material is placed in a sufficiently strong magnetic field a magnetization **M** is induced.

$$\mathbf{M} = \frac{1}{\mu_0}\chi\mathbf{B}, \tag{20}$$

where $\mathbf{B}$ is the magnetic induction and χ is the magnetic susceptibility. In a uniaxial liquid crystal two components $\chi_{ZZ} \equiv \chi_{\parallel}$ and $\chi_{XX} = \chi_{YY} \equiv \chi_{\perp}$ corresponding to the director parallel or perpendicular to the magnetic field direction, can in principle be determined. The trace of χ is essentially temperature independent and can be taken to be the isotropic value

$\bar{\chi} = \mathrm{Tr}\chi/3$. However, the difference between the parallel and perpendicular components changes significantly with temperature and is related to the ordering existing in the system. It is reasonable, for this property, to assume that the macroscopic diamagnetic susceptibility χ is the average of independent molecular contributions corresponding to molecular magnetic polarizability χ^{MOL}. Thus

$$\chi_{\parallel} = \langle \chi_{ZZ}^{\mathrm{LAB}} \rangle \tag{21}$$

$$= \sum_{a,b} \langle R_{Za} \chi_{ab}^{\mathrm{MOL}} \tilde{R}_{bZ} \rangle \tag{22}$$

$$= \langle R_{Zz}^2 \rangle \chi_{\parallel}^{\mathrm{MOL}} + (\langle R_{Zx}^2 \rangle + \langle R_{Zy}^2 \rangle) \chi_{\perp}^{\mathrm{MOL}} \tag{23}$$

$$= \langle \cos^2 \beta \rangle \chi_{\parallel}^{\mathrm{MOL}} + \langle \sin^2 \beta \rangle \chi_{\perp}^{\mathrm{MOL}} \tag{24}$$

$$= \bar{\chi} + \frac{2}{3} \Delta\chi \langle P_2 \rangle, \tag{25}$$

where R_{ab} are elements of the cartesian rotation matrix [1] and $\bar{\chi}^{\mathrm{MOL}} = \frac{1}{3}(\chi_{\parallel}^{\mathrm{MOL}} + 2\chi_{\perp}^{\mathrm{MOL}})$ equals $\bar{\chi}$. Thus determining $\chi_{\parallel}$ and the isotropic value $\bar{\chi}$ gives $\langle P_2 \rangle$ if the molecular anisotropy $\Delta\chi \equiv \chi_{\parallel}^{\mathrm{MOL}} - \chi_{\perp}^{\mathrm{MOL}}$ is known. DeJeu and coworkers [9] have measured the diamagnetic susceptivity anisotropy in a series of Schiff's base nematics which include the popular mesogens n –(4 –methoxy benzylidene) –$4'$ –n –Butyl aniline (MBBA), p –methoxy benzylidene p –cyananiline (MBCA), anisylidene –p –aminophenyl acetate (APAPA), as well as o –hydroxy –p –methoxy benzylidene p' –Butyl aniline (OHMBBA). In fig. 3 we report the temperature dependence of $\langle P_2 \rangle$ for a few cases. We notice that the order decreases with increasing temperature and then suddenly jumps to zero, as expected for this first order phase transition. Plotting the results in terms of reduced temperature T/T_{NI} shows a similar trend for the different compounds, even though the detailed behaviour is not really universal. The temperature dependence of the order parameter is often found to be well represented by the so called Haller equation

$$\Delta\chi(T/T_{NI}) = \Delta\chi(0)(1 - T/T_{NI})^{\beta}, \tag{26}$$

where $\Delta\chi(0)$ and the exponent β are fitting parameters. The exponent β that describes the temperature dependence of $\langle P_2 \rangle$ when approaching the transition has values $\beta = 0.17 - 0.22$ for many liquid crystals [10]. For the materials in fig. 3 $\beta = 0.170$ (MBBA), 0.185 (APAPA), 0.198 (OHMBBA) while for MBCA the value is rather different: $\beta = 0.134$.

The experimental determination of second and fourth rank order parameter can now be achieved with a number of techniques, as discussed in detail in the various chapters of [11].

24

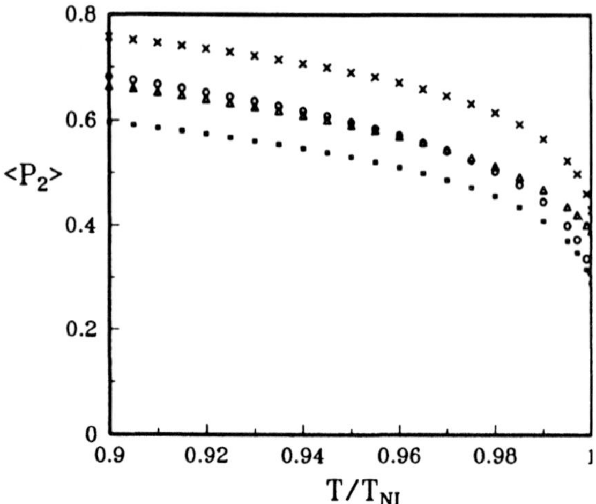

Figure 3. The second rank order parameter $\langle P_2 \rangle$ for MBBA (squares), OHMBBA (circles), MBCA (crosses), APAPA (triangles) obtained from diamagnetic anisotropy measurements as a function of temperature T/T_{NI} scaled with respect to the nematic–isotropic transition temperature [9].

2.3. GETTING ORDER PARAMETERS FROM SIMULATIONS

The calculation of orientational order parameters is clearly of particular importance in computer simulations of model liquid crystals. It also requires the development of some new algorithms as compared to simulations of isotropic fluids. We start by considering for simplicity a uniaxial mesophase formed of cylindrically symmetric particles, so that the description of ordering can be realized in terms of Legendre polynomial averages $\langle P_L \rangle$. In practice second and fourth rank parameters are particularly relevant and we shall consider their calculation in some detail.

2.4. SECOND RANK

The second rank order parameter $\langle P_2 \rangle$ can ideally be calculated by averaging $P_2(\cos \beta)$ over the probability $P(\beta)$ of finding the molecule at an orientation β with respect to the director, eq. 16.

This method can be used if the director orientation $\hat{n}$ is known. This could in principle be achieved in MD, preparing a system with the director along Z, say, and exploiting the much slower time scale for director tumbling compared to molecular reorientation. Alternatively an attempt can be made, even in MC, of pinning the director along the z direction by an external field, for instance adding to the potential energy a term

$$U_{ext} = -\zeta \sum_{i=1}^{N} P_2(\cos\beta_i). \tag{27}$$

where the positive coupling parameter ζ determines the strength of interaction with the field and β_i measures the angle between the axis of the i-th molecule and the field. In these rather special cases $\langle P_2 \rangle_{\text{LAB}}$ can be simply calculated from an average over M equilibrium configurations of the sample order parameter $\langle P_2 \rangle_S$. We have

$$\langle P_2 \rangle_{\text{LAB}} = \frac{1}{M} \sum_{\mathcal{J}}^{M} \langle P_2 \rangle_S^{(\mathcal{J})}, \tag{28}$$

where

$$\langle P_2 \rangle_S^{(\mathcal{J})} \equiv \frac{1}{N} \sum_{i=1}^{N} P_2(\hat{\mathbf{u}}_i \cdot \hat{\mathbf{n}})^{(\mathcal{J})} \tag{29}$$

is the order parameter computed for the $\mathcal{J}$-th configuration and $\hat{\mathbf{u}}_i \equiv (\sin\beta_i\cos\alpha_i, \sin\beta_i\sin\alpha_i, \cos\beta_i)$, defines the orientation of the i-th molecule in the laboratory frame. In this particular case it may be even simpler to calculate a histogram for the singlet orientational probability $P(\beta)$ and subsequently determine all the desired order parameters $\langle P_L \rangle_{\text{LAB}}$ by integration.

The problem in using the simple method just described is that, at least in the ordinary MC method, we work in an arbitrary laboratory frame and we have no applied field. Thus we do not know the orientation of the director in each configuration and we have no a priori guarantee that it will remain the same during the simulation. This in turn means that we cannot normally calculate $\langle P_2 \rangle$ as in eq. 16 in a configuration and then average the result over cycles. To find a way out [13] it is helpful to remember that computer simulations can be considered as experimental techniques where we can choose our observable at will. Thus we introduce a simple single molecule matrix property $\mathbf{A}$, whose only non vanishing component is along the molecule symmetry axis $\hat{\mathbf{u}}_i$:

$$A_{ab}^{\text{MOL}} = \delta_{az}\delta_{bz}. \tag{30}$$

The sample average of $\mathbf{A}$ in our arbitrary laboratory frame is obtained relating the components of $\mathbf{A}$ to the molecule fixed components and summing over all particles:

$$\langle A_{ab}^{\text{LAB}} \rangle_S = \frac{1}{N} \sum_{i=1}^{N} \{ \sum_{a'b'} (R_i)_{aa'}(A_i)_{a'b'}^{\text{MOL}}(\tilde{R}_i)_{b'b} \} \tag{31}$$

$$= Q_{ab} + \frac{1}{3}\delta_{ab}. \tag{32}$$

Here we have defined the ordering matrix $\mathbf{Q}$ averaged over the sample (configuration) as

$$\mathbf{Q} = \frac{1}{N}\sum_{i=1}^{N} \begin{pmatrix} u_{i,x}^2 - \frac{1}{3} & u_{i,x}u_{i,y} & u_{i,x}u_{i,z} \\ u_{i,x}u_{i,y} & u_{i,y}^2 - \frac{1}{3} & u_{i,y}u_{i,z} \\ u_{i,x}u_{i,z} & u_{i,y}u_{i,z} & u_{i,z}^2 - \frac{1}{3} \end{pmatrix}. \tag{33}$$

since $(R_i)_{aZ} = u_{i,a}$. Notice that $\mathbf{Q}$ is symmetric and traceless. Diagonalization of $\langle \mathbf{A}^{\text{LAB}}\rangle_S$ with the rotation matrix $\mathbf{U}$ identifies the director frame where

$$\langle A_{ZZ}^{\text{DIR}}\rangle_S = \sum U_{aZ}U_{bZ}\langle A_{ab}^{\text{LAB}}\rangle_S \tag{34}$$

$$= \frac{2}{3}\langle P_2\rangle_S + \frac{1}{3}. \tag{35}$$

The diagonalization procedure is equivalent to determining the order by maximizing the expression

$$\langle P_2\rangle_S^{(\mathcal{J})} = \frac{1}{N}\sum_{i=1}^{N} P_2(\hat{\mathbf{u}}_i \cdot \hat{\mathbf{n}}^{(\mathcal{J})}), \tag{36}$$

with respect to the unit vector $\hat{\mathbf{n}}^{(\mathcal{J})}$. Indeed in the special case that the director $\hat{\mathbf{n}}^{(\mathcal{J})}$ is parallel to the Z axis we see immediately that

$$\mathbf{Q} = \begin{pmatrix} -\frac{1}{3}\langle P_2\rangle_S - \xi & 0 & 0 \\ 0 & -\frac{1}{3}\langle P_2\rangle_S + \xi & 0 \\ 0 & 0 & \frac{2}{3}\langle P_2\rangle_S \end{pmatrix}. \tag{37}$$

The sample biaxiality parameter ξ corresponds to different ordering with respect to the laboratory X and Y axis and will tend to zero at large sample sizes if the mesophase has uniaxial symmetry [14]. It is now obvious that the rotation diagonalizing $\langle \mathbf{A}^{\text{LAB}}\rangle$ or equivalently $\mathbf{Q}$ defines the orientation of the director frame in our laboratory frame. The director itself is defined by the eigenvector corresponding to the largest eigenvalue, λ_{max}, of $\mathbf{Q}$ [13, 15]. The second rank order parameter referred to the director in the sample, $\langle P_2\rangle_\lambda$ is obtained from this λ_{max} as $\langle P_2\rangle_{\lambda,S} = \frac{3}{2}\lambda_{max}$. Thus we can define a $\mathbf{Q}$ tensor for every configuration, say $\mathbf{Q}^{(\mathcal{J})}$ for the $\mathcal{J}$-th one. By diagonalizing $\mathbf{Q}^{(\mathcal{J})}$, we obtain an order parameter $P_2^{(\mathcal{J})}$ and a director $\hat{\mathbf{n}}^{(\mathcal{J})}$, that can change from one configuration to the next. Since $P_2^{(\mathcal{J})}$ is

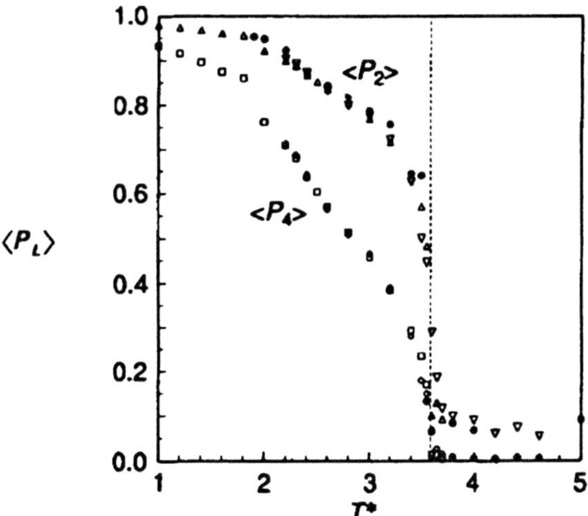

Figure 4. Orientational order parameters $\langle P_2 \rangle$, $\langle P_4 \rangle$ as a function of scaled temperature for a Gay–Berne system [17, 18]. We show results for a for N=512 (•) and N=1000 system in cooling (∇, $\diamond$) and heating ($\triangle$, $\square$) sequences. The vertical dashed line indicates the nematic–isotropic transition.

obtained as an eigenvalue and the eigenvalues of a matrix are rotationally invariant (scalar), we then calculate

$$\langle P_2 \rangle_\lambda = \frac{3}{2M} \sum_{J}^{M} (\lambda_{max})^{(J)}, \tag{38}$$

where $(\lambda_{max})_J$ is the largest eigenvalue of the average tensor in configuration J. The calculation of the orientational distribution with respect to the director strictly involves transforming the orientations, after each diagonalization, to the new director frame.

2.5. FOURTH RANK

We generalize the frame independent procedure by defining [16] a fourth rank molecular property as the direct square of the matrix $\mathbf{A}$ defined in eq. 30:

$$\mathbf{F} = \mathbf{A} \otimes \mathbf{A}. \tag{39}$$

We notice that in the molecule fixed frame the fourth rank property defined in this way has only one non vanishing component, F_{zzzz}, i.e.

$$F_{ijkl}^{\mathrm{MOL}} = \delta_{iz}\delta_{jz}\delta_{kz}\delta_{lz}. \tag{40}$$

28

The sample average of F is

$$\langle \mathbf{F}^{\text{LAB}} \rangle_S = \langle \mathbf{A}^{\text{LAB}} \otimes \mathbf{A}^{\text{LAB}} \rangle_S \tag{41}$$

$$= \langle (\mathbf{U}\mathbf{A}^{\text{DIR}}\tilde{\mathbf{U}}) \otimes (\mathbf{U}\mathbf{A}^{\text{DIR}}\tilde{\mathbf{U}}) \rangle_S \tag{42}$$

$$= (\mathbf{U} \otimes \mathbf{U})\langle \mathbf{A}^{\text{DIR}} \otimes \mathbf{A}^{\text{DIR}} \rangle_S (\tilde{\mathbf{U}} \otimes \tilde{\mathbf{U}}) \tag{43}$$

$$= (\mathbf{U} \otimes \mathbf{U})\langle \mathbf{F}^{\text{DIR}} \rangle_S (\tilde{\mathbf{U}} \otimes \tilde{\mathbf{U}}), \tag{44}$$

where LAB,DIR indicate laboratory and director frame. We then relate the components in the director frame to those in the molecular frame

$$\langle \mathbf{F}^{\text{DIR}} \rangle_S = \langle (\mathbf{R}\mathbf{A}^{\text{MOL}}\tilde{\mathbf{R}}) \otimes (\mathbf{R}\mathbf{A}^{\text{MOL}}\tilde{\mathbf{R}}) \rangle_S \tag{45}$$

$$= \langle (\mathbf{R} \otimes \mathbf{R})\mathbf{A}^{\text{MOL}} \otimes \mathbf{A}^{\text{MOL}}(\tilde{\mathbf{R}} \otimes \tilde{\mathbf{R}}) \rangle_S \tag{46}$$

$$= \langle (\mathbf{R} \otimes \mathbf{R})\mathbf{F}^{\text{MOL}}(\tilde{\mathbf{R}} \otimes \tilde{\mathbf{R}}) \rangle_S. \tag{47}$$

The fourth rank order parameter $\langle P_4 \rangle$ can be obtained from

$$\langle F^{\text{DIR}}_{ZZZZ} \rangle_S = \langle (R_{ZZ})^4 \rangle_S \tag{48}$$

$$= \langle \cos^4 \beta \rangle_S \tag{49}$$

$$= \sum_{a,b,c,d} U_{aZ}U_{bZ}U_{cZ}U_{dZ}\langle F^{\text{LAB}}_{abcd} \rangle_S \tag{50}$$

as

$$\langle P_4 \rangle_S = \frac{35}{8}\langle \cos^4 \beta \rangle_S - \frac{30}{8}\langle \cos^2 \beta \rangle_S + \frac{3}{8}. \tag{51}$$

In fig. 4 we show, as an application of these techniques, the order parameters $\langle P_2 \rangle$, $\langle P_4 \rangle$ obtained from MC simulations of a Gay–Berne liquid crystal [18].

3. Biaxial order parameters

The determination of order parameters for different molecular and phase symmetries from computer simulations is far from simple. However, the idea, outlined above, of introducing suitable molecular observables and determining their average in the laboratory system as in a virtual experiment is quite helpful and has been used to determine the biaxial and uniaxial order parameters [19]. The minimum set of order parameters required to describe biaxial molecules in a biaxial phase is [20, 21, 12] $\langle P_2 \rangle$, $\langle R^2_{02} \rangle$, $\langle R^2_{20} \rangle$, $\langle R^2_{22} \rangle$. The R^L_{mn} are combinations of Wigner functions D^L_{mn} symmetry-adapted for the D_{2h} group of the particles and phase [4]. Their explicit expressions, for the even terms, are:

$$R^L_{mn} \equiv \frac{1}{2}\mathcal{R}e(D^L_{mn} + D^L_{m-n}) \tag{52}$$

$$= \frac{1}{2}\Big[\cos(m\alpha)\cos(n\gamma)[d^L_{-mn}(\beta) + d^L_{mn}(\beta)] + $$

$$\sin(m\alpha)\sin(n\gamma)[d^L_{-mn}(\beta) - d^L_{mn}(\beta)] \Big]. \tag{53}$$

Experimentally one would try to select a sufficiently high number of molecular properties A_{ij}^{MOL} and measure their average values $\langle A_{ij}^{\text{LAB}} \rangle$. Then, through a diagonalization of these average tensors $\langle \mathbf{A}^{\text{LAB}} \rangle$, one could determine the order parameters. The requirement that these order parameters are the same for different observables helps in assigning the correct principal laboratory frame. As an illustration the explicit expressions for the eigenvalues of a tensor $F_{ab} = \delta_{az}\delta_{bz}$ are

$$f_{XX} = \frac{1}{3} - \frac{1}{3}\langle P_2 \rangle + \sqrt{\frac{2}{3}}\langle R_{20}^2 \rangle \tag{54}$$

$$f_{YY} = \frac{1}{3} - \frac{1}{3}\langle P_2 \rangle - \sqrt{\frac{2}{3}}\langle R_{20}^2 \rangle \tag{55}$$

$$f_{ZZ} = \frac{1}{3} + \frac{2}{3}\langle P_2 \rangle. \tag{56}$$

The non-trivial problem is finding a consistent way of assigning the three eigenvalues f_1, f_2, f_3 of the matrix $\mathbf{F}$ to the X,Y,Z axis. In the uniaxial phase or anyway when $\langle R_{20}^2 \rangle \simeq 0$, and taking $\langle P_2 \rangle > 0$, we have $f_{ZZ} > f_{XX} \simeq f_{YY}$ and letting $f_{ZZ} = \max(f_1, f_2, f_3)$ is sufficient to assign the axes except for an irrelevant exchange of X and Y. However, in the biaxial phase it may well happen that $f_{XX} > f_{ZZ}$, e.g. when $\langle P_2 \rangle / \langle R_{20}^2 \rangle < \sqrt{\frac{2}{3}}$, and, even if we assume that $\langle P_2 \rangle$ and $\langle R_{20}^2 \rangle$ are always positive, there is not a unique choice of axes other than assigning Y using $f_{YY} = \min(f_1, f_2, f_3)$. In particular the basic assumption used to calculate $\langle P_2 \rangle$ in the uniaxial case, i.e. $\max(f_1, f_2, f_3) = f_{ZZ}$, breaks down. Fortunately in simulations we can perform more virtual experiments, determining the average of other probe properties sensitive to the alignment of the two other molecular axes. In practice, equations containing $\langle R_{02}^2 \rangle$ and $\langle R_{22}^2 \rangle$ as well as $\langle P_2 \rangle$, $\langle R_{20}^2 \rangle$ are constructed from the average of two other matrices, say $G_{ab} = \delta_{ax}\delta_{bx}$ and $H_{ab} = \delta_{ay}\delta_{by}$. The resulting expressions of the order parameters in terms of the average tensors are:

$$\langle P_2 \rangle \;=\; \frac{3}{2}f_{ZZ} - \frac{1}{2} \tag{57}$$

$$=\; 1 - \frac{3}{2}(g_{ZZ} + h_{ZZ}) \tag{58}$$

$$\langle R_{20}^2 \rangle \;=\; \sqrt{\frac{3}{8}}(f_{XX} - f_{YY}) \tag{59}$$

$$=\; \sqrt{\frac{3}{8}}(g_{YY} - g_{XX} + h_{YY} - h_{XX}) \tag{60}$$

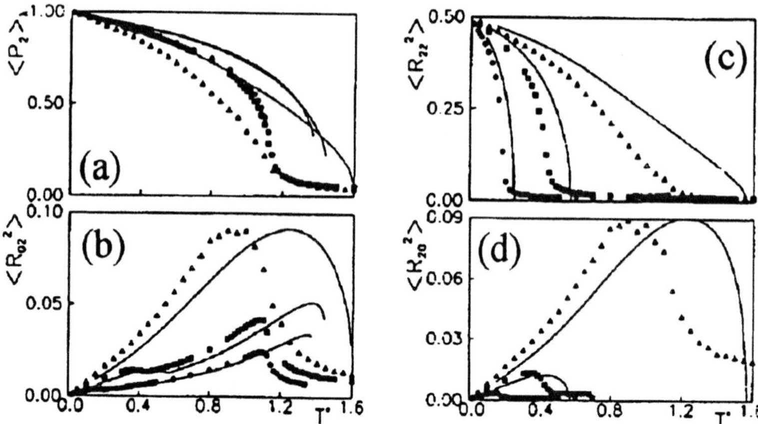

Figure 5. The second rank order parameter $\langle P_2 \rangle$ (a), $\langle R_{02}^2 \rangle$ (b), $\langle R_{20}^2 \rangle$ (c), $\langle R_{22}^2 \rangle$ (d) vs. temperature obtained from simulation: $\lambda = 0.2$ (circles), $\lambda = 0.3$ (squares) $\lambda = 0.40825$ (triangles) and from Mean Field Theory (lines) [19]

$$\langle R_{02}^2 \rangle = \sqrt{\frac{3}{8}}(g_{ZZ} - h_{ZZ}) \tag{61}$$

$$= \sqrt{\frac{3}{8}}(h_{XX} + h_{YY} - g_{XX} - g_{YY}) \tag{62}$$

$$\langle R_{22}^2 \rangle = \frac{1}{2}(g_{XX} - g_{YY} + h_{YY} - h_{XX}). \tag{63}$$

The normalized eigenvectors of the matrices give the axes of the reference system, except for the sign, so there are $3! = 6$ different systems corresponding to the eigenvalue permutations. In [19] we have chosen the eigenvalue permutation which satisfies the following conditions: a) $\langle P_2 \rangle > 0$; b) the same order parameters must have the same values in all the ways they are computed (here, e.g., $\langle P_2 \rangle$ and $\langle R_{20}^2 \rangle$ are computed in two different ways); c) for each configuration at one temperature the order parameters must be as close as possible to the mean value of the order parameters of the previous temperature (the sum of the differences is minimized). The above procedure effectively assigns the X and Y axes when the phase is biaxial. In the uniaxial phase X and Y are undistinguishable and the method, even though not needed, is not applicable because it forces a difference that is completely spurious. In a similar way, application of the algorithm previously described for finding $\langle P_2 \rangle$ to an isotropic phase will give a spurious non-zero order parameter (decreasing with size). In fig. 5 we show a set of order parameters obtained from MC simulation of the simple second rank attractive pair potential [19, 22]:

$$U(\omega_{ij}) = -\epsilon_{ij}\{P_2(\cos\beta_{ij}) + 2\lambda[R^2_{02}(\omega_{ij}) + R^2_{20}(\omega_{ij})]$$
$$+ 4\lambda^2 R^2_{22}(\omega_{ij})\} \tag{64}$$

with the biaxial molecules, or "spins", fixed at the sites of a three dimensional cubic lattice. The coupling parameter, ϵ_{ij}, is taken to be a positive constant, ϵ when particles i and j are nearest neighbors and zero otherwise. ω_{ij} is the relative orientation of the molecular pair. The biaxiality parameter λ accounts for the deviation from cylindrical molecular symmetry: when λ is zero, the potential reduces to the Lebwohl-Lasher P_2 potential, while for λ different from zero the particles tend to align not only their major axis, but also their faces. The value $\lambda = 1/\sqrt{6}$ marks the boundary between a system of prolate ($\lambda < 1/\sqrt{6}$) and oblate molecules ($\lambda > 1/\sqrt{6}$) [20].

We notice the different temperature dependence and the different magnitude of the four order parameters. Given the numerical errors inevitable in computer simulation results, e.g. those associated with finite size, the order parameter $\langle R^2_{22} \rangle$ can be recommended as a particularly effective monitor of the biaxial transition.

4. Pair properties

4.1. PAIR DISTRIBUTION

We can define a positional–orientational pair distribution using once more delta functions as counting devices:

$$P^{(2)}(\mathbf{r}_1,\omega_1,\mathbf{r}_2,\omega_2)/[N(N-1)] = \langle\delta(\mathbf{r}_1 - \mathbf{r}'_1)\delta(\omega_1 - \omega'_1)\delta(\mathbf{r}_2 - \mathbf{r}'_2)\delta(\omega_2 - \omega'_2)\rangle. \tag{65}$$

As the separation between the particles becomes very large the probability of finding molecule 1 at $(\mathbf{r}_1,\omega_1)$ and molecule 2 at $(\mathbf{r}_2,\omega_2)$ will be the product of these two independent events and the pair distribution will tend to the product of two single particle ones.

$$\lim_{r\to\infty} \langle\delta(\mathbf{r}_1 - \mathbf{r}'_1)\delta(\omega_1 - \omega'_1)\delta(\mathbf{r}_2 - \mathbf{r}'_2)\delta(\omega_2 - \omega'_2)\rangle$$
$$= \langle\delta(\mathbf{r}_1 - \mathbf{r}'_1)\delta(\omega_1 - \omega'_1)\rangle \langle\delta(\mathbf{r}_2 - \mathbf{r}'_2)\delta(\omega_2 - \omega'_2)\rangle \tag{66}$$

Thus it is convenient to write, for uniform systems

$$P^{(2)}(\mathbf{r}_1,\omega_1,\mathbf{r}_2,\omega_2) \equiv \rho^2 G(\mathbf{r}_{12},\omega_1,\omega_2)$$
$$= \rho^2 P(\omega_1)P(\omega_2)g(\mathbf{r}_{12},\omega_1,\omega_2). \tag{67}$$

32

The reduced pair distribution function $g(\mathbf{r}_{12}, \omega_1, \omega_2)$ introduced in eq.67 expresses a spatial - orientational correlation function or simply a *pair correlation function*. We have

$$\lim_{r_{12} \to \infty} g(\mathbf{r}_{12}, \omega_1, \omega_2) = 1 \qquad (68)$$

i.e. the density of particles at large distances just becomes that of the bulk. This limiting value is often subtracted from the $g(\mathbf{r}_{12}, \omega_1, \omega_2)$ to define the *total correlation function*

$$h(\mathbf{r}_{12}, \omega_1, \omega_2) = g(\mathbf{r}_{12}, \omega_1, \omega_2) - 1 \qquad (69)$$

or more generally

$$h(\mathbf{r}_1, \omega_1, \mathbf{r}_2, \omega_2) = g(\mathbf{r}_1, \omega_1, \mathbf{r}_2, \omega_2) - 1. \qquad (70)$$

It is also clear that at large separations the only orientational correlation between particles will be that indirectly coming from the fact that both molecule 1 and 2 are separately parallel to the same director, if that exists. In particular no long range orientational correlations exist in a normal fluid.

Another limiting situation is obtained for very short distances. If the molecules have a hard impenetrable core, there is vanishing probability of finding a second particle nearer than a minimum approach distance $\sigma(\mathbf{r}_{12}, \omega_1, \omega_2)$ from the first one. Thus

$$g(\mathbf{r}_{12}, \omega_1, \omega_2) = 0, \quad \text{if} \quad r_{12} < \sigma(\mathbf{r}_{12}, \omega_1, \omega_2). \qquad (71)$$

4.2. STONE EXPANSION OF THE PAIR DISTRIBUTION

We have seen earlier on, when discussing the calculation of $\langle P_2 \rangle$ from MC simulations, the advantages of using a rotational invariant description when calculating order parameters for a sample where no field is applied and the director can fluctuate from a configuration to the next. Here we wish to discuss the calculation of suitable rotationally invariant pair properties. The pair distribution can depend on orientations of the two molecules ω_1, ω_2 and on the intermolecular vector orientation (i.e. $\mathbf{r}_{12}$, not just r_{12}) but only through rotationally invariant combinations [23, 24, 25, 26, 4]. For example if we have linear molecules with orientation defined by unit vectors $\hat{\mathbf{u}}_1, \hat{\mathbf{u}}_2$ the distribution could depend on $\hat{\mathbf{u}}_1 \cdot \hat{\mathbf{u}}_2$, $\hat{\mathbf{u}}_1 \cdot \mathbf{r}_{12}$, $\hat{\mathbf{u}}_2 \cdot \mathbf{r}_{12}$ and their powers, but not on $\hat{\mathbf{u}}_1$, $\hat{\mathbf{u}}_2$, $\mathbf{r}_{12}$ by themselves. Let us now examine the general case

$$G(\mathbf{r}, \omega_1, \omega_2) = g(\mathbf{r}, \omega_1, \omega_2) P(\omega_1) P(\omega_2), \qquad (72)$$

where $G(\mathbf{r}, \omega_1, \omega_2)$, $P(\omega_i)$ are the pair (cf. 67) and single particle distribution functions and $\mathbf{r} = \mathbf{r}_2 - \mathbf{r}_1 = \mathbf{r}_{12}$ is the intermolecular vector with

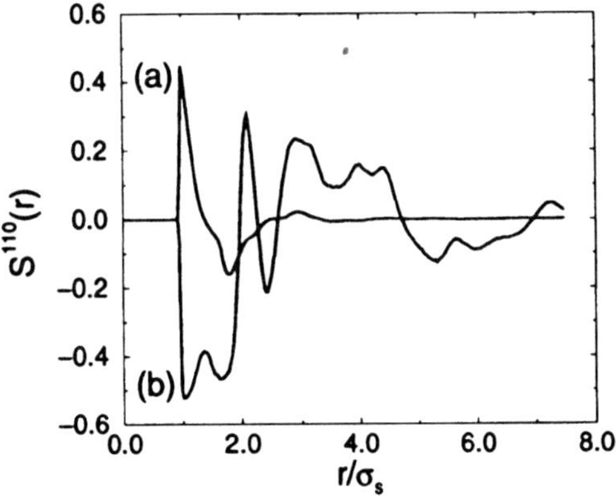

Figure 6. The dipole space correlation $S^{110}(r) = -\frac{1}{\sqrt{3}}\langle[\hat{u}_1 \cdot \hat{u}_2]\rangle_r$ for a system of elongated Gay-Berne particles with dipoles. The curves refer to central (a) and shifted (b) axial dipoles in the smectic[27].

length r and orientation $\omega_r = (\alpha_r, \beta_r)$. The pair correlation (e.g. for uniaxial molecules) can be expanded in Stone invariants (cf. Appendix)

$$g(\omega_1, \omega_2, \mathbf{r}) = \sum_{L_1,L_2,L_3} g^{L_1,L_2,L_3}(r) S^{L_1,L_2,L_3}(\omega_1, \omega_2, \omega_r), \qquad (73)$$

where the expansion coefficients are defined as

$$g^{L_1,L_2,L_3}(r) = \int d\omega_1\, d\omega_2\, g(\omega_1, \omega_2, \mathbf{r}) S^{L_1,L_2,L_3\,*}(\omega_1, \omega_2, \omega_r)$$

$$= \frac{(2L_1+1)(2L_2+1)(2L_3+1)}{256\pi^5} S^{L_1,L_2,L_3\,*}(r). \qquad (74)$$

Average rotational invariants as a function of inter-particle separation r are extremely useful to describe the molecular organization in a liquid crystal. As an example we show in fig. 6 the orientationally averaged rotational invariant $S^{110}(r)$ for a system of elongated Gay-Berne particles with an axial dipole at the center or near the end of the molecule [27]. The invariant shows that for central dipoles neighbouring dipoles tend to be antiparallel, while the opposite is true for shifted dipoles. Indeed monolayer smectic liquid crystals and modulated antiferroelectric bilayer stripe domains similar to the experimentally observed "antiphase" structures [28] are obtained in the two cases.

4.3. INTERMOLECULAR VECTOR CORRELATION FUNCTIONS

Given the anisotropy of liquid crystals, we can expect that pair distributions along the director or transversal to it, for instance, can be quite different.

34

More generally it is useful to define distributions along different orientations $\omega_{r_{12}}$ of the intermolecular vector relative to the director [29, 30, 18]:

$$g(r_{12}, \omega_{r_{12}}) = \int d\mathbf{r}_1 \, d\omega_1 \, d\omega_2 \, g(\mathbf{r}_1, \omega_1, \mathbf{r}_1 + \mathbf{r}_{12}, \omega_2) / \int d\mathbf{r}_1 d\omega_1 d\omega_2. \qquad (75)$$

This quantity gives the probability of finding a particle at a certain distance r_{12} from a particle chosen as origin when their intermolecular vector has an orientation $\omega_{r_{12}} = (\alpha_{r_{12}}, \beta_{r_{12}})$. For systems that are at most uniaxial, we can consider orientations defined with respect to a laboratory system with Z axis parallel to the director (perhaps after a suitable overall rotation of the sample) and we do not need to consider the angle $\alpha_{r_{12}}$, so that the intermolecular vector distribution reduces to $g(\mathbf{r}_{12}) = g(r_{12}, \beta_{r_{12}})/2\pi$. In fig. 7 we see an example of $g(r_{12}, \beta_{r_{12}})$ for a Gay-Berne system [18]. We notice that the radial distribution is not isotropic even in the nematic phase and that it changes quite significantly with temperature. The very low temperature one ($T^* = 1.8$) shows that as we move from a molecule along the z laboratory axis ($\cos\beta_r = 1$) a second molecule is found slightly below the particle length σ_e. However if we move transversally to the director ($\cos\beta_r = 0$) very sharp, well defined peaks appear, indicating structuredness in the layer. The characteristic double structure of the second peak indicates for this structure an hexagonal ordering, as expected in a smectic B or crystalline layer structure. It is convenient to expand $g(r_{12}, \beta_{r_{12}})$ as

$$g(r_{12}, \beta_{r_{12}}) = g_0(r_{12}) \sum_L (2L + 1) g_L^+(r_{12}) P_L(\cos\beta_{r_{12}}), \qquad (76)$$

where we have the standard radial distribution

$$g_0(r_{12}) = \frac{1}{2} \int d\beta_{r_{12}} \sin\beta_{r_{12}} g(r_{12}, \beta_{r_{12}}). \qquad (77)$$

This simple centre of mass, radial, distribution $g_0(r_{12})$ is very similar in the isotropic and in the nematic phase. This is because the $g_0(r_{12})$ are mainly structured at short range, and at short range liquid crystals are just like normal liquids.

The set of quantities $g_L^+(r_{12})$ associated to the intermolecular vector correlation function represent a sort of order parameters [29, 30]

$$\begin{aligned} g_L^+(r_{12}) &= \frac{1}{2\,g_0(r_{12})} \int d\beta_{r_{12}} \sin\beta_{r_{12}} g(r_{12}, \beta_{r_{12}}) P_L(\cos\beta_{r_{12}}) && (78) \\ &= \langle P_L(\cos\beta_{r_{12}})\rangle_{r_{12}}. && (79) \end{aligned}$$

We can also define a pair density function along the director

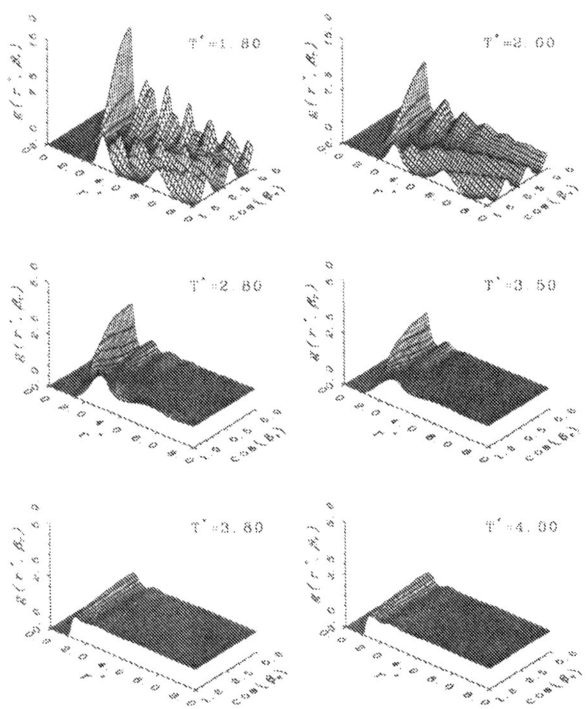

Figure 7. The intermolecular vector distribution $g(r^*, \beta_r)$ for a Gay-Berne potential at various dimensionless temperatures T^* in the crystalline, smectic: $T^* = 1.80$, 2.00, nematic: $T^* = 2.80$, 3.50 and isotropic: $T^* = 3.80$, 4.00 phase [18].

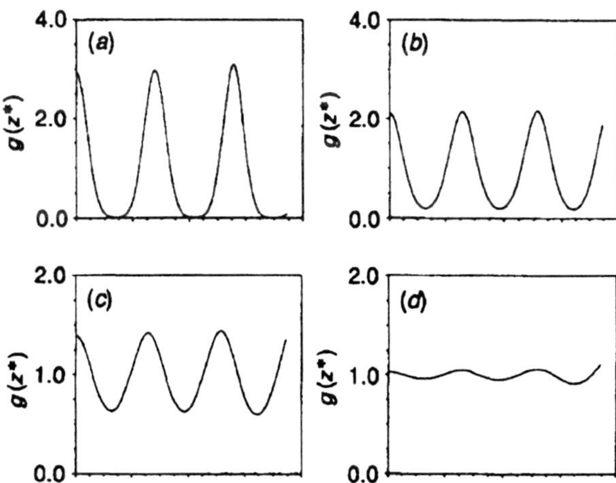

Figure 8. The pair density $g(z^*)$ for a Gay-Berne system at $T^* = 1.80$ (a), 2.00 (b), 2.30 (c) and 2.40 (d) [18].

36

$$g(z) = \frac{\int \, \mathrm{d}r \, r^2 \mathrm{d}\beta_r \, \sin\beta_r \delta(r\cos\beta_r - z)g(r,\beta_r)}{\int \, \mathrm{d}r \, r^2 \mathrm{d}\beta_r \, \sin\beta_r \delta(r\cos\beta_r - z)}. \tag{80}$$

The density $g(z)$ will be sinusoidally varying in a smectic and essentially constant in a nematic, as we see for a Gay-Berne fluid of elongated particles in fig. 8 [18].

5. Thermodynamic observables

The distribution functions introduced earlier on can be used to write down expressions for the various thermodynamic functions. Quite often these will be too complicated to be practically applicable as such but they nevertheless constitute the basis for approximate formulations or for algorithms to be used in computer simulations. Some relevant formulas are given here.

5.1. ENERGY

The total configuration energy of a system of N particles is most often assumed to be a sum of pairwise intermolecular contributions, and in turn the observed average can be written in terms of the pair distribution. Using the definition of pair correlation

$$U = \sum_{i=1}^{N-1} \sum_{j=i+1}^{N} \langle U(\mathbf{r}_i, \omega_i, \mathbf{r}_j, \omega_j) \rangle \tag{81}$$

$$= \frac{1}{2} \int \mathrm{d}\mathbf{r}_1 \mathrm{d}\omega_1 \mathrm{d}\mathbf{r}_2 \mathrm{d}\omega_2 P^{(2)}(\mathbf{r}_1\omega_1, \mathbf{r}_2, \omega_2) U(\mathbf{r}_1, \omega_1, \mathbf{r}_2, \omega_2) \tag{82}$$

and for a uniform system

$$U = \frac{V}{2}\rho^2 \int \mathrm{d}\mathbf{r}_{12} \mathrm{d}\omega_1 \mathrm{d}\omega_2 P(\omega_1) P(\omega_2) g(\mathbf{r}_{12}, \omega_1, \omega_2) U(\mathbf{r}_{12}, \omega_1, \omega_2). \tag{83}$$

5.2. HEAT CAPACITY

If we differentiate the microscopic expression for the energy we can show that the constant volume heat capacity is related to the mean square fluctuations in the energy:

$$C_V = \left(\frac{\partial U}{\partial T} \right)_V \tag{84}$$

$$= \frac{1}{k_B T^2} \left(\langle U^2 \rangle - \langle U \rangle^2 \right) \tag{85}$$

$$= \frac{1}{2} \int \mathrm{d}\mathbf{r}_1 \mathrm{d}\omega_1 \mathrm{d}\mathbf{r}_2 \mathrm{d}\omega_2 U(\mathbf{r}_1, \omega_1, \mathbf{r}_2, \omega_2) \frac{\partial}{\partial T} P^{(2)}(\mathbf{r}_1, \omega_1, \mathbf{r}_2, \omega_2), \tag{86}$$

where k_B is the Boltzmann constant and the last equation applies to temperature independent potentials. Eq. 85 shows that the heat capacity is a non-negative quantity.

Notice that the heat capacity is not a pairwise quantity, even if the potential is a pairwise one. C_V does not depend only on the pair distribution at a given temperature but on its derivative. If we try to perform the derivative we see that the microscopic expression depends on more than two particles simultaneously. Thus the specific heat is really a collective property and it is reasonable that it can change significantly and diverge at a phase transition where the collective organization changes.

5.3. PRESSURE

The calculation of pressure provides an important observable and is also essential for implementing proper isobaric control, e.g. in NPT MC simulations. To derive a molecular expression for the pressure we start from the thermodynamic definition as a volume derivative of the free energy A [31, 32]

$$P = -\left(\frac{\partial A}{\partial V}\right)_T \tag{87}$$

$$= k_B T \left(\frac{\partial \ln Q_N}{\partial V}\right)_T. \tag{88}$$

We can render the volume dependence of the configurational integral Q_N an explicit one by changing the positional variable $\mathbf{r}_i$ to dimensionless units $\mathbf{s}_i$. Thus we let

$$\mathbf{r}_i = V^{\frac{1}{3}} \mathbf{s}_i. \tag{89}$$

and find

$$Q_N = \frac{V^N}{N!} \int \{ds\}^N \{d\omega\}^N e^{-U(\{s,\omega\}^N)/k_B T}, \tag{90}$$

which gives

$$P = \frac{N k_B T}{V} - \frac{V^N N!}{Q_N} \int \{ds\}^N \{d\omega\}^N \frac{\partial U(\{s,\omega\}^N)}{\partial V} e^{-U(\{s,\omega\}^N)/k_B T}$$

$$= \frac{N k_B T}{V} - \langle \frac{\partial U(\{r,\omega\}^N)}{\partial V} \rangle. \tag{91}$$

The volume derivative of the potential energy is

$$\frac{\partial U(\{\mathbf{r},\omega\}^N)}{\partial V} = \sum_i \frac{\partial}{\partial \mathbf{r}_i} U(\{\mathbf{r},\omega\}^N) \cdot \frac{\partial \mathbf{r}_i}{\partial V}, \tag{92}$$

38

where $\partial/\partial \mathbf{r}_i U$ is the potential gradient. Since

$$\frac{\partial \mathbf{r}_i}{\partial V} = \frac{1}{3V} \mathbf{r}_i, \tag{93}$$

we find the *virial* equation [31, 4]

$$P = \frac{Nk_BT}{V} - \frac{1}{3V} \langle \sum_i \mathbf{r}_i \cdot \nabla_i U(\{\mathbf{r},\omega\}^N) \rangle. \tag{94}$$

For the special case of a pairwise potential the volume derivative of the potential energy in eq. 91 is

$$\frac{\partial U(\{\mathbf{r},\omega\}^N)}{\partial V} = \frac{1}{3V} \sum_{i<j} \frac{\partial}{\partial \mathbf{r}_{ij}} U(\mathbf{r}_{ij}, \omega_i, \omega_j) \cdot \mathbf{r}_{ij}. \tag{95}$$

Notice that these expressions have to be modified if the potential energy also depends directly on the volume, as is the case for instance of long range Coulomb interactions summed with Ewald or reaction field formulas [33].

5.4. SURFACE TENSION

We consider two coexisting phases with a plane interface separating them localized around $z = 0$. The thermodynamic definition relates the surface tension γ to the variation in free energy as the area A is changed [3, 34]:

$$\gamma = -\left(\frac{\partial \mathcal{A}}{\partial A}\right)_{TV} \tag{96}$$

$$= k_BT \left(\frac{\partial \ln Q_N}{\partial A}\right)_{TV}. \tag{97}$$

To derive a molecular expression for the surface tension we can proceed as we did for the pressure to get a virial type expression, by making the surface dependence of the configurational integral Q_N an explicit one by changing the positional variable to dimensionless units. In practice we can use

$$r_x = A^{\frac{1}{2}} s_x, \qquad \frac{\partial r_x}{\partial A} = \frac{r_x}{2A} \tag{98}$$

$$r_y = A^{\frac{1}{2}} s_y, \qquad \frac{\partial r_y}{\partial A} = \frac{r_y}{2A} \tag{99}$$

$$r_z = \frac{V}{A} s_z, \qquad \frac{\partial r_z}{\partial A} = -\frac{r_z}{A}. \tag{100}$$

The surface tension, which can be written in terms of the average derivative of the potential energy with respect to the area, is, for a pairwise potential:

$$\gamma = \left\langle \frac{\partial U(\{\mathbf{r},\omega\}^N)}{\partial A} \right\rangle_{TV} \tag{101}$$

$$= \frac{1}{2A}\sum_{i<j}\left\langle \mathbf{r}_{ij}\cdot\frac{\partial U(\mathbf{r}_{ij},\omega_i,\omega_j)}{\partial \mathbf{r}_{ij}} - 3z_{ij}\frac{\partial U(\mathbf{r}_{ij},\omega_i,\omega_j)}{\partial z_{ij}}\right\rangle_{TV} \tag{102}$$

$$= \frac{1}{4V_z}\int_{-z}^{z} dz \sum_{i<j}\left\langle \left(x_{ij}\frac{\partial U_{ij}}{\partial x_{ij}} + y_{ij}\frac{\partial U_{ij}}{\partial y_{ij}}\right) - 2z_{ij}\frac{\partial U_{ij}}{\partial z_{ij}}\right\rangle_{TV}, \tag{103}$$

where $U_{ij} = U(\mathbf{r}_{ij},\omega_i,\omega_j)$ and V_z is the volume of a thin layer $\mathcal{L}$ containing the interface of thickness $2z$ and with area A equal to the sample cross section [35].

6. Dynamic evolution of a molecular property

We now turn to briefly considering the calculation of dynamic properties [36, 37, 38, 39, 40]. We start considering once more a system of N molecules at equilibrium in a volume V at temperature T. The full distribution of the system is

$$\rho_0(\mathbf{\Gamma}) \equiv \frac{1}{Z_N}\exp[-\beta\mathcal{H}_0(\mathbf{\Gamma})], \tag{104}$$

where $\beta \equiv 1/(k_BT)$ and $\mathbf{\Gamma} \equiv (\{\mathbf{q},\mathbf{p}\}^N) = (\mathbf{q}_1,\mathbf{p}_1,\mathbf{q}_2,\mathbf{p}_2,\ldots,\mathbf{q}_N,\mathbf{p}_N)$ is the set of coordinates and momenta needed to specify a point in phase space and the unperturbed hamiltonian $\mathcal{H}_0$, is the sum of the potential $U(\{\mathbf{q}\}^N)$ and kinetic $K(\{\mathbf{p}\}^N)$ contributions:

$$\mathcal{H}_0(\{\mathbf{q}\}^N,\{\mathbf{p}\}^N) = U(\{\mathbf{q}\}^N) + K(\{\mathbf{p}\}^N) \tag{105}$$

and correspondingly

$$\rho_0(\mathbf{\Gamma}) = \frac{1}{Z_N}\exp[-\beta U(\{\mathbf{q}\}^N)]\exp[-\beta K(\{\mathbf{p}\}^N)] \tag{106}$$

$$= P(\{\mathbf{q}\}^N)P_M(\{\mathbf{p}\}^N), \tag{107}$$

where $P(\{\mathbf{q}\}^N)$ is the configurational probability studied earlier in this chapter, where we have considered, $\{\mathbf{q}\}^N = (\mathbf{q}_1,\mathbf{q}_2,\ldots,\mathbf{q}_N)$, and each $\mathbf{q}_i$ can depend on a set of positions and orientations, $(\mathbf{r}_i,\omega_i)$. $P_M(\{\mathbf{p}\}^N)$ is the Maxwell distribution of linear and angular momenta.

For a system described by an hamiltonian $\mathcal{H}$ the equations of motion for coordinates and momenta are [41]

$$\frac{\partial \mathcal{H}}{\partial \mathbf{p}_i} = \dot{\mathbf{q}}_i, \quad \frac{\partial \mathcal{H}}{\partial \mathbf{q}_i} = -\dot{\mathbf{p}}_i. \tag{108}$$

The time derivative of a function A that depends on generalized coordinates $\mathbf{q}_i$ and momenta $\mathbf{p}_i$ i.e. of the state point $\mathbf{\Gamma}$, will be

$$\dot{A}(t) \quad = \frac{\partial A}{\partial t} + \sum_i \left(\frac{\partial A}{\partial \mathbf{p}_i} \cdot \dot{\mathbf{p}}_i + \frac{\partial A}{\partial \mathbf{q}_i} \cdot \dot{\mathbf{q}}_i \right) \tag{109}$$

$$= \frac{\partial A}{\partial t} + \mathcal{H}^\times A. \tag{110}$$

The differential Liouville operator or evolution operator $\mathcal{H}^\times$ is defined as

$$\mathcal{H}^\times \quad \equiv - \sum_i \left(\frac{\partial \mathcal{H}}{\partial \mathbf{q}_i} \cdot \frac{\partial}{\partial \mathbf{p}_i} - \frac{\partial \mathcal{H}}{\partial \mathbf{p}_i} \cdot \frac{\partial}{\partial \mathbf{q}_i} \right) \tag{111}$$

$$= \sum_i \left(\mathbf{F}_i \cdot \frac{\partial}{\partial \mathbf{p}_i} + \frac{1}{m_i} \mathbf{p}_i \cdot \frac{\partial}{\partial \mathbf{q}_i} \right), \tag{112}$$

where $\mathbf{F}_i$ is the force acting on molecule i. If $\partial A / \partial t = 0$, the evolution of $A(t)$ is formally given by the so called Heisenberg evolution equation

$$A(t) = e^{\mathcal{H}^\times t} A(0). \tag{113}$$

In particular the evolution of the probability distribution $\rho(\mathbf{\Gamma})$ at equilibrium is

$$\frac{d\rho}{dt} = \frac{\partial \rho}{\partial t} + \mathcal{H}^\times \rho, \tag{114}$$

but the total density must be conserved in time $(\dot{\rho}_0(\mathbf{\Gamma}) = 0)$ so that evolution takes place through the Liouville equation

$$\frac{\partial \rho}{\partial t} = -\mathcal{H}^\times \rho, \tag{115}$$

where we have written

$$\rho \equiv \rho(\mathbf{\Gamma}) = \frac{1}{Z_N} \exp[-\beta \mathcal{H}(\mathbf{\Gamma})]. \tag{116}$$

Thus if the hamiltonian is the unperturbed one $\mathcal{H}_0$, it does not modify the equilibrium distribution ρ_0

$$\mathcal{H}_0^\times \rho_0 = \mathcal{H}_0^\times \frac{1}{Z_N} e^{-\beta \mathcal{H}_0} = -\beta \frac{1}{Z_N} e^{-\beta \mathcal{H}_0} \mathcal{H}_0^\times \mathcal{H}_0 = 0. \tag{117}$$

In other words ρ_0 is an eigenvector of the operator $\mathcal{H}_0^\times$ corresponding to a zero eigenvalue. Correspondingly for such a stationary system $\mathcal{H}_0^\times \langle A \rangle = 0$. The evolution operator is a combination of derivatives and thus it follows the standard rules of derivatives. For instance

$$\mathcal{H}^\times (AB) = (\mathcal{H}^\times A)B + A(\mathcal{H}^\times B). \tag{118}$$

In particular, given the average of two properties A,B, we have $\mathcal{H}_0^\times \langle AB \rangle = 0$ and the antisymmetric relation

$$
\begin{aligned}
\langle A\mathcal{H}_0^\times B \rangle &= \int d\Gamma \rho_0 A \mathcal{H}_0^\times B \\
&= -\langle (\mathcal{H}_0^\times A) B \rangle.
\end{aligned}
\tag{119}
$$

More generally

$$
\langle A e^{t\mathcal{H}_0^\times} B \rangle = \langle (e^{-t\mathcal{H}_0^\times} A) B \rangle.
\tag{120}
$$

The quantity just introduced is called an equilibrium time correlation function for properties A and B, $C_{AB}(t)$. In general it corresponds to an average of a property A taken at a certain time t_0 with the complex conjugate of a property B taken at time $t_0 + t$

$$
\begin{aligned}
C_{AB}(t) &= \langle A(t_0) B^*(t_0 + t) \rangle \tag{121} \\
&= \langle A(0) B^*(t) \rangle \tag{122} \\
&= \langle A(0) e^{t\mathcal{H}_0^\times} B^*(0) \rangle \tag{123} \\
&= \int d\Gamma \rho_0(\Gamma) A(\Gamma) e^{t\mathcal{H}_0^\times} B^*(\Gamma). \tag{124}
\end{aligned}
$$

Time correlation functions can be calculated from the sequence of instantaneous values of the observables taken along a Molecular Dynamics generated trajectory. Thus

$$
C_{AB}(t) \approx \frac{1}{M-n} \sum_{k=1}^{M-n} A(k\Delta t) B^*([k+n]\Delta t), \quad t = n\Delta t.
\tag{125}
$$

Since our system should be independent on where we start measuring the time, there is no dependence on t_0. We shall see in the next section that correlation functions are directly linked to observable dynamic properties. In this context it is useful to introduce also the Fourier-Laplace transform of a correlation function, which is called a *spectral density*

$$
\begin{aligned}
j_{AB}(\hat{\omega}) &= \int dt\, e^{i\hat{\omega}t} C_{AB}(t) \tag{126} \\
&= \langle A(0) \frac{1}{\mathcal{H}^\times - i\hat{\omega}} B^*(0) \rangle. \tag{127}
\end{aligned}
$$

Here for simplicity we assume real quantities A, B.

7. Contact with experiment. Linear response theory

Let us consider the measurement of a property A of a molecular system through the application of a weak measuring field $f(t)$ [36, 37, 39, 40]. We assume that the field is switched on a time $t = 0$ and that it interacts with the system through a perturbation hamiltonian

$$\mathcal{H}_1 = -B(\mathbf{\Gamma})f(t). \tag{128}$$

The property coupling to the field B depends in the most general case on the coordinates and the momenta of all molecules, $B \equiv B(\mathbf{\Gamma})$. We assume that the observed value of property A changes from its static equilibrium value in the absence of the field $\langle A \rangle_0$. Since the applied field is weak the observed non equilibrium value in the presence of the field at time t, $\langle \delta A(t) \rangle_f$, should be linear in the field strength. Considering that the system may not react instantaneously to the field, what we observe at time t is a sum of the contributions from all possible time lags τ between application and observation

$$\langle \delta A(t) \rangle_f \;\; = \;\; \sum_{\tau_i} K_{AB}(\tau_i) f(t - \tau_i) \tag{129}$$

$$= \;\; \int_0^\infty \mathrm{d}\tau K_{AB}(\tau) f(t - \tau). \tag{130}$$

The observed value is a *convolution* of the field function f with a " kernel" K_{AB} whose functional form depends on the type of applied field and the observable property. The dynamics of molecular phenomena is most often explored as a frequency dependence of a certain observable property rather than a direct time evolution from a given starting event (there are exceptions of course, e.g. time domain fluorescence depolarization experiments). We can write the Fourier-Laplace transform of the time dependent response as

$$\langle \delta \tilde{A}(\hat{\omega}) \rangle_f \;\; \equiv \;\; \int_0^\infty \mathrm{d}t \, e^{i\hat{\omega}t} \langle \delta A(t) \rangle_f \tag{131}$$

$$= \chi_{AB}(\hat{\omega}) \tilde{f}(\hat{\omega}), \tag{132}$$

where the Fourier-Laplace transform

$$\chi_{AB}(\hat{\omega}) = \int_0^\infty \mathrm{d}\tau K_{AB}(\tau) e^{i\hat{\omega}\tau} \tag{133}$$

is called a *susceptibility*. Thus the Fourier transform of a convolution integral of two functions is the product of the Fourier transform of the functions, and this useful result is called the *convolution theorem*. The equation we have just seen is macroscopic, but it is clear that if we could obtain a microscopic expression for the susceptibility we could be able to calculate the response to a measuring field. The importance of Linear Response Theory is that it gives a molecular interpretation to the susceptibility in terms of fluctuations of the unperturbed system.

8. Evolution in the presence of a perturbation

When the time dependent perturbation $\mathcal{H}_1(t) = -B(\Gamma)f(t)$ is added, so that

$$\mathcal{H}(t) = \mathcal{H}_0 + \mathcal{H}_1(t), \tag{134}$$

the perturbation produces an evolution of ρ_0:

$$\mathcal{H}_1^{\times}(t)\rho_0 \;=\; -\beta \frac{1}{Z_N} e^{-\beta \mathcal{H}_0} \mathcal{H}_1^{\times}(t)\mathcal{H}_0 \tag{135}$$

$$= -\beta \rho_0 \dot{B} f(t) \tag{136}$$

since using the definition of $\mathcal{H}^{\times}$

$$\mathcal{H}_1^{\times}(t)\mathcal{H}_0 \;=\; -\mathcal{H}_0^{\times}\mathcal{H}_1(t) \tag{137}$$

$$= f(t)\mathcal{H}_0^{\times} B(\Gamma)$$

$$= f(t)\dot{B}. \tag{138}$$

In the presence of $\mathcal{H}_1(t)$ the distribution becomes at first order

$$\rho(t) = \rho_0 + \delta\rho(t) \tag{139}$$

and the non equilibrium value is

$$\langle \delta A(t)\rangle_f = \int d\Gamma\, A(\Gamma)\delta\rho(t). \tag{140}$$

From the Liouville equation we have, substituting eq. 139 and keeping only linear terms,

$$\delta\dot{\rho}(t) \;=\; -\mathcal{H}_0^{\times}\delta\rho - \mathcal{H}_1^{\times}(t)\rho_0$$

$$= -\mathcal{H}_0^{\times}\delta\rho + \beta\rho_0 f(t)\dot{B}. \tag{141}$$

This is a simple first order linear equation whose general solution is available[1]

---

[1] *The known differential equation is* $\dot{y}(t) + Py = Q(t)$ *if* $P = \mathcal{H}_0^{\times}$, $Q(t) = -\beta\rho_0\dot{\mathcal{H}}_1$ *that with* $y = y_0$ *when* $t = t_0$ *has the solution*

$$e^{Pt}y - e^{Pt_0}y_0 = \int_{t_0}^{t} e^{Pt'} Q(t')dt'$$

$$\delta\rho(t) = \beta \int_{-\infty}^{t} dt' e^{[(t'-t)\mathcal{H}_0^{\times}]} \rho_0 \dot{B} f(t'). \tag{142}$$

Substituting in eq. 140

$$\begin{aligned}
\langle \delta A(t) \rangle_f &= \beta \int_{-\infty}^{t} dt' f(t') \int d\boldsymbol{\Gamma} A(\boldsymbol{\Gamma}) e^{[(t'-t)\mathcal{H}_0^{\times}]} \dot{B}(\boldsymbol{\Gamma}) \rho_0(\boldsymbol{\Gamma}) \\
&= \beta \int_{-\infty}^{t} dt' f(t') \int d\boldsymbol{\Gamma} A(0) \dot{B}(t'-t) \rho_0(\boldsymbol{\Gamma}) \\
&= \beta \int_{-\infty}^{0} d\tau f(\tau + t) C_{A\dot{B}}(\tau),
\end{aligned} \tag{143}$$

where we have shifted the time origin $(\tau = t' - t)$. We can do one further manipulation noticing that

$$C_{A\dot{B}}(t) = \frac{d}{dt} C_{AB}(t) \tag{144}$$

and that the correlation $C_{AB}(t)$ is invariant for time reversal

$$C_{AB}(t) = C_{AB}^{*}(-t) \tag{145}$$

and then

$$C_{A\dot{B}}(t) = -C_{A\dot{B}}(-t). \tag{146}$$

We can then write our final result

$$\langle \delta A(t) \rangle_f = -\beta \int_{0}^{\infty} d\tau f(t - \tau) C_{A\dot{B}}(\tau). \tag{147}$$

Thus Linear Response theory shows that the change observed in a property A when the perturbation $-Bf(t)$ is switched on at time 0 can be obtained simply from equilibrium time correlation functions that we can calculate in the absence of the perturbation, for instance from molecular dynamics simulations of the unperturbed system. Taking the Fourier transform and comparing with eq. 132

$$\chi_{AB}(\hat{\omega}) = -\beta \int_{0}^{\infty} C_{A\dot{B}}(\tau) e^{i\hat{\omega}\tau} d\tau. \tag{148}$$

The resulting expression for the susceptivity is quite simple:

$$\chi_{AB}(\hat{\omega}) = \beta C_{AB}(0) + i\hat{\omega}\beta \int_{0}^{\infty} C_{AB}(\tau) e^{i\hat{\omega}\tau} d\tau, \tag{149}$$

where $C_{AB}(\tau)$ is the *correlation function* for the two properties A, B.

9. Theory of dielectric response

As an example of application of Linear Response theory we obtain equations for the dielectric response of a material [42], for an idealized case where the problem of connecting the applied external field with that felt by the molecules [43] can be ignored. Let us consider a system of N molecules with permanent dipole moments μ_i to which a uniform external electric field i.e. $\mathbf{E}(t) = \mathbf{E}_0 \exp(i\hat{\omega}t)$ is applied. The system has a total dipole moment

$$\mathbf{M} = \sum_i^N \mu_i, \tag{150}$$

while the general perturbation hamiltonian $\mathcal{H}_1 = -B(\mathbf{\Gamma})f(t)$ becomes

$$\mathcal{H}_1 = -\mathbf{M}(\{\omega_i\}^N) \cdot \mathbf{E}(t). \tag{151}$$

Notice that molecular positions have not been included since the field is assumed to be the same at every position. The expression for the dielectric susceptivity follows from general linear response theory as:

$$\chi_{\mathbf{MM}}(\hat{\omega}) = -\frac{1}{kT_B} \int_0^\infty C_{\mathbf{M\dot{M}}}(\tau)e^{i\hat{\omega}\tau}\,\mathrm{d}\tau \tag{152}$$

$$\chi_{\mathbf{MM}}(\hat{\omega}) = \frac{1}{kT_B}\left(\langle\mathbf{MM}\rangle + i\hat{\omega}\int_0^\infty C_{\mathbf{MM}}(\tau)e^{i\hat{\omega}\tau}\,\mathrm{d}\tau\right), \tag{153}$$

where $C_{\mathbf{MM}}(\tau)$ is the dipole correlation function

$$C_{\mathbf{MM}}(t) = \langle\mathbf{M}(0)\mathbf{M}(t)\rangle \tag{154}$$

and we assume $\langle\mathbf{M}(t)\rangle = 0$.
A large number of dynamic properties and transport coefficients can be formulated in a similar way in terms of equilibrium correlation functions [44, 37, 38].

Acknowledgments

This work was supported by University of Bologna, CNR and MURST. I would like to thank all the members of the group in Bologna and in particular Roberto Berardi, Silvia Orlandi, Francesco Spinozzi and Paolo Pasini for many enjoyable and stimulating discussions on the topics of this chapter.

46

Appendix

ROTATIONAL INVARIANTS

We often need to consider invariant functions of the position and orientation of two molecules $f(\mathbf{r}_1, \omega_1, \mathbf{r}_2, \omega_2)$. We assume that the function should be translationally invariant

$$f(\mathbf{r}_1, \omega_1, \mathbf{r}_2, \omega_2) = f(\mathbf{R} + \mathbf{r}_1, \omega_1, \mathbf{R} + \mathbf{r}_2, \omega_2). \tag{155}$$

This implies that the most general function will be

$$f(\mathbf{r}_{12}, \omega_1, \omega_2) = f(r_{12}, \omega_r, \omega_1, \omega_2), \tag{156}$$

where $\mathbf{r}_{12} \equiv \mathbf{r}_2 - \mathbf{r}_1$ is the inter-centre vector with orientation ω_r. The function should also be invariant for an arbitrary rotation of the laboratory frame. General rotationally invariant combinations can be constructed following Blum and Torruella [24] and Stone [25]. $f(r_{12}, \omega_r, \omega_1, \omega_2)$ can be expanded in a basis of products of three Wigner rotation matrices

$$D_{m_1 n_1}^{J_1 *}(\omega_{1L'}) D_{m_2 n_2}^{J_2 *}(\omega_{2L'}) D_{m0}^{J}(\omega_{rL'}), \tag{157}$$

where the last subscript is zero because only the two angle α_r, β_r are needed to specify the intermolecular vector orientation. Thus we proceed to a symmetrization of these products by performing first an arbitrary rotation $\omega_{LL'}$ from L' to L. Using the closure relation of Wigner matrices [1] the original rotation from the laboratory can be rewritten starting from an arbitrary auxiliary frame L

$$D_{m,n}^{J *}(\omega_{1L'}) = \sum_{q=-J}^{J} D_{qm}^{J}(\omega_{L'L}) D_{q,n}^{J *}(\omega_{1L}). \tag{158}$$

Summation on all possible orientations $(\omega_{LL'})$ using the integral of three Wigner rotation matrices (Gaunt formula)[1] :

$$\int d\omega_{LL'} D_{qm}^{J *}(\omega_{LL'}) D_{q_2 m_2}^{J_2}(\omega_{LL'}) D_{q_1 m_1}^{J_1}(\omega_{LL'})$$

$$= \frac{8\pi^2}{(2J+1)} \delta_{q_1+q_2,q} \delta_{m_1+m_2,m} C(J_1 J_2 J; q_1 q_2) C(J J' J''; m_1 m_2) \tag{159}$$

gives

$$\mathcal{R} D_{m0}^{J}(\omega_{rL}) D_{m_2,n_2}^{J_2 *}(\omega_{2L}) D_{m_1,n_1}^{J_1 *}(\omega_{1L})$$

$$= \frac{8\pi^2}{(2J+1)} C(J J' J''; m_1 m_2)$$

$$\sum_{q_1,q_2} C(J_1 J_2 J; q_1 q_2) D_{q_1,n_1}^{J_1 *}(\omega_{1L}) D_{q_2,n_2}^{J_2 *}(\omega_{2L}) D_{q,0}^{J}(\omega_{rL}). \tag{160}$$

The quantity on the right is a rotationally invariant combination of the three orientations and could be used to define, apart from a constant, a useful invariant basis. In particular we use the rotationally invariant functions $S^{k_1 k_2}_{J_1 J_2 J}(\omega_1, \omega_2, \omega_r)$, as defined by Stone [25], which read in our notation

$$
\begin{aligned}
S^{k_1,k_2}_{J_1,J_2 J}(\omega_1, \omega_2, \omega_r) = {} & \frac{(i)^{J_1-J_2-J}}{\sqrt{2J+1}} \\
& \times \sum_{m_1,m_2} C(J_1 J_2 J; -m_1 - m_2) D^{J_1 *}_{m_1,k_1}(\omega_1) D^{J_2 *}_{m_2,k_2}(\omega_2) D^{J}_{-m_1-m_2,0}(\omega_r).
\end{aligned}
$$

$$(161)$$

Using this definition and the properties of Clebsch - Gordan coefficients, we find the complex conjugate of the rotational invariant as

$$
S^{k_1,k_2 *}_{J_1,J_2 J}(\omega_1, \omega_2, \omega_r) = (-1)^{k_1+k_2} S^{-k_1,-k_2}_{J_1,J_2 J}(\omega_1, \omega_2, \omega_r). \tag{162}
$$

The rotational invariants form an orthogonal basis of functions for the space $\{\omega_1, \omega_2, \omega_r\}$:

$$
\begin{aligned}
\int d\omega_1 d\omega_2 d\omega_r \quad & S^{k_1 k_2 *}_{J_1, J_2, J}(\omega_1, \omega_2, \omega_r) S^{k'_1 k'_2}_{J'_1, J'_2, J'}(\omega_1, \omega_2, \omega_r) \\
& = \frac{256 \pi^5 \delta_{J_1,J'_1} \delta_{J_2,J'_2} \delta_{J,J'} \delta_{k_1,k'_1} \delta_{k_2,k'_2}}{(2J_1+1)(2J_2+1)(2J+1)}
\end{aligned} \tag{163}
$$

and can be used to expand the intermolecular potential [25, 26] or the pair correlation function[24, 15]. The expansion of an arbitrary function will be

$$
f(r_{12}, \omega_1, \omega_2, \omega_r) = \sum_{\substack{J_1, J_2, J \\ k_1, k_2}} f^{J_1, J_2 J}_{k_1, k_2}(r_{12}) S^{k_1, k_2}_{J_1, J_2 J}(\omega_1, \omega_2, \omega_r). \tag{164}
$$

Since the expansion is valid in an arbitrary frame, we can in particular adopt the intermolecular frame, with z axis along the inter-centre axis and $\omega_r = (000)$. Thus $D^{J *}_{-m_1-m_2,0}(\omega_r) = \delta_{m_1,m_2}$

$$
\begin{aligned}
f(r_{12}, \omega_1, \omega_2, 0) = {} & \sum_{\substack{J_1, J_2 J \\ k_1, k_2}} f^{J_1, J_2 J}_{k_1, k_2}(r_{12}) \frac{(i)^{J_1-J_2-J}}{\sqrt{2J+1}} \\
& \times \sum_{m_1} C(J_1 J_2 J; -m_1 - m_1) D^{J_1 *}_{m_1,k_1}(\omega_1) D^{J_2 *}_{m_1,k_2}(\omega_2) \\
\equiv {} & \sum_{\substack{J_1, J_2, J \\ k_1, k_2, m_1}} f^{J_1, J_2, J}_{k_1, k_2, m_1}(r_{12}) D^{J_1 *}_{m_1,k_1}(\omega_1) D^{J_2 *}_{m_1,k_2}(\omega_2).
\end{aligned} \tag{165}
$$

48

Thus we can switch from a space fixed to a molecule fixed expansion writing the expansion coefficients of one representation in terms of those of the other:

$$f^{J_1,J_2\,J}_{k_1,k_2}(r_{12}) = \sum_{m_1} \frac{(i)^{J_1-J_2-J}}{\sqrt{2J+1}} C(J_1 J_2 J; -m_1 - m_1) f^{J_1,J_2}_{k_1,k_2,m_1}(r_{12}). \tag{166}$$

For uniaxial molecules we only need the subset

$$
\begin{aligned}
S^{J_1,J_2,J}(\omega_1,\omega_2,\omega_r) &\equiv & S^{00}_{J_1,J_2,J}(\omega_1,\omega_2,\omega_r) \\
&=& \frac{i^{J_2-J_1-J}}{(2J+1)^{1/2}} \sum_{m_1,m_2} (-)^{m_1+m_2} C(J_1,J_2,J;m_1,m_2) \\
&\times& D^{J_1\,*}_{m_1,0}(\omega_1) D^{J_2\,*}_{m_2,0}(\omega_2) D^{J\,*}_{-m_1-m_2,0}(\omega_r).
\end{aligned}
\tag{167}
$$

The first few rotational invariants for uniaxial molecules are, in cartesian terms, and using the unit vectors $\hat{\mathbf{u}}_1$, $\hat{\mathbf{u}}_2$ along the z axis of the two molecules [25]

$$S^{000} = 1 \tag{168}$$

$$S^{110} = -\frac{1}{\sqrt{3}} \hat{\mathbf{u}}_1 \cdot \hat{\mathbf{u}}_2 \tag{169}$$

$$S^{101} = -\frac{1}{\sqrt{3}} \hat{\mathbf{u}}_1 \cdot \hat{\mathbf{r}}_{12} \tag{170}$$

$$S^{011} = +\frac{1}{\sqrt{3}} \hat{\mathbf{u}}_2 \cdot \hat{\mathbf{r}}_{12} \tag{171}$$

$$S^{112} = \frac{1}{\sqrt{30}} [\hat{\mathbf{u}}_1 \cdot \mathbf{z}_2 - 3(\hat{\mathbf{u}}_1 \cdot \hat{\mathbf{r}}_{12})(\hat{\mathbf{u}}_2 \cdot \hat{\mathbf{r}}_{12})] \tag{172}$$

$$S^{121} = \frac{1}{\sqrt{30}} [(\hat{\mathbf{u}}_1 \cdot \hat{\mathbf{r}}_{12}) - 3(\hat{\mathbf{u}}_1 \cdot \hat{\mathbf{u}}_2)(\hat{\mathbf{u}}_2 \cdot \hat{\mathbf{r}}_{12})] \tag{173}$$

$$S^{211} = -\frac{1}{\sqrt{30}} [(\hat{\mathbf{u}}_2 \cdot \hat{\mathbf{r}}_{12}) - 3(\hat{\mathbf{u}}_1 \cdot \hat{\mathbf{u}}_2)(\hat{\mathbf{u}}_1 \cdot \hat{\mathbf{r}}_{12})] \tag{174}$$

$$S^{220} = \frac{1}{2\sqrt{5}} [3(\hat{\mathbf{u}}_1 \cdot \hat{\mathbf{u}}_2)^2 - 1] \tag{175}$$

$$S^{202} = \frac{1}{2\sqrt{5}} [3(\hat{\mathbf{u}}_1 \cdot \hat{\mathbf{r}}_{12})^2 - 1] \tag{176}$$

$$S^{022} = \frac{1}{2\sqrt{5}} [3(\hat{\mathbf{u}}_2 \cdot \hat{\mathbf{r}}_{12})^2 - 1] \tag{177}$$

$$
\begin{aligned}
S^{222} = \frac{1}{\sqrt{70}} [& 2 - 3(\hat{\mathbf{u}}_1 \cdot \hat{\mathbf{u}}_2)^2 - 3(\hat{\mathbf{u}}_1 \cdot \hat{\mathbf{r}}_{12})^2 - 3(\hat{\mathbf{u}}_2 \cdot \hat{\mathbf{r}}_{12})^2 \\
& +9(\hat{\mathbf{u}}_1 \cdot \hat{\mathbf{u}}_2)(\hat{\mathbf{u}}_1 \cdot \hat{\mathbf{r}}_{12})(\hat{\mathbf{u}}_2 \cdot \hat{\mathbf{r}}_{12})]
\end{aligned}
\tag{178}
$$

$$S^{224} = \frac{1}{4\sqrt{70}}[1 + 2(\hat{\mathbf{u}}_1 \cdot \hat{\mathbf{u}}_2)^2 - 5(\hat{\mathbf{u}}_1 \cdot \hat{\mathbf{r}}_{12})^2$$
$$-5(\hat{\mathbf{u}}_2 \cdot \hat{\mathbf{r}}_{12})^2 - 20(\hat{\mathbf{u}}_1 \cdot \hat{\mathbf{u}}_2)(\hat{\mathbf{u}}_1 \cdot \hat{\mathbf{r}}_{12})(\hat{\mathbf{u}}_2 \cdot \hat{\mathbf{r}}_{12})$$
$$+35(\hat{\mathbf{u}}_1 \cdot \hat{\mathbf{r}}_{12})^2(\hat{\mathbf{u}}_2 \cdot \hat{\mathbf{r}}_{12})^2] \tag{179}$$
$$S^{242} = \frac{1}{4\sqrt{70}}[1 - 5(\hat{\mathbf{u}}_1 \cdot \hat{\mathbf{u}}_2)^2 + 2(\hat{\mathbf{u}}_1 \cdot \hat{\mathbf{r}}_{12})^2 - 5(\hat{\mathbf{u}}_2 \cdot \hat{\mathbf{r}}_{12})^2$$
$$-20(\hat{\mathbf{u}}_1 \cdot \hat{\mathbf{u}}_2)(\hat{\mathbf{u}}_1 \cdot \hat{\mathbf{r}}_{12})(\hat{\mathbf{u}}_2 \cdot \hat{\mathbf{r}}_{12}) + 35(\hat{\mathbf{u}}_1 \cdot \hat{\mathbf{u}}_2)^2(\hat{\mathbf{u}}_2 \cdot \hat{\mathbf{r}}_{12})^2]. \tag{180}$$

Higher invariants can be generated e.g. using the coupling formula of two rotational invariants, obtained from the coupling formulae for Wigner rotation matrices [26].

References

1. Rose, M.E. (1957) *Elementary Theory of Angular Momentum*. Wiley, New York.
2. Chandrasekhar, S. (1992) *Liquid Crystals*. 2nd ed., Cambridge U.P., Cambridge.
3. Allen, M.P. and Tildesley, D.J. (1987) *Computer Simulation of Liquids*. Clarendon Press, Oxford.
4. Zannoni, C. (1979) in *The Molecular Physics of Liquid Crystals*. Luckhurst, G.R. and Gray, G.W. (eds.), Academic Press, London, Chap. 3, p. 51.
5. Zannoni, C. and Guerra, M. (1981) *Molec. Phys.*, **44**, 849.
6. Biscarini, F., Chiccoli, C., Pasini, P. and Zannoni, C. (1991) *Molec. Phys.*, **73**, 439.
7. Abramowitz, M. and Stegun, I.A. (eds.) (1964) *Handbook of Mathematical Functions*. Dover.
8. Tsvetkov, V. (1939) *Acta Physicoch. U.S.S.R.*, **10**, 557.
9. Leenhouts, F., de Jeu, W. and Dekker, A.J. (1979) *J. Physique*, **40**, 989.
10. Wu, S.T. and Cox, R.J. (1988) *J. Appl. Phys.*, **64**, 821.
11. Luckhurst, G.R. and Veracini, C.A. (eds.) (1994) *The Molecular Dynamics of Liquid Crystals*. Kluwer, Dordrecht.
12. Zannoni, C. (1994) in *The Molecular Dynamics of Liquid Crystals*. Luckhurst, G.R. and Veracini, C.A. (eds.), Kluwer, Dordrecht.
13. Viellard - Baron, J. (1974) *Molec. Phys.*, **28**, 809.
14. Eppenga, R. and Frenkel, D. (1984) *Molec. Phys.*, **52**, 1303.
15. Zannoni, C. (1979) in *The Molecular Physics of Liquid Crystals*. Luckhurst, G.R. and Gray, G.W. (eds.), Academic Press, London, Chap. 5, p. 191.
16. Fabbri, U. and Zannoni, C. (1986) *Molec. Phys.*, **58**, 763.
17. Gay, J.G. and Berne, B.J. (1981) *J. Chem. Phys.*, **74**, 3316.
18. Berardi, R., Emerson, A.P.J and Zannoni, C. (1993) *J. Chem. Soc. Faraday Trans.*, **89**, 4069.
19. Biscarini, F., Chiccoli, C., Pasini, P., Semeria, F. and Zannoni, C. (1995) *Phys. Rev. Lett.*, **75**, 1803.
20. Straley, J.P. (1974) *Phys. Rev. A*, **10**, 1881.
21. Allen, M.P. (1990) *Liq. Cryst.* **8**, 499.
22. Luckhurst, G.R. and Romano, S. (1980) *Molec. Phys.*, **40**, 129.
23. Jepsen, D.W. and Friedman, H.L. (1963) *J. Chem. Phys.*, **38**, 846.
24. Blum, L. and Torruella, A.J. (1972) *J. Chem. Phys.*, **56**, 303.
25. Stone, A.J. (1978) *Molec. Phys.*, **36**, 241.

26. Price, S.L., Stone, A.J. and Alderton, M. (1984) *Molec. Phys.*, **52**, 987.
27. Berardi, R., Orlandi, S. and Zannoni, C. (1996) *Chem. Phys. Lett.*, **261**, 357.
28. Levelut, A.M., Tarento, R.J., Hardouin, F., Achard, M.F. and Sigaud, G. (1981) *Phys. Rev. A*, **24**, 2180; Prost, J. and Barois, P. (1983) *J. Chim. Physique*, **80**, 65.
29. Emerson, A.P.J., Hashim, R. and Luckhurst, G.R. (1992) *Molec. Phys.*, **76**, 241.
30. Humphries, R.L., James, P.G. and Luckhurst, G.R. (1972) *J. Chem. Soc. Faraday Trans. 2*, **68**, 1031.
31. Barker, J.A. and Henderson, D. (1976) *Rev. Mod. Phys.*, **48**, 587.
32. Hansen, J.P. and McDonald, I.R. (1976) *Theory of simple liquids.* Academic Press, London.
33. Hummer, G., Grønbech-Jensen, N. and Neumann, M. (1998) *J. Chem. Phys.*, **109**, 2791.
34. Evans, R. (1989) in *Liquids at interfaces.* Charvolin, J., Joanny, J.F. and Zinn-Justin, J. (eds.), Elsevier, Amsterdam.
35. del Río, E.M., de Miguel E. and Rull, L.F., (1995) *Physica A*, **213**, 138.
36. Kubo, R. (1966) *Rep. Progr. Phys.*, **29**, 255.
37. Steele, W.A. (1969) in *Transport Phenomena in Fluids.* Hanley, H.J.M. (ed.), Dekker, New York, p. 209.
38. Berne, B.J. (1971) *Physical Chemistry, an Advanced Treatise.* Eyring, H., Henderson, D. and Jost, W. (eds.), Academic Press, **8B**, p. 539.
39. Berne, B.J. and Pecora, R. (1976) *Dynamic Light Scattering.* Wiley, N.Y.
40. Friedman, H.L. (1985) *A Course in Statistical mechanics.* Prentice Hall.
41. Goldstein, H. (1953) *Classical Mechanics.* Addison-Wesley.
42. Böttcher, C.J.F. and Bordewijk, P. (1978) *Theory of Electric Polarization.* vol. II, Elsevier.
43. Luckhurst, G.R. and Zannoni, C. (1975) *Proc. Roy. Soc.*, **A343**.
44. Gordon, R.G. (1968) *Advances in Magnetic Resonance.* Waugh, J.S. (ed.), Academic Press, **3**, p. 1.

PHASE BEHAVIOR OF LYOTROPIC LIQUID CRYSTALS

D. FRENKEL

FOM Institute for Atomic and Molecular Physics
Kruislaan 407
1098 SJ Amsterdam
The Netherlands.

Abstract. In these lectures, I discuss techniques to compute the phase diagram of lyotropic liquid crystals. I review the standard techniques to compute the free-energy of various phases. Subsequently, I focus on phase transitions in liquid crystals. Recent techniques to determine the dependence of phase boundaries on the shape and flexibility of the constituent molecules, are discussed. Finally, I devote attention to Monte Carlo techniques that are particularly suited to study the phase behavior of flexible molecules.

1. Introduction

These lectures focus on some of the technical aspects of the simulation of lyotropic liquid crystals. First, I review several techniques that can be used to locate first-order phase transitions. The availability of such techniques is of particular relevance for liquid-crystal simulations because many of the standard techniques to study phase-coexistence in computer simulations of simple atomic or molecular systems cannot be applied to liquid crystals. A key quantity that must be computed in order to determine the point where two phases coexist, is the chemical potential μ of the molecules in either phase. Most techniques to compute the chemical potential are limited to model systems consisting of small molecules at low densities. Actual liquid-crystal formers (both thermotropic and lyotropic) rarely meet this specification. I shall therefore also discuss techniques to study systems consisting of large, flexible molecules.

P. Pasini and C. Zannoni (eds.), Advances in the Computer Simulations of Liquid Crystals, 51–72.

2. Phase transitions and free energy

The most direct way to study phase coexistence in a computer simulation would be to simply change the temperature or pressure of the system under study until a phase transformation occurs. In the real world it is often (but by no means always) possible to ensure that such a phase change takes place reversibly. The coexistence point is defined as the point where the reversible phase transformation occurs. At coexistence, the temperature and pressure of the coexisting phases are equal. In addition, the chemical potential of every individual species must have the same value in every phase. Following the seminal work of Panagiotopoulos [1], much progress has been made in the direct simulation of phase coexistence of moderately dense fluid phases using the "Gibbs-ensemble" method. This method relies on the fact that it is possible to satisfy the above conditions for coexistence between two bulk phases (or, to be more precise, homogeneous phases with periodic boundary conditions) by allowing them to exchange both volume and molecules. Unfortunately, such a direct simulation method is of limited value in computer simulations of transitions involving dense phases that have some translational order. The reason why the Gibbs-ensemble method breaks down under those circumstances is twofold. First of all, pronounced hysteresis effects are usually observed in computer simulations of a strong first order phase transition, such as melting. This implies that it is difficult for the molecules in the system to spontaneously rearrange from a configuration belonging to the 'old' phase, to one that corresponds to the 'new' phase. But even if the two different phases have somehow been prepared, it is usually impossible to exchange particles between them. As a consequence, we cannot ensure the equality of the chemical potential in the two phases.

Under those circumstances, it is still possible to locate the point where the two phases coexist. But in order to do so, we must explicitly compute the chemical potential of the homogeneous phases at the same temperature and pressure and find the point where the two μ's are equal. The direct calculation of chemical potentials of dense phases is the first topic that I address in these lectures. In practice, it is often the Helmholtz free energy F, rather than the chemical potential μ, that is computed. In what follows, I shall use the terms chemical-potential calculations and free-energy calculations interchangeably, as F and μ are simply related (e.g., for a one-component system of N particles in a volume V at pressure P, we have $F = N\mu - PV$).

2.0.1. *The natural way*
When discussing techniques to measure free energies, it is useful to recall how such quantities are measured experimentally. In the real world, free

energies cannot be obtained from a single measurement either. What can be measured, however, is the derivative of the free energy with respect to volume V and temperature T:

$$\left(\frac{\partial F}{\partial V}\right)_{NT} = -P \qquad (1)$$

and

$$\left(\frac{\partial F/T}{\partial 1/T}\right)_{NV} = E. \qquad (2)$$

Here P is the pressure and E the energy of the system under consideration. The trick is now to find a reversible path that links the state under consideration to a state of known free energy. The change in F along that path can then simply be evaluated by integration of Eqns. 1 and 2. In the real world the free energy of a substance can only be evaluated directly for a very limited number of thermodynamic states. One such state is the ideal gas phase, the other is the perfectly ordered ground state at $T = 0K$. In computer simulations, the situation is quite similar. In order to compute the free energy of a dense liquid, one may construct a reversible path to the very dilute gas phase. It is not really necessary to go all the way to the ideal gas. But at least one should reach a state that is sufficiently dilute that the free energy can be computed accurately, either from knowledge of the first few terms in the virial expansion of the compressibility factor PV/Nk_BT, or that the chemical potential can be computed by other means (see below). For the solid, the ideal gas reference state is less useful (although techniques have been developed to construct a reversible path from a dense solid to a dilute (lattice-) gas [2]). The obvious reference state for solids is the harmonic lattice. Computing the absolute free energy of a harmonic solid is relatively straightforward, at least for atomic and simple molecular solids. However, not all solid phases can be reached by a reversible route from a harmonic reference state. For instance, in molecular systems it is quite common to find a strongly an-harmonic plastic phase just below the melting line. This plastic phase is not (meta-) stable at low temperatures.

2.0.2. *Artificial reversible paths*

Fortunately, in computer simulations we do not have to rely on the presence of a 'natural' reversible path between the phase under study and a reference state of known free energy. If such a path does not exist, we can construct an artificial path. This is in fact a standard trick in statistical mechanics (see e.g. [3]). It works as follows: Consider a case where we need to know the free energy $F(V,T)$ of a system with a potential energy function U_1, where U_1 is such that no 'natural' reversible path exists to a state of known free energy. Suppose now that we can find another model system with a

54

potential energy function U_0 for which the free energy *can* be computed exactly. Now let us define a generalized potential energy function $U(\lambda)$, such that $U(\lambda = 0) = U_0$ and $U(\lambda = 1) = U_1$. The free energy of a system with this generalized potential is denoted by $F(\lambda)$. Although $F(\lambda)$ itself cannot be measured directly in a simulation, we can measure its derivative with respect to λ:

$$\left(\frac{\partial F}{\partial \lambda}\right)_{NVT\lambda} = \left\langle \frac{\partial U(\lambda)}{\partial \lambda} \right\rangle_{NVT\lambda}. \tag{3}$$

If the path from $\lambda = 0$ to $\lambda = 1$ is reversible, we can use Eqn. 3 to compute the desired $F(V,T)$. We simply measure $< \partial U/\partial \lambda >$ for a number of values of λ between 0 and 1. Typically, 10 quadrature points will be sufficient to get the absolute free energy per particle accurate to within $0.01\, k_B T$. It is however important to select a reasonable reference system. One of the safest approaches is to choose as a reference system an Einstein crystal with the same structure as the phase under study [4]. This choice of reference system makes it extremely improbable that the path connecting $\lambda = 0$ and $\lambda = 1$ will cross an (irreversible) first order phase transition from the initial structure to another, only to go back to its original structure for still larger values of λ. Nevertheless, it is important that the parametrization of $U(\lambda)$ be chosen carefully. Usually, a linear parametrization (i.e. $U(\lambda) = \lambda U_1 + (1-\lambda)U_0$) is quite satisfactory. But occasionally such a parametrization may lead to weak (and relatively harmless) singularities in Eqn. 3 for $\lambda \to 0$. More details about such free energy computations can be found in refs. [5, 6].

2.1. PHASE TRANSITIONS IN LIQUID CRYSTALS

2.1.1. *Isotropic and nematic phases*
For the isotropic phase we can take the ideal gas as a reference and integrate along the equation of state using the density ρ as the integration parameter

$$\frac{F(\rho, L)}{Nk_B T} - \frac{F_{id}(\rho)}{Nk_B T} = \int_0^\rho \frac{P(\rho', L) - \rho' k_B T}{\rho'^2 k_B T} d\rho'. \tag{4}$$

The isotropic-nematic transition usually exhibits some hysteresis. As a consequence, direct integration of the equation of state through the transition region is subject to statistical errors. This problem can be alleviated by switching on a strong ordering field. In the presence of such a field, the first-order isotropic-nematic transition is suppressed and a reversible expansion to the dilute gas becomes possible [7].

A second method that can be used to compute the free energy of the nematic phase is based on the particle-insertion method of Widom [8]. This method was first applied to the evaluation of the free energy of nematics by Eppenga and Frenkel [9]. As I shall discuss particle-insertion schemes in some detail below, I defer the discussion of this technique to section 4.1. Suffice it to say that this scheme works best for strongly anisometric molecules that undergo a transition to the nematic phase at low density.

2.1.2. *Solid phase*

The strong first order transition separating the solid phase from the other phases rules out the integration along the equation of state. Instead, we choose as reference system for the solid an Einstein crystal with the same structure [4]. Now the reversible path transforms the original system to an Einstein crystal with fixed center-of-mass, by gradually coupling the atoms to their equilibrium lattice position. For a system of anisometric particles, the orientation also needs to be coupled to an aligning field. The Hamiltonian that can be used to achieve the coupling is

$$\beta H_{\mu,\lambda} = \mu \sum_i \left(\mathbf{r}_i - \mathbf{r}_i^0 \right)^2 + \lambda \sum_i \sin^2 \theta_i, \tag{5}$$

where μ and λ are the coupling constants which determine the strength of the harmonic forces. The free energy of the system can be related to the (known) free energy of an Einstein crystal by thermodynamic integration

$$\frac{\beta F(\rho^*)}{N} = \frac{\beta F_{ein}}{N} - \int_0^{\mu_{max}} d\mu \left\langle \Delta r^2 \right\rangle_\mu - \int_0^{\lambda_{max}} d\lambda \left\langle \sin^2 \theta \right\rangle_\lambda - \frac{\ln V}{N} . \tag{6}$$

Here $\left\langle \Delta r^2 \right\rangle_\mu$ is the mean-square displacement and $\left\langle \sin^2 \theta \right\rangle_\lambda$ the mean square sine of the angle between a particle and the aligning field in a simulation with Hamiltonian $H_{\mu,\lambda}$. The free energy of the Einstein crystal (with fixed center-of-mass) in the limit of large coupling constants is given by

$$\beta F_{ein} = \frac{3}{2} \ln N - \frac{3}{2}(N-1) \ln \frac{\pi}{\mu} - N \ln \frac{2\pi}{\lambda} . \tag{7}$$

By performing several simulations at different values of μ and λ one can numerically evaluate the integrals in Eqn. 6. As the values μ and λ at which the integrand is evaluated can be chosen freely, the error in the integration can be minimized by using Gauss-Legendre quadrature. Occurrence of any first order transition was avoided by performing two Gauss-Legendre integrations in succession. The first fixes the positions while leaving $\lambda = 0$, the second aligns all spherocylinders while keeping $\mu = \mu_{max}$. It is convenient to choose the maximum values of λ and μ such that in a simulation at these maximum values, there are essentially no overlaps between the particles. Otherwise it is necessary to correct Eqn. 7 for the occurrence of overlaps [4].

56

2.1.3. *Smectic phase*

The smectic phase does not have an obvious reference state for which the free energy is known. In the case of hard spherocylinders, Veerman and Frenkel [10] used the fully aligned system as a reference system. However, the free energy of the aligned parallel smectic itself is subject to numerical error. An alternative is to couple the mesogenic molecules with an harmonic spring to the smectic layer to which they belong and subsequently align them. In this way, the smectic phase of hard spherocylinders can be transformed into what is essentially a 2D hard disk fluid for which the free energy is well known [2]. In principle, one could apply the Einstein integration method used in the previous section with one difference: the position field couples only the z-coordinates of the particles to the layer positions and leaves the x, y coordinates completely free. If we consider the first part of the integration, where the particle are confined to their layers, the free energy of smectic phase can be related to this planar system by

$$\frac{\beta F_{\mu=0}}{N} = \frac{\beta F_{\mu=\mu_0}^{planar}}{N} - \int_0^{\mu_0} d\mu \left\langle \Delta r^2 \right\rangle_\mu - \frac{\ln V}{N} . \tag{8}$$

In the second integration, the difficulty arises that a infinite amount of aligning energy is needed to get all spherocylinders completely parallel.

$$\frac{\beta F_{\lambda=0,\mu=\mu_0}^{planar}}{N} = \frac{\beta F_{\lambda=\infty,\mu=\mu_0}^{planar,aligned}}{N} - \int_0^\infty d\lambda \left\langle \sin^2 \theta \right\rangle_\lambda . \tag{9}$$

To keep the energy values finite, we subtract on both sides of this equation the free energy of an ideal rotator in the same field.

$$\frac{\beta F_{\lambda=\lambda_0,\mu=\mu_0}^{planar,id}}{N} = \frac{\beta F_{\lambda=\infty,\mu=\mu_0}^{planar,aligned,id}}{N} - \int_{\lambda_0}^\infty d\lambda \left\langle \sin^2 \theta \right\rangle_{id,\lambda} \tag{10}$$

which results in

$$\frac{\beta F_{\lambda=0,\mu=\mu_0}^{planar}}{N} = \frac{\beta F_{\lambda=\lambda_0,\mu=\mu_0}^{planar,id}}{N} + \frac{\beta F_{\lambda=\infty,\mu=\mu_0}^{planar,aligned,ex}}{N}$$
$$- \int_0^{\lambda_0} d\lambda \left\langle \sin^2 \theta \right\rangle_\lambda - \int_{\lambda_0}^\infty d\lambda \left[\left\langle \sin^2 \theta \right\rangle_\lambda - \left\langle \sin^2 \theta \right\rangle_{id,\lambda} \right] . \tag{11}$$

The excess free energy of the completely aligned planar system, that is $\beta F_{\lambda=\infty,\mu=\mu_0}^{planar,aligned,ex}$, is equal to the excess free energy of a 2D hard disk fluid. The free energy of the ideal planar system (with fixed center-of-mass) in the limit of large coupling constants is given by

$$\beta F_{\lambda=\lambda_0,\mu=\mu_0}^{planar,id} = \frac{1}{2} \ln N - \frac{1}{2}(N-1) \ln \frac{\pi}{\mu} - N \ln \frac{2\pi}{\lambda} . \tag{12}$$

The integral over the difference of the $\sin^2$ terms in Eqn. 11 is finite. We can change the integration boundaries by substituting $\lambda = 1/\xi^2$.

$$\int_{\lambda_0}^{\infty} d\lambda \left[\left\langle \sin^2 \theta \right\rangle_{\lambda} - \left\langle \sin^2 \theta \right\rangle_{id,\lambda} \right] = \int_0^{\frac{1}{\sqrt{\lambda_0}}} d\xi\, 2\lambda^{\frac{2}{3}} \left[\left\langle \sin^2 \theta \right\rangle_{\lambda} - \left\langle \sin^2 \theta \right\rangle_{id,\lambda} \right].$$
(13)

In conventional MC sampling, the statistical error of both terms in the integrand is larger than the difference itself. Under those circumstances, the following approach is useful: instead of rotating a spherocylinder i around an angle $d\theta_i$ we choose a completely new trial value of θ_i from the probability distribution

$$P(\theta) \sim \exp(-\beta\lambda \sin^2 \theta) \,.$$
(14)

This is the equilibrium distribution for an ideal rotator with a Hamiltonian according to Eqn. 5 and results in the correct value for $\left\langle \sin^2 \theta \right\rangle_{id,\lambda}$. If no overlap occurs the trial move will be accepted and we will have

$$\sin_\lambda^2 \theta_i - \sin_{id,\lambda}^2 \theta_i = 0 \,.$$
(15)

If an overlap does occur the trial move will be rejected and the particle will retain its old value. The difference now will be

$$\sin_\lambda^2 \theta_i - \sin_{id,\lambda}^2 \theta_i = \sin_\lambda^2 \theta_i^{old} - \sin_{id,\lambda}^2 \theta_i^{new} \,.$$
(16)

The statistical error in the average of the difference is always smaller than the average itself. This will enable us to determine the integrand more accurately. By combining Eqns. 8,11 and 13 the complete expression for free energy of the smectic phase follows

$$\frac{\beta F_{\mu=0}}{N} = \frac{\beta F_{disk}^{ex}}{N} - \frac{\beta F_{\lambda=\lambda_0,\mu=\mu_0}^{planar,id}}{N} - \frac{\ln V}{N} - \int_0^{\mu_0} d\mu \left\langle \Delta r^2 \right\rangle_{\mu}$$

$$- \int_0^{\lambda_0} d\lambda \left\langle \sin^2 \theta \right\rangle_{\lambda} - \int_0^{\frac{1}{\sqrt{\lambda_0}}} d\xi\, 2\lambda^{\frac{2}{3}} \left[\left\langle \sin_\lambda^2 \theta - \sin_{id,\lambda}^2 \theta \right\rangle \right] \,.$$
(17)

The excess hard disk free energy can be obtained by subtracting the ideal term $\beta F_{disk}^{id} = \ln \rho$ from the free energy in Ref [2]. Another problem that may arise is that, as the smectic phase forms from the nematic, the fluctuations in the number of particles per layer get frozen in. As a result, different layers may have different $(2D)$ densities. To ensure that the $2D$ densities in the smectic layers are equal throughout the system one can use shifted periodic boundaries: the periodic boundaries in the x direction are shifted exactly one layer period along the z-axis, while leaving them the same in the y and z direction. In this way, a particle leaving the simulation

58

box at the left side will reenter the box at the right *one layer higher*. This particle can diffuse through the whole system, as there is effectively only one layer. This ensures that fluctuations in the number of particles per smectic layer can relax, even at high density where normal inter-layer diffusion is completely frozen out [11].

2.1.4. *Nematic-smectic free energy difference*

Even when all tricks to compute the absolute free energies of the different phases are used, it remains difficult to locate the nematic-smectic coexistence region with reasonable accuracy. However, we are hardly ever interested in the absolute free energies themselves. It is often attractive to calculate the free energy difference between a stable nematic and a stable smectic directly. In order to find a reversible path from the nematic to smectic one can use the following Hamiltonian

$$H_\lambda = \lambda \left(\sum_i \cos(\frac{2\pi n r_{i,z}}{L_z}) + 1 \right), \tag{18}$$

where n is the number of smectic layers, L_z the box length in the z-direction, $r_{i,z}$ the z-coordinate of particle i and λ the coupling parameter determining the strength of the smectic ordering. At low density this Hamiltonian will produce, by increasing λ, a gradual transition from a nematic to a smectic phase. We start with a smectic phase and applied a cosine field at large enough λ. Subsequently, the smectic is expanded to lower density, while measuring the pressure. Finally, the cosine field is slowly turned off. The free energy difference now simply is

$$\frac{\Delta F_{ns}}{N} = \frac{F_{smec}}{N} - \frac{F_{nem}}{N} = \int_0^{\lambda_{max}} d\lambda \left\langle \sum_i \cos(\frac{2\pi n r_{i,z}}{L_z}) + 1 \right\rangle_{smec}$$

$$- \int_{\rho_n}^{\rho_s} \frac{P(\rho)}{\rho^2} d\rho - \int_0^{\lambda_{max}} d\lambda \left\langle \sum_i \cos(\frac{2\pi n r_{i,z}}{L_z}) + 1 \right\rangle_{nem}. \tag{19}$$

Of course, the value of λ_{max} should be chosen large enough that the first-order S-N transition is completely suppressed. For more details, see ref. [11]

2.1.5. *Changing the particle shape*

Usually, we are interested in the phase diagram of a model system, for a whole range of model parameters. In the case of spherocylinders, the relevant parameter is the length-to-width ratio L/D (strictly speaking, L is the length of the cylindrical part – the total length is $L + D$). Once the free energy of a phase for a given value of L/D has been determined at some density ρ, the free energy at other values of L/D can be obtained by a simple thermodynamic integration scheme. We can compute the reversible work involved in changing the aspect ratio of the spherocylinders from L_0 to L and subsequently changing the density from ρ_0 to ρ (for convenience, we have chosen $D = 1$):

$$\frac{F(\rho, L)}{N} = \frac{F_0(\rho_0, L_0)}{N} + \int_{L_0}^{L} \left(\frac{\partial F}{\partial L}\right)_{\rho_0} dL + \int_{\rho_0}^{\rho} \frac{P(\rho, L)}{\rho^2} d\rho . \qquad (20)$$

The pressure is obtained from an MD simulation in the usual way, by time averaging the virial.

$$\frac{\beta P}{\rho} - 1 = \frac{1}{3}\beta \sum_{i \leq j} \langle \mathbf{f}_{ij} \cdot \mathbf{r}_{ij} \rangle , \qquad (21)$$

where r_{ij} is the vector joining the centers of mass of particles i and j, and f_{ij} denotes the (impulsive) force on j due to i. The derivative $\kappa = (\partial F/\partial L)_\rho$ can be measured at the same time by taking the projection of the intermolecular force along the particle axis.

$$\kappa = \left(\frac{\partial F}{\partial L}\right)_\rho = \frac{1}{2} \sum_{ij} \langle \mathbf{f}_{ij} \cdot (\mathbf{u}_i + \mathbf{u}_j) \rangle . \qquad (22)$$

The average κ is calculated at constant number density ρ. However, it is more convenient to measure it at constant reduced density ρ^* (i.e. at a constant fraction of the close-packing density). If we denote this derivative by κ', we get

$$\kappa' = \left(\frac{\partial F}{\partial L}\right)_{\rho^*} = \left(\frac{\partial F}{\partial L}\right)_\rho + \left(\frac{\partial F}{\partial \rho}\right)_L \left(\frac{\partial \rho}{\partial L}\right)_{\rho^*} = \left(\frac{\partial F}{\partial L}\right)_\rho - \frac{\sqrt{3}}{2\rho^*} P(\rho^*, L) \quad (23)$$

and Eqn. 20 becomes

$$\frac{F(\rho^*, L)}{N} = \frac{F_0(\rho_0, L_0)}{N} + \qquad (24)$$

$$\int_{L_0}^{L} dL \left(\left(\frac{\partial F}{\partial L}\right)_\rho - \frac{\sqrt{3}}{2\rho^*} P(\rho^*, L) \right) dL + \int_{\rho_0}^{\rho^*} \frac{1}{\rho_{cp}(L)} \frac{P(\rho^*, L)}{\rho^{*2}} d\rho^* .$$

2.2. GIBBS-DUHEM INTEGRATION

The location of a fluid-solid coexistence curve can be determined by performing several free-energy calculations and measurements of the equation-of-state for a large number of L/D values. However, this approach is computationally rather expensive. To avoid this problem, we use a modification of a method that was recently developed by Kofke to trace coexistence curves [12]. The advantage of this method is that only equation-of-state information *at the coexistence curve* is required to follow the L/D-dependence of the melting curve. In its original form, the Kofke scheme is based on the Clapeyron equation which describes the temperature-dependence of the pressure at which two phases coexist:

$$\frac{dP}{dT} = \frac{\Delta H}{T\Delta V} , \qquad (25)$$

where ΔH is the molar enthalpy difference and ΔV the molar volume difference of the two phases. This equation is not self starting, in the sense that one point on the coexistence curve must be known before the rest of the curve can be computed by integration of Eqn. 25.

For hard-core systems, we are not interested in the (trivial) temperature dependence of the coexistence curve, but in the dependence of the coexistence pressure on L/D, the shape anisotropy of the spherocylinders. In order to obtain a Clapeyron-like equation relating the coexistence pressure to L/D, we should first write down the explicit dependence of the (Gibbs) free energy of the system on L/D:

$$dG = N\mu = VdP + \kappa dL , \qquad (26)$$

where κ is the derivative $(\partial F/\partial L)_\rho$ defined in Eqn. 22 and where we have used the fact that D is our unit of length. Along the coexistence curve, the difference in chemical potential of the two phases is always equal to zero. Hence,

$$\Delta\mu = \Delta v dP + \frac{1}{N}\Delta\kappa dL = 0 , \qquad (27)$$

where Δv is the difference in molar volume of the two phases at coexistence and $\Delta\kappa = \kappa_1 - \kappa_2$. From Eqn. 27 we can immediately deduce the equivalent of the Clausius-Clapeyron equation

$$\frac{dP}{dL} = -\frac{1}{N}\frac{\Delta\kappa}{\Delta v} . \qquad (28)$$

In Kofke's application of the Gibbs-Duhem method, the MC simulations are carried out in the isothermal-isobaric (NPT) ensemble. However, in the present case (hard-core particles), it is more efficient to use Molecular Dynamics to compute the derivative κ. In practice, we use a hybrid approach where MD simulations are embedded in a constant NPT-MC scheme. True constant-pressure MD is not an attractive option for hard-core models.

3. Simulations at infinite aspect ratio

3.1. SCALING

Most theoretical information about lyotropic liquid crystals has been obtained in the limit of infinite aspect ratio. Clearly, it would be interesting to perform simulations in the same limit. At first sight this seems impossible because in general the system size scales with L^3. However, at finite reduced density ρ^*, i.e. not in the isotropic phase or low ρ^* nematic phase, the average angle θ that a particle makes with the director scales as $1/L$, which means that the particles are (almost) completely aligned. In this regime, we can bring the volume down to finite sizes by scaling the system along the director (chosen to be along the z-axis) with a factor L [11]. This will change the shape of the particle from a spherocylinder to a shifted cylinder of height 1 and diameter D. The height of the cylinder is always 1 because the angle $\theta \propto 1/L$ and the difference in height $1 - \cos(1/L) \approx 1/L^2$ vanishes as $L/D \to \infty$. The shift of the cylinder in the xy plane perpendicular to the director is finite because it is given by $L \sin \theta = \mathcal{O}(\mathcal{D})$ in the limit $L/D \to \infty$. The top and bottom end of the cylinder are flat and always perpendicular to the director. The hemispheres of the spherocylinder have completely disappeared by the scaling procedure.

Because the shape of the particle is different from a spherocylinder we need a new overlap criterion. This is given by the shortest distance between two particle axes in the xy plane. In the xy plane a cut through the cylinder results in a circle of diameter D. Therefore, if the shortest distance is smaller than the diameter D, an overlap will occur.

Scaling of the box in this particular way will not effect the reduced density because the close packing density will scale in the same way as the number density. The pressure will be multiplied by a factor L, whereas P/ρ remains unaffected by the scaling. We can therefore measure the equation of state in this limit using normal NPT-MC simulations.

Because the particles are free to shift any arbitrary amount in the xy plane it is convenient to keep the nematic director always along the z-axis. That is, we keep the total amount of shift in the xy plane equal to zero. This can be achieved by starting with a completely aligned system and shift two particles with the same amount in opposite direction at every MC trial move. In order to avoid multiple overlaps, shifts larger than half the box-length are forbidden. Standard MC trial moves are not very effective in reproducing the collective motion of tilted layers. In the smectic phase, we therefore allowed for two neighboring layers to tilt collectively by equal but opposite amounts, so that the constraint of a constant director is satisfied.

A similar scaling technique can be (and has been [13]) applied to oblate hard particles. In this case, the method can be used to study the nematic,

62

columnar and crystalline phases of disklike particles.

3.2. MIXTURES

Almost all liquid crystals of practical interest are mixtures. Either mixtures of different mesogenic molecules, or mixtures of a mesogenic and non-mesogenic molecules. In simulations, the fact that we deal with mixtures, rather than pure compounds, poses no special problems. For instance, Camp et al. [14] have studied the phase diagram of mixtures of prolate and oblate ellipsoids. The simulations reveal the presence of an isotropic phase, two nematics and one biaxial nematic.

An example of a simulation of a mixture of a mesogen and a non-mesogen is the work by Bolhuis et al. [15] on mixtures of hard spherocylinders and polymers. In this case, the presence of the polymers leads to the appearance of fluid-fluid phase transitions in the various liquid-crystalline phases.

More interesting from a technical point of view are simulations of mixtures of particles with a continuous size or shape distribution. The general approach in this case is discussed in a paper by Bolhuis and Kofke, who studied the freezing of polydisperse hard-sphere mixtures [16]. This approach was extended to polydisperse lyotropic liquid crystals by Bates and Frenkel [17]. The interesting point is that, in this case, poly-dispersity may induce phases that are absent in the phase diagram of mono-disperse particles. For example: mono-disperse spherocylinders do not exhibit a columnar phase, but polydisperse spherocylinders do.

4. Chemical potential of flexible molecules

Before discussing techniques to measure the chemical potential of chain molecules, I first review the particle-insertion method of Widom [8].

4.1. THE PARTICLE INSERTION METHOD

A particularly simple and elegant method to measure the chemical potential μ of a species in a pure fluid or in a mixture is the 'particle-insertion' method (often referred to as the Widom-method [8]). The statistical mechanics that is the basis for this method is quite simple. Consider the definition of the chemical potential μ_α of a species α. From thermodynamics we know that μ is defined as:

$$\mu = \left(\frac{\partial G}{\partial N} \right)_{PT}$$

$$\begin{aligned}
&= \left(\frac{\partial F}{\partial N}\right)_{VT} \\
&= -T\left(\frac{\partial S}{\partial N}\right)_{VE} .
\end{aligned} \tag{29}$$

Where G, F and S are the Gibbs free energy, the Helmholtz free energy and the entropy, respectively. Here, and in the next few paragraphs we focus on a one-component system, and hence we drop the subscript α. Let us first consider the situation at constant NVT. If we express the Helmholtz free energy of an N-particle system in terms of the partition function Q_N

$$\begin{aligned}
F(N,V,T) &= -k_B T \ln Q_N \\
&= -k_B T \ln\left(\frac{(q(T)V)^N}{N!}\right) - k_B T \ln\left(\int ds^N \exp[-\beta U(s^N;L)]\right) \\
&= F_{id}(N,V,T) + F_{ex}(N,V,T) ,
\end{aligned} \tag{30}$$

then it is obvious from Eqn. 29 that, for sufficiently large N the chemical potential is given by: $\mu = -k_B T \ln(Q_{N+1}/Q_N)$. If we use the explicit form (Eqn. 30) for Q_N, we find:

$$\begin{aligned}
\mu &= -k_B T \ln(Q_{N+1}/Q_N) \\
&= -k_B T \ln\left(\frac{qV}{(N+1)}\right) - k_B T \ln\left(\frac{\int ds^{N+1} \exp(-\beta U(s^{N+1})))}{\int ds^N \exp(-\beta U(s^N))}\right) \\
&= \mu_{id}(V) + \mu_{ex} .
\end{aligned} \tag{31}$$

In the first line of Eqn. 31, we have assumed that the system is contained in a cubic box with diameter $L = V^{\frac{1}{3}}$ and have defined scaled coordinates s^N, by:

$$\mathbf{q}_i = L \mathbf{s}_i$$

for $i = 1, 2, \cdots, N$. In the last line of Eqn. 31, we have indicated the separation in the ideal-gas contribution to the chemical potential, and the excess part. As $\mu_{id}(V)$ can be evaluated analytically, we focus on μ_{ex}. We now separate the potential energy of the $N+1$-particle system into the potential energy function of the N-particle system, $U(s^N)$, and the interaction energy of the $N+1$-th particle with the rest: $\Delta U \equiv U(s^{N+1}) - U(s^N)$. Using this separation, we can write μ_{ex} as:

$$\mu_{ex} = -k_B T \ln < \int ds_{N+1} \exp(-\beta \Delta U) >_N , \tag{32}$$

where $< \cdots >_N$ denotes canonical ensemble averaging over the configuration space of the N-particle system. The important point to note is that where $< \cdots >_N$ denotes canonical ensemble averaging over the configuration space of the N-particle system. The important point to note is that equation 32 expresses μ_{ex} as an ensemble average that can be sampled by the conventional Metropolis scheme [18]. There is only one aspect of this equation that makes it different form the averages that we considered before, namely the fact that we compute the average of an *integral* over the position of particle $N + 1$. This last integral can be sampled by brute-force (unweighted) Monte Carlo sampling. In practice the procedure is as follows: we carry out a perfectly normal constant-NVT Monte Carlo simulation on the system of N particles. At frequent intervals during this simulation (for instance, after every MC trial move) we randomly generate a coordinate s_{N+1}, uniformly over the unit cube. With this value of s_{N+1}, we then compute $\exp(-\beta \Delta U)$. By averaging the latter quantity over all generated trial positions, we obtain the average that appears in Eqn. 32. So, in effect, we are computing the average of the Boltzmann factor associated with the random insertion of an additional particle in an N-particle system, *but we never accept any such trial insertions*, because then we would no longer be sampling the average needed in Eqn. 32. The Widom method provides us with a very powerful scheme to compute the chemical potential of (not too dense) atomic and simple molecular liquids.

The particle insertion scheme fails when the probability of 'accepting' a trial insertion becomes very small. One consequence is that the simple particle insertion method is less suited for molecular than for atomic systems. This is so because the probability of accepting the random trial insertion of a large molecule in a fluid is usually extremely small.

4.2. CHEMICAL POTENTIAL OF MACRO-MOLECULES WITH DISCRETE CONFORMATIONS

In order to understand the methods that have been devised to calculate the chemical potential of chain molecules, it is instructive to first consider how we would compute μ_{ex} of a chain molecule with the Widom technique. To this end, I introduce the following notation: the position of the first segment of the chain molecule is denoted by $\mathbf{q}$ and the conformation of the molecule is described by Γ. The configurational part of the partition function of a system of chain molecules can be written as

$$Q_{chain}(N, V, T) = \frac{1}{N!} \int d\mathbf{q}^N \sum_{\Gamma_1, \cdots, \Gamma_n} \exp(-\beta U(\mathbf{q}^N, \Gamma^N)) . \qquad (33)$$

where of the $Q(N+1,V,T)$ is the (configurational part of) the partition function of a system of $N+1$ interacting chain molecules and $Q(N,V,T) \times Q_{non-interacting}(1,V,T)$ the partition function for a system consisting of N interacting chains and one chain that does not interact with the others. The latter chain plays the role of the ideal gas molecule in the previous sections. Note, however, that although this molecule does not interact with any of the other molecules it *does* interact with itself, both through bonded and through non-bonded interactions. Unfortunately, this is not a particularly useful reference state, as we do not, in general, know the partition function of an isolated self-avoiding chain.

We therefore use another reference state, namely that of the isolated non-selfavoiding chain. To be specific, let us consider the case of a molecule that consists of ℓ segments. Starting from segment 1, we can add segment 2 in b_2 equivalent directions, and so on. Clearly, the total number of non-selfavoiding conformations is $\Omega_{id} = \prod_{i=1}^{\ell} b_i$. For convenience, I have assumed that for a given i, all b_i directions are equally likely (i.e. I ignore *gauche-trans* potential energy differences and I even allow the ideal chain to fold back on itself). These limitations are not essential but they simplify the notation. Finally, I assume that all b_i are the same. Hence, for the simple model that we consider, $\Omega_{id} = b^{\ell}$. If we use such an ideal chain as our reference system, the expression for the excess chemical potential becomes

$$
\begin{aligned}
\beta\mu_{ex} &= -k_BT \ln\left(\frac{Q_{chain}(N+1,V,T)}{Q(N,V,T)Q_{ideal}(1,V,T)}\right) \\
&= -k_BT \ln < \exp[-\beta\Delta U(\mathbf{q}^N,\mathbf{\Gamma}^N;\mathbf{q}_{N+1},\mathbf{\Gamma}_{N+1})] > , \quad (34)
\end{aligned}
$$

where ΔU denotes the interaction of the test chain with the N chains that are already present in the system *and with itself*, while $< \cdots >$ indicates averaging over all starting positions and all ideal-chain conformations of a randomly inserted chain.

The problem with the Widom approach to Eqn. 34 is that almost all randomly inserted ideal chain conformations will overlap either with particles already present in the system, or internally. The most important contributions to μ_{ex} will come from the extremely rare cases where the trial chain happens to be in just the right conformation to fit into the available space in the fluid. Clearly, it would be desirable if we could restrict our sampling to those conformations that satisfy this condition. If we do that, we introduce a bias in our computation of the insertion probability and we must somehow correct for that bias. In practice, the scheme involves two steps: in the first step a chain conformation is generated in such a way that 'acceptable' conformations are created with a high probability. The next step corrects for this bias by multiplying with a weight factor. A scheme that generates 'acceptable' chain conformations with a high probability was developed by Rosenbluth and Rosenbluth in the early fifties [19]. In the Rosenbluth scheme, a conformation of a chain molecule is constructed segment-by-segment. For every segment, we have a choice of b possible directions. In the Rosenbluth scheme, this choice is not random but favors the direction with the largest Boltzmann factor. To be specific, the probability (P) to generate a polymer with a conformation Γ using the Rosenbluth algorithm is given by

$$P_\Gamma = \prod_{i=1}^{\ell} \frac{\exp\left[-\beta u^{(i)}(\Gamma_i)\right]}{Z_i} \, , \tag{35}$$

where $u^{(i)}(\Gamma_i)$ denotes the energy of segment i of the chain with conformation Γ (note that this energy excludes the contributions of segments $i+1$ to l, so the total energy of the chain is given by: $U_\Gamma = \sum_{i=1}^{\ell} u^{(i)}(\Gamma_i)$). Z_i in equation 35 is shorthand for

$$Z_i \equiv \sum_{j=1}^{b} \exp\left[-\beta u^{(i)}(\Gamma_j)\right] \, .$$

where j enumerates all possible orientations from which the i-th segment of the chain can be chosen and $u^{(i)}(\Gamma_j)$ denotes the potential energy of the i-th segment in orientation j. An important property of the probability given by Eqn. 35 is that it is normalized, i.e

$$\sum_{\Gamma} P_\Gamma = 1 \, .$$

The Rosenbluth weight factor that corrects for the bias in the selection of conformation Γ is given by

$$W_\Gamma \equiv \prod_{i=1}^{\ell} \frac{Z_i}{b} \, . \tag{36}$$

Now let us assume that we use the Rosenbluth scheme to generate a large number of chain conformations while keeping the coordinates of all other particles in the system fixed. For this set of conformations, we compute the average of the Rosenbluth weight factor W, $\overline{W}$. If we also perform an ensemble average over all coordinates and conformations of the N particles in the system, we obtain

$$\langle W \rangle = \left\langle \sum_{\Gamma} P_\Gamma(\mathbf{q}^N, \Gamma^N) W_\Gamma(\mathbf{q}^N, \Gamma^N) \right\rangle \, , \tag{37}$$

where the angular brackets denote the ensemble average over all configurations of the system $\{\mathbf{q}^N, \mathbf{\Gamma}^N\}$ of the 'solvent'. Note that the test polymer does not form part of the N-particle system. Therefore the probability to find the remaining particles in a configuration $\{\mathbf{q}^N, \mathbf{\Gamma}^N\}$ does not depend on the conformation $\mathbf{\Gamma}$ of the polymer.

In order to simplify the expression for the average in Eqn. 37, we first consider the average of the Rosenbluth factor for a given configuration $\{\mathbf{q}^N, \mathbf{\Gamma}^N\}$ of the solvent.

$$\overline{W}(\{\mathbf{q}^N, \mathbf{\Gamma}^N\}) = \sum_{\mathbf{\Gamma}} P_{\mathbf{\Gamma}}(\mathbf{q}^N) W_{\mathbf{\Gamma}}(\{\mathbf{q}^N, \mathbf{\Gamma}^N\}) . \tag{38}$$

Substitution of equations (35) and (36) yields

$$
\begin{aligned}
\overline{W} &= \sum_{\mathbf{\Gamma}} \left[\prod_{i=1}^{\ell} \frac{\exp\left[-\beta u^{(i)}(\mathbf{\Gamma}_i)\right]}{Z_i} \right] \left[\prod_{i=1}^{\ell} \frac{Z_i}{b} \right] \\
&= \sum_{\mathbf{\Gamma}} \prod_{i=1}^{\ell} \frac{1}{b} \exp\left[-\beta u^{(i)}(\mathbf{\Gamma}_i)\right] \\
&= \sum_{\mathbf{\Gamma}} \frac{1}{b^{\ell}} \exp\left[-\beta U_{\mathbf{\Gamma}}\right] ,
\end{aligned}
\tag{39}
$$

where we have dropped all explicit reference to the solvent coordinates $\{\mathbf{q}^N, \mathbf{\Gamma}^N\}$. Note that Eqn. 39 can be interpreted as an average over all *ideal* chain conformations of the Boltzmann factor $\exp\left[-\beta U_{\mathbf{\Gamma}}\right]$. If we now substitute Eqn. 39 in Eqn. 38 we obtain

$$\langle W \rangle = \frac{\sum_{\mathbf{\Gamma}} < \exp[-\beta \Delta U(\mathbf{q}^N, \mathbf{\Gamma}^N; \mathbf{q}_{N+1}, \mathbf{\Gamma}_{N+1})] >}{\sum_{\mathbf{\Gamma}}} . \tag{40}$$

If we compare Eqn. 40 with Eqn. 34, we see that the ensemble average of the Rosenbluth factor is directly related to the excess chemical potential of the chain molecule.

$$\beta \mu_{ex} = -k_B T \ln \langle W \rangle , \tag{41}$$

The above method to measure the chemical potential is not limited to chain molecules on a lattice. What *is* essential is that the number of possible directions for each segment (b) relative to the previous one is finite.

4.3. EXTENSION TO CONTINUOUSLY DEFORMABLE MOLECULES

The numerical computation of the (excess) chemical potential of a flexible chain (with or without elastic forces that counteract bending), is rather

68

different from the corresponding calculation for a chain molecule that has a large but fixed number of undeformable conformations. Below, I shall consider the case of a flexible molecule *with* internal energy. Consider a 'worm-like' chain of ℓ linear segments. The potential energy of a given conformation has two contributions:

1. The internal potential energy U_{int} is equal to the sum of the contributions of the individual joints. A joint between segments i and $i+1$ (say) has a potential energy $u(\theta_i)$ that depends on the angle θ_i between the successive segments. For instance, $u(\theta_i)$ could be of the form $u(\theta)=\alpha\theta^2$. For realistic models for poly-atomic molecules, U_{int} would account for all local internal potential energy changes due to bending and torsion.

2. The 'external' potential energy U_{ext}. This energy accounts for all interactions with other molecules and for the non-bonded intra-molecular interactions. In addition, interactions with any external field that may be present are also included in U_{ext}.

In what follows I shall denote the chain in the absence of the 'external' interactions as the *ideal* chain. Clearly, the conformational partition function of the ideal chain is equal to

$$Z_{id} = c \int \cdots \int d\Gamma_1 \cdots d\Gamma_\ell \prod_{i=1}^{\ell} \exp(-\beta u_{id}(\theta_i)) \qquad (42)$$

where c is a numerical constant. Our aim is to compute the effect of the external interactions on the conformational partition function. Hence, we wish to evaluate Z/Z_{id}, where Z denotes the partition function of the interacting chain. The excess chemical potential of the interacting chain is given by

$$\mu_{ex} = -k_B T \ln(Z/Z_{id}) \, .$$

The numerical procedure to compute the chemical potential is similar to the scheme to compute the excess chemical potential of a chain molecule with fixed conformations 4.2. Yet, there is an important difference precisely because the number of conformations is now, in principle, infinite. We can never hope to sample over *all* possible orientations of a new segment as we grow a chain. Hence, we generate a random sample of possible segment directions and use these in a modified Rosenbluth scheme. To compute μ_{ex}, we apply the following 'recipe' to construct a conformation of a chain of ℓ segments. The construction of chain conformations proceeds segment by segment. Let us consider the addition of one such segment. To be specific, let us assume that we have already grown i segments, and that we are trying to add segment $i+1$. This is done as follows:

1. Generate a fixed number (say b) trial segments. The orientations of the trial segments are distributed according to the Boltzmann weight associated with the internal energy $u(\theta)$. We denote the different trial segment by indices $1, 2, \cdots b$.
2. For all b trial segments, we compute the 'external' Boltzmann factor $\exp(-\beta u_{ext}(j))$.
3. Select one of the trial segments, say j, with a probability

$$P_j = \frac{\exp(-\beta u_{ext}(j))}{Z_i} \, ,$$

where we have defined

$$Z_i \equiv \sum_{j'=1}^{b} \exp(-\beta u_{ext}(j')) \, .$$

4. Add this segment as segment $i + 1$ to the chain and store the corresponding partial Rosenbluth weight $w_i = Z_i / b$.

The desired ratio Z/Z_{id} is than equal to the average value (over many trial chains) of the product of the partial Rosenbluth weights:

$$Z/Z_{id} = < \prod_{i=1}^{\ell} w_i > \, . \tag{43}$$

The advantage of this scheme is that step 3 biases the sampling towards energetically favorable conformations. However, it still remains to be shown that equation 43 is, in fact, correct. To show this, we consider the probability to generate a given chain conformation. This probability is the product of a number of factors. Let us first consider these factors for one segment, and then later extend the result to the complete chain. The probability to generate a given set of b trial segments with orientations Γ_1 through Γ_b is

$$P_{id}(\Gamma_1).P_{id}(\Gamma_2) \cdots P_{id}(\Gamma_b) d\Gamma_1 \cdots d\Gamma_b.$$

The probability of selecting any one of these trial segments, say segment j, is

$$\frac{\exp(-\beta u_{ext}(j))}{Z_i} \, .$$

We wish to compute the average of a quantity, say w, over all possible sets of trial segments and all possible choices of the segment. To this end, we must sum over all j and integrate over all orientations $\prod_{j'=1}^{b} d\Gamma_{j'}$ (i.e, we

We wish to compute the average of a quantity, say w, over all possible sets of trial segments and all possible choices of the segment. To this end, we must sum over all j and integrate over all orientations $\prod_{j'=1}^{b} d\Gamma_{j'}$ (i.e, we average over the normalized probability distribution for the orientation of segment $i+1$):

$$< w >= \int [\prod_{j'=1}^{b} d\Gamma_{j'} P_{id}(\Gamma_{j'})] \sum_{j''=1}^{b} \frac{\exp(-\beta u_{ext}(j''))}{Z_i} w(1,2,\cdots,b) \ . \quad (44)$$

Now we make use of the fact that $w_i(1,2,\cdots,b)$ is equal to $Z_i)/b$ (see step 4 of the 'recipe' above). Inserting this expression in Eqn. 44, we obtain:

$$< w >= \int [\prod_{j'=1}^{b} d\Gamma_{j'} P_{id}(\Gamma_{j'})] \sum_{j''=1}^{b} \frac{\exp(-\beta u_{ext}(j''))}{b} \ . \quad (45)$$

As the labeling of the trial segments is arbitrary, all b terms in the sum in Eqn. 45 yield the same contribution, and Eqn. 45 simplifies to

$$< w > \ = \ \int d\Gamma P_{id}(\Gamma) \exp(-\beta u_{ext}(\Gamma)) \quad (46)$$

$$= \ \frac{\int d\Gamma \exp(-\beta [u_{id}(\Gamma) + u_{ext}(\Gamma)])}{\int d\Gamma \exp(-\beta u_{id}(\Gamma))} \quad (47)$$

$$= \ \frac{Z^{(1)}}{Z_{id}^{(1)}} \ , \quad (48)$$

which is indeed the desired result, but for the fact that the expression in Eqn, 46 refers to one segment (as indicated by the superscript in $Z^{(1)}$). The extension to a chain of ℓ segments is straightforward, be it that the intermediate expressions become a little unwieldy.

5. Configurational bias Monte Carlo scheme

Up to this point, I have been speaking about techniques to estimate the chemical potential of flexible molecules. However, the Rosenbluth trial insertion scheme can be used as a starting point for a Monte Carlo scheme to sample equilibrium configurations of systems consisting of chain molecules. At first sight, this may not appear to be a new result but a very old one. After all, the original Rosenbluth scheme itself was designed as a method to sample polymer conformations. However, the Rosenbluth scheme suffers from the drawback that it generates an unrepresentative sample of all polymer conformations: i.e. the probability to generate a particular conformation Γ using the Rosenbluth scheme, is *not* proportional to the Boltzmann weight of that conformation. The Rosenbluth weight W, discussed in works for relatively short chains. This drawback of the Rosenbluth sampling scheme is, in fact, well known (see, e.g. [20]). The solution of this problem is to bias the Rosenbluth sampling in such a way that the correct (Boltzmann) distribution of chain conformations is generated in a Monte Carlo sequence. In the configurational bias scheme, the Rosenbluth weight is used to bias the *acceptance* of trial conformations that are generated with the Rosenbluth procedure. As a consequence, all conformations are generated with their correct Boltzmann weight. This removes the main drawback of the original Rosenbluth scheme. For details, I refer the reader to ref. [21].

The CBMC scheme has been applied to several models for lyotropic liquid crystals. Dijkstra and Frenkel studied the effect of flexibility on the I-N transition of semi-flexible hard rods [22] and subsequently, Polson and Frenkel combined the scaling approach described above with CBMC to study the effect of flexibility on the N-Sm transition [23].

Acknowledgments

The work of the FOM Institute is part of the research program of FOM and is made possible with financial help from the Nederlandse Organisatie voor Wetenschappelijk Onderzoek (NWO).

References

1. Lectures presented at this School and e.g. Panagiotopoulos, A.Z. (1995) *Observation, Prediction and Simulation of Phase Transitions in Complex Fluids.* Baus, M., Rull, L.F. and Ryckaert, J.-P. (eds.), volume 460 of *NATO ASI Series C*, Kluwer Academic Publishers, Dordrecht, p. 463.
2. Hoover, W.G. and Ree, F.H. (1967) *J. Chem. Phys.*, **47**, 4873.
3. Hansen, J.P. and McDonald, I.R. (1986) *Theory of Simple Liquids.* 2nd edition, Academic Press, London.
4. Frenkel, D. and Ladd, A.J.C. (1984) *J. Chem. Phys.*, **81**. 3188.
5. Frenkel, D. (1985) in: *Molecular Dynamics Simulations of Statistical Mechanical Systems.* Proceedings of the 97th International School of Physics 'Enrico Fermi', Ciccotti, G. and Hoover, W.G. (eds.), North-Holland, Amsterdam, p. 151.
6. Meijer, E.J., Frenkel, D., LeSar, R.A. and Ladd, A.J.C. (1990) *J. Chem. Phys.*, **92**, 7570.
7. Frenkel, D. and Mulder, B.M. (1985) *Mol. Phys.*, **55**, 1171.
8. Widom, B., (1963) *J. Chem. Phys.*, **39**, 2808.
9. Eppenga, R. and Frenkel, D., (1984) *Mol. Phys.*, **52**, 1303.
10. Veerman, J.A.C. and Frenkel, D. (1990) *Phys. Rev.*, **A41**, 3237.
11. Bolhuis, P.G. and Frenkel, D., (1997) *J. Chem. Phys.*, **106**, 666.
12. Kofke, D.A.J. (1993) *Chem. Phys.*, **98**, 4149.
13. Bates, M.A. and Frenkel, D. (1998) *Phys. Rev. E*, **57**, 4824.
14. Camp, P.J., Allen, M.P., Bolhuis, P.G. and Frenkel, D. (1997) *J. Chem. Phys.*, **106**, 9270.
15. Bolhuis, P.G., Stroobants, A., Frenkel, D. **and** Lekkerkerker, H.N.W. (1997) *J. Chem. Phys.*, **107**, 1551.
16. Bolhuis, P.G. and Kofke, D.A. (1996) *Phys. Rev. E*, **54**, 634.
17. Bates, M.A. and Frenkel, D. (1998) *J. Chem. Phys.*, **109**, 6193.

18. Metropolis, N., Rosenbluth, A.W., Rosenbluth, M.N. Teller, A.H. and Teller, E. (1953) *J. Chem. Phys.*, **21**, 1087.
19. Rosenbluth, M.N. and Rosenbluth, A.W. (1955) *J. Chem. Phys.*, **23**, 356.
20. Kremer, K. and Binder, K. (1988) *Computer Physics Reports*, **7**, 259.
21. Frenkel, D. and Smit, B. (1996) *Understanding Molecular Simulation. From Algorithms to Applications.* Academic Press, Boston.
22. Dijkstra, M. and Frenkel, D. (1995) *Phys. Rev. E*, **51**, 5891.
23. Polson, J.M. and Frenkel, D. (1997) *Phys. Rev. E*, **56**, 6260.

MODELLING LIQUID CRYSTAL STRUCTURE, PHASE BEHAVIOUR AND LARGE-SCALE PHENOMENA

MICHAEL P. ALLEN
University of Bristol
H. H. Wills Physics Laboratory
Royal Fort
Tyndall Avenue
Bristol BS8 1TL
United Kingdom

Abstract. This chapter summarizes recent simulation work aimed at investigating the effects of changing molecular shape and interaction parameters on the phase behaviour and structure of liquid crystalline phases. The focus will be on some simple test cases: hard particles of various kinds, and the Gay-Berne family of models. In some of these cases, it is possible to compare directly with molecular-scale theories, such as Onsager theory. Then some examples will be given of the use of computer simulation to study larger-scale phenomena, with the ultimate intention of testing continuum theories of liquid crystals, and bridging the gap with molecular-scale effects that are hard to model using theories of this kind. Examples will include recent work on structure and dynamics near the isotropic-nematic transition; a study of the smectic-A* twist-grain-boundary phase; and the calculation of orientational elastic coefficients.

1. Introduction

The computer simulation of liquid crystals is a rapidly expanding field: many of the techniques needed to measure the properties of interest, and to study the relevant phase transitions, are still being developed. In principle, computer simulation combined with liquid-state theories should give us an insight into the link between molecular structure and liquid crystalline behaviour. In practice, this approach complements, and to an extent competes with, the very well established experimental method of synthesizing

P. Pasini and C. Zannoni (eds.), Advances in the Computer Simulations of Liquid Crystals, 73–97.

and testing large numbers of compounds. The sensitivity of liquid crystal properties to molecular details means that simulation has a tough task ahead. Nonetheless, the first elements of a coherent picture are beginning to emerge from simulations, and the prospects look good for further progress. Onsager's theory is an early example of the now-popular density-functional theories of the liquid state: such theories directly link molecular properties with bulk-phase behaviour. Computer simulation can make a valuable contribution by testing theories of this kind, perhaps pointing the way to improving them. The first part of this chapter gives examples of this kind of study, emphasizing the special simulation methods that sometimes need to be employed.

Continuum models of liquid crystals, based on phenomenological elastic constants and hydrodynamic transport coefficients, have been outstandingly successful in modelling behaviour on the length and time scales of most interest to device manufacturers. Simulation can contribute by providing methods of calculating the relevant coefficients for a given molecular model. Moreover, interest is developing in molecular-scale effects, such as behaviour near surfaces and defects, which cannot be modelled properly in the continuum picture. To bridge the gap, it is necessary to simulate very large systems, and some recent work in this area will be described in the second part of this chapter.

Here I am only able to give selected details, due to limitations of space; more information may be found in the original papers.

2. Onsager theory and hard-particle phase diagrams

Onsager's theory [1], a forerunner of modern density functional theories, is based on an expression for the free energy in terms of the single-particle density, $\varrho(\mathbf{r}, \mathbf{u})$, itself a function of position $\mathbf{r}$ and orientation $\mathbf{u}$. In the nematic and isotropic phases, the position dependence is trivial: $\varrho(\mathbf{r}, \mathbf{u}) = \rho f(\mathbf{u})$ where $f(\mathbf{u})$ is the orientational distribution function, and $\rho = N/V$ the density. The Helmholtz free energy $\mathcal{F} = \mathcal{F}^{\mathrm{id}} + \mathcal{F}^{\mathrm{ex}}$ may then be written

$$\frac{\beta \mathcal{F}^{\mathrm{id}}[\varrho]}{N} = \left(\ln \rho \Lambda^3 - 1\right) + \int d\mathbf{u}_1 \, f(\mathbf{u}_1) \ln 4\pi f(\mathbf{u}_1)$$

$$\frac{\beta \mathcal{F}^{\mathrm{ex}}[\varrho]}{N} = \frac{1}{2}\rho \int d\mathbf{u}_1 \int d\mathbf{u}_2 \, f(\mathbf{u}_1) f(\mathbf{u}_2) \, E(\mathbf{u}_1, \mathbf{u}_2) + \ldots$$

where $\beta = 1/k_{\mathrm{B}}T$ and Λ is the thermal de Broglie wavelength. The term $\mathcal{F}^{\mathrm{id}}$ is the free energy of an ideal mixture of different species (molecular orientations); interest here is restricted to uniaxial molecules, and the factor 4π is included to make the integral vanish for the isotropic case $f(\mathbf{u}) = 1/4\pi$. The excess free energy $\mathcal{F}^{\mathrm{ex}}$ is written as a virial expansion, truncated at the leading, pairwise, term: it depends on $E(\mathbf{u}_1, \mathbf{u}_2)$, the excluded volume of a pair of molecules with specified orientations. The approximation that three-body and higher terms are neglected, becomes increasingly good at low densities and for high elongations. Variational minimization of $\mathcal{F}$ subject to the normalization condition $\int d\mathbf{u}\, f(\mathbf{u}) = 1$ leads to a self-consistency equation

$$\ln f(\mathbf{u}_1) = \text{constant} - \rho \int d\mathbf{u}_2\, f(\mathbf{u}_2)\, E(\mathbf{u}_1, \mathbf{u}_2)$$

which may be solved iteratively to give $f(\mathbf{u})$ and hence $\mathcal{F}$. For suitably high elongations, two types of solutions may be found: an isotropic phase, stable at low density, and a nematic phase which becomes stable at higher density.

An empirical improvement of Onsager's theory, due to Parsons and Lee [2–4], introduces a density-dependent prefactor into the two-body term:

$$\frac{\beta \mathcal{F}^{\text{ex}}[\varrho]}{N} = \frac{1}{2} \frac{\eta(4 - 3\eta)}{4v_0(1 - \eta)^2} \int d\mathbf{u}_1 \int d\mathbf{u}_2\, f(\mathbf{u}_1) f(\mathbf{u}_2)\, E(\mathbf{u}_1, \mathbf{u}_2) + \ldots \, .$$

Here $\eta = \rho v_0$ is the packing fraction and v_0 is the molecular volume. Variational minimization proceeds as before to give $f(\mathbf{u})$ and $\mathcal{F}$. The form of the prefactor is chosen so that for hard spheres, when $E(\mathbf{u}_1, \mathbf{u}_2) \to 8v_0$, the free energy becomes $\beta \mathcal{F}^{\text{ex}}/N \to \eta(4 - 3\eta)/(1 - \eta)^2$ which is the well-known, and very accurate, Carnahan-Starling equation of state. Another interpretation of the Parsons improvement, following from functional differentiation of the above expressions, is that the Onsager theory sets the direct correlation function c equal to the Mayer f-function

$$c(\mathbf{r}_1 - \mathbf{r}_2, \mathbf{u}_1, \mathbf{u}_2) = f(\mathbf{r}_1 - \mathbf{r}_2, \mathbf{u}_1, \mathbf{u}_2) = \begin{cases} -1 & \text{overlap} \\ 0 & \text{no overlap} \end{cases}$$

which is true at low density; while in the Parsons theory, c is just f multiplied by the density-dependent scaling factor. Thermodynamic properties then follow by inserting c into the 'compressibility' equation of state. So, the Parsons theory reduces to that of Onsager at high elongation, while giving a good equation of state in the hard sphere limit.

2.1. HIGHLY ELONGATED MOLECULES

To test the Onsager and Parsons theories, we have carried out computer simulations of hard ellipsoids of revolution of elongation $e = a/b$ where $a = \text{length}$, $b = \text{width}$, in the range $5 \leq e \leq 20$ [5]. To locate the phase transition, it is necessary to determine the state points which satisfy the conditions for thermodynamic equilibrium: $T_1 = T_2$, $P_1 = P_2$, $\mu_1 = \mu_2$. For

76

particles interacting only with infinitely repulsive potentials, the temperature is not a significant thermodynamic variable, and the first equality is always satisfied. The last equation is the most difficult to determine.

2.1.1. *Determining the chemical potential*

The chemical potential at a point (T, P) in each phase may be found by Widom test particle insertion. Since $\mu^{\text{ex}} = F_{N+1}^{\text{ex}} - F_N^{\text{ex}}$,

$$\beta\mu^{\text{ex}} = -\ln\langle\exp(-\beta\mathcal{V}_{\text{test}})\rangle$$

where $\mathcal{V}_{\text{test}}$ is the potential energy of interaction of a randomly inserted test particle with the N particles in the system. Then one may use thermodynamic integration

$$\mu(P) = \mu(P_0) + \int_{P_0}^{P} \rho^{-1}\,\mathrm{d}P$$

to solve $\mu_1(P_1) = \mu_2(P_2)$. Automatic equilibration of two phases, with each phase in its own simulation box, may be achieved using the Gibbs ensemble method of Panagiotopoulos (see [6, 7]. Briefly, this technique involves standard Monte Carlo moves in each box; volume exchange moves between boxes, which guarantee $P_1 = P_2$; and particle transfer moves arranged to ensure $\mu_1 = \mu_2$. Agreement between the Widom method and the Gibbs ensemble method for the I-N transition for $e = 20$ hard ellipsoids is demonstrated in Fig. 1.

For dense fluids, both the Widom method and the Gibbs ensemble become less reliable, due to the low acceptance rates for particle insertion. One solution which we have adopted is to use an *expanded ensemble* in which N-particle and $(N+1)$-particle systems are linked by intermediate fractional-particle states [8, 9]. One particle is scaled in size by a parameter κ, taking values $0 \leq \kappa \leq 1$. In the Monte Carlo procedure, we allow transitions between species $\kappa \rightleftharpoons \kappa'$. We determine the probability histogram $\mathcal{P}(\kappa)$, and hence the relative free energies of species $\mathcal{F}(\kappa)$; then $\mathcal{F}(1) - \mathcal{F}(0)$ gives μ^{ex}. To ensure uniform sampling of species, we apply an (iteratively refined) weighting function $\mathcal{W}(\kappa)$ to the κ-moves. Details are given in Ref. [5]. It is essential that the scaled particle samples fluid configurations efficiently. To assist, we introduce two additional types of move which dramatically relocate it. Firstly, we attempt to move the scaled particle to a completely random position, as in the (random-insertion) Widom method. Such moves are accepted with high probability when κ is small. Secondly, we attempt to exchange the scaled particle with full-size particles. This works best when κ is large.

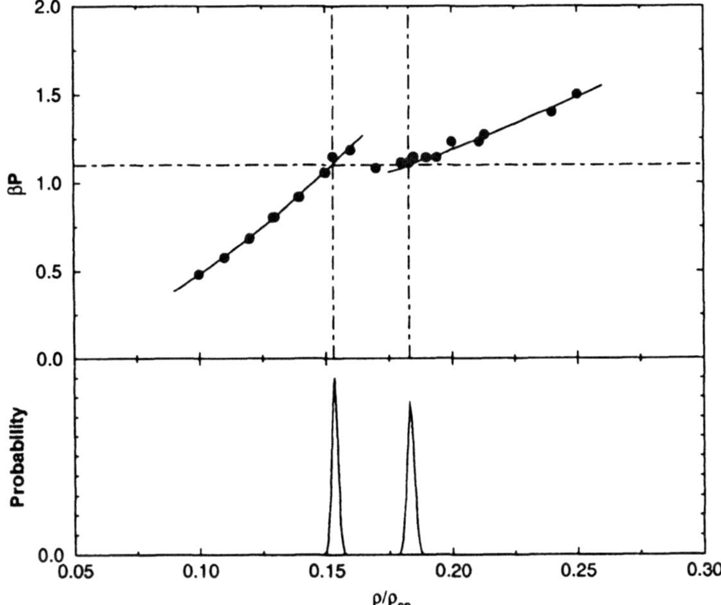

Figure 1. I-N coexistence for $e = 20$ ellipsoids. The data points with solid lines are the equations of state $P(\rho)$ for isotropic and nematic phases, with the coexistence value P_{NI} determined by Widom test-particle insertion plus thermodynamic integration, indicated by dot-dashed lines. Below this is the probability distribution function for density ρ averaged over both boxes in a Gibbs simulation.

2.1.2. *Tracing out the phase boundary*

An essential part of a study of this kind is to determine how the transition pressure changes as we slowly vary molecular shape, or some other parameter. The idea is to avoid computing chemical potentials from scratch, for each case. A useful approach, due to Kofke [10], is to derive and then use a variant of the Clapeyron equation. Suppose we characterize the molecular shape by a parameter λ; here we choose $\lambda = \ln a$ ($a =$ length) or $\ln b$ ($b =$ width) for convenience. Then we define a thermodynamic variable $\Gamma = (\partial \mu / \partial \lambda)_{TP}$ conjugate to λ; this means that we may write $\mathrm{d}\mu = \Gamma \mathrm{d}\lambda + v\mathrm{d}P$. It follows that

$$\frac{\mathrm{d}P}{\mathrm{d}\lambda} = -\frac{\Delta\Gamma}{\Delta v} \quad \text{or} \quad \frac{\mathrm{d}\ln P}{\mathrm{d}\lambda} = -\frac{\Delta\Gamma}{P\Delta v}.$$

Kofke's method consists of evaluating the terms on the right, and integrating step-by-step with respect to λ along the coexistence curve.

For hard particles, Γ may be calculated by making a 'phantom' increase in molecular dimensions, $\delta\lambda$, and counting the resultant number of pair overlaps, $\mathcal{N}^{\text{overlap}}$, as illustrated in Fig. 2. Γ may be obtained from the

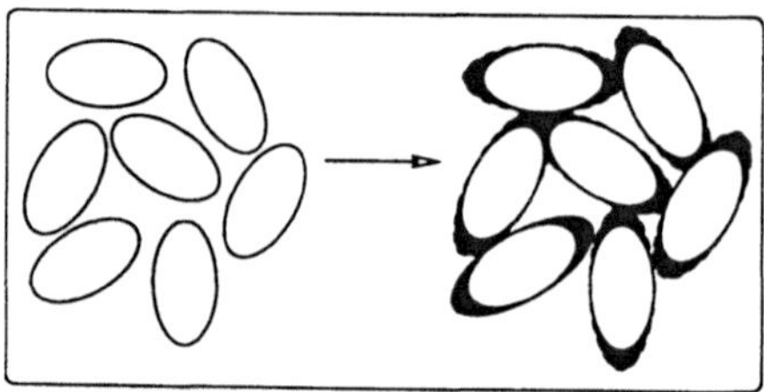

Figure 2. Phantom particle shape changes

expression

$$\Gamma = \lim_{\delta\lambda\to 0} k_{\mathrm{B}}T \frac{\left\langle \mathcal{N}^{\text{overlap}}\right\rangle}{N\delta\lambda} \, .$$

Applying these techniques to the hard ellipsoid system, we were able to show the extent of applicability of both Onsager and Parsons theories, and the expected decrease in strength of transition on decreasing elongation, as shown in Fig. 3. The effects of molecular biaxiality on the phase diagram have recently been investigated in the same way using a hard spheroid model [11].

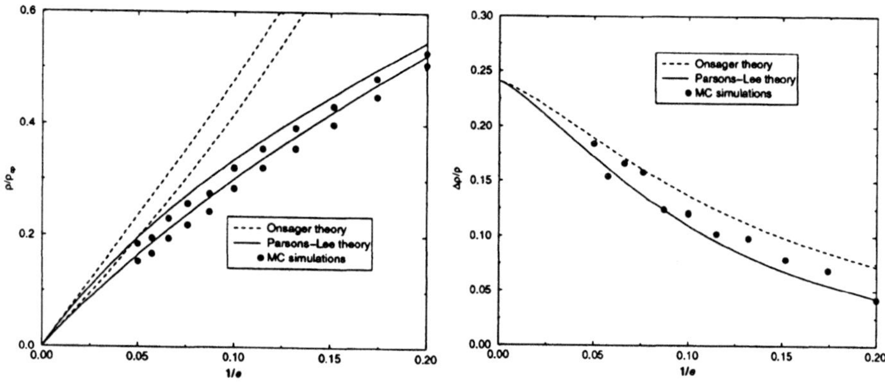

Figure 3. Coexistence densities (left) and fractional density change (right) at the I-N transition vs elongation for hard ellipsoids of revolution. For details see Ref. [5].

2.2. ROD-PLATE MIXTURES

Mixtures of rod-like and plate-like molecules may form isotropic, uniaxial and biaxial phases as shown in Fig. 4. The biaxial phase competes with demixing into two uniaxial phases: this is an example of phase separation due to completely repulsive interactions. We have studied mixtures of ellipsoids of the same molecular volume, and two conjugate pairs of shapes:

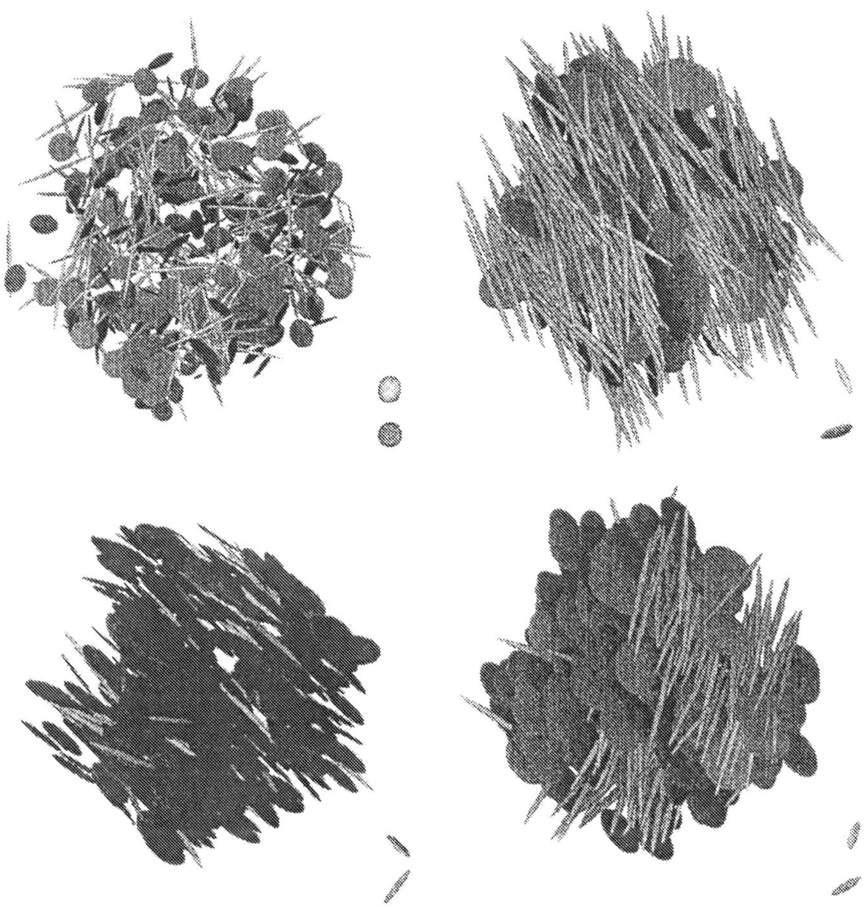

Figure 4. Rod and plate configurations for $e = 20, 1/20$ ellipsoid mixtures, showing isotropic, uniaxial nematic, discotic and biaxial phases. The biaxial configuration also shows some demixing. At the bottom right corner of each configuration, two spheroids give a visual representation of the orientational ordering of rods (upper) and plates (lower).

$e = 15, 1/15$ and $e = 20, 1/20$ to investigate this competition. Once again it was possible to compare with Onsager theory and the Parsons-Lee rescaling. The phase transitions were located by performing Gibbs ensemble simulations, and observing the behaviour of order parameters characterizing both phases; full details are given elsewhere [12]. The phase diagram for the $e = 15, 1/15$ system is shown in Fig. 5 (that for $e = 20, 1/20$ is similar). A striking feature is the asymmetric line separating the biaxial phase from the two-phase demixing region. For these conjugate shapes, a symmetry of the second virial coefficient under the exchange $e \leftrightarrow 1/e$ means that the phase

80

diagram predicted by the Onsager or Parsons theory is symmetric about the 50:50 composition. Most of the phase boundaries exhibit this symmetry, approximately, but the dividing line between biaxial and two-phase regions is clearly very sensitive to three-body, or perhaps higher, interactions.

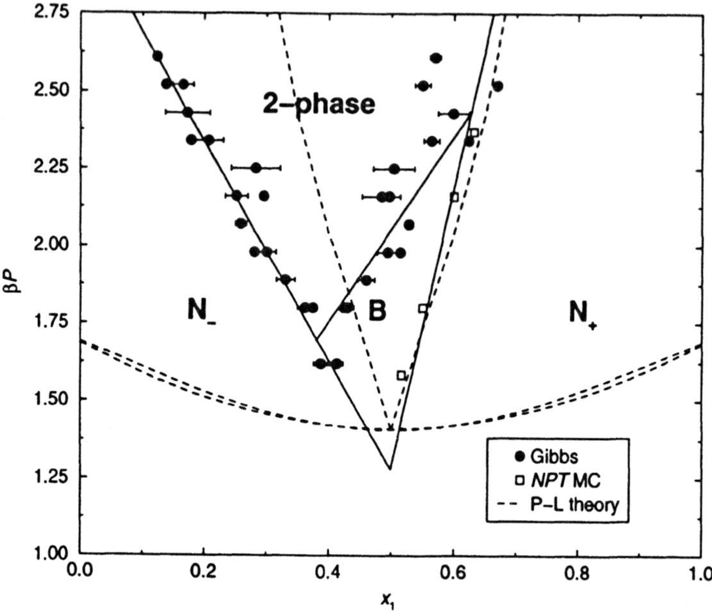

Figure 5. Phase diagram for $e = 15, 1/15$ rod-plate mixtures in the pressure-composition plane: x_1 is the mole fraction of rods. The points are simulation results, the solid lines mark approximate phase boundaries, and the dashed-lines are the predictions of Parsons-Lee theory. For further details see Ref. [12].

3. The Gay-Berne model

The introduction of attractive forces into the molecular model makes the phase diagram a little more realistic, and more complicated. The well-known Gay-Berne potential has become a standard model for the study of liquid crystalline phases, and it is discussed by several others in this volume. The molecular pair potential

$$U = 4\varepsilon(\hat{\mathbf{r}}, \mathbf{u}_1, \mathbf{u}_2)\left[\left(\frac{\sigma_0}{r - \sigma(\hat{\mathbf{r}}, \mathbf{u}_1, \mathbf{u}_2) + \sigma_0}\right)^{12} - \left(\frac{\sigma_0}{r - \sigma(\hat{\mathbf{r}}, \mathbf{u}_1, \mathbf{u}_2) + \sigma_0}\right)^{6}\right]$$

depends upon the molecular axis vectors $\mathbf{u}_1$ and $\mathbf{u}_2$, and on the direction $\hat{\mathbf{r}}$ and magnitude r of the centre-centre vector $\mathbf{r} = \mathbf{r}_1 - \mathbf{r}_2$. The parameter σ_0 determines the smallest molecular diameter and there are two orientation-dependent quantities: a diameter $\sigma\left(\hat{\mathbf{r}}, \mathbf{u}_1, \mathbf{u}_2 \mid \kappa, \kappa', \mu, \nu\right)$ and an energy $\varepsilon\left(\hat{\mathbf{r}}, \mathbf{u}_1, \mathbf{u}_2 \mid \kappa, \kappa', \mu, \nu\right)$. Each quantity depends in a complicated way (not given here) on κ, the shape anisotropy parameter, κ', the energy anisotropy parameter, and two exponents μ, ν. The original 'reference' model has $\mu = 2$, $\nu = 1$, $\kappa = 3$, $\kappa' = 5$. Jointly with E. de Miguel and E. Martín del Rio, we have been studying the effects of varying the attractive anisotropy and the elongation.

3.1. VARYING THE ATTRACTIVE ANISOTROPY

The effects of varying κ' in the range $1 \leq \kappa' \leq 25$, with fixed $\kappa = 3$, have been reported in detail elsewhere [13, 14]. Perhaps the most interesting effect is the growth of the liquid-vapour coexistence envelope on reducing κ', as illustrated in Fig. 6 The envelope was determined by Gibbs ensemble simulation, but this became less efficient at lower temperatures as the liquid density became higher; just at this point, evidence emerged of an I-N transition on the liquid side, raising the prospect of being able to simulate a nematic-vapour interface. The coexisting vapour density is very low at these temperatures, so a constant-pressure simulation with $P = 0$ provides a good guide to the behaviour of the liquid with changing temperature; however, it was also possible to conduct thermodynamic integration using the Kofke method, to determine both vapour and liquid densities at coexistence. Studies of the nematic-vapour interface for this model have been undertaken by two groups [13, 15, 16].

3.2. VARYING THE ELONGATION

Values of κ in the range $3 \leq \kappa \leq 4$ with fixed $\kappa' = 5$ have been studied [17]. Here, there is some interest in characterizing different types of positional order so as to characterize the smectic phases. Smectic ordering is typically identified through the longitudinal pair distribution function $g_{\parallel}(r_{\parallel})$ where $r_{\parallel} = \mathbf{r} \cdot \mathbf{n}$, $\mathbf{r}$ is the separation vector, and we project along the director $\mathbf{n}$ which is assumed normal to the smectic layers. This function shows a regular set of peaks corresponding to the smectic layer spacing: examples are shown in Fig. 7.

Sometimes the transverse structure is represented by a function $g_{\perp}(r_{\perp})$ where $r_{\perp} = |\mathbf{r} - r_{\parallel}\mathbf{n}|$ is the transverse separation of a pair of molecules, but we found this inconvenient as it depends on the choice of a cutoff on pair separations in the 'parallel' direction. Instead, we define functions based on the assignment of particles to layers, which is straightforward given the phase and period of the density wave (this may be easily obtained from the structure factor). The simplest examples, illustrated in Fig. 8, are the transverse *within-layer* distribution function $g_{\perp}^{(0)}(r_{\perp})$ which only examines

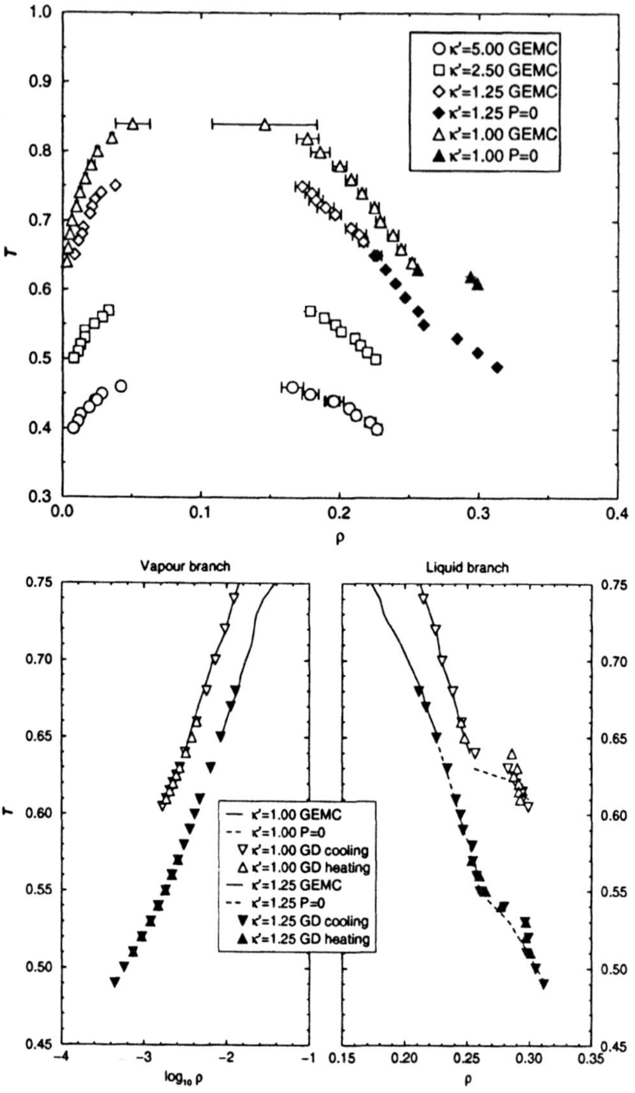

Figure 6. The effects of varying the anisotropy of the attractive forces in the Gay-Berne potential on the liquid-vapour coexistence envelope. Above: results of Gibbs simulations and zero-pressure Monte Carlo. There is an I-N transition on the liquid branch for $\kappa' = 1.00$ and 1.25. Below: results of Gibbs-Duhem integration through the I-N transition (note the logarithmic density scale on the vapour side).

correlations for pairs within the same layer; and the transverse *between-layer* distribution function $g_{\perp}^{(1)}(r_{\perp})$, defined for pairs in adjacent layers. Systematic extension to next-nearest and more remote layers is straightforward. In a smectic-A phase, $g_{\perp}^{(0)}(r_{\perp})$ shows 2-d liquid-like behaviour,

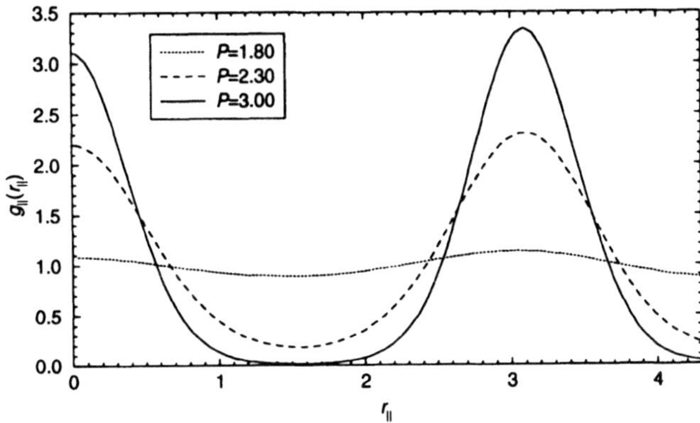

Figure 7. Longitudinal pair distribution function for Gay-Berne model with $\kappa = 3.6$ at different pressures along the isotherm $T = 1.00$: $P = 1.80$, nematic phase; $P = 2.30$, smectic-A phase; $P = 3.00$, smectic-B phase.

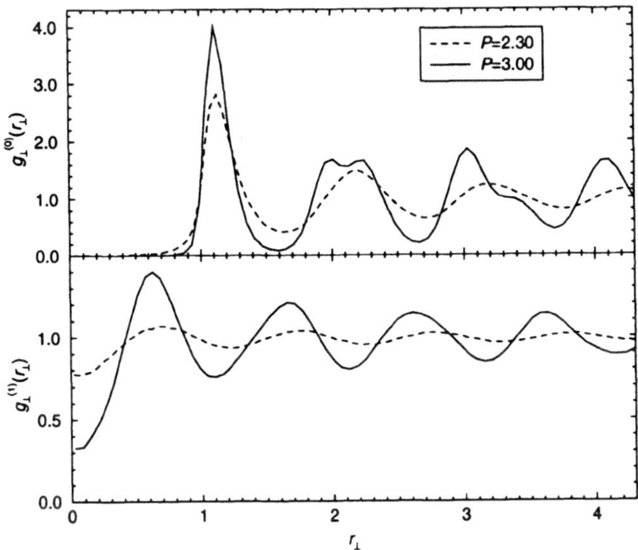

Figure 8. Transverse pair distribution functions for Gay-Berne model with $\kappa = 3.6$ at different pressures along the isotherm $T = 1.00$: $P = 2.30$, smectic-A phase; $P = 3.00$, smectic-B phase. Above: within-layer function; below: between-layer function.

while very little structure is seen in $g_{\perp}^{(1)}(r_{\perp})$, since adjacent layers are not registered. In a smectic-B phase, $g_{\perp}^{(0)}(r_{\perp})$ develops long-ranged peaks characteristic of ordering within the layers, and $g_{\perp}^{(1)}(r_{\perp})$ may show some short-range correlation between layers. (So-called 'hexatic' and 'crystal' variants of the smectic-B phase are distinguished by quasi-long-ranged and true long-ranged order in positional correlation functions, but we cannot properly examine this in our simulations, due to system-size limitations). In a crystal, all these functions show well-developed long-ranged structure.

The effects of increasing κ from 3 to 4 are indicated in the phase diagrams of Fig. 9. The liquid-vapour envelope contracts and disappears, while the smectic-A phase becomes stable, but bounded at high and low T. In the course of this study we failed to find any transition between smectic-B and crystal on lowering T; this suggests that the smectic-B phase is actually a crystal. Further details may be found in Ref. [17].

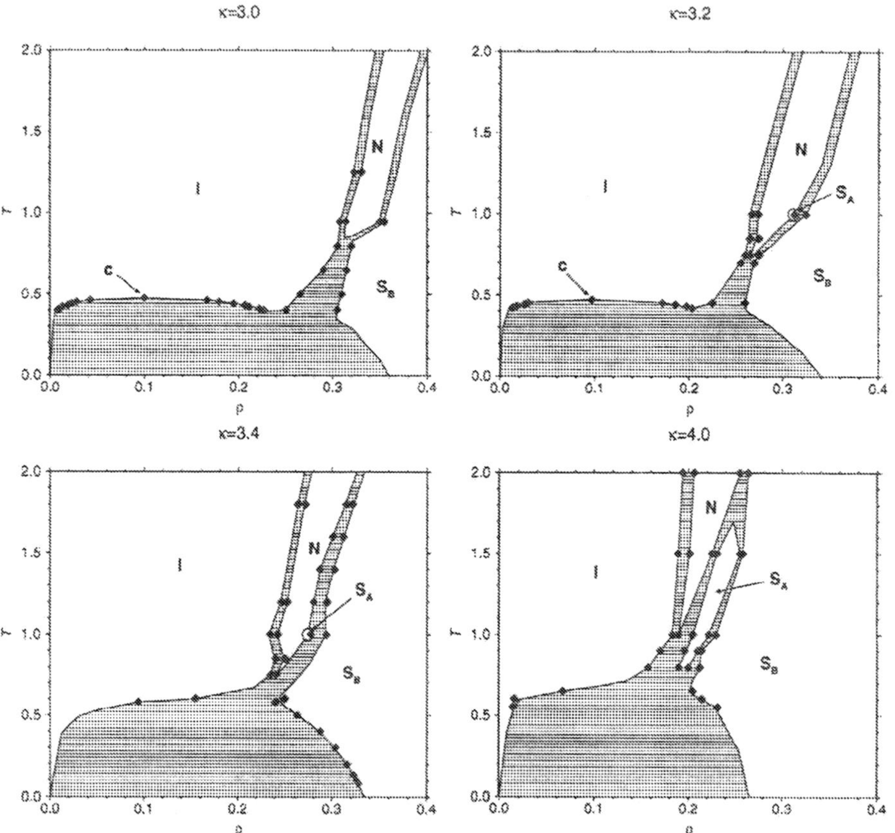

Figure 9. Schematic phase diagrams for Gay-Berne systems at various elongations. For more details see Ref. [17].

4. Diffusion and rotation in smectic-A phases

Some properties of liquid crystals require special simulation techniques, because the timescales or length scales are too long to study in the conventional way. The first of these concerns the relatively slow processes of translational diffusion between layers, and end-over-end rotation, in a smectic-A phase. It has been predicted by theory and subsequently confirmed by simulation [18] that, while most molecules lie within the smectic layers, a small proportion of so-called *transverse interlayer* molecules can be found. Could these play a role in either of the above processes? If we treat these as barrier-crossing events, the first stage in answering this question is to determine the free energy barrier in each case: this can be quite high, and hard to determine in a conventional simulation.

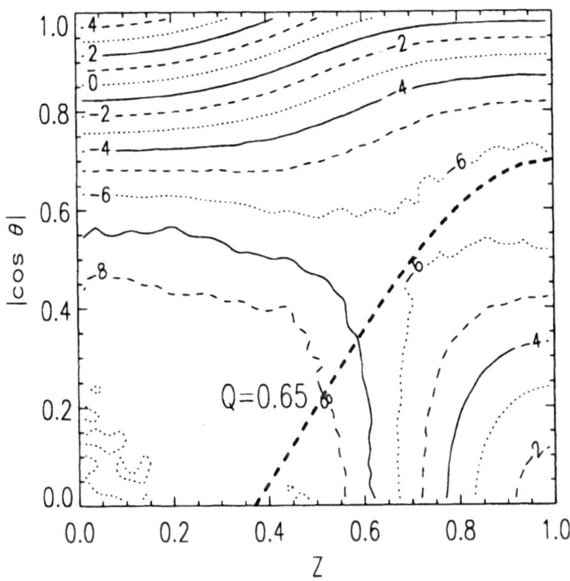

Figure 10. Single-particle probability density landscape, showing $\ln \rho(z, |\cos\theta|)$ as a function of position z and orientation $\cos\theta$. The boundary defined by $Q = 0.65$ is indicated.

To investigate this, we have studied systems of hard spherocylinders, with elongations $L/D = 3.8$ and 5 [19]. First it is necessary to define a probability function, and hence a free energy, as a function of a suitably-defined order parameter. Consider the single-particle density function $\rho(z, |\cos\theta|)$ which describes both positional and orientational ordering: this is normalized so that $\int\int \rho(z, |\cos\theta|) \mathrm{d}|\cos\theta| \mathrm{d}z = 1$. For convenience, the z coordinate normal to layers is scaled so that $z = 0$ withins a smectic layer, $z = 1$ between smectic layers. θ is the angle with the layer normal. Most molecules have $z = 0, |\cos\theta| = 1$; transverse interlayer particles have

86

$z \approx 1, |\cos\theta| \approx 0$; we are interested in particle motion between these two limits. The two-dimensional contour map of $\rho(z, |\cos\theta|)$, for $L/D = 3.8$, determined by conventional Monte Carlo, is illustrated in Fig. 10.

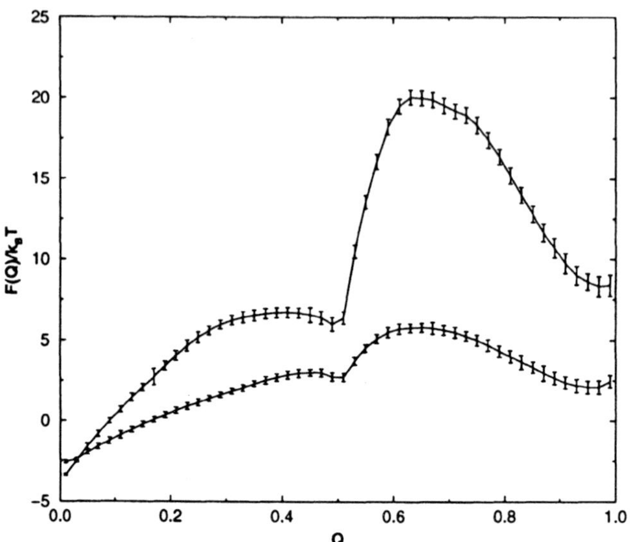

Figure 11.   Free-energy barriers for $L/D = 5$ (higher) and $L/D = 3.8$ (lower) as functions of the order parameter Q.

We found it useful to define a single order parameter Q

$$Q = \frac{1}{2}\left\{1 + \sin^2(\pi z/2) - |\cos\theta|\right\}$$

to discriminate between normal intralayer ($Q = 0$), and transverse inter-layer ($Q = 1$) particles. Then a Landau free energy $\mathcal{F}(Q) = -k_B T \ln \mathcal{P}(Q)$ could be defined in terms of the probability distribution function $\mathcal{P}(Q)$, and this was estimated efficiently by umbrella sampling (see Ref. [19] for details). Results for $L/D = 5$ and $L/D = 3.8$ are shown in Fig. 11. For $L/D = 3.8$, the maximum occurs at $Q \approx 0.65$, and this is also indicated on the two-dimensional free energy landscape shown in Fig. 10. For the higher elongation, the free energy free energy barrier is very high ($\sim 20 k_B T$), and would be difficult to determine without umbrella sampling.

Although no dynamical properties were calculated here, one may draw some preliminary conclusions on the basis of the free energy barriers alone: a high barrier exists to translation into the interlayer region, but an even higher one inhibits rotation in this region; other results show that the barrier to rotation within the layers is higher still. Thus, it is likely that trans-verse interlayer particles play some role in the end-over-end rotation mech-anism, but probably not in the process of translation.

5. Large-scale phenomena

The simulation of truly large-scale phenomena requires parallel computing resources, and the programs used for the work described in the following sections were developed as a consortium project within the U.K. High Performance Computing Initiative. The parallel MD code developed in this project is described in the chapter of Wilson, and elsewhere [20].

5.1. STRUCTURE AND DYNAMICS NEAR THE I-N TRANSITION

Approaching the isotropic-nematic transition temperature T_{NI} from the isotropic side, the correlation length for second-rank orientational fluctuations ξ_2 diverges and collective orientational dynamics become slow, rather like density fluctuations near the liquid-vapour critical point. The extrapolated divergence temperature T^* is very slightly lower than the thermodynamic transition temperature, and the transition is weakly first-order rather than being continuous. Although a proper system-size dependent study of this phase transition would be very expensive, and is only really feasible for a lattice model [21, 22], there is some interest in observing pre-transitional fluctuations using a molecular model, and comparing with simple theoretical predictions. We have studied a system of $N = 8000$ molecules, interacting with the version of the Gay-Berne potential proposed by Berardi et $al.$ [23], using typical run lengths ~ 20 ns. For temperatures just above the transition temperature we have studied structural correlations and dynamical correlation functions [24]. The pair correlation function, $h(1,2) \equiv g(1,2) - 1 = g(\mathbf{r}_1 - \mathbf{r}_2, \mathbf{u}_1, \mathbf{u}_2) - 1$, may be expanded in a set of rotationally invariant functions [25]

$$h(1,2) = \sum_{mnl} h^{mnl}(r)\Phi^{mnl}(\mathbf{u}_1, \mathbf{u}_2, \hat{\mathbf{r}}) = 4\pi \sum_{mn\chi} h_{mn\chi}(r)Y_\chi^m(\mathbf{u}_1)Y_{-\chi}^n(\mathbf{u}_2)$$

where, as before, $\hat{\mathbf{r}}$ is the direction and r the magnitude of the centre-centre vector $\mathbf{r}_1 - \mathbf{r}_2$. The first set of expansion coefficients refers to a space-fixed laboratory frame, and the second to a coordinate system based on the molecular centre-centre vector. The two sets of coefficients are easily interconverted. The direct correlation function $c(1,2)$ may be similarly expanded, and the Ornstein-Zernike equation linking h with c

$$h(1,2) = c(1,2) + \frac{\rho}{4\pi} \int d\mathbf{r}_3 d\mathbf{u}_3 \, h(1,3)c(3,2)$$

expressed as a matrix of integral equations in the coefficients $h_{mn\chi}(r)$ and $c_{mn\chi}(r)$. Thus, it is possible to invert the simulation data, obtaining $c(1,2)$. $c(1,2)$ may be well behaved near T_{NI} even though $h(1,2)$ becomes long ranged. Specifically, defining integrals of the space-fixed coefficients

$$h^{(m)} \equiv \rho(2m+1)^{-1/2}4\pi \int_0^{\infty} dr \ r^2 \ h^{mm0}(r)$$

$$c^{(m)} \equiv \rho(2m+1)^{-1/2}4\pi \int_0^{\infty} dr \ r^2 \ c^{mm0}(r)$$

it may be shown that there is an exact relation: $h^{(m)} = c^{(m)}/(1 - c^{(m)})$. $c^{(m)} \to 1$ implies that the isotropic phase is unstable relative to the nematic phase [26, 27]. Our results, shown in Fig. 12 suggest that, indeed, the appropriate components of $c(1,2)$ remain short-ranged (at least, for the temperatures investigated here) while the corresponding components of $h(1,2)$ develop long-ranged correlations. Also, the instability criterion $c^{(m)} \to 1$ was observed to hold on the approach to the transition.

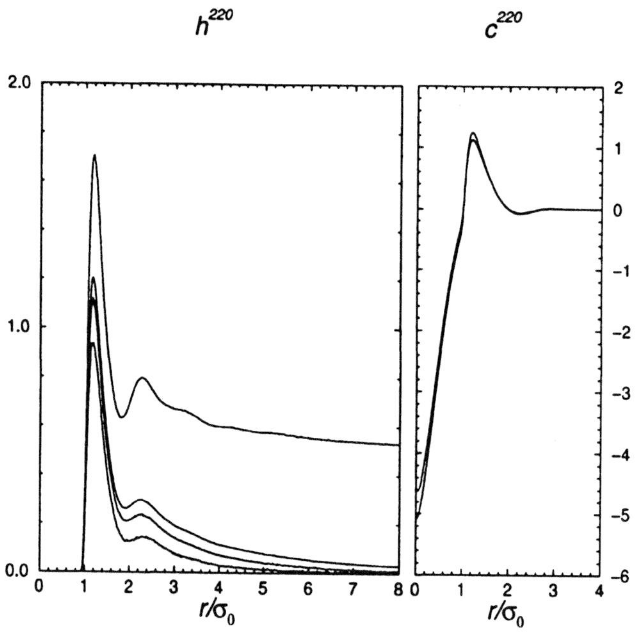

Figure 12. On the left we show the component of the total correlation function $h^{220}(r)$ which shows pretransitional growth of orientational correlations, approaching the I-N transition at $T = 4.0, 3.6, 3.5$ and at $T = 3.45$, below the transition, showing long-ranged order. On the right we show the corresponding component of the direct correlation function, $c^{220}(r)$, approaching the transition at $T = 4.0, 3.5$.

To study the collective dynamics we defined a variable

$$Q_{\alpha\beta}(\mathbf{k}, t) = \sum_i \left(\frac{3}{2}u_{i\alpha}(t)u_{i\beta}(t) - \frac{1}{2}\delta_{\alpha\beta} \right) e^{i\mathbf{k}\cdot\mathbf{r}_i(t)} \qquad \alpha, \beta = x, y, z \qquad (1)$$

and measured the time correlation function

$$C(k,t) = \sum_{\alpha\beta} \langle Q_{\alpha\beta}(-\mathbf{k},0)Q_{\alpha\beta}(\mathbf{k},t)\rangle \propto \sum_{ij} \left\langle P_2(\mathbf{u}_i(0)\cdot\mathbf{u}_j(t))\; e^{i\mathbf{k}\cdot(\mathbf{r}_i(0)-\mathbf{r}_j(t))}\right\rangle$$

From the simulation results we observed roughly exponential decay $C(k,t) = A(k)\exp\{-t/\tau(k)\}$ with $\tau(k) \sim k^{-2}$. At $k = 0$, $\tau^{-1} \propto \xi_2^{-2} \propto (T - T^*)$, in agreement with a simple Landau-de Gennes description [28]. For further details see Ref. [24].

5.2. STRUCTURE OF THE S_A^* TGB PHASE

In a liquid crystal, director twist may arise through molecular chirality or a mechanical torque. In chiral nematic (N^*) phases, the director $\mathbf{n}$ twists smoothly through space. In smectic-A (S_A) phases, the director $\mathbf{n}$ is parallel to the layer normal $\mathbf{N}$; in chiral smectic-A* (S_A^*) phases, $\mathbf{N}$ (and hence $\mathbf{n}$) cannot twist smoothly. Instead, the twist is concentrated in *twist grain boundaries* between domains of untwisted smectic phase.

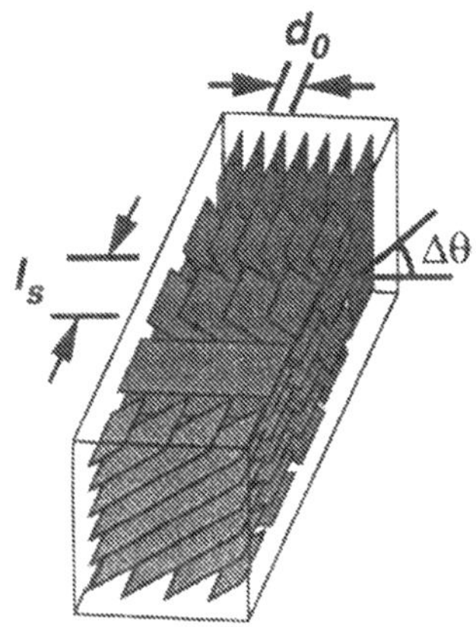

Figure 13. Structure of the S_A^* TGB phase, showing domains of untwisted smectic separated by twist grain boundaries.

The structure of the S_A^* TGB phase is characterized by several lengths and angles. The smectic layer spacing, d_0, is approximately the same as the molecular length. Smectic domains of width l_s are separated by planar twist-grain boundaries, as illustrated in Fig. 13. Within each grain boundary there is an array of screw dislocation line defects separated by distance

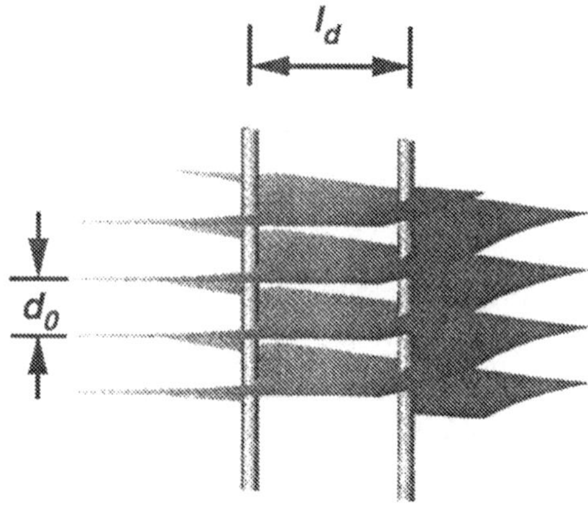

Figure 14. Two screw dislocation defects within a grain boundary. The surfaces represent smectic layers.

l_d, as shown in Fig. 14. The rotation angle $\Delta\theta$ between one domain and the next is given by

$$\Delta\theta = (l_s/p)2\pi = 2\arctan(d_0/2l_d) \approx d_0/l_d$$

where p is the pitch of the twist.

To investigate this phase, we adopted the Gay-Berne pair potential with $\kappa = 4.4$, $\kappa' = 20$, $\mu = 1$, $\nu = 1$. The phase behaviour of this system has been investigated by Luckhurst and Bates, and it is known to exhibit a smectic-A phase [29]. Large system sizes ($N = 21000, 84000$ molecules) are required, both to allow a large enough twist pitch, and to ensure that the transverse periodic boundary conditions do not influence the smectic layer structure too much. Typical run lengths $t \approx 10$ ns allowed the system to evolve from an initial, equilibrated, chiral nematic N* phase at temperature $T^* = 1.4$, to various temperatures in the smectic region [30]. Special twisted periodic boundary conditions were employed to allow the simulation box to induce a pitch equal to one quarter of the box length; these are illustrated in Fig. 15. The screw dislocation defects were located by a simulated annealing technique.

Despite the simplicity of the model, it proved possible to model domain sizes of the same order of magnitude as those seen experimentally [31]; a comparison is given in Table 1. Undoubtedly many caveats remain concerning the equilibration times for defects of this size, large-scale defect

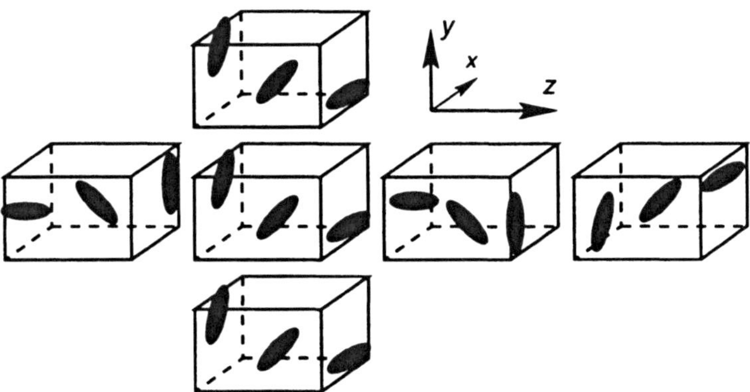

Figure 15.   Twisted periodic boundary conditions

TABLE 1. Comparison of simulation results with experiment [31] for the twist grain boundary phase.

Property	Simulation $/\sigma_0$	Experiment /nm
Smectic layer spacing d_0	4.3	4.1
Helix pitch p	320	500
Smectic domain size l_s	20	24
Dislocation line separation l_d	11	15

motion, and the influence of periodicity on the defect structure, but these early results are quite encouraging.

5.3. BULK ELASTIC CONSTANTS

Long-wavelength deformations of the director field $\mathbf{n}(\mathbf{r})$ are described by the Frank free energy

$$\mathcal{F}_{\text{elastic}} = \int_V \mathrm{d}\mathbf{r}\, f_b + \int_S \mathrm{d}\sigma\, f_s$$

$$f_b = \frac{1}{2}K_{11}\left(\nabla \cdot \mathbf{n}\right)^2 + \frac{1}{2}K_{22}\left(\mathbf{n} \cdot \nabla \wedge \mathbf{n}\right)^2 + \frac{1}{2}K_{33}\left(\mathbf{n} \wedge \nabla \wedge \mathbf{n}\right)^2$$

$$f_s = \frac{1}{2}W_\theta \sin^2\theta\,.$$

There are two classes of terms: an integral, over the sample volume, of a bulk free energy density which depends on the splay (K_{11}), twist (K_{22}) and bend (K_{33}) elastic constants; and an integral, over the bounding surface, of a surface free energy density, which has been written here simply as a surface

92

anchoring term W_θ. This involves deformations of the director by an angle θ away from the preferred anchoring orientation at the surface. Omitted here are the surface-like elastic terms K_{13} and K_{24} which have been of recent experimental and theoretical interest (see e.g. [32–34]). The question to be addressed here is 'Can simulation provide values of the coefficients in the above expression?'.

In the nematic phase, for sufficiently long wavelengths that the above free energy expression holds, equilibrium thermal fluctuations of the Fourier-transformed orientation density $\hat{Q}_{\alpha\beta}(\mathbf{k})$ (defined by Eqn. (1)) are determined by the elastic constants. It is convenient to define the following:

$$W_{13}(k_1, k_3) \quad \propto \quad \left\langle \hat{Q}_{13}(\mathbf{k})\hat{Q}_{13}(-\mathbf{k}) \right\rangle^{-1} \equiv \langle |\hat{Q}_{13}(\mathbf{k})|^2 \rangle^{-1} \propto K_1 k_1^2 + K_3 k_3^2$$

$$W_{23}(k_1, k_3) \quad \propto \quad \left\langle \hat{Q}_{23}(\mathbf{k})\hat{Q}_{23}(-\mathbf{k}) \right\rangle^{-1} \equiv \langle |\hat{Q}_{23}(\mathbf{k})|^2 \rangle^{-1} \propto K_2 k_1^2 + K_3 k_3^2$$

Here, the director is chosen to lie in the 3-direction; and $\mathbf{k} = (k_1, 0, k_3)$ is in the 1-3 plane. These quantities may be determined in a standard computer simulation, and a fit made to W_{13} and W_{23} as functions of k_1^2 and k_3^2. Taking the low-k limit is the crucial part of this process, so it is essential to have a large enough simulation box size L to guarantee the accuracy of the extrapolation. It is convenient to constrain the 123 axis system to coincide with the xyz coordinate system of the simulation box, otherwise the components k_1, k_3 will change as the director drifts; this can be done using Lagrangian constraints in molecular dynamics. We have studied [35] the Gay-Berne systems with $\mu = 2, \nu = 1, \kappa = 3, \kappa' = 5$ [36] and $\mu = 1, \nu = 3, \kappa = 3, \kappa' = 5$ [23]. Slow fluctuations of the long-wavelength modes necessitated long simulation runs, while system sizes up to $N = 8000$ particles were used to guarantee the correctness of the low-k extrapolation. Typical slices through the fitted surface, with the corresponding data points, are shown in Fig. 16.

The key conclusions of this work were that elastic constants can indeed be measured quite accurately by this method, and that the absolute magnitudes of the elastic constants, and their ratios, are consistent with those expected for rod-like molecules. In particular the bend elastic constant K_{33} was, typically, rather larger than the other two, and the values were all reasonably close to those found for hard ellipsoids of similar shape, at comparable densities. It should be noted that an alternative method of estimating elastic constants, via the direct correlation function, has been proposed [37–41]. The advantage of this method is the possibility of calculating these quantities from short-ranged functions, without the need to conduct a low-k extrapolation. The disadvantage is that some approximations must be introduced to make the method tractable.

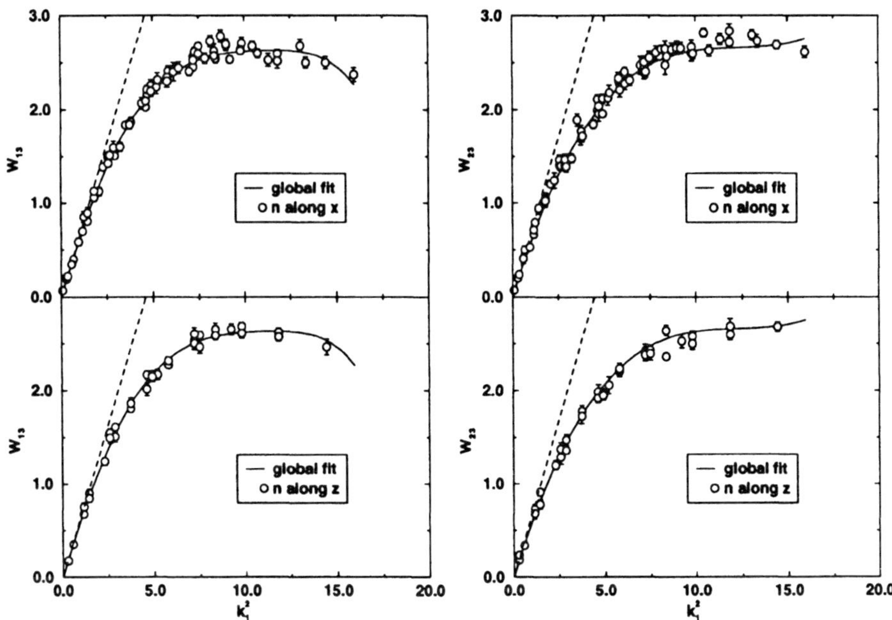

Figure 16. Splay (left) and twist (right) fluctuations as functions of wavelength for the Gay-Berne model with $\mu = 2, \nu = 1, \kappa = 3, \kappa' = 5$ at $\rho = 0.32$, $T = 0.90$. Here the wave-vector component along the director k_3 is taken to be zero, and the inverse fluctuations quantities W_{13} and W_{23} are plotted as functions of k_1^2. The solid lines show sections through the globally-fitted surface and the dashed lines indicate the low-k limiting gradient. For more details see Ref. [35].

5.4. SURFACE ANCHORING STRENGTHS

In the measurement of the surface anchoring coefficient, the first problem is the simulation of a properly equilibrated and well-characterized surface. Here are some preliminary results [42] for a very simple, idealized model: hard ellipsoids, of elongation $e = 15$, confined between parallel hard walls in a planar slab geometry. The wall potential is taken to act on the ellipsoid centres, not on their surfaces, so they may partially penetrate the wall; packing considerations lead to homeotropic anchoring (i.e. the preferred alignment is normal to the walls). Advantages of studying this system are that the bulk nematic phase is stable at relatively low density, that the confining surfaces are well defined, and that the liquid crystal forms an adsorbed layer at the wall which is relatively unstructured. Measurement of the transverse centre-centre pair correlation function in the adsorbed layer confirms that it is liquid-like. The anchoring strength at one of the walls is measured by applying an orienting perturbation at the opposite wall, which generates a director profile across the slab.

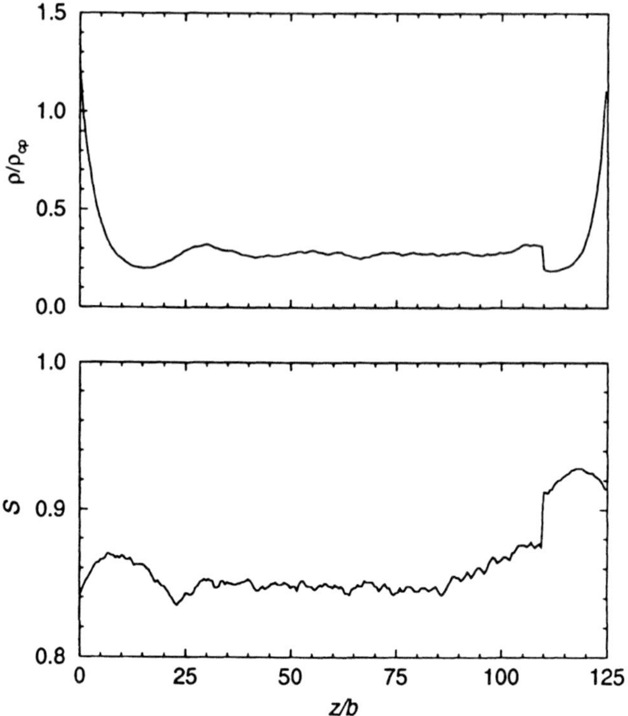

Figure 17. Density profile ρ/ρ_{cp} and nematic order parameter profile S for hard ellipsoids with homeotropic anchoring in slab geometry. An orientational perturbing field is applied at the right wall; the left wall is unperturbed. Distances are measured in units of the semi-axis length b. See Ref. [42] for details.

A variational treatment of the elastic free energy leads to the following approximate expression for the director angle $\theta(z)$ at the wall:

$$\frac{W_\theta}{K_{33}} \approx \left.\frac{d\theta(z)/dz}{\theta(z)}\right|_{z=0} \equiv \lambda^{-1}$$

The expression yields a characteristic extrapolation length λ. Preliminary results, for a range of different ordering fields, are shown in Figs 17 and 18. For this model, the orientation profiles are quite well described by the elastic theory, and the extrapolation length (and hence the surface anchoring coefficient) are easily measurable. For this system we see 'strong' anchoring, i.e. λ is comparable with molecular dimensions (but recall that the 'molecules' here are quite long). A surface exhibiting 'weak' anchoring would respond to an orientational perturbation with a large displacement $\theta(z = 0)$ at the wall, and a correspondingly large extrapolation length λ; this would present a different set of problems for a computer simulation.

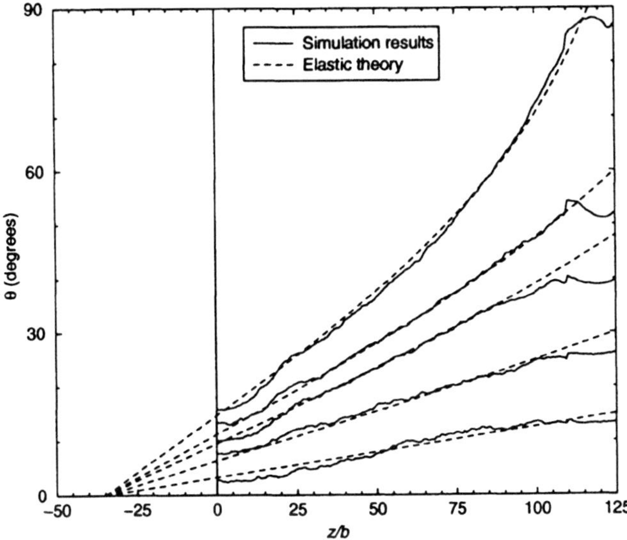

Figure 18. Measuring a surface anchoring coefficient. An orienting field is applied in a region near the right wall; the resulting profile of director angle θ is shown, and the limiting behaviour near the left wall fitted to an elastic description. Results are shown for various magnitudes of orientational perturbation. Simulation results: solid lines; elastic theory: dashed lines. See Ref. [42] for details.

Looking ahead, it may be anticipated that molecular simulations will contribute most in situations where the simple elastic theory breaks down, specifically where the density and order parameter vary rapidly with position. This applies near surfaces and in liquid crystal defects. How should we describe this behaviour theoretically? One possibility is to use versions of the Landau-de Gennes continuum theory, in which the free energy is expressed as a function of the order tensor $\mathbf{Q}$. However this introduces many more phenomenological coefficients, which may be difficult to relate to molecular properties; moreover the theory is essentially an expansion about the isotropic state, which restricts its applicability, and it is typically still limited to small gradients. An alternative is to develop density functional descriptions, starting with direct minimization of the Onsager free energy functional in the given boundary conditions for an inhomogeneous density function $\varrho(\mathbf{r}, \mathbf{u})$. A similar approach has been used [43] in the context of calculating the depletion force in colloid-polymer mixtures. Our preliminary results using the Onsager free energy seem quite encouraging, and will be reported elsewhere.

96

Acknowledgements

Much of the work described here was carried out by postgraduate students and postdoctoral researchers within my group: Muataz Al-Barwani, Julian Brown, Philip Camp, Jeroen van Duijneveldt, Alexey Lyulin, Carl Mason, and Mark Warren. It is a pleasure also to acknowledge collaborations with Peter Bolhuis, Glenn Evans, Daan Frenkel, Dave Kofke, Elvira Martin del Rio, Andrew Masters, Enrique de Miguel, Bela Mulder, Igor Neelov, plus my colleagues in the High Performance Computing Initiative project, Chris Care, Doug Cleaver, George Jackson, Geoffrey Luckhurst, Andrew Masters, Maureen Neal, and Mark Wilson. Computing support for that project was provided by Nick Allsopp, Nick Floros, Denis Nicole and Kenji Takeda at Southampton University. Roberto Berardi and Claudio Zannoni provided valuable advice on their simulations, not to mention assistance and hospitality, during my visits to the CINECA supercomputer centre; Martin Bates also provided valuable information concerning his simulations. The support of Bristol University, particularly our departmental computing officer Jean-Pierre Melot, is gratefully acknowledged, and, finally, funding for this research was provided by EPSRC.

References

1. Onsager, L. (1949) *Ann. N.Y. Acad. Sci.*, **51**, 627.
2. Parsons, J.D. (1979) *Phys. Rev. A*, **19**, 1225.
3. Lee, S.-D. (1987) *J. Chem. Phys.*, **87**, 4972
4. Lee, S.-D. (1989) *J. Chem. Phys.*, **89**, 7036.
5. Camp, P.J., Mason, C.P., Allen, M.P., Khare, A.A. and Kofke, D.A. (1996) *J. Chem. Phys.*, **105**, 2837.
6. Panagiotopoulos, A.Z. (1987) *Molec. Phys.*, **61**, 813.
7. Panagiotopoulos, A.Z. (1995), in *Observation, Prediction and Simulation of Phase Transitions in Complex Fluids*. Baus, M., Rull, L.F. and Ryckaert, J.-P. (eds.), volume 460 of *NATO ASI Series C*, Kluwer Academic Publishers, Dordrecht, pages 463–501. Proceedings of the NATO Advanced Study Institute on 'Observation, Prediction and Simulation of Phase Transitions in Complex Fluids', Varenna, Italy, July 25–August 5, 1994.
8. Nezbeda, I. and Kolafa, J. (1991) *Molec. Simul.*, **5**, 391.
9. Attard, P. (1993) *J. Chem. Phys.*, **98**, 2225.
10. Kofke, D.A. (1993) *Molec. Phys.*, **78**, 1331.
11. Camp, P.J. and Allen, M.P. (1997) *J. Chem. Phys.*, **106**, 6681.
12. Camp, P.J., Allen, M.P., Bolhuis, P.G. and Frenkel, D. (1997) *J. Chem. Phys.*, **106**, 9270.
13. Martín del Río, E. (1996) *Estudio de las propiedades interfaciales de cristales liquidos nematicos*. PhD thesis, Seville University,
14. de Miguel, E., Martín del Río, E., Brown, J.T. and Allen, M.P. (1996) *J. Chem. Phys.*, **105**, 4234.
15. Martín del Río, E. and de Miguel, E. (1997) *Phys. Rev. E*, **55**, 2916.
16. Emerson, A.P.J., Faetti, S. and Zannoni, C. (1997) *Chem. Phys. Lett.*, **271**, 241,
17. Brown, J.T., Allen, M.P., de Miguel, E. and Martín del Río, E. (1998) *Phys. Rev. E*, **57**, 6685.

18. van Roij, R., Bolhuis, P., Mulder, B. and Frenkel, D. (1995) *Phys. Rev. E*, **52**, R1277.

19. van Duijneveldt, J.S. and Allen, M.P. (1997) *Molec. Phys.*, **90**, 243.

20. Wilson, M.R., Allen, M.P., Warren, M.A., Sauron, A. and Smith, W. (1997) *J. Comput. Chem.*, **18**, 478.

21. Zhang, Z., Mouritsen, O.G. and Zuckermann, M.J. (1992) *Phys. Rev. Lett.*, **69**, 2803.

22. Zhang, Z., Zuckermann, M.J. and Mouritsen, O.G. (1993) *Molec. Phys.*, **80**, 1195.

23. Berardi, R., Emerson, A.P.J. and Zannoni, C. (1993) *J. Chem. Soc. Faraday Trans.*, **89**, 4069.

24. Allen, M.P. and Warren, M.A. (1997) *Phys. Rev. Lett.*, **78**, 1291.

25. Gray, C. and Gubbins, K.E. (1984) *Theory of Molecular Fluids*. Clarendon Press, Oxford.

26. Stecki, J. and Kloczkowski, A. (1979) *J. Phys., Paris*, **40** (C3), 360.

27. Perera, A., Patey, G.N. and Weis, J.J. (1988) *J. Chem. Phys.*, **89**, 6941.

28. de Gennes, P.G. and Prost, J. (1995) *The Physics of Liquid Crystals*. Clarendon Press, Oxford, second, paperback edition.

29. Bates, M. *private communication.*

30. Allen, M.P., Warren, M.A. and Wilson, M.R. (1998) *Phys. Rev. E*, **57**, 5585.

31. Ihn, K.J., Zasadzinski, J.A.N., Pindak, R., Slaney, A.J. and Goodby, J. (1992) *Science*, **258**, 275.

32. Crawford, G.P. and Žumer, S. (1995) *Int. J. Mod. Phys. B*, **9**, 331.

33. Crawford, G.P. and Žumer, S. (eds.) (1996) *Liquid Crystals in Complex Geometries Formed by Polymer and Porous Networks*. London, Taylor and Francis.

34. Yokoyama, H. (1997) *Phys. Rev. E*, **55**, 2938.

35. Allen, M.P., Warren, M.A., Wilson, M.R., Sauron, A. and Smith, W. (1996) *J. Chem. Phys.*, **105**, 2850.

36. de Miguel, E., Rull, L.F., Chalam, M.K. and Gubbins, K.E. (1991) *Molec. Phys.*, **74**, 405.

37. Poniewierski, A. and Stecki, J. (1979) *Molec. Phys.*, **38**, 1931.

38. Lipkin, M.D., Rice, S.A. and Mohanty, U. (1985) *J. Chem. Phys.*, **82**, 472.

39. Stelzer, J., Longa, L. and Trebin, H.-R. (1995) *J. Chem. Phys.*, **103**, 3098.

40. Stelzer, J., Longa, L. and Trebin, H.-R. (1995) *Mol. Cryst. Liq. Cryst.*, **262**, 455.

41. Stelzer, J., Trebin, H.-R. and Longa, L. (1997) *J. Chem. Phys.*, **107**, 1295 Erratum.

42. Allen, M.P. *Molec. Phys.*, submitted for publication.

43. Mao, Y., Cates, M.E. and Lekkerkerker, H.N.W. (1997) *J. Chem. Phys.*, **106**, 3721.

LIQUID CRYSTAL LATTICE MODELS I.
BULK SYSTEMS

PAOLO PASINI, CESARE CHICCOLI
Istituto Nazionale di Fisica Nucleare, Sezione di Bologna
Via Irnerio 46, 40126, Bologna, ITALY

AND

CLAUDIO ZANNONI
Dipartimento di Chimica Fisica ed Inorganica, Università di Bologna
Viale Risorgimento 4, 40136, Bologna, ITALY

Abstract. Monte Carlo simulations of spin models for liquid crystal bulk systems are described. A concise description of second rank Lebwohl-Lasher models of various dimensionality and of models containing in addition interactions of first and fourth rank is provided. Biaxial lattice models are also briefly discussed.

1. Introduction

Lattice spin models consist of systems of interacting centres (*"spins"*) placed at the sites of a certain regular lattice. The spins can be thought of as idealized unit vectors assuming discrete or continuously varying orientations in a space of given "spin dimensionality" s . The lattice of positions will have its own, possibly different, dimensionality d. Classical examples are the Ising and Heisenberg models [1] that have played and still play a key role in the study of magnetism. Indeed, despite their simplicity, spin models have proved to be extremely important in the study of phase transitions and critical phenomena in many fields of physics ranging from liquids to polymers [1–3]. Although lattice systems with their intrinsic positional order are in some sense the very antithesis of liquid crystals, they have also been succesfully employed in investigating nematics since the pioneering work of Lebwohl and Lasher (LL) [4]. In this case the spins that represent

99

P. Pasini and C. Zannoni (eds.), Advances in the Computer Simulations of Liquid Crystals, 99–119.

the molecules or groups of molecules should possess full rotational freedom, rather than a discrete set of orientations, so as not to affect the long range orientational behavior. A large amount of work has been and is currently done on generalizations of the LL hamiltonian even though in the last few years more realistic potentials with full translational freedom, like the Gay-Berne one [5], have become increasingly popular thanks also to the continuous increase in computing power. It is, in any case, fair to say that as long as the properties of interest are purely orientational, there are several advantages in using simple lattice models, with respect to potentials with translational freedom, the foremost of which is probably the possibility of performing simulations on a larger (often $10^2 - 10^3$ times larger!) number of particles while conserving the essence of the physics. As an alternative, when using smaller lattices, it is possible to investigate potentials for relatively complicated systems depending on additional parameters, for example associated with varying boundary conditions and field strengths, over a wide range of state points.

Here we wish to present and briefly review some lattice models of bulk liquid crystals and their computer simulations to show how these simple potentials can be useful in investigating the orientational properties of nematics.

2. Periodic boundary conditions

Before going into the details of the various models we wish to mention the ubiquitous problem of the choice of boundary conditions, i.e. of what to surround the simulated sample with. Tackling it is unavoidable since computer simulations are usually performed on a relatively limited number of particles N. Even for lattice models N is of the order of $10^3 - 10^6$ in comparison with a bulk system for which the interacting particles are of the order of the Avogadro number. Then, apart from choosing a lattice size as large as possible, it is very important to adopt some artifact at the sample surfaces so as to minimize the effects of the finite size of the system. The appropriate choice of the boundary conditions becomes then essential especially when small systems are investigated.

The most often used boundary conditions are the so called periodic ones (PBC), where the sample box is surrounded by exact replicas of itself (see Fig. 1). Although this kind of boundary conditions introduces a non existent periodicity and thus some spurious correlation, PBC effectively reduce the effect of the finite size and of the sample surfaces. Due to the greater correlation between sites it is expected that periodic boundary conditions will overestimate the transition temperature T_C. The opposite case arises when the correlation between sites is underestimated as in the case of free

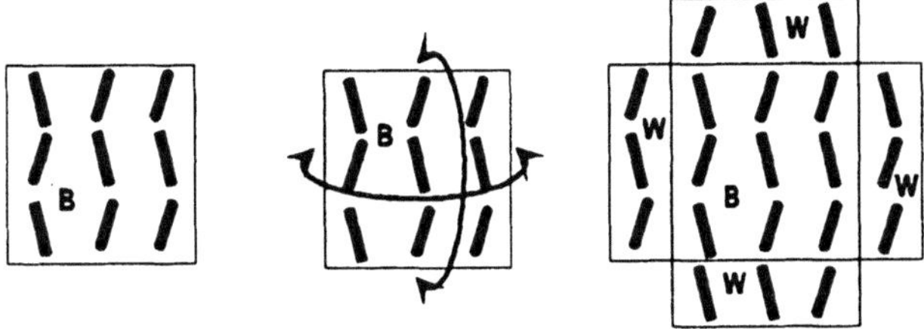

Figure 1. A schematic representation of boundary conditions in a 2D lattice system: empty or free (left), periodic (middle) and cluster (right).

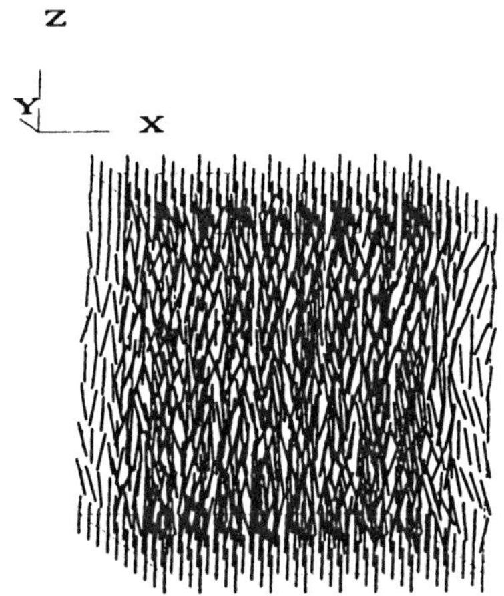

Figure 2. The LL lattice model in the nematic phase.

boundary surfaces. Consequentely periodic and free boundaries should give, respectively, an upper and a lower bound for the transition temperature:

$$T_C(free) < T_C < T_C(periodic), \tag{1}$$

and this has been checked at least for 2d lattices [6]. In any case PBC represent the standard approach to simulating bulk phases and we shall adopt it here, although we shall see later that other and sometimes more effective approaches can also be employed.

3. The Lebwohl-Lasher model

The prototype lattice model for modelling liquid crystals was devised many years ago by Lebwohl and Lasher (LL) [4] and is the simplest one with the correct symmetry for nematics (in particular the potential is invariant for an head-tail flip of the molecules). The particles, assumed to have uniaxial

102

symmetry and represented by three dimensional spins located at the sites of a $L \times L \times L$ cubic lattice (see Fig. 2), interact through a pair potential of the form:

$$U_{ij} = -\epsilon_{ij} P_2(\cos \beta_{ij}), \qquad (2)$$

where ϵ_{ij} is a positive constant, ϵ, for nearest neighbour spins i and j and zero otherwise, P_2 is the second Legendre polynomial and β_{ij} is the angle between the molecules. The interaction tends to bring molecules parallel to one another and effectively models whatever underlying intermolecular interaction either attractive or repulsive that does that. The model has been studied [7–12] and generalized by many authors [13–16] and we shall see later a few relevant examples.

3.1. OBSERVABLES

While Monte Carlo computer simulations of LC lattice models typically proceed following the standard Metropolis [17] procedure (see Chapter 1) two issues require special attention and will be covered here starting with the LL model: one is the determination of phase transitions and the other the calculation of orientational order and other anisotropic observables.

3.1.1. *Energy and heat capacity*
The phase transition of the model is located by monitoring as a function of temperature the constant volume specific heat defined as:

$$C_V^* = \partial U^* / \partial T^*, \qquad (3)$$

and obtained by a numerical differentiation of energy with respect to temperature [7]. We use the star for dimensionless variables, e.g. $T^* = kT/\epsilon$ with k the Boltzmann constant.

This is a only seemingly simple task since numerical differentiation typically requires smoothing and this in turn masks the transition. Indeed at times the derivative is best calculated solving an integral equation! [18] The specific heat can also be calculated from the energy fluctuations:

$$C_V^*/k = (\langle U^{*2} \rangle - \langle U^* \rangle^2)/(kT^*)^2, \qquad (4)$$

although this tends to be somewhat noisy.

The dimensionless energy per particle of a system with N spins, $U^* = \langle U \rangle / N\epsilon$, is in turn calculated by the sum:

$$U^* = \frac{1}{N(N-1)\epsilon} \sum_{i=1}^{N} \sum_{j=i+1}^{N} U_{ij}. \qquad (5)$$

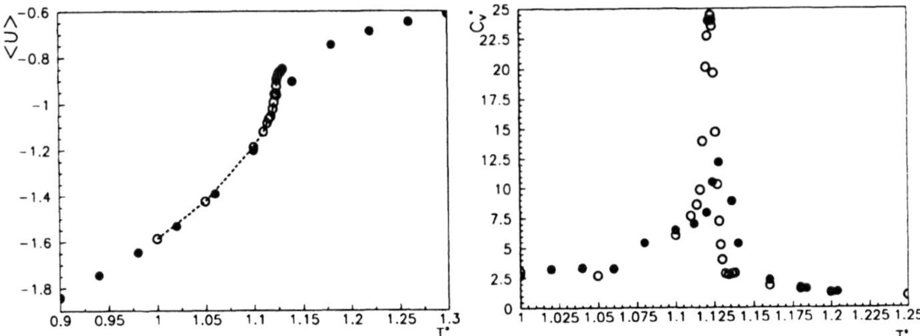

Figure 3. The single particle energy U^* (left) and heat capacity C_V^* (right)vs dimensionless temperature $T^* = kT/\epsilon$ for the Monte Carlo simulation of a $10 \times 10 \times 10$ (full circles) and a $30 \times 30 \times 30$ (empty circles) [9] Lebwohl-Lasher systems.

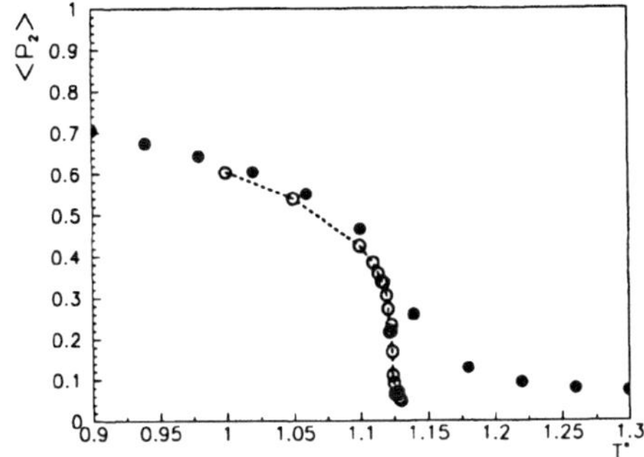

Figure 4. The second rank order parameter $\langle P_2 \rangle$ vs dimensionless temperature $T^* = kT/\epsilon$ as obtained from MC simulations on a $10 \times 10 \times 10$ (full circles) and a $30 \times 30 \times 30$ (empty circles) [9] LL systems.

For the LL system the values range then from $U^* = -3$ for a perfectly aligned system to $U^* = 0$ for an isotropic phase.

First order transitions are characterized by a singularity in the heat capacity in the thermodynamic limit. In a finite system, however, the transition region is broadened and the heat capacity just peaks at a phase transition. Then C_V^* is used in locating the phase transition temperature T_c^*. For the LL model T_{NI}^* was determined [9] in this way to be $T_{NI}^* = 1.1232 \pm 0.0006$. Furthermore the maximum of the peak increases with the system size [19, 20, 22]:

$$C_V^{max}(L) = a + bL^3, \tag{6}$$

where a and b are size independent parameters.

The energy and the heat capacity variation with temperature as obtained from Monte Carlo simulations on two lattices with different sizes

104

with PBC can be seen in Figure 3. The peaks sharpens with the increase
of L and actually scales as expected from (6). The first order character
of the transition is confirmed also by an analysis of the distribution of energy values collected, as histograms, during the simulation runs, that shows
the double peak behaviour expected for coexisting ordered and disordered
states [21]. A more modern analysis [46] based on Ferrenberg and Swendsen
[23] reweighting method confirms the results of [9].

3.1.2. *Order parameters*

The second rank orientational order parameter can be defined as :

$$\langle P_2 \rangle = \frac{1}{N} \sum_{i=1}^{N} P_2(\mathbf{u}_i \cdot \mathbf{n}), \tag{7}$$

where $\mathbf{u}_i$ is the molecular axis of the i-th particle and $\mathbf{n}$ the director. However, in these Monte Carlo simulations there is no external field applied to
pin the director and $\mathbf{n}$ can change during the system evolution. A description of the determination of the order parameters in computer simulations is
reported in Chapter 2 of this book. Here we recall only that the problem of
determining the order parameter reduces to that of finding the unit vector
$\mathbf{n}$ which renders the order in a certain configuration $\langle P_2 \rangle_S$ a maximum and
this in turn amounts to calculating and diagonalizing the ordering matrix
$\mathbf{Q}$ defined as:

$$\mathbf{Q} = \frac{1}{2N} \sum_{i=1}^{N} 3u_{i\alpha} u_{i\beta} - \frac{1}{2}\delta_{\alpha\beta}, \tag{8}$$

with $u_{i\alpha}$ the direction cosines of the *i-th* molecule, and identifying the order
in the S configuration $\langle P_2 \rangle_S$ with its largest eigenvalue [7]. The eigenvalues of a matrix are scalars, independent on an overall frame rotation and
thus on the orientation of the instantaneous director $\mathbf{n}$ with respect to the
laboratory frame. Then a global average, $\langle P_2 \rangle_\lambda$, can be obtained by first
calculating the order parameter by diagonalization of $\mathbf{Q}$ at each desired
configuration and then averaging over a sufficiently large number of configurations. A related procedure has been given for $\langle P_4 \rangle_\lambda$ in ref. [9]. The
second rank order parameters obtained in this way from the simulations
of two LL lattice systems are shown in Figure 4. As mentioned before the
LL model reproduces rather well the temperature dependence of the order
parameter versus temperature observed in real nematics, that can normally
be written as:

$$\langle P_2 \rangle = (1 - T/T_{NI})^\beta + \langle P_2 \rangle_{iso} \qquad T < T_{NI}. \tag{9}$$

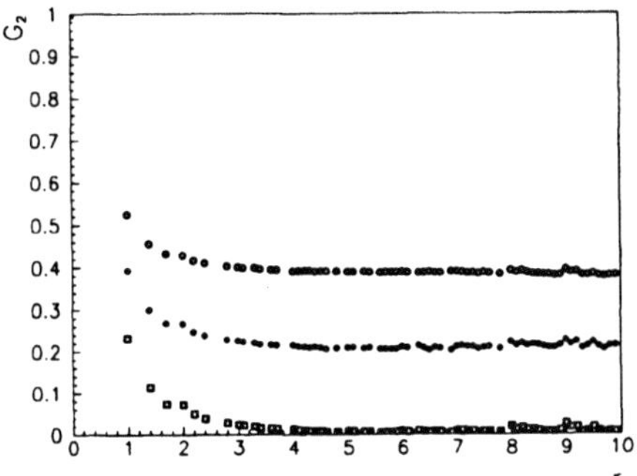

Figure 5. The second rank orientational correlation function G_2 vs distance r as obtained from MC simulations on a $10 \times 10 \times 10$ LL systems. The results are for $T^* = 1.0$ (upper curve), $T^* = 1.1$ (middle curve) and $T^* = 1.2$ (bottom curve),

In fact it has been found that the exponent ranges approximately from 0.17 up to 0.25 for a large series of both Schiff base and cyanobiphenyl nematics [24, 25] while the LL model gives 0.22 ± 0.01 [16]. In real experiments $\langle P_2 \rangle_{iso} = 0$ while for the finite samples used in simulations $\langle P_2 \rangle_{iso} \approx O(\sqrt{N})$.

This behavior can be obtained also using small lattices (a few thousand spins) and possibly the model works so well because a "spin" can be thought to represent, rather than a single particle, a closely packed group of molecules, that maintains its local structure at various temperatures and even across the nematic/isotropic phase transition [26]. As a special case these domains could comprise just one molecule but it seems more realistic to assume that they typically include a few tens of particles [27].

3.1.3. *Orientational correlation functions*

While $\langle P_2 \rangle$ and the higher $\langle P_L \rangle$ offer a description of the orientational long range order in the case it exists, the problem of deciding if true long range order does indeed exist remains to be tackled and orientational correlation functions $G_L(r)$ are particularly useful to this effect. The set of correlations $G_L(r)$ can be defined as expansion coefficients of the rotationally invariant pair distribution [7]:

$$G(r, \beta_{12}) = G_0^{00}(r) \sum_L \frac{2L + 1}{64\pi^2} G_L(r) P_L(\cos \beta_{12}). \tag{10}$$

106

$G_0^{00}(r)$ is the particle centre distribution that, for a cubic lattice, is just:

$$G_0^{00}(r) = \frac{1}{4\pi\rho r^2} \sum_k z_k \delta(r - r_k), \tag{11}$$

where ρ is the density and z_k the number of neighbours at r_k. Thus $G_L(r)$ are a sort of two particle order parameters , which give the correlation between the orientations of two particles separated by a distance r:

$$G_L(r) = \langle P_L(\cos\beta_{12})\rangle_r, \tag{12}$$

where $\langle.....\rangle_r$ is a normalized average over all spins falling in a thin spherical shell centred at r and of width corresponding to the chosen resolution Δ.

The pair coefficients $G_L(r)$ should start from one and tail off to essentially $\langle P_L\rangle^2$ [7]. Thus we expect $G_L(r)$ to decay to a plateau only if long range order exists, as in the nematic phase. As we can gather from the above formulas the calculation runs on particle pairs and can be quite time consuming when the size of the lattice is large, representing a relevant percentage of the total time spent in the simulation. Usually the first two angular pair correlation coefficients G_2 and G_4 are calculated.

In fig. 5 we show $G_2(r)$ at some selected temperature below and above the phase transition for the LL model.

3.2. LOW DIMENSIONAL SYSTEMS

Low dimensional systems present interesting and challenging problems such as the very existence and nature of their phase transitions and have received a lot of interest from many authors [28–34]. We notice that the planar LL model is of interest also from a purely theoretical point of view to test the existence of topological phase transitions in non-abelian two dimensional systems [35, 36].

We here consider systems with reduced space dimensionality: $d = 2$ and $d = 1$, while keeping $s = 3$, in other words a planar and a linear LL lattice where molecules can still reorient in three dimensions.

Looking at the heat capacity behavior of the $d = 2$ lattice as a function of size it is clear from the simulation results that the scenario is very different in comparison with the 3D system and that in particular the heat capacity is insensitive to the increase in the number of particles as can be seen in Fig. 6 where the results for four increasing sizes are shown.

Although no true phase transition is expected in a two dimensional LL system [37] the heat capacity anomaly divides the temperature range in two regions which shows a different behavior in the decay of orientational pair correlation function (Fig. 7 *left*). The analysis of these quantities [30]

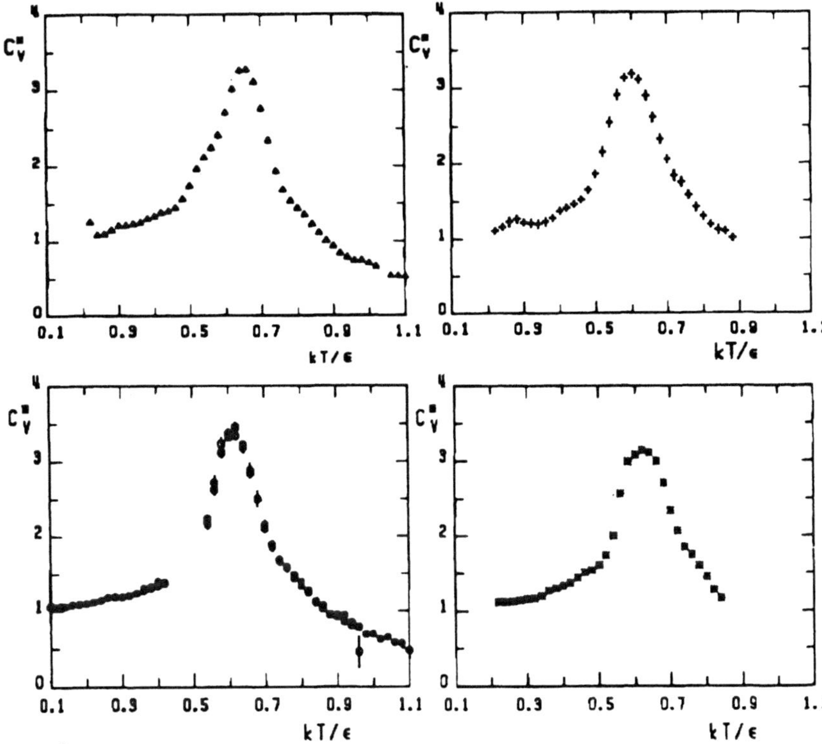

Figure 6. The heat capacity C_V^* obtained from differentiation of energy plotted versus dimensionless temperature $T^* = kT/\epsilon$ for the Monte Carlo simulations of four size planar Lebwohl-Lasher systems, i.e. 10×10 (*top left*), 20×20 (*top right*), 60×60 (*bottom left*) and 80×80 (*bottom right*).

indicates that a power law decay of the type:

$$G_2(r) = A_p/r^{k_p} \tag{13}$$

is the best fit for the simulation data in the ordered phase while an exponential decay:

$$G_2(r) = (1 - A_e)e^{-k_e r} + A_e \tag{14}$$

describes the correlation function behavior above the pseudo-transition temperature. Thus the planar system can present large ordered domains for $T^* < T_c^*$ but not for $T^* > T_c^*$.

To conclude this short overview of space dimensionality effects it is interesting to look now at a one dimensional, $d = 1$, $s = 3$ lattice. This 1D system is constituted by a chain of particles which are free to rotate in a 3D space. An analytical solution for this model, given by Vuillermot and Romerio [38], exists and shows that no phase transition occurs and that the

108

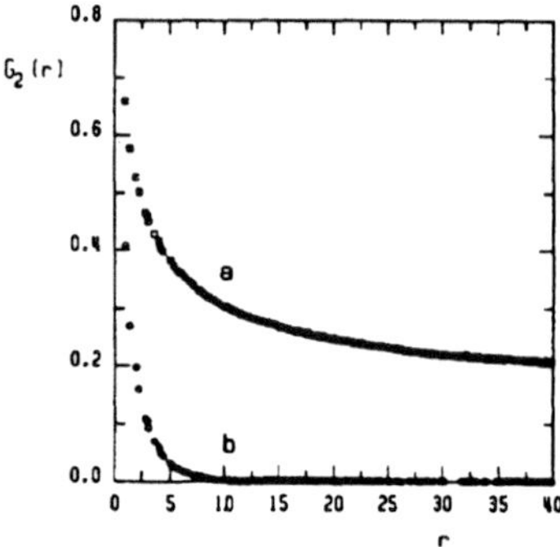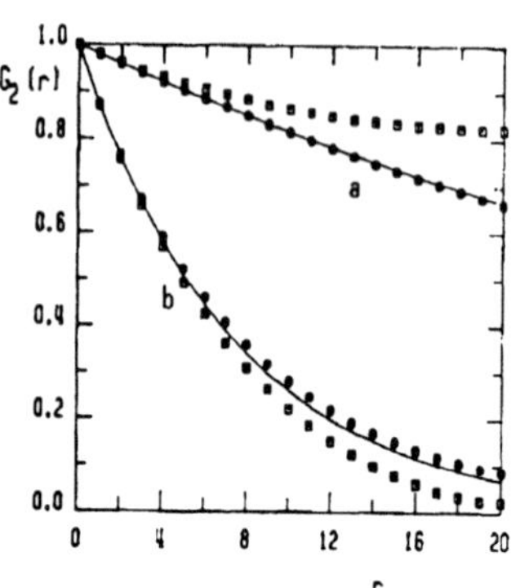

Figure 7. The second rank orientational correlation function G_2 versus distance r (lattice units) at two selected temperatures, i.e. below (a) and above (b) the heat capacity anomaly. *Left:* G_2 for the planar LL model. *Right:* G_2 for the monodimensional case for $L = 40$ (squares) and $L = 100$ (circles) compared with the analytic solution [38] (lines).

system is ordered only at zero temperature (see Fig. 8). It is instructive

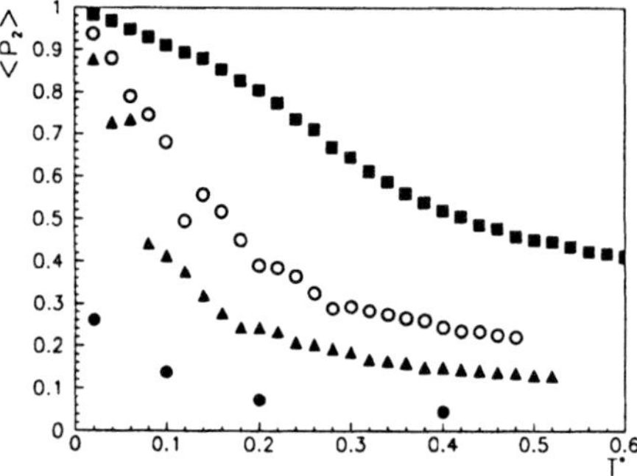

Figure 8. The second rank order parameter $\langle P_2 \rangle$ vs dimensionless temperature $T^* = kT/\epsilon$ for a 1D LL system of various lengths: L=10 (squares), L=40 (empty circles), L=100 (triangles), L=1000 (full circles).

to see that $\langle P_2 \rangle$ can be quite misleading as an indicator of the existence of true long range order, especially if a size dependent study is not performed (Fig. 8). On the contrary a fit of $G_2(r)$ reveals very clearly the exponential decay of orientational correlations [31] (Fig. 7 *right*).

In summary we see that the different behavior of $G_2(r)$ can help in assessing the presence and particularly the absence of long range order.

3.3. CLUSTER BOUNDARY CONDITIONS

Although quite satisfactory when far from a phase transition, the use of periodic boundary conditions (PBC) leads to large smearing and broadening of the heat capacity and order parameter vs. temperature curves. This complicates the location of the transition and demands the use of very large samples. There is therefore an interest in looking for alternative schemes and, for instance, another type of boundary condition was proposed within the Cluster Monte Carlo (CMC) method [39]. In this approach the simulation sample is surrounded by an additional layer of particles (*ghosts*) which have on average the same properties as the particles inside. In the CMC method, the desired bulk or global average of a quantity A is written as an average over all the external "world" configurations $[W]$ of the values $< A >_{[W]}$ calculated for a fixed configuration of the "world" outside the sample box [39]. Thus the global average is

$$< A >_G \quad = << A >_{[W]}>_W \tag{15}$$

$$\approx (1/M_W) \sum_{[W]} < A >_{[W]} . \tag{16}$$

In practice a MC simulation is run to obtain $< A >_{[W]}$ and the outside world configurations needed are obtained by creating a layer of ghost particles outside the sample box having the same one-particle distribution of the system inside the box. The orientations of the virtual neighbours are sampled from an orientational distribution function constructed, using maximum entropy principles,[40] from the order parameters calculated inside the sample, i.e.

$$P(\cos \beta) = \exp[\sum_{L=0}^{L'} a_L P_L(\cos \beta)], \tag{17}$$

where the coefficients a_L are determined from the constraint that the available $< P_L >$ can be reobtained by averaging $P_L(x)$ over the distribution. For example $<P_2>$ and $<P_4>$ have been used for the LL model, and the coefficients a_2, a_4 have been determined by solving the non linear system

$$<P_L> = \frac{\int_0^\pi d\beta \sin \beta P_L(\cos \beta) \exp[a_2 P_2(\cos \beta) + a_4 P_4(\cos \beta)]}{\int_0^\pi d\beta \sin \beta \exp[a_2 P_2(\cos \beta) + a_4 P_4(\cos \beta)]}, \quad L = 2, 4 \tag{18}$$

During the simulation the order parameters $< P_2 >$, $< P_4 >$ inside the sample are calculated and a_2 and a_4 are determined. The orientations for the ghost particles outside the box are then sampled from the distribution

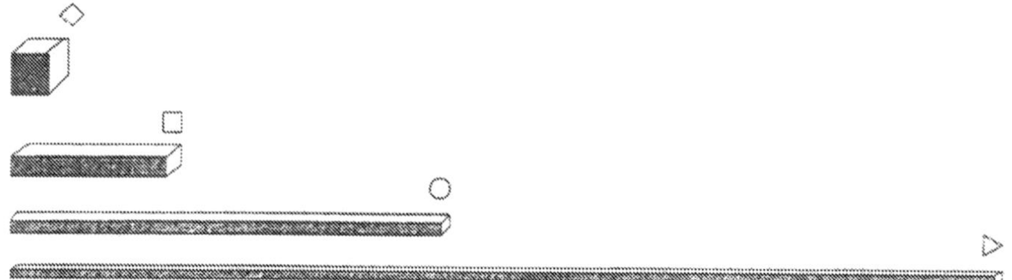

Figure 9. The various shape systems with similar volume, i.e., from top to bottom: $10 \times 10 \times 10$, $5 \times 5 \times 40$, $3 \times 3 \times 110$ and $2 \times 2 \times 250$.

in Eq. 18, the energy of the system is then recalculated and evolution proceeds. In the subsequent cycles the order parameters with respect to the Z laboratory direction P_L^J for the spins inside the box are still calculated. After a certain number of cycles M an average is calculated for this K trajectory segment together with the attendant standard deviation σ_K. These $< P_2 >_{in}$ and $< P_4 >_{in}$ parameters are then compared to the ones outside and if the difference is statistically significant a new set of orientations for the ghost molecules is generated using the new order parameters. The other parts of the Monte Carlo simulation method and particularly lattice updates proceed as usual. A more detailed description of the method is given in [39].

The method, that avoids the spurious correlations between particles separated by more than half the box size, has been successfully tested for various lattice models [13, 39, 41, 42] where it has given results comparable with those obtained employing PBC on lattices up to 2^d times larger in d dimensional systems. The CMC boundaries are particularly useful when potentials with one or more additional parameters have to be studied and a set of independent simulations has to be performed to obtain a phase diagram [13] as we shall show later on.

3.4. SAMPLE SHAPE

Another possibly significant advantage in using non-periodic CMC boundaries is the complete freedom over the shape of the sample that it allows us to simulate, for example, spherical samples [16, 42] and that different shapes with similar N can be used without affecting the results. Since computer simulations are numerical experiments performed on finite and rather small systems with the aim of reproducing the bulk it is to be hoped that the simulations should not depend very significantly on the sample size and shape: indeed these are accessories to the calculation not relevant in a truly bulk sample. Ideally the choice of the boundary conditions should ensure that the size and the shape employed do not affect the behavior of the system under study. On the other hand employing Periodic Boundary Conditions in non cubic systems may give results very different from a bulk behavior [43], as we can see from the results of simulations performed on LL systems of very different shapes [43] (see Fig. 9), containing approximately the same number of spins (Figs.10 and 11) We see that changing the shape of the sample the results show a pronounced variation for PBC but not for CMC. Upon decreasing the breadth and width of the sample while keeping the volume constant, the system tends to approach, using PBC, the limit of one dimensional model (cf. Fig. 9). Also the second rank order parameter results confirm that PBC tends to induce a one dimensional behavior increasing the length to breadth ratio of the sample and, in this latter case $\langle P_2 \rangle$ (see Fig. 11 *left plate*) can be compared with the 1D simulation data (Fig. 8).

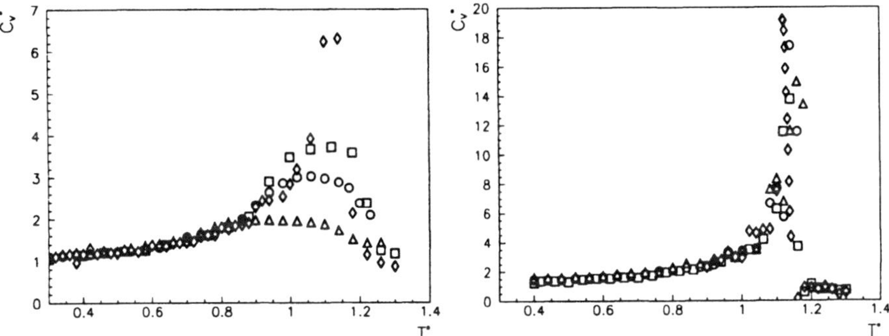

Figure 10. The heat capacity C_V versus dimensionless temperature $T^* = kT/\epsilon$ as obtained from PBC (left) and CMC (right) simulations of LL systems with different shapes, i.e. $2 \times 2 \times 250$ (squares), $3 \times 3 \times 110$ (empty circles), $5 \times 5 \times 40$ (triangles) and $10 \times 10 \times 10$ (full circles).

4. Some other nematic lattice spin models

The simplicity of lattice models allows to study in detail potentials depending on more than one relevant parameter. In these cases the simulation has to be repeated for various values of these physical parameters and the determination of transition temperatures and transition behavior implies a challenging exercise in computer simulations [1, 2, 39]. The choice of boundary conditions like the CMC ones is of considerable importance because a

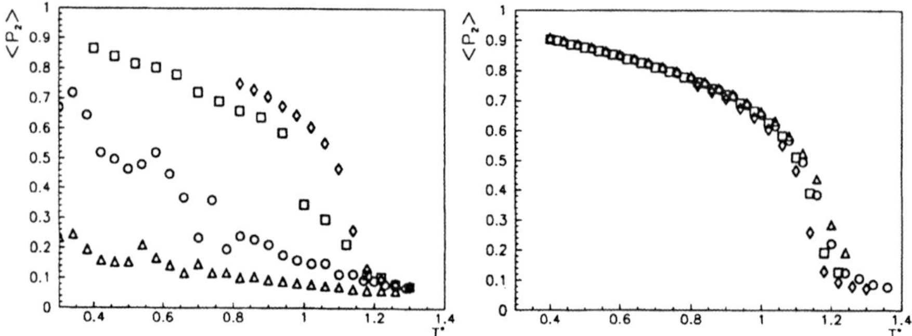

Figure 11. The second rank order parameter $\langle P_2 \rangle$ vs. dimensionless temperature $T^* = kT/\epsilon$ as obtained from PBC (left) and CMC (right) simulations of LL systems with different shapes. System sizes and symbols are as in Fig. 9.

large number of simulations can be performed using smaller lattices in order to obtain a phase diagram for the model. Here we consider two examples involving generalizations of the LL model.

4.1. A P_2P_4 MODEL

As mentioned above the temperature dependence of the orientational order for the LL model is in quite a good agreement with the esperimental results for nematics. However, it is interesting to examine the effects of a fourth rank contribution, easily the first neglected term in a general expansion of the pair interaction and examine how strictly the observed experimental results are related to the specific second rank nature of the potential. A fourth rank contribution has often been invoked in interpreting experimental results in nematics [44] and in membrane vesicles [45] and some simulation studies of the mixed P_2P_4 interaction potential have been performed [14, 16, 46].

The P_2P_4 hamiltonian can be written as:

$$U_{ij} = -\epsilon_{ij}[P_2(\mathbf{u}_i \cdot \mathbf{u}_j) + C_4 P_4(\mathbf{u}_i \cdot \mathbf{u}_j)] \quad ; \quad \text{with } i \neq j, \qquad (19)$$

where C_4 designates the relative strength of the interactions. A three dimensional representation of the potential as a function of β_{ij} and C_4 (Fig. 12) shows how the location of the potential minima changes with the fourth rank contribution. The MC simulation results for the specific heat for various values of C_4 are reported in Fig. 13. We see that the fourth rank term has a profound effect on the transition. A positive contribution shifts T^*_{NI} to higher temperature and makes the transition much more pronouncedly first order, while a negative C_4 has the opposite effect, weakening the transition and shifting it to a lower temperature.

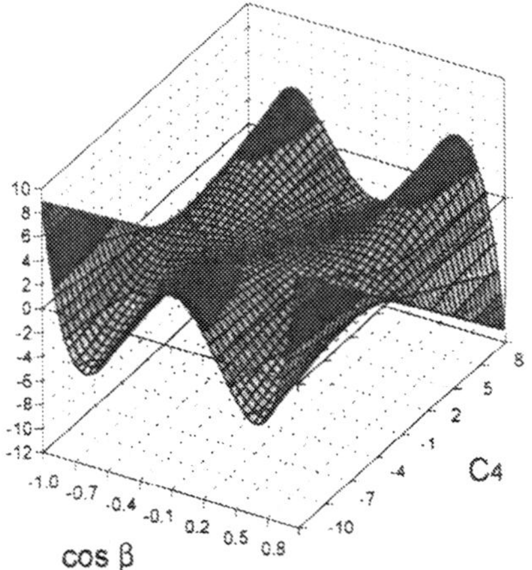

Figure 12. A plot of the $P_2 P_4$ potential between two spins as a function of their relative orientation $\cos\beta = \mathbf{u}_i \cdot \mathbf{u}_j$ for various fourth rank contributions C_4.

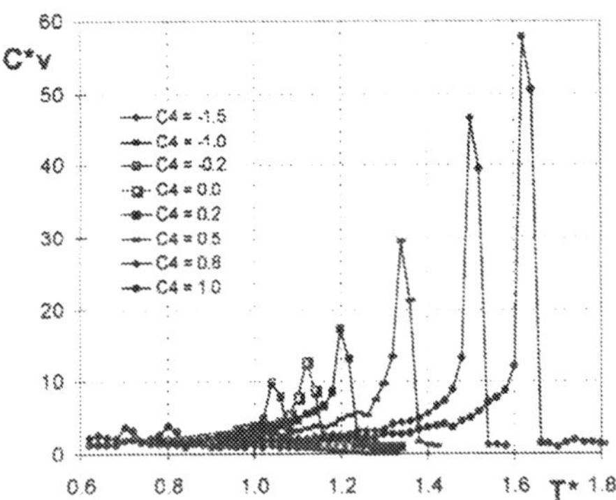

Figure 13. Heat capacity $C_V^* \equiv C_V/k$ dependence on reduced temperature $T^* \equiv kT/\epsilon$ for various values of C_4 as obtained from MC simulation on an approximate spherical lattice with Cluster boundary conditions. The lines are a guide for the eye.

The order parameter versus reduced temperature curve (Fig. 14) shows that a fourth rank contribution can vary the effective exponent β in eq. 9 and that only a limited range of C_4 can yield a β value compatible with experiment.

4.2. A $P_1 P_2$ MODEL

Another interesting mixed rank potential contains a simple combination of first and second rank interactions proposed by Krieger and James [47]

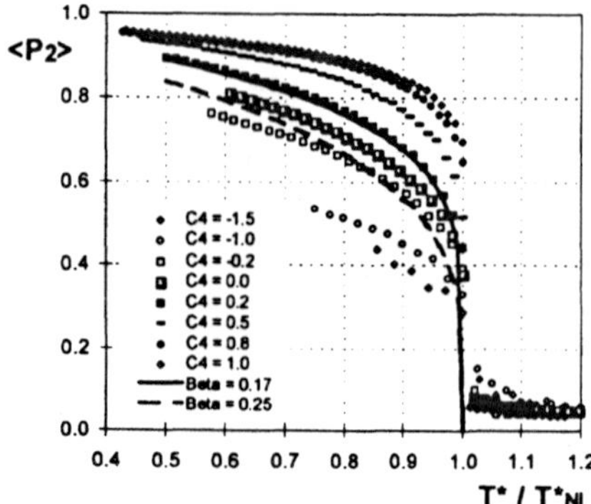

Figure 14. The second rank order parameter $<P_2>$ dependence on scaled temperature $T^+ \equiv T^*/T^*_{NI}$ for various values of C_4. The continuous curves denote the region limited by the experimental exponents β in the Haller law as obtained for real liquid crystals.

and later by Lin Lei [48]. The first rank term simulates the head - tail asymmetry and the potential can be viewed as a prototype model for bowlic and ferroelectric liquid crystals. The hamiltonian reads:

$$U_{ij} = -\epsilon_{ij}[P_2(\cos\beta_{ij}) + \xi P_1(\cos\beta_{ij})], \tag{20}$$

where the parameter ξ determines the relative importance of the first rank term (Heisenberg model) with respect to the second one (Lebwohl-Lasher model), while its sign determines ferroelectric or antiferroelectric type interactions. Realization of a molecular system with ferroelectric type ordering is actively seeked and could be made possible by a combination of steric and dipolar interactions as, e.g., in pyramidic systems [49, 50].

The simulations [13] confirm the Mean Field predictions [47] about the phase diagram, shown in Fig. 15, with three phases: polar, nematic and isotropic. In the polar phase (P) both the first and the second rank order parameters, $<P_1>$ and $<P_2>$ respectively, are non zero. In the nematic region (N) $<P_1>$ is zero while, as usual, $<P_2>$ survives and both of them vanish in the isotropic phase (I). The tricritical point occurs at a value of $\xi = 0.3578$.

4.3. A BIAXIAL MODEL

In all the lattice models presented above, as in the large majority of theoretical calculations and computer simulations of liquid crystals, the mesogenic molecules are assumed to be cylindrically symmetric. However it is important to recall that nematogen molecules are invariably not cylindrically symmetric and that a much more realistic approximation is to treat

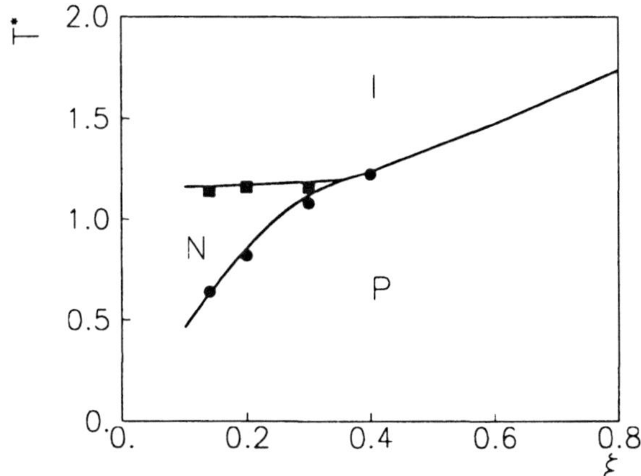

Figure 15. The $P_1 P_2$ model phase diagram showing the reduced transition temperature versus the relative strength parameter ξ. The simulations results (points) are reported together with the Two Site Cluster predictions (curves).

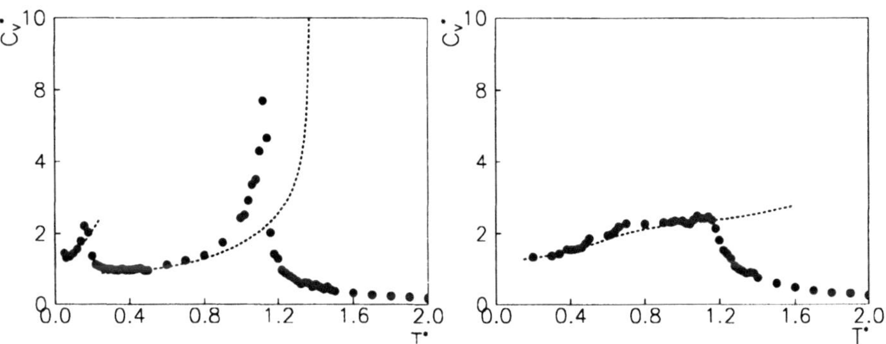

Figure 16. The heat capacity C_V^* versus temperature for two different molecular biaxialities, i.e. $\lambda = 0.2$ (left) and $\lambda \approx 1/\sqrt{6}$ (right). The results are from the simulations (dots) and MF Theory (lines).

them at least as biaxial objects. Ordered phases formed by biaxial particles have indeed been studied using a number of theoretical methods: Mean Field Theory (MFT) [51–55], counting methods [56], Landau-deGennes theory [57], bifurcation analysis [58, 59]. Both attractive interactions and hard particle models have been investigated. It should be stressed that typical nematic phases have uniaxial symmetry around the preferred direction, the director, even if the constituent molecules are themselves biaxial. However, the possibility of a biaxial nematic mesophase has been predicted by all these studies. The existence of this phase has also been confirmed by Monte Carlo simulations of a lattice system of biaxial particles [15, 60] and of a fluid system of biaxial spherocylinders [61]. A simple lattice model of a biaxial system is described by the second rank attractive pair potential:

$$U(\omega_{ij}) = -\epsilon_{ij}\{P_2(\cos\beta_{ij}) + 2\lambda[R_{02}^2(\omega_{ij}) + R_{20}^2(\omega_{ij})] + 4\lambda^2 R_{22}^2(\omega_{ij})\}, \quad (21)$$

where λ is the biaxiality parameter that accounts for the deviation from cylindrical molecular symmetry: when λ is zero, the biaxial potential reduces to the Lebwohl - Lasher P_2 potential, while for λ different from zero the particles tend to align not only their major axis, but also their short axis. $\omega \equiv (\alpha, \beta, \gamma)$ is the set of Euler angles specifying the orientation of a molecule. The potential depends on the relative orientation ω_{ij} of the molecular pair, R_{mn}^L are combinations of Wigner functions symmetry - adapted for the D_{2h} group of the two particles:

$$R_{00}^2 = \tfrac{3}{2}\cos^2\beta - \tfrac{1}{2} \quad (22)$$

$$R_{20}^2 = \tfrac{1}{2}\sqrt{\tfrac{3}{2}}\sin^2\beta\cos 2\alpha \quad (23)$$

$$R_{02}^2 = \tfrac{1}{2}\sqrt{\tfrac{3}{2}}\sin^2\beta\cos 2\gamma \quad (24)$$

$$R_{22}^2 = \tfrac{1}{4}\left(\cos^2\beta + 1\right)\cos 2\alpha\cos 2\gamma - \tfrac{1}{2}\cos\beta\sin 2\alpha. \quad (25)$$

The model has been studied on a fcc lattice by Luckhurst and Romano for $\lambda = 0.2$ [15] and on a cubic lattice for a fairly large set of biaxialities by Chiccoli et al. [62]. The largest value for λ, $\lambda = \frac{1}{\sqrt{6}}$, separates the region of distorted rods from that of distorted disks that can be mapped into one another [52]. This means that for $\lambda > \frac{1}{\sqrt{6}}$, that is for discotic molecules, one can change the y and z axes of the molecules and use the potential with the corresponding $\lambda' < \frac{1}{\sqrt{6}}$ and ϵ'. In other words for $\lambda > \frac{1}{\sqrt{6}}$ there is a mapping of the system to another system with $\lambda < \frac{1}{\sqrt{6}}$, and all the thermodynamic results should be the same (of course the temperature $T = kT/\epsilon$ will correspond to $T' = kT/\epsilon'$).

The simulations has proved very useful in investigating the thermodynamics of this biaxial model and improving the Mean Field Theory prediction. At low values of biaxiality two transitions occur as clearly visible from the two peaks in the heat capacity curve (see Fig. 16), peaks which coalesce approaching the $\lambda = \frac{1}{\sqrt{6}}$, self dual, case. The phase diagram, obtained from a set of MC simulations at various molecular biaxiality is shown in Fig. 17 together with the MFT prediction. The lower transition lines identify the biaxial - nematic phase transition and have a second order character. The upper curve denotes the nematic isotropic phase transition: it is first order for low values of the molecular biaxiality and becomes more second order approaching the dual point. This uniaxial-isotropic transition highlights some differences between MC and MFT. In fact, while the MFT

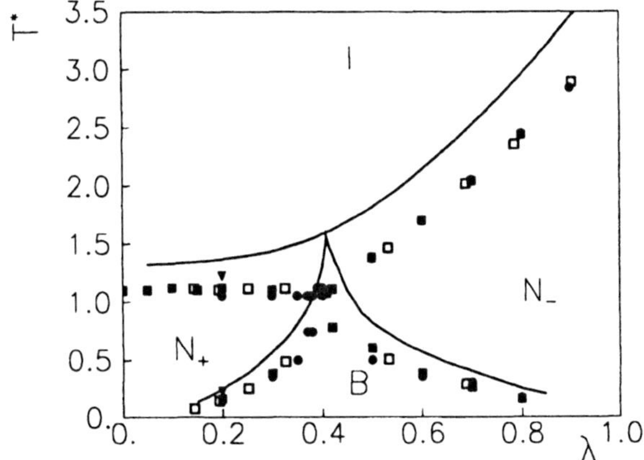

Figure 17. The biaxial model phase diagram showing the reduced transition temperature versus molecular biaxiality λ. The points are simulation results while the continuous curves are the mean field predictions. The tricritical point corresponds to a value $\lambda = 1/\sqrt{6}$) [60].

curve increases with λ the MC results are nearly constant or even show an opposite behavior. This could be relevant in understanding the difficulties in observing a true thermotropic nematic which would be competing in real systems with potential smectic or crystal phases. If, as observed in [63] the typical range of a nematic is of the order of 10% from T_{NI} then the MC phase diagram shows that the biaxial region (B) accessible to experiments is reduced further at lower temperatures and at higher values of λ with respect to the MFT expectations.

The simulations have also been used to calculate the biaxial order parameters. A method to obtain the order parameters is described in Chapter 2 of the present book where a full set of this second rank quantities is also presented. The availability of these biaxial data has allowed other researchers to calculate elastic constants for the model [64].

5. Conclusions

Computer simulations of lattice spin models for liquid crystals have been around for many years but still offer interesting opportunities for investigating anisotropic materials. Their simplicity allows to easily modify the models by adding terms that try to keep into account various effects such as contributions of different rank and symmetry. In particular they are useful in studying phase transitions, transition temperatures and collective properties to a precision not too lower than experiment something that requires in turn very large simulation sample sizes. They have also revealed useful in studying relatively small lattices under a variety of different conditions to investigate confined nematic liquid crystals as we shall see in the next Chapter.

Acknowledgments

We thank MURST, CNR, University of Bologna, INFN (grant I.S. BO12) for support and our coworkers over many years, particularly F. Biscarini and F. Semeria.

References

1. Binder, K.(ed.), (1984) *Applications of the Monte Carlo Method in Statistical Physics* (Springer - Verlag).
2. Allen, M.P. and Tildesley, D.J. (1987) *Computer Simulation of Liquids*, Clarendon Press, Oxford.
3. Kremer, K. and Binder, K. (1988) *Computer Phys. Rep.*, **7**, 259.
4. Lebwohl, P.A. and Lasher G., (1972) *Phys Rev. A*, **6**, 426.
5. Gay, J.G. and Berne, B.J. (1981) *J. Chem. Phys.*, **74**, 3316.
6. Binder, K., (1974) *Adv. Phys.*, **23**, 917.
7. Zannoni, C., (1979) in: *The Molecular Physics of Liquid Crystals*, eds. Luckhurst, G.R. and Gray, G.W., ch. 9. Academic Press, New York.
8. Luckhurst, G.R. and Simpson, P., (1982) *Mol. Phys.*, **47**, 251.
9. Fabbri, U. and Zannoni, C., (1986) *Molec. Phys.*, **58**, 763.
10. Zhang, Z., Zuckermann, M.J. and Mouritsen, O.G., (1991) *Phys. Rev. Lett.*, **69**, 2803.
11. Cleaver, D.J. and Allen, M.P., (1991) *Phys. Rev. A*, **43**, 1918.
12. Greef,C.W. and Lee, M.A., (1994) *Phys. Rev. E*, **49**, 3225.
13. Biscarini, F., Zannoni, C., Chiccoli, C. and Pasini, P., (1991) *Molec. Phys.*, **73**, 439.
14. Fuller, G.J., Luckhurst, G.R. and Zannoni, C. (1985) *Chem. Phys.*, **92**, 105.
15. Luckhurst, G.R. and Romano, S. (1980) *Mol. Phys.*, **40**, 129.
16. Chiccoli, C., Pasini, P. and Zannoni, C., (1997) *Int. J. Mod. Phys. B.*, **11**, 1937.
17. Metropolis N., Rosenbluth A.W., Rosenbluth M.N., Teller A.H. and Teller E., (1953) *J. Chem. Phys.*, **21**, 1087.
18. Chiccoli, C., Pasini, P., and Zannoni, C., (1987) *Liq. Cryst.*, **2**, 39.
19. Fisher, M.E. (1971) in *Proceedings of the International School E. Fermi*, Course 51, Varenna, Green, M.S. (ed), Academic, New York.
20. Binder, K. and Landau, D.P. (1984) *Phys. Rev. B*, **30**, 1477.
21. Mouritsen, O.G. (1984), *Computer Studies of Phase Transitions and Critical Phenomena* Springer, Berlin.
22. Zhang, Z., Mouritsen, O. G. and Zuckermann, M. (1993) *Mod. Phys. Lett. B*, **7**, 217.
23. Ferrenberg A.M. and Swendsen, R.H. (1988) *Phys. Rev. Lett.*, **61**, 2635.
24. Leenhouts, F. , de Jeu, W.H. and Dekker, A.J. (1979) *J. de Physique*, **40**, 989.
25. Wu, S.T. and Cox, R.J. (1988) *J. Appl. Phys.*, **64**, 821.
26. Luckhurst, G.R. and Zannoni, C. (1977) Nature, **267**, 412.
27. Berggren, E., Chiccoli, C., Pasini, P., Semeria, F. and Zannoni, C. (1994) *Phys. Rev. E*, **50**, 2929.
28. Mountain, R. and Rujgrok, Th.W. (1977) *Physica*, **89A**, 522.
29. Tobochnik, J. and Chester, G.V. (1979) *Phys. Rev B*, **20**, 3761.
30. Chiccoli, C., Pasini, P., and Zannoni, C., (1988) *Physica*, **148A**, 298.
31. Chiccoli, C., Pasini, P., and Zannoni, C., (1988) *Liq. Cryst.*, **3**, 363.
32. Romano, S. (1987) *Nuovo Cim.*, **B100**, 447.
33. Romano, S. (1988) *Nuovo Cim.*, **D10**, 1459.

34. Romano, S. (1991) *Liq. Cryst.*, **10**, 73.
35. Kunz, H and Zumbach, G., (1991) *Phys. Lett.* , **257B**, 299.
36. Caracciolo, S., Edwards, R.G., Pellissetto, A. and Sokal, A.D. (1993) *Nucl. Phys. B (Proc. Suppl.)*, **30**, 815.
37. Vuillermot, P.A. and Romerio, M.V. (1975) *Commun. Math. Phys.*, **41**, 281.
38. Vuillermot, P.A. and Romerio, M.V. (1973) *J. Phys. C*, **6**, 2922.
39. Zannoni, C., (1986) *J. Chem. Phys.*, **84** , 424.
40. Levine, R.D. and Tribus, M. (eds.) (1979) *The Maximum Entropy Formalism* MIT Press, Boston.
41. Chiccoli, C. , Pasini, P., Biscarini, F., Zannoni, C. (1988) *Molec. Phys.*, **65**, 1505.
42. Chiccoli, C., Pasini, P., Semeria, F. and Zannoni, C. (1993) . *Int. J. Mod. Phys. C*, **4**, 1041.
43. Chiccoli, C., Pasini, P., Semeria, F. and Zannoni, C., (1993) *Phys. Lett. A*, **176**, 428.
44. Pottel, H., Herreman, W., van der Meer, B.W. and Ameloot, M. (1986) *Chem. Phys.*, **102**, 37.
45. Ameloot, M., Hendrickx, H., Herreman, W., Pottel, H., van Cauwelaert, F. and van der Meer, W. (1984) *Biophys. J.*, **46**, 525.
46. Zhang, Z., Zuckermann, M. and Mouritsen, O. G. *Mol. Phys.* (1993), **80**, 1195.
47. Krieger, T.J. and James, H.M., (1954) *J. Chem. Phys.*, **22** , 796.
48. Lin Lei, (1987) *Mol. Cryst. Liq. Cryst.*, **146** , 41.
49. Malthête, J. and Collet, A., (1985) *Nouv. J. de Chimie*, **9** , 151.
50. Zimmermann, H., Poupko, R., Luz, Z. and J. Billard, (1985) *Z.Naturforsch.*, **40a** , 149.
51. Freiser, M.J. (1970) *Phys. Rev. Lett.*, **24**, 1041.
52. Straley, J.P. (1974) *Phys. Rev. A*, **10**, 1881.
53. Luckhurst, G.R., Zannoni, C., Nordio, P.L., Segre, U. (1975) *Mol. Phys.*, **30**, 1345.
54. Priest, R.G. (1975) *Solid State Comm.*, **17**, 519.
55. Remler, D.K. and Haymet, A.D.J. (1986) *J. Phys. Chem.*, **90**, 5426.
56. Shih, C.S. and Alben, R. (1972) *J. Chem. Phys.*, **57**, 3055.
57. Gramsbergen, E.F., Longa, L. and de Jeu, W.H. (1986) *Phys. Reports*, **135**, 195.
58. Mulder, B.M. (1986) *Liq. Cryst.*, **1**, 539.
59. Holyst, R. and Ponierewski, A. (1990) *Mol. Phys.*, **69**, 193.
60. Biscarini, F., Chiccoli, C., Pasini, P., Semeria, F., and Zannoni, C. (1995) *Phys. Rev. Lett.*, **75**, 1803.
61. Allen, M.P. (1990) *Liq. Cryst.*, **8**, 499.
62. Chiccoli, C., Pasini, P., Semeria, F., and Zannoni, C. (1999) *Int. J. Mod. Phys. C*, **10**, 469.
63. Ferrarini A., Nordio P.L., Spolaore E. and Luckhurst G.R., (1995) *J. Chem. Soc. Faraday Trans.*, **91**, 3177.
64. Longa L., Stelzer J. and Dunmur D., (1998), *J. Chem. Phys.*, **109**, 1555.

LIQUID CRYSTAL LATTICE MODELS II.
CONFINED SYSTEMS

PAOLO PASINI, CESARE CHICCOLI
Istituto Nazionale di Fisica Nucleare, Sezione di Bologna
Via Irnerio 46, 40126, Bologna, ITALY

AND

CLAUDIO ZANNONI
Dipartimento di Chimica Fisica ed Inorganica, Università di Bologna
Viale Risorgimento 4, 40136, Bologna, ITALY

Abstract. Monte Carlo simulations of lattice spin models represent a powerful method for the investigation of confined nematic liquid crystals and allow a study of the molecular organization and thermodynamics of these systems. Here some models of confined liquid crystals, such as polymer dispersed liquid crystals, twisted nematic, in-plane switching liquid crystal displays and hybrid aligned films are described together with their computer simulations.

1. Introduction

Confined nematic systems are a class of materials of wide interest both from the technological and basic research point of view [1]. The first aspect is obvious since a large number of electroptical devices is based on the properties of nematics confined in suitable geometries and boundaries. The academic interest is related to the effects that confinement induces on the phase transitions and on the molecular organization of these systems. This organization in turn stems from a competition between the effects due to surface boundary conditions, to the nematic ordering inside the system and to the disordering caused by temperature. Many experiments and theories have been employed to improve our understanding of these phenomena, but Monte Carlo (MC) simulations seem to be a particularly useful method in

P. Pasini and C. Zannoni (eds.), Advances in the Computer Simulations of Liquid Crystals, 121–138.
© 2000 *Kluwer Academic Publishers. Printed in the Netherlands.*

studying relatively small lattices of confined nematics, particularly in the presence of complex geometries or boundary conditions not amenable to analytic solutions. The need of understanding and predicting experiments where orientational ordering plays the key role makes the simple spin models, introduced in the previous Chapter, a convenient and flexible tool to simulate fairly realistic experimental conditions. In particular this technique has proved useful in investigating droplets with fixed surface anchoring [2] mimicking polymer dispersed liquid crystals [3], nematic displays [4, 5] and hybrid aligned nematic cells [6]. Here we wish to present some examples of applications of the model systems.

2. Polymer dispersed liquid crystals

Polymer dispersed liquid crystals (PDLC) [3] are composite materials that consist of microscopic nematic droplets, with typical radii from a few hundred Angström to more than a micron, embedded in a polymer matrix. These systems are interesting for technical applications [3] but PDLC also represent practical realizations of systems exhibiting topological defects of interest in many fields of physics [7]. A number of experimental works have considered different boundary conditions at the droplet surface, for example radial [8, 9], axial [9], toroidal [10] and bipolar [8, 9, 11] that can be obtained by choosing the polymer matrix and the preparation methods. Additional effects of interest come from the application of external, electric or magnetic, fields [8]. MC simulations have been used to study PDLC in a variety of these physical situations: different boundary conditions [12, 14, 16], influence of the anchoring strength at the nematic/polymer interface [13] and the effect of an external applied field [15]. Particular attention has been devoted to simulating quantities that can be directly observed in real experiments in an attempt to bridge the gap with experimental investigations performed on the same systems. For instance, methodologies to calculate powder deuterium NMR lineshapes and textures observable in polarized light experiments corresponding to the microscopic configurations obtained from computer simulations have been developed [2, 15, 16].

The PDLC model used in simulations concentrates on a single droplet and consists of an approximately spherical sample $\mathcal{S}$ carved from a cubic lattice with spins interacting with the Lebwohl-Lasher (LL) potential described in the previous Chapter, while the surface effects are modelled with an external layer of "ghost" spins, $\mathcal{G}$, with fixed orientations chosen to mimick the desired boundary conditions. The boundary layer acts on the inside particles according to the simple pair interaction:

$$U_{i,j} = -\epsilon_{ij} J[\frac{3}{2}(\mathbf{u}_i \cdot \mathbf{u}_j)^2 - \frac{1}{2}], \qquad for \qquad i \in \mathcal{S}, j \in \mathcal{G}, \qquad (1)$$

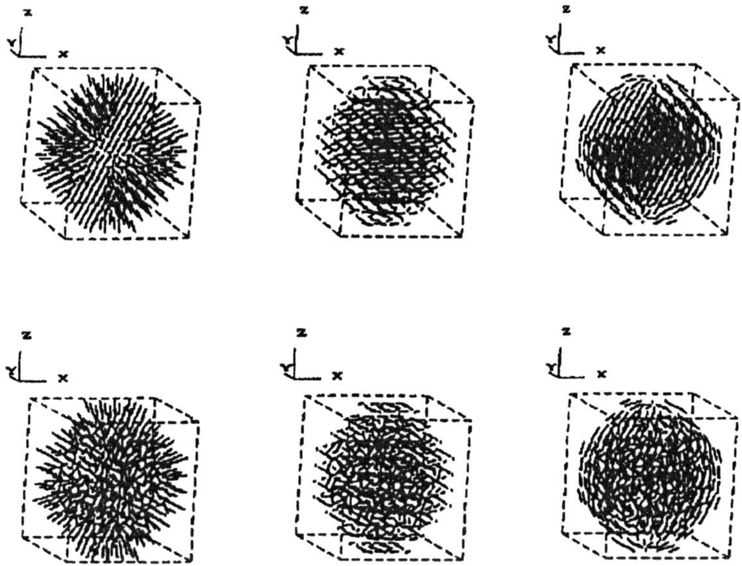

Figure 1. Monte Carlo configurations for a $N = 304$ droplet with radial (left), toroidal (middle) and bipolar (right) surface alignement at a temperature below, $T^* = 0.2$ (top), and one above, $T^* = 1.4$ (bottom) the nematic/isotropic transition are shown [2].

where the sign of the parameter J determines the main direction of anchoring (parallel or perpendicular to the ghost spins) and $\mid J \mid$ its strength at the polymer surface. When $J = 1$ the interaction between two neighbors, one on the surface of the nematic droplet and one belonging to the outside matrix, is the same as that between two liquid crystal spins, while $J = 0$ would correspond to a droplet in vacuum. In Figure 1 sample configurations corresponding to the following three different boundary conditions at the interface nematic/polymer are presented:

$i)$ Radial boundary conditions (RBC), that are imposed by orienting the spins in the matrix normally to the local surface, so that they point towards the center of the droplet.

$ii)$ Toroidal boundary conditions. (TBC) obtained when the spins in the polymer interface lie in planes perpendicular to the z axis and are oriented tangentially to the droplet surface.

$iii)$ Bipolar boundary conditions (BBC) for which the ghost spins are oriented tangentially to the droplet surface and belong to planes parallel to the z axis.

In Figure 1 we have included for each case the outer layer of oriented "ghost" spins appropriate to these boundary conditions.

124

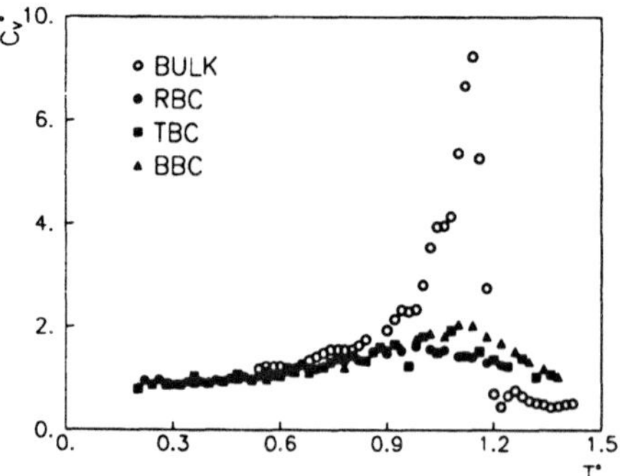

Figure 2. The heat capacity C_V^* versus reduced temperature, T^*, as obtained from Monte Carlo simulations of lattice model droplets with different boundary conditions: radial (RBC), bipolar (BBC) and Toroidal (TBC) and with a bulk simulation. The simulations has been performed on a small lattice ($N = 304$) and with $J = 1$.

From the heat capacity behavior reported in Fig. 2 [2] we see that the nematic-isotropic phase transition is suppressed for small enough confined systems.

To examine the ordering inside the microdroplet various second rank order parameters have been calculated for the systems investigated. The ordinary second rank order parameter, $\langle P_2 \rangle_\lambda$, obtained from diagonalization of the ordering matrix [17] (see previous Chapter), is however not always appropriate as it quantifies the nematic order with respect to an hypothetical global director which may not exist as such. However, MC simulations offer the possibility of evaluating some other order parameters more appropriate to each special case. For example, in case of RBC, it is not possible to distinguish between a perfect ordered radial configuration and a completely disordered system just from the value of $\langle P_2 \rangle_\lambda$ which would vanish in both cases. It is then more useful to define a radial order parameter, $\langle P_2 \rangle_R$ [12]:

$$\langle P_2 \rangle_R = \frac{1}{N} \sum_{i=1}^{N} P_2(\mathbf{u}_i \cdot \mathbf{r}_i), \tag{2}$$

where $\mathbf{r}_i$ is the radial vector of the ith spin. For a perfect hedgehog configuration $\langle P_2 \rangle_R = 1$, while for a truly disordered system $\langle P_2 \rangle_R = 0$. Following the same reasoning it is possible to define a configurational order parameter, $\langle P_2 \rangle_C$, which tends to one for a configuration perfectly ordered according to the idealized structure induced by the boundary conditions used. Thus

$$\langle P_2 \rangle_C = \frac{1}{N} \sum_{i=1}^{N} P_2(\mathbf{u}_i \cdot \mathbf{c}_i), \tag{3}$$

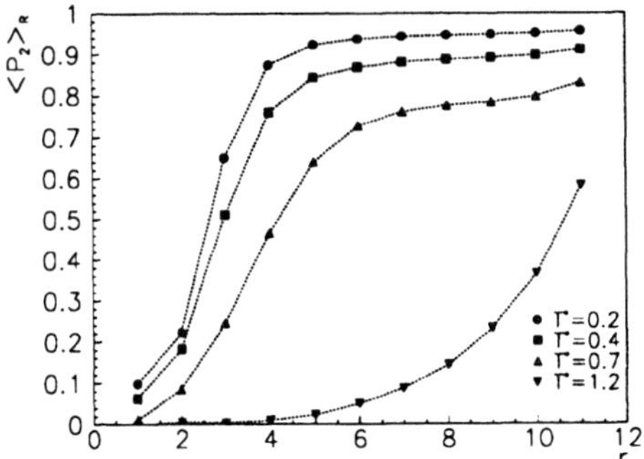

Figure 3. The radial order parameter, $\langle P_2 \rangle_R$, versus distance r starting from the center of the droplet. The results are obtained from Monte Carlo simulations of a RBC lattice model droplet with 5832 spins at some seletcted temperatures.

where c_i is the direction corresponding to the local surface induced alignment. For example in the bipolar case c_i is a local meridian that lies on the plane defined by the droplet axis (z axis) and the radial vector r_i of the particle while being perpendicular to r_i itself.

An investigation of these configurational order parameters across the sample is interesting to test theories of the molecular organisation inside the droplet. In MC this can be achieved dividing the droplet in concentric shells and calculating the relevant quantities in each region so as to have the variation of the ordering going from the center to the border of the system. As an example, the behavior of $\langle P_2 \rangle_R$ with respect to the distance from the center is reported in Fig. 3 at some selected temperatures. These results show, in the nematic region, a ordered core at the center of the droplet, consistent with a ring disclination [18, 19], with a radius which becomes larger as the temperature increases.

The standard nematic order parameter $\langle P_2 \rangle_\lambda$ shows, quite reasonably, an opposite behavior; i.e. it is a maximum at the center of the droplet where the aligned core is found and decreases approaching the surface where the spins are radially oriented. In Figure 4 the order parameters $\langle P_2 \rangle_\lambda$, for the three different boundary conditions here considered, are shown as a function of the distance from the droplet center in lattice units. In the right hand side plates of Fig. 4 $\langle P_2 \rangle_\lambda$ is plotted against a scaled distance r/r_{max}, where r_{max} is the radius of the sphere, to investigate if the ordering inside the droplet depends on the system size or if the behavior is just the same in these reduced units. As mentioned earlier we expect, for the radial case, the nematic order to be greater near the core of the droplet (inner shell) and to decrease in the other shells. It is interesting to notice that in the RBC case the size of the aligned core does not depend on the droplet size and has a radius of about 3-4 lattice units [19] (see Figure 4 *top*) for all the

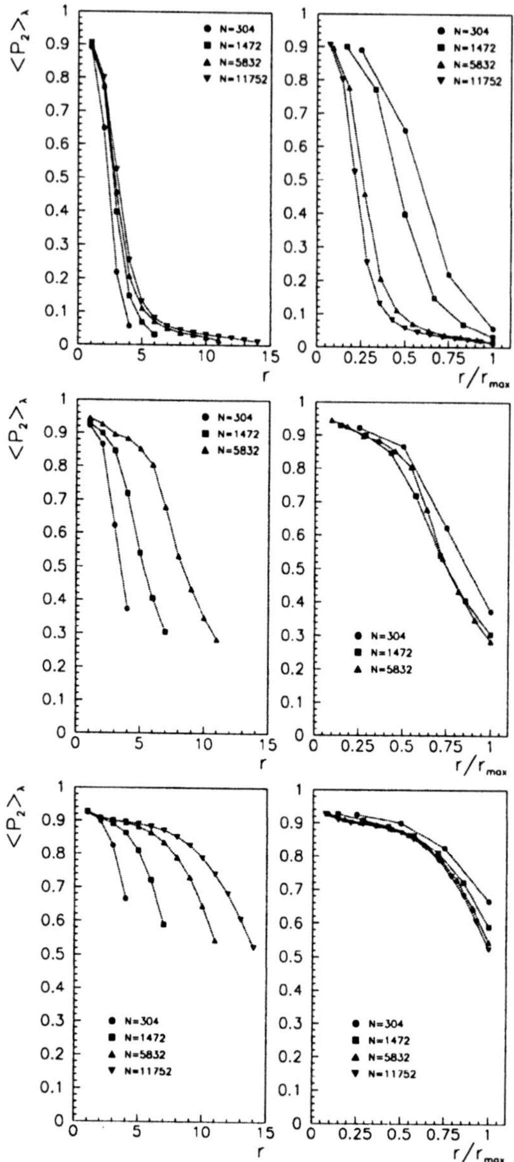

Figure 4. The nematic order parameter, $\langle P_2 \rangle_\lambda$, versus distance r starting from the center of the droplet. The results are obtained from Monte Carlo simulations of various sizes with radial (top), toroidal (middle) and bipolar (bottom) boundary conditions. The right plates show the curves plotted against a normalized distance r/r_{max}.

system sizes studied. This hints that the core size is a true material property [19] rather than being dependent on the droplet size. The picture changes completely when going to a planar surface achoring: the nematic ordering at the surface becomes larger for the toroidal boundary conditions and then even larger for the bipolar case with respect to the radial one (Fig.4 *middle* and *bottom*). Moreover the aligned region at the center becomes larger and, above all, its size increases linearly with the droplet radius yielding almost superimposed curves for the different system sizes, as shown on the right in Fig. 4 *middle* and *bottom*). We can then say that the behavior of the planar (toroidal and bipolar) boundary condition systems is the same in scaled units, differently from that of RBC.

The similarity in the behavior of properties calculated for different sample sizes strengthens the argument that each of our spins could really be considered to represent a microdomain of some tens of particles, and that our results also are applicable to droplets in the micron size, that have been investigated experimentally [9] by optical techniques.

A particularly interesting case comes from the application of an external field, a situation that corresponds to many real experiments. At the microscopic level this is modeled by adding an extra term to the hamiltonian of the system (1). Assuming second rank interactions ($L = 2$) the total configuration energy is written as:

$$U = -\sum_i^N \{\sum_j^N \epsilon_{ij} P_2(\mathbf{u}_i \cdot \mathbf{u}_j) + \epsilon \xi P_2(\mathbf{u}_i \cdot \mathbf{B})\} \qquad , j > i, \qquad (4)$$

where $\mathbf{B}$ is a unit vector along the field direction and the parameter ξ depends on the anisotropy of the electric or magnetic susceptivity and on the field intensity. In this case too we have defined an appropriate second rank order parameter, which now expresses the molecular alignment with respect to the field, $\langle P_2 \rangle_B$:

$$\langle P_2 \rangle_B = \frac{1}{N} \sum_{i=1}^N P_2(\mathbf{u}_i \cdot \mathbf{B}). \qquad (5)$$

Examples of these thermodynamic observables have been presented for various physical situations for the three boundary conditions listed above in our works on PDLC systems [15, 16].

2.1. MOLECULAR ORGANIZATION AND DEUTERIUM NMR SPECTRA

Monte Carlo simulations allow us to generate, apart from averages of thermodynamic observables, full sets of coordinates and angles representing instantaneous configurations of the lattice that can be used for visualization (cf. Fig. 1) or to calculate other quantities of interest such as, for example, polydomain deuterium NMR lineshapes for the model system of fictitious molecules.

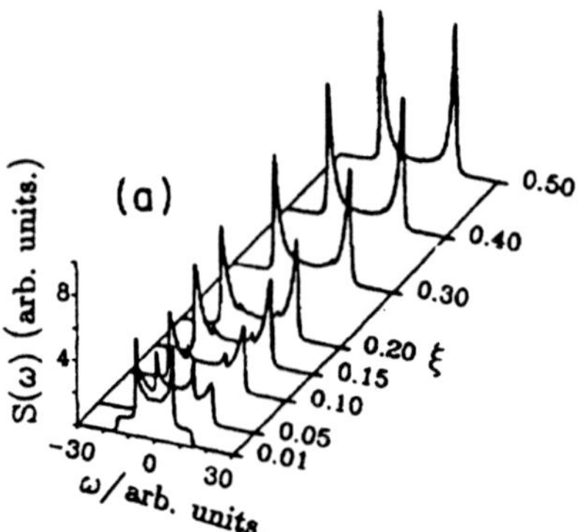

Figure 5. An example of NMR lineshapes as obtained from Monte Carlo simulations of a model droplet with radial boundary conditions at a reduced temperature $T^* = 0.4$. The applied external field is directed along the z axis and its intensity is modulated by the parameter ξ [15].

Deuterium NMR of deuterated liquid crystals has been frequently used in studying PDLC droplets [9], particularly when the droplets are so small that standard optical methods are not viable. The use of ^2HNMR allows focusing on the molecules inside the droplet (the only deuterated ones) thus giving in principle a direct handle on their properties. Each deuteron with quadrupole coupling constant ν_Q and angle θ between effective quadrupole axis and molecular axis provides a couple of lines at frequency

$$\omega_Q(\cos\beta_i) = \pm\frac{3}{4}\nu_Q P_2(\cos\beta_i)P_2(\cos\theta), \qquad (6)$$

where β_i is the angle between molecule and field axis, and uniaxial symmetry of the tensor and of the molecule are assumed. If the effect of the NMR spectrometer magnetic field on the configuration is negligible, as it is the case at least for sub-micron droplets [20] then field effects due to the applied external field can be examined.

In order to calculate simulated lineshapes from the Monte Carlo configurations we have assumed a system of fictitious deuterated molecules with axis of effective molecular uniaxial symmetry corresponding to that of the spins [15] as obtained for our configurations. Moreover, if molecular diffusion can be assumed to be negligible at the chosen experimental conditions, then the deuterium NMR spectrum becomes a powder like one and can give information on the director distribution or more generally on the molecular organization. These apparently rather stringent conditions have been shown to hold in various experimental studies [8, 11].

In practice the total spectrum for a configuration is calculated as the sample average

$$\mathcal{S}(\omega) = \left\langle \mathcal{S}[\omega, \omega_Q(\cos \beta_i), T_2^{-1}] \right\rangle_S \qquad (7)$$

$$= \frac{1}{N} \sum_{i=1}^{N} \mathcal{S}[\omega, \omega_Q(\cos \beta_i), T_2^{-1}] \qquad (8)$$

where N is the number of molecules in the droplet. Every particle provides a line shape contribution.

$$\mathcal{S}[\omega, \omega_Q(\cos \beta_i), T_2^{-1}] = \sum_{p=\pm 1} \frac{T_2^{-1}}{[\omega - p\omega_Q(\cos \beta_i)]^2 + [T_2^{-1}]^2}. \qquad (9)$$

In Ref. [15] we have used data appropriate to $4' - methoxy - 4 - cyanobiphenyl - d_3$ (10CB): $\nu_Q = 175 kHz$ and $\theta = 59.45$ degrees corresponding to the angle between CD_3 axis and molecular axis and consistent with the assumption of fast rotation of the CD_3 group, as from Ref. [8] and an intrinsic line width $T_2^{-1} = 200$Hz. This establishes a correspondence between kHz and the arbitrary units used for the frequency scale. The resulting spectrum is then further averaged over a number of configurations to improve the signal to noise ratio. An example of NMR line shapes calculated from an average over droplet configurations of a RBC droplet with N=5832 at different field strengths, ξ, is shown in Fig. 5 [15]. The deuterium splitting is a maximum when the molecules have their principal axis parallel to the applied field. At the lowest field, the droplet configurations is characterized by a near perfect hedgehog configuration and this gives rise to a lineshape characteristics of an essentially three dimensionally isotropic distribution of molecules with respect to the direction of the applied field. At stronger fields the population of molecules parallel to the field increases and the lineshape progressively reduces to a doublet corresponding to the parallel splitting.

In the case of RBC droplets the simulation results are compatible with a first order transition in the microscopic organization inside the droplet as the strength of the applied field increases as predicted by Dubois-Violette and Parodi [21] for a similar system.

2.2. POLARIZED LIGHT TEXTURES

Another experimental technique used to investigate micrometer size droplets is polarized light microscopy [9]. Also this kind of experimental observables can be calculated starting from the Monte Carlo configurations of the lattice spin model [16] exploiting a standard matrix approach which has been

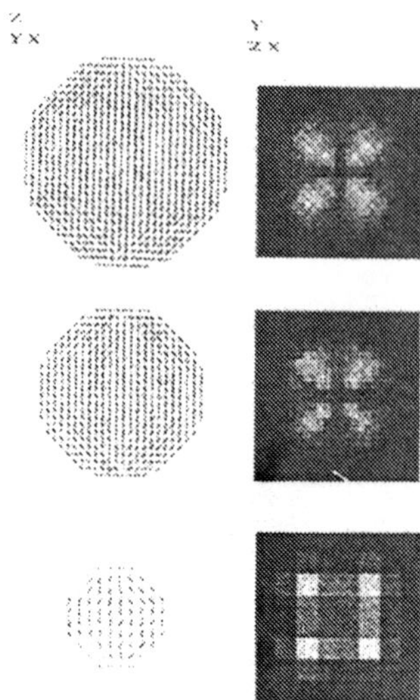

Figure 6. An example of snapshots and polarised optical images as obtained from Monte Carlo simulations of model droplets with bipolar boundary conditions. Three different sizes with 11752 (top), 5832 (middle) and 304 (bottom) spins at a reduced temperature $T^* = 0.4$ are shown.

employed in calculations based on continuum theory [9, 22, 23]. The basic idea in the matrix approach is that ray optics can be used and that each site in the droplet is described by a Müller matrix [24]. Then the light ray passing through a row of particles across the droplet is retarded by the matrix resulting from the product of the Müller matrices corresponding to each site in the light path. Each matrix involves the angles ϕ_j and θ_j, describing the orientation of a domain j, taken from the simulation data, and the phase difference which depends on the thickness of the layer, h, the wave length, λ, and the refractive indices, n_0 and n_e. In Ref. [16] we have used $h = 5.3 \mu m / (2 r_{max})$ (r_{max} is the radius of the droplet in lattice units), $\lambda = 545 nm$, $n_0 = 1.5$ and $n_e = 1.7$, similar to those of the nematic liquid crystal 5CB [9]. Since we assume the local domain to be basically unchanged throughout the simulations, thus only describing the disordering of each domain with respect to the others, we have arbitrarily taken this intrinsic refractivity to be constant with temperature.

To observe the light retarded by the droplet we assume to have crossed polarizers placed at each side of the sample cell, $\mathbf{P}_{in}$ and $\mathbf{P}_{out}$, and the resulting Stokes vector of the polarized and retarded light beam is thus given by [22, 23]:

$$s = P_{out} \prod_j M_j P_{in} s_{in}, \tag{10}$$

where s_{in} corresponds to the Stokes vector of unpolarized light. The intensity is proportional to the first element in the output Stokes vector s. To improve the quality of the optical image we further average over a number (typically around 20) equilibrated configurations. A texture obtained from configurations defined by a lattice of $22 \times 22 \times 22$, corresponding to a droplet of 5832 particles, provides a projection of 22×22 pixels perpendicular to the direction of the retarded and polarized lightbeams. The intensity of each pixel is grey coded for each picture with a normalized scale going from black, lowest intensity of light, to white, highest intensity, with 32 different grey levels. As an example the simulated optical patterns as obtained from Monte Carlo simulations of BBC droplets of different sizes are shown in Fig. 6. It is clear from these images that the basic features of the optical textures, as obtained experimentally by Doane group [9], are reproduced even with the smallest droplet size with only 304 spins.

3. Liquid crystal displays

Other quite different types of confined systems are the one dimensional ones, where a thin nematic film is confined between two surfaces as in the cells employed in Liquid Crystal Displays (LCD). Although these devices have been popular for two decades a large part of the know-how on them seems to be empirical or based on macroscopical continuum models. However, the range and scale of computer modelling have now grown to the point where it is possible to try and attempt a complete simulation of a model display starting from microscopic interactions. The techniques described in the previous section allow to simulate LCD images while at the same time MC can provide a unique tool for understanding and predicting ordering and microscopic organization inside the display cell. We briefly describe here the lattice simulations of two types of LCD: the Twisted Nematic (TN) [25] and a more recently proposed one based on the In-Plane Switching effect [26].

3.1. TWISTED NEMATIC DISPLAY

The spin model we employ tries to catch the essential features of the well known TN cell. The fixed "ghosts" on the top of the cell are oriented perpendicularly to those on the bottom while both are parallel to the cell surfaces (cf. Fig. 7 *left*). The alignment induced by these surfaces tends to propagate inside the liquid crystal cell producing a twisted nematic con-

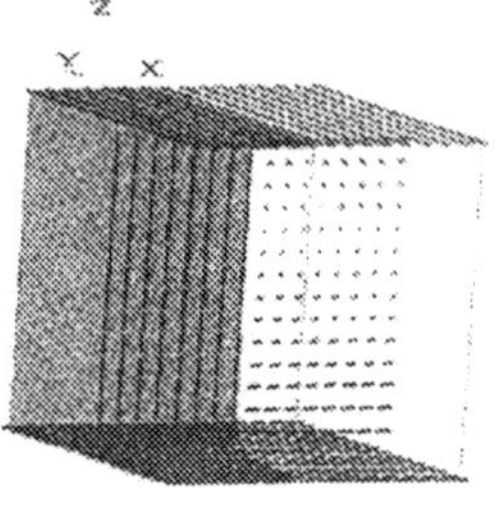

Figure 7. The twisted nematic lattice model. (*Left*): The simulation cell with the ordered surfaces and the regions in presence (grey) and absence of an applied field. (*Right:*) An example of a simulated optical image.

figuration. Periodic boundary conditions are employed around the other four faces of the cell. Moreover the lattice is divided in a regular array of sublattices where a field can be applied or not. At a microscopic level this is realized adding a local second rank term to the LL hamiltonian as in Equation (4). The field **B** is directed along the z-axis of the display, i.e. perpendicular to the oriented surfaces, and a factor $F = 1$ or 0 in front of the second term in Equation (4) acts as a switch to turn on or off the local external field. We assume that an electric field is applied and we take $\xi > 0$, corresponding to a material with positive dielectric anisotropy. When a sufficiently strong field is applied (*on* region), the molecules on which it acts align on average along the field direction, while the helical structure is conserved in the rest of the cell (*off* region), as schematically shown in Figure 7 (*left plate*). An helical order parameter appropriate to determin-

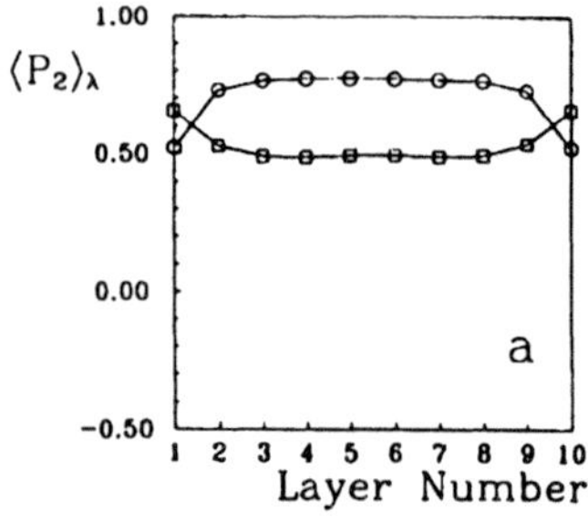

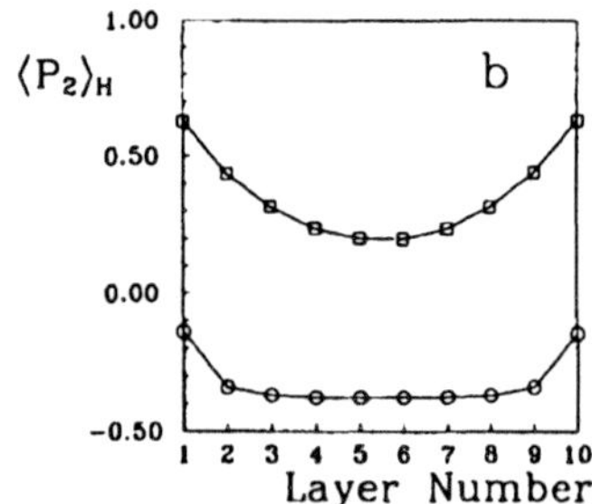

Figure 8. The standard nematic order parameter $\langle P_2 \rangle_\lambda$ (left) and helical order parameter $\langle P_2 \rangle_H$ (right) in regions with (squares) and without (circles) the applied field are illustrated for each layer of the display. The simulation was made at a scaled temperature, $T^* = 1.0$, and with the applied field strength, $\xi = 1.0$.

ing the order in the different layers and in the two regions: with external

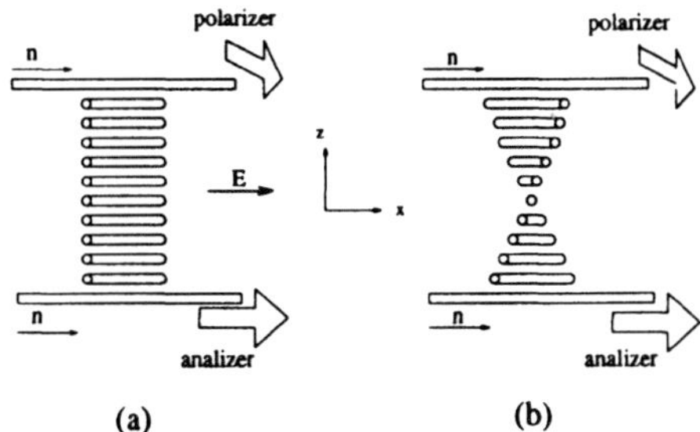

Figure 9.   A sketch of the operation mode of the In-Plane Switching LC display with the light propagation direction along the z axis. The alignment direction of the surfaces is indicated by **n**. (a) Field off: the polarized light is not transmitted through the cell; (b) Field on: a layer of nematic molecules rotates and the director twist changes the polarization of the incoming light which now passes through the analyzer.

field *off* or *on* can be introduced. This more specific microscopic quantity expressing the deviation from the ideal twist configuration, is the *helical order parameter*, $\langle P_2 \rangle_H$ defined [4] as follows:

$$\langle P_2 \rangle_H = \frac{1}{N_L} \sum_{i=1}^{N_L} P_2(\mathbf{u}_i \cdot \mathbf{t}_i), \tag{11}$$

where N_L is the number of particles contained in the Lth layer and $\mathbf{t}_i$ is the ideal twist direction at point i. In the limiting case that all the particles lie in the direction defined by the discretized helix between the bottom and top surfaces $\langle P_2 \rangle_H = 1$. In the field *on* regions $\langle P_2 \rangle_H$ becomes negative corresponding to the molecules being on average perpendicular to the ideal helix axis. In Fig. 8 the $\langle P_2 \rangle_\lambda$ and $\langle P_2 \rangle_H$, in the regions with and without the applied field, are shown. $\langle P_2 \rangle_\lambda$ is, except near the surface, higher in the *on* than in the *off* regions.

The link between simulation and experiment can be made more direct calculating the appearance of the display (Figure 7, *right*). For the perfectly twisted organization we obtain the largest intensity of transmitted light, and the *off* regions become light grey while the *on* regions, corresponding to written symbols, are darker, or ideally black. The appearance of display hints that simulations could start to be employed in modeling complex liquid crystal devices.

3.2. IN-PLANE SWITCHING EFFECT DISPLAY

In this device the top and bottom transparent cell surfaces are treated to induce homogeneous, i.e. surface parallel, alignment along the same direction (x in Fig. 9). A polarizer and an analyzer, with orthogonal polarization directions (for example along the x and y axis) are placed respectively above and below the cell. Thus a display element, a pixel say, does not let light through with no field applied and is black. Differently from the twisted nematic case, the liquid crystalline material filling the cell is chosen to have a negative dielectric anisotropy. The lateral switching effect is due to the application of an external electric field (see Fig. 9) across two electrodes placed at a certain distance from the surfaces and in a plane orthogonal to the light direction. The field is applied across the cell, and ideally it acts only on the molecules belonging to a thin intermediate layer of the liquid crystal sample. Notice that the direction of the applied field is parallel to the surface alignment direction but, due to the negative dielectric anisotropy, the molecules subjected to the field tend to rotate by 90 degrees so that a twisted alignment is induced between this intermediate layer and the two aligned surfaces.

 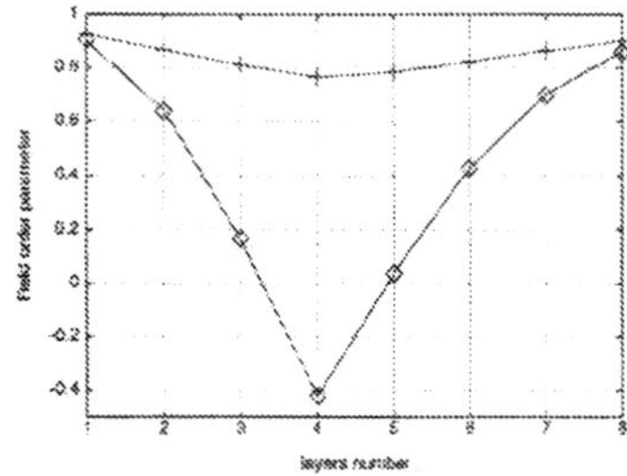

Figure 10. (*Left:*) An example of simulated optical image of a in-plane switching display as obtained by a lattice spin model simulation. (*Right:*) The order parameter with respect to the field direction (cf. Eq. 5) calculated for the same sample in the region where the field is active (diamonds) or off (plus).

Contrary to the usual twisted nematic display the light is thus transmitted only where the effect of the field is sufficiently strong and the background of the image is black. A simulated in-plane liquid crystal display image is shown in Fig. 10 together with the field order parameters calculated at each layer of the cell.

4. Hybrid aligned nematic film

The third type of confined system we consider here is a model of an hybrid cell, with random planar orientation on the bottom surface and

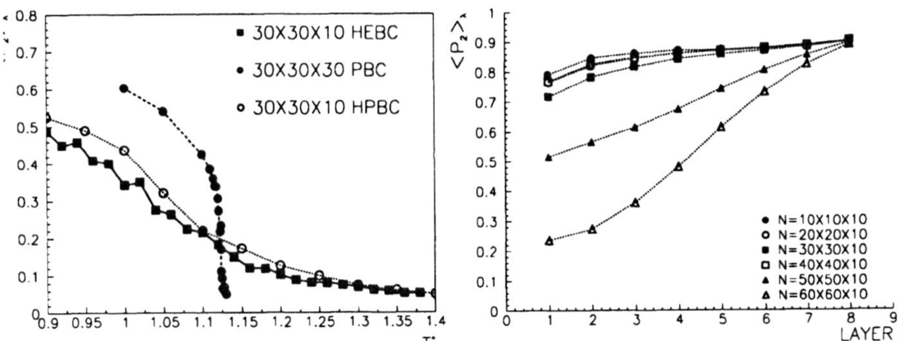

Figure 11. (*Left*): The order parameter dependence on temperature in an HAND system with empty (HEBC) or periodic (HPBC) lateral boundary conditions. As a comparison the bulk behavior is also reported. (*Right*): The order parameter calculated at each layer of the cell for different lateral size systems.

homeotropic, normal orientation at the top. These conditions have been experimentally realized [6] e.g. placing a liquid crystal film on top of an isotropic liquid substrate such as polyethylenglycol or glycerine and leaving a free air/liquid crystal surface. The interest on these systems is related to one of the most important and telling properties of liquid crystals: the structure of their topological defects, that is points or lines along which it is impossible to define an order parameter, under different conditions [7]. Lavrentovich [6] has demonstrated that this hybrid nematic liquid crystal films produce very interesting polarized-microscopy textures that are probably due to the presence of the two competing boundary conditions. Also in this case the MC method proves quite useful in studying these patterns and allows a precise control over the factors involved, such as thickness and anchoring strength.

The hybrid aligned nematic (HAN) cell [27] is mimicked assuming a $L \times L \times h$ lattice with suitable boundary conditions [28]. The spins of the bottom layer, $z = 0$, have random fixed orientations in the horizontal (x, y) plane, while those of the top layer, $z = h$, are fixed along the surface normal. Open, i.e. empty space, boundary conditions are assumed on the four planes surrounding the cell instead of the usually employed periodic BC with identical replicas surrounding the sample [28]. This artificial periodicity causes no fundamental artefacts in the modelling of uniform states.

136

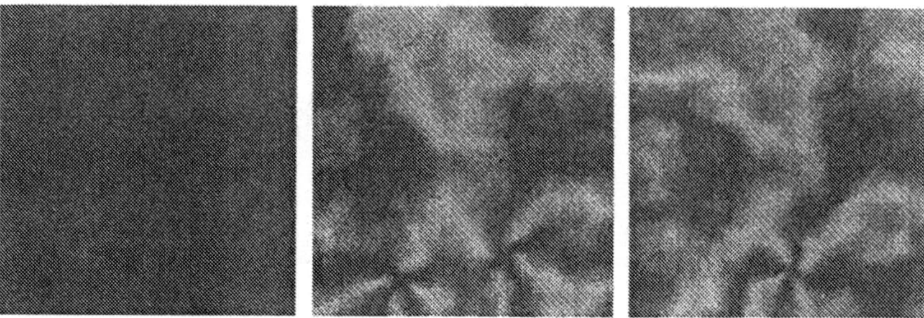

Figure 12. The appearance of a stable point defect in an hybrid nematic film as obtained by MC simulations of a $100 \times 100 \times 12$ lattice system [28]. The images correspond to 100, 6000 and 80000 MC evolution cycles from left to right respectively.

However, when the ground state contains topological defects, they might be incompatible with periodic BC and hence our reason for not using them. The molecular organization resulting from this conflicting surface boundaries is then inhomogeneous across the sample going from the disordered configuration of the first layer to the aligned one of the last layer. The system goes from an ordered to an isotropic phase at a temperature which is approximately from 5 to 10% lower than that exibited by the LL model [29], as we see in Fig 11 (*left*). Moreover there are no evident changes in the phase transition behavior increasing the horizontal dimensions of the cell. The defects do not appear until the lateral size L becomes much larger than h (e.g. $L \geq 50$ for $h = 10$). For films with small and large L/h, the director is strongly deformed in the vertical plane, obviously following the antagonistic boundary conditions at $z = 0$ and $z = h$. The horizontal director field is uniform, although not perfectly, and only small variations of the director exist. The situation dramatically changes for large L/h where there are strong and stable horizontal deformations associated with topological defects. The texture develops from black to a set of brushes, with each defect marked by four brushes emerging from its core [30]. The defects are of strength $m = \pm 1$, i.e., the director field undergoes a $\pm 2\pi$ rotation as one goes once around the defect core. The absolute value $|m| = 1$ is the lowest possible topological charge of a defect in a HAN film. The texture evolves as the system anneals, but the defects do not disappear even in the longest runs performed (120000 cycles, where a cycle is a full lattice update) even though they occasionally migrate outside the sample. The core of the defect is located near the lower surface; the distortions vanish as one moves towards the upper plate and the molecules reorient along the z axis. Note that the results described above are specifically related to the hybrid alignment of the film. In this case the nematic order parameter $< P_2 >_\lambda$ (see, e.g. [17]) across the film changes substantially for the larger lattices but not for the small ones (see Fig. 11, *right plate*).

A striking result of the MC simulations is that the model based exclusively on pure nearest-neighbors molecular interactions mimics the long-range deformations with topologically stable defects in agreement with continuum theory predictions.

5. Conclusions

We have described lattice spin models for the simulation of various confined nematic systems. The biggest advantage of Monte Carlo simulations is the possibility of investigating the system at a microscopic level. Moreover, apart from the usual thermodynamic properties it is possible to define and calculate specific configurational order parameters suitable for the different types of systems under investigation. Molecular organizations can be visualized as snapshots but the optics of a simulated device can also be calculated as shown for applications to models of nematic liquid crystal displays (LCD), polymer dispersed liquid crystals (PDLC) , and hybrid aligned films.

Acknowledgments

We thank University of Bologna, MURST, CNR and INFN (grant I.S. BO12) for support and our collaborators and students for their invaluable help.

References

1. Crawford, G.P. and Žumer, S. (eds.) (1995) *Liquid Crystals in Complex Geometries Formed by Polymer and Porous Networks*, Taylor and Francis, London.
2. Chiccoli, C., Pasini, P., Semeria, F., Berggren, E. and Zannoni, C. (1995) *Mol. Cryst. Liq. Cryst.* **266**, 241 *and references therein*.
3. Crawford, G.P. and Doane, J.W. (1992) *Condens. Matter News*, 1, 5.
4. Berggren, E., Zannoni, C., Chiccoli, C., Pasini, P. and Semeria, F. (1995) *Int. J. Mod. Phys. C*, **6**, 135.
5. Chiccoli, C., Guzzetti, S., Pasini, P. and Zannoni, C. (1998) *Int. J. Mod. Phys. C*, **9**, 409.
6. Lavrentovich, O.D. (1992) *Liq. Cryst. Today*, **2**, 3 *and references therein*.
7. Mermin, N.D (1976) *Rev. Mod. Phys.*, **51**, 591.
8. Golemme, A., Žumer, S., Doane, J.W. and Neubert, M.E. (1988) *Phys. Rev. A*, **37**, 559.

9. Ondris-Crawford, R., Boyko, E.P., Erdmann, B.G., Žumer S. and Doane, J.W. (1991) *J. Appl. Phys.*, **69**, 6380.
10. Drzaic, P. (1988) *Mol. Cryst. Liq. Cryst.*, **154**, 289.
11. Aloe, R., Chidichimo, G. and Golemme, A. (1991) *Mol. Cryst. Liq. Cryst.*, **203**, 1155.
12. Chiccoli, C., Pasini, P., Semeria F. and Zannoni, C. (1990) *Phys. Lett.*, **150A**, 311.
13. Chiccoli, C., Pasini, P., Semeria F. and Zannoni, C. (1992) *Mol. Cryst. Liq. Cryst.*, **212**, 197.
14. Chiccoli, C., Pasini, P., Semeria F. and Zannoni, C. (1992) *Mol. Cryst. Liq. Cryst.*, **221**, 19.
15. (a) Berggren, E., Zannoni, C., Chiccoli, C., Pasini, P. and Semeria, F. (1992) *Chem. Phys. Lett.*, **197**, 224. (b) (1994) *Phys. Rev. E*, **49**, 614.
16. Berggren, E., Zannoni, C., Chiccoli, C., Pasini, P. and Semeria, F. (1994) *Phys. Rev. E*, **50**, 2929.
17. Zannoni, C. (1999) Chapter 2 of this volume.
18. Schopol N. and T.J. Sluckin (1988) *J. Phys. France* , **49**, 1097.
19. Chiccoli, C., Pasini, P., Semeria, F., Sluckin, T.J. and Zannoni, C. (1995) *J. de Physique II*, **5**, 427.
20. Crawford, G.P., Ondris-Crawford, R., Žumer, S. and Doane, J.W. (1993) *Phys. Rev. Lett*, **70**, 1838 .
21. Dubois-Violette, E. and Parodi, O. (1969) *J. de Physique* , Colloq. **30**, C4-57.
22. Xu, F. , Kitzerow, H.-S. and Crooker, P.P. (1992) *Phys. Rev. A*, **46**, 6535.
23. Kilian, A. (1993) *Liq. Cryst.*, **14**, 1189.
24. Schellman, J.A. (1988) in *Polarized Spectroscopy of Ordered Systems*, Samori', B. and Thulstrup, E.W.(eds.), Kluwer, Dordrecht, p. 231.
25. Schadt, M. (1989) *Liq. Cryst.*, **5**, 57.
26. Baur, G. Kiefer, R., Klausmann, H. and Windscheid, F. (1995) *Liquid Crystals Today*, **5**, 13 *and References therein.*
27. O.D. Lavrentovich and V.M. Pergamenshchik, (1995) *Int. J. Mod. Phys. B*, **9**, 2839.
28. Chiccoli, C. O.D. Lavrentovich, Pasini, P. and Zannoni C. (1997) *Phys. Rev. Lett.*, **79**, 4401.
29. Chiccoli, C., Pasini, P. and Zannoni, C. (1999) *Mol. Cryst. Liq. Cryst.* (in press).
30. Kleman, M (1983) *Points, Lines and Walls* Wiley, New York.
31. Fabbri, U. and Zannoni, C. (1986) *Mol. Phys.*, **58**, 763.

COMPUTER SIMULATION OF LYOTROPIC LIQUID CRYSTA AS MODELS OF BIOLOGICAL MEMBRANES

OLE G. MOURITSEN

Department of Chemistry
Technical University of Denmark
Building 206, DK-2800 Lyngby
Denmark

1. Introduction: membranes, lipid bilayers, and smectics

During evolution, Nature has evolved biological membran structure [1] as an optimal form of micro–encapsulation technology which on the one hand imparts the necessary durability to the particularsoft condensed matter that membranes are made of, and at the other hand sustains the lively dynamics that are needed to support and control the mechanisms of the many essential cellular functions associated with membranes, such as transport, growth, and enzymatic activity [2]. Biological membranes are generally very stratified and composite structures whose central element is the fluid lipid bilayer [3] as illustrated in Fig. 1. An important function of the lipid bilayer is to act as a passive permeability barrier to ions and other molecular substances and leave the trans–membrane transport to active carriers and channels.

The lipid–bilayer of the cell membrane is a *lyotropic smectic liquid crystal* who owns its existence to the amphiphilic nature of lipid molecules and the omnipresence of the biological solvent, water. Hence, all life as we know it depends on the liquid–crystalline state of condensed matter: the lipid bilayer is *Nature's preferred liquid crystal* [4].

The lamellar smectic state of lipids is only one of a large family of supramolecular aggregates that may form spontaneously when lipids are mixed with water. Other aggregation states, as illustrated in Fig. 2, include monolayers, micelles, hexagonal phases, cubic phases, as well as structures of more disordered and complex form, such as sponge phases and microemulsions [5].

139

P. Pasini and C. Zannoni (eds.), Advances in the Computer Simulations of Liquid Crystals, 139–187.
© 2000 *Kluwer Academic Publishers. Printed in the Netherlands.*

140

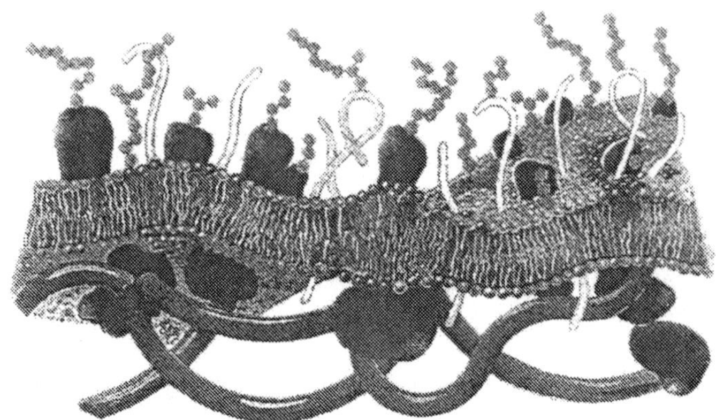

Figure 1. Nature's preferred liquid crystal: Schematic illustration of a biological eucaryotic cell membrane which highlights the membrane as a composite of the smectic, fluid lipid bilayer matrix sandwiched between the carbohydrate glycocalyx on the outside and the cytoskeleton on the inside. Intercalated in the lipid bilayer are shown schematically various integral and peripheral proteins and polypeptides. The lipid bilayer component is seen to be subject to a substantial degree of heterogeneity both with respect to molecular composition, bilayer thickness, and local curvature. (Illustration by O. Broo Sørensen, Technical University of Denmark.)

The self-assembly process of amphiphilic molecules involves a number of complex phenomena and implies a subtle competition between forces of different nature [6]. Since these forces are of physical origin with no covalent bonding involved, and since many of the forces are of a colloidal and entropic nature, the relative stability of the resulting structures and phases is intimately dependent on temperature. The self-assembly process is hierarchical in the sense that aggregates, e.g. lipid bilayers, can form at some basic level and these aggregates can then on a higher level arrange among themselves to form superstructures, such as multi–lamellar bilayer stacks, cf. Fig. 2f. Furthermore, within each basic aggregate, organization on different levels can take place, e.g. formation of various crystalline and liquid-crystalline phases within a lipid bilayer or various morphological structures involving the large–scale conformational complexity of the whole bilayer sheet. The different levels in the hierarchy are usually connected, e.g. the large–scale conformation of a lipid bilayer as well as the stabilizing forces between adjacent lipid bilayers are controlled by the in–plane bilayer structure and visa versa. Which aggregate form in Fig. 2 is the stable one depends on composition, thermodynamic conditions, and the structure of the lipid molecules in question. Formation of non–lamellar structures, such as the hexagonal phase in Fig. 2g, requires lipids of non-cylindrical shape which

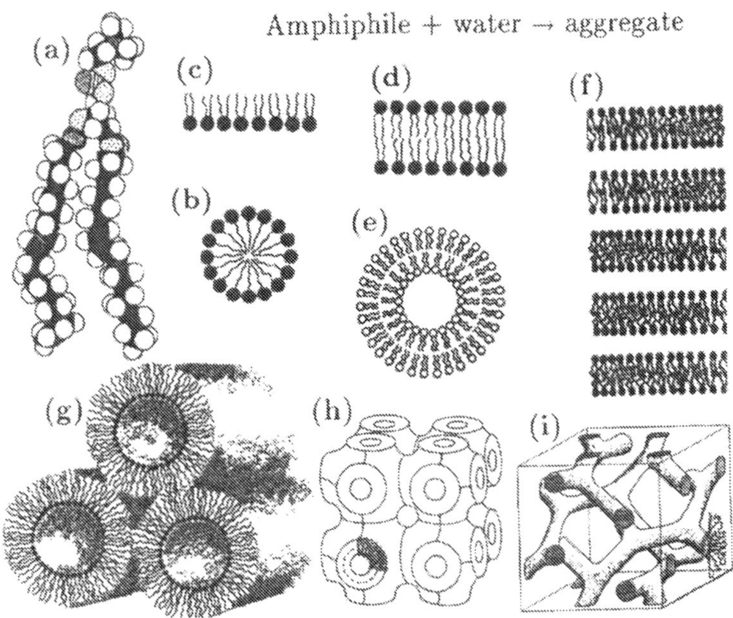

Figure 2.   Schematic illustration of a lipid molecule (a) and the self–organization of lipids into supra–molecular aggregates in association with water. (b): micelle, (c): lipid monolayer, (d): lipid bilayer, (e): vesicle/liposome (closed lipid bilayer), (f): multi-lamellar bilayer phase, (g): inverse hexagonal phase, (h): cubic phase, and (i): sponge phase.

introduces a curvature stress field in the aggregate. Despite their exotic appearance, some of the non–lamellar structures in Fig. 2 may be relevant for the biological functioning of membranes in connection with various membrane processes, such as fusion, cytosis, cell–cell interactions, as well as cell motility. A lamellar bilayer tends to close onto itself and form a closed object, a so–called vesicle or liposome (Fig. 2e), which is the prototype of a cell–membrane model. Often, such closed structures are formed within themselves leading to multi–lamellar structures with a considerable degree of three-dimensional ordering among the lamellae, cf. Fig. 2f.

It is instructive to keep in mind the spatial dimensions of liposomes. Whereas the lipid bilayer itself is only about 50Å thick, the diameters of vesicles and liposomes are orders of magnitude larger, typically in the range from 500Å to 50000Å. In comparison, typical procaryotic and eucaryotic cells have diameters that are one and two orders of magnitude larger than the largest liposomes. Hence, lipid bilayers are very thin organic films associated with unique material characteristics. It is the peculiar physical properties of soft lipid-bilayer membranes, which are so unique to ensure effective encapsulation of cells and cell organelles, that lead to the opportunities but also the many limitations and problems encountered in the use of liposomes for drug–delivery purposes [7]. It is the hypothesis, that the lipid–bilayer softness, the dynamic structure of the membrane, and the corresponding lipid organization are important regulators of membrane function and the ability of the membrane to support biological activity. A consequence of this hypothesis is that the generic effects of peptides, proteins, and drugs on membrane structure and function on the one side, and the influence of bilayer structure on these compounds on the other side may be understood in part by the ability of these compounds to alter the lipid–bilayer organization [8]. The picture of lipid bilayers to be advocated in the present set of lecture notes takes its starting point in the bilayer as a physical entity [3], a complex and structured fluid, with a considerable dynamics, a special trans–bilayer molecular profile, and a highly non–trivial lateral organization of the membrane components on many different length- and time scales. This picture anticipates the physical fact that the molecular constituents of the bilayer are large amphiphilic molecules with many internal degrees of freedom, and that the membrane assembly is a many–particle system which, by basic laws of Nature, displays correlated and cooperative dynamical modes involving many molecules. The amphiphilic nature of the lipid molecules, which is the main reason for their spontaneous self–assembly into lipid–bilayer aggregates in an aqueous milieu (cf. Fig. 2), implies that the bilayer and the adjacent parts of the aqueous embedding medium have a particular molecular interfacial structure and cannot be considered as separable entities. The special interfacial structure is of paramount importance when it comes to the interaction of the membrane with foreign molecular components, such as proteins and peptides. On the mesoscopic scale the lipid bilayer can be considered as a soft and flexible interface that is characterized by continuum-mechanic physical properties like interfacial tension and elastic bending modulii. The bilayer exhibits a substantial degree of in–plane fluctuations and nano–scale heterogeneity which in turn control the mechanical modulii of the membrane and hence the repulsive entropic interactions with approaching macromolecules and other neighboring membranes and surfaces. And this is where *computer–simulation techniques* come in as an indispensable set of tools. By these techniques it is possible to study the complexity of membranes and membrane models, both with respect to molecular conformation, self–organization and molecular organization, thermodynamic and thermomechanic properties, as well as aspects of relationships between membrane structure and function. In the following we shall provide a perspective of this type of approach to membrane science with an emphasis on phase transitions and in–plane molecular organization in lamellar lipid bilayers. Throughout the lecture notes references will mostly be given to recent review papers and books from which the present text to a large extent is composed. The interested reader has to consult these sources for detailed references to original research papers.

2. Computer simulations: molecular dynamics or Monte Carlo?

The use of molecular dynamics methods to elucidate aspects of membrane structure and dynamics has virtually exploded over the last few years [9–13]. Modern computers and new fast simulation algorithms have made it possible to carry out dynamic simulations on model membranes characterized by very realistic inter–atomic potentials, including interactions with molecular water and ions. Simulations of this type have recently lead to detailed results regarding, e.g., trans-bilayer structure in the different phases, bilayer surface tension, the role of water and hydration forces, interaction of peptides with bilayers, small molecules interacting with bilayers, interaction of peripheral and integral proteins with lipid bilayers, as well as aspects of trans–bilayer transport of water and ions. Due to the very detailed level of description entailed by molecular dynamics simulations they are subject to severe limitations in both the number of molecules that can be dealt with as well as the time range over which the molecular system can be studied. Typically, a state–of–the–art simulation of this type can treat of the order of a hundred lipid molecules and some thousand water molecules over a time span in the sub-nanosecond regime. In the case of proteins and peptides, usually only a single macromolecule is considered. These limitations currently exclude molecular dynamics studies of lipid membrane structure beyond the spatial scale of a single protein molecule. Lateral lipid–bilayer structure, not to speak about phase equilibria, morphology, and out–of–the–plane structure, which involve a very large number of molecules and their diffusional characteristics, would require simulations of thousands of molecules on time–scales up to micro-seconds and longer, considering that it typically takes a lipid molecule several nano–seconds to travel its own diameter.

Monte Carlo computer–simulation methods applied to lipid membranes [14–16] circumvent some of these problems by introducing stochastic elements in the simulation and by usually invoking simplified and semi–phenomenological and coarse–grained models for the molecular interactions. These are very severe restrictions on their part and this is probably why Monte Carlo simulation has appealed much less to those biologically oriented scientists who favor molecular detail. Consequently, the membrane literature on Monte Carlo simulations is much more scarce than is the case

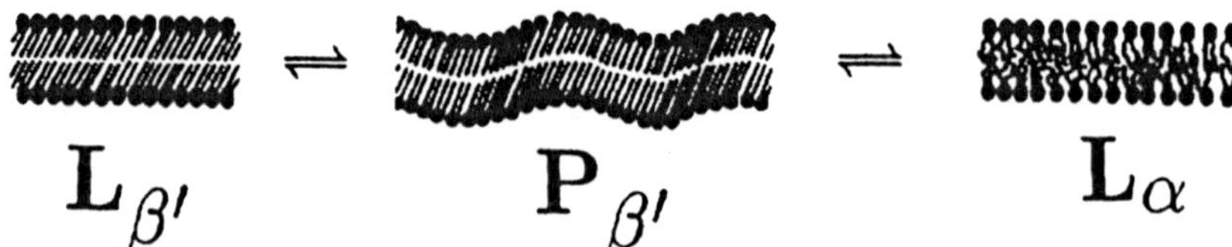

Figure 3. Schematic illustration of three lamellar smectic phases of lipid bilayers in water and the associated phase transitions. The second transition to the right is the so–called main phase transition. This transition implies acyl–chain melting as well as melting of the crystalline lattice of the solid $P_{\beta'}$ phase.

for molecular dynamic studies. It is mostly physicists with an interest in biophysics who have applied these methods to study general aspects of membrane structure and organization. The viewpoint of physicists in this context is very flavored by the success in physics of powerful concepts like universality, scaling, and generic behavior, i.e. that the many–bodiness of a system imparts to the system certain highly non–trivial properties that are robust to details. These properties can then conveniently be studied by simple models and sturdy Monte Carlo simulation techniques.

Membrane organization and formation of lipid domains in the nano–meter range are difficult to investigate experimentally, and to date only indirect evidence of their existence is available [9, 17, 18]. In this situation, Monte Carlo computer–simulation calculations have proved particularly useful to investigate, within simple model–membrane systems, under which conditions small–scale membrane structure arises, how it can be characterized, and how it can be modulated.

3. Phase transitions in lipid bilayers

Under physiological conditions, the lipid–bilayer component of cell membranes is in a so–called fluid phase, the L_α–phase, which is a smectic A phase. Depending on the lipid species in question, the bilayer symmetry may itself correspond to a number of different thermodynamic smectic phases separated by different thermotropic phase transitions, as illustrated in Fig. 3. We shall here be concerned with one of these transitions, the main phase transition which takes the bilayer from a low–temperature so–called *gel* phase to a high–temperature *fluid* phase.

We shall neglect the ripple phase ($P_{\beta'}$) and basically consider the gel phase as a type of smectic H structure. When characterizing the gel and fluid phases, one is confronted with the requirement of at least two variables to describe the nature of the phase, one that denotes the order in internal conformational degrees of freedom of the lipid acyl chains and one which refers to the translational order. Hence one can perceive the main transition to proceed in terms of two different degrees of freedom. It is usually assumed that this thermotropic transition involves two distinct, but coupled phase transitions: two–dimensional lattice melting and chain melting associated with the translational- and chain conformational degrees of freedom, respectively [19]. In other words the transition takes the bilayer from a low–temperature (gel) phase, which is a solid and has (quasi) long–range translational order and a high degree of conformational order within the lipid chains (the so–called solid–ordered, SO, phase), to a high–temperature (fluid or liquid) phase, which displays disorder in both the translational and chain conformational degrees of freedom (the so–called liquid-disordered, LD, phase). A number of theoretical and experimental problems are associated with the main transition with regard to the detailed properties of the transition [19]. Moreover, there is still some controversy as to the very question of the nature of the transition, i.e. whether it is continuous or of first order, or whether there is a phase transition at all. Finally, the study of the main transition is of interest since it appears that this transition is strongly influenced by thermal density fluctuations which may control biological function. The density fluctuations, which are consequences of the cooperative phenomena in the membrane, lead to dynamically heterogeneous membrane states. It is a question of key interest as to how various membrane components, such as cholesterol, proteins, enzymes, and receptors as well as other foreign molecular compounds interacting with membranes, such as drugs, couple to the heterogeneous membrane states and change the lateral organization and hence the membrane function.

4. Models of lipid bilayer phase transitions

The formulation of a useful theoretical model for lipid bilayers is a balance between physical realism and computational feasibility [16]. This balance becomes quite delicate when phase transitions are at issue because it requires that large assemblies of molecules are taken into account. With our present knowledge about interatomic potentials in hydrocarbon systems it is possible to write down a rather realistic model of a lipid bilayer membrane. However, it would be virtually impossible to calculate any properties near the phase transition of such a model due to its complexity. By contrast, it is possible to design highly simplified models which are analytically tractable though this is sometimes at the expense of physical realism. The virtue of the analytically solvable models is that they have provided us with valuable information about those features of the interaction potentials which are necessary in order to produce the characteristics of the

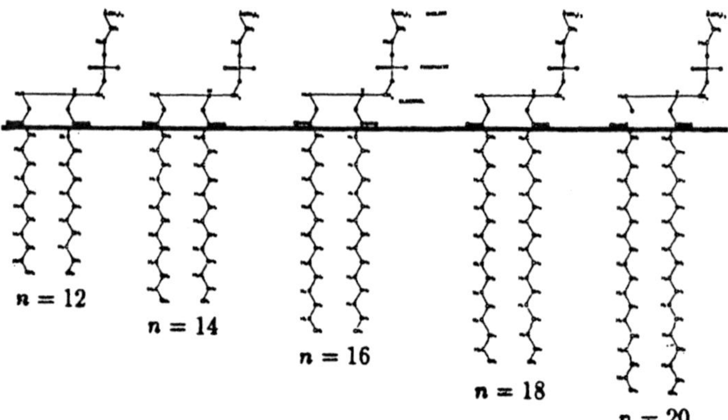

Figure 4. Schematic illustration of the chemical structure of a series of saturated di-acyl–glycero phosphatidylcholine lipids denoted as DC_nPC, with n carbon atoms in each acyl chain. The molecules are arranged in a manner that reflects the approximate position of the glycerol backbone at the hydrophobic–hydrophilic interface of the membrane.

transition and those which are only marginally relevant. Such information is indispensable when the next level of more realistic but still computationally tractable models is approached.

Most of the computer–simulation calculations to be described in these lecture notes have been performed on models for saturated di–acyl glycero phosphatidylcholine lipids denoted as DC_nPC [1] with varying number of carbon atoms, n, in the acyl chains. A collection of the lipids to be referred to in the following is shown in Fig. 4.

4.1. THE FUNDAMENTAL VARIABLES

A theoretical treatment of any complex molecular system requires an identification of those mechanical variables which are relevant for the phenomenon under consideration. Obviously it is usually impossible as well as undesirable to treat any given phenomenon using the full set of variables. Some of the variables have therefore to be 'frozen out' in the calculation. In models of the main transition, a planar lamellar membrane geometry is assumed and the lipid molecules are confined to a plane by anchoring the level of the glycerol backbones. Secondly, all but the acyl–chain degrees of freedom are usually 'frozen out', i.e. the flexibility of the polar head group

---

[1]Abbreviation used: DC_nPC: di-acyl phosphatidylcholine with n carbon atoms in each acyl chain.

is neglected. Thirdly, the two acyl chains of a diacyl lipid molecule are often assumed to be independent. This leaves us with a set of fundamental variables, $\{m_i, \vec{r}_i\}$, where m denotes the conformational state of the acyl chain and $\vec{r}$ is the translational variables of the chain.

Even this dramatic reduction in the degrees of freedom of the membrane is not sufficient to permit the construction of models which are computationally tractable in the transition region. This may sound surprising, but it can be appreciated by considering the many unsolved problems regarding the transition in as simple a system as a two–dimensional particle system with purely translational degrees of freedom. Hence, most approaches introduce dramatic assumptions about at least one of the two variables, m or $\vec{r}$. Different strategies which have been followed put emphasis on different variables:

Strategy (i): realistic conformational variables and frozen translational variables (Sec. 4.2)

Strategy (ii): realistic conformational variables and approximate translational variables

Strategy (iii): realistic translational variables and approximate conformational variables (Sec. 4.3).

The simplest possible way of freezing out the translational variables is to adopt a regular lattice approximation in which each acyl chain is positioned on a regular (e.g. triangular) two–dimensional lattice. The multi–state Pink model [14] to be described below in Sec. 4.2 builds on such a lattice approximation. Within the description of the Pink model it is furthermore assumed that the conformational properties of a single acyl chain can be described by a small number of selected conformational states corresponding to the mapping of the three–dimensional acyl–chain conformations onto a finite, discrete set of projected two–dimensional coarse-grained variables. It is possible to proceed from this level of description to strategy (ii) by introducing, on the lattice, a new set of variables [14], formally like in the multi–state Potts model, which in a very approximate manner account for the crystalline coherence of the solid–ordered phase. A much more satisfactory way of treating the translational variables is to introduce a dynamic random lattice, while maintaining an approximate description of the conformational variables (Strategy (iii)). We shall describe an example of this approach in Sec. 4.3.

Having identified the fundamental variables of the problem, the next step is to couple the variables by means of a 'model' in which the physical interactions are embodied. By a model we mean a Hamiltonian function defined in terms of the chosen relevant microscopic mechanical variables. The Hamiltonian includes the proper interaction potentials, possibly parameterized in terms of interaction constants which are often unknown. The microscopic model calls for a full statistical mechanical treatment of the related many-body problem whose solution usually requires a computer–simulation calculation.

4.2. STRATEGY (I): THE MULTI-STATE PINK MODEL

In the multi–state Pink lattice model [14] the conformational chain variables are coupled by anisotropic van der Waals interactions in the spirit of anisotropic liquid crystals. The interaction between the hydrophilic moieties are modeled by a Coulomb–like force or more simply by an effective intrinsic lateral pressure. The Pink model allows for a series of ten conformational states of the acyl chains of the lipid molecules. The ten–state model provides a reasonably accurate description of the phase behavior of pure lipid bilayers and the associated density fluctuations since it accounts for the most important conformational states of the lipid chains as well as their mutual interactions and statistics. In the Pink model the bilayer is considered to be composed of two monolayer sheets which are independent of each other. Each monolayer is represented by a triangular lattice. The model is therefore a pseudo–two–dimensional lattice model which neglects the translational modes of the lipid molecules and focuses on the conformational degrees of freedom of the acyl chains.

Since we will be extending the formalism to include several molecular species, we label the lipid variables corresponding to a particular lipid species by A. Each acyl chain can take on one of ten conformational states m, each of which is characterized by an internal energy E_m^A, a hydrocarbon chain length d_m^A (corresponding to half a bilayer, $2d_m^A = d_L$), and a degeneracy D_m^A, which accounts for the number of conformations that have the same area A_m^A and the same energy E_m^A, where $m = 1, 2, \ldots, 10$. The ten states can be derived from the all-*trans* state in terms of *trans–gauche* isomerism. The state $m = 1$ is the non-degenerate gel–like ground state, representing the all-*trans* conformation, while the state $m = 10$ is a highly degenerate excited state characteristic of the melted or fluid phase. The eight intermediate states are gel–like states containing kink and jogs excitations satisfying the requirement of low conformational energy and optimal packing. The conformational energies, E_m^A, are obtained from the energy required for a *gauche* rotation relative to the all-*trans* conformation. The values of D_m^A are determined by combinatorial considerations. The chain cross-sectional areas, A_m^A, are reciprocally related to the values of d_m^A since the volume of an acyl chain varies only slightly under temperature changes.

The saturated hydrocarbon chains are coupled by nearest-neighbor anisotropic forces which represent both van der Waals and steric interactions. These interactions are formulated in terms of products of shape-dependent nematic factors. The lattice approximation automatically accounts for the excluded volume effects and to some extent for that part of the interaction with the aqueous medium which assures bilayer integrity. Finally, an effective two-dimensional lateral pressure, Π, is included in the model. The Hamiltonian for the pure lipid bilayer can then be written in terms of site occupation variables, $\mathcal{L}_{im}^{A} : \mathcal{L}_{im}^{A} = 1$ if the chain on site i is in state m, otherwise $\mathcal{L}_{im}^{A} = 0$, in the following form

$$\mathcal{H}^{A} = \sum_{i} \sum_{m=1}^{10} \left(E_{m}^{A} + \Pi A_{m}^{A} \right) \mathcal{L}_{im}^{A} - \frac{J_{A}}{2} \sum_{<i,j>} \sum_{m,n=1}^{10} I_{m}^{A} I_{n}^{A} \mathcal{L}_{im}^{A} \mathcal{L}_{jn}^{A}, \quad (1)$$

where J_{A} is the strength of the van der Waals interaction between neighboring chains, and $I_{m}^{A} I_{n}^{A}$ is an interaction matrix which involves both distance and shape dependence. $\langle i, j \rangle$ denote nearest-neighbor indices on the triangular lattice. The interaction matrix can be expressed as

$$I_{m}^{A} I_{n}^{A} = V_{mn}^{A} S_{m}^{A} S_{n}^{A}, \quad (2)$$

where V_{mn}^{A} describes the distance dependence of the potential, and S_{m}^{A} is the nematic shape factor or average acyl–chain order parameter (segmental order parameter)

$$S_{m}^{A} = \frac{1}{2(N-1)} \sum_{i=2}^{N} (3 \cos^{2} \vartheta_{mi} - 1). \quad (3)$$

The summation extends over all segments of the chain and ϑ_{mi} is the angle between the bilayer normal and the normal to the plane spanned by the ith CH_2-group of the mth conformational state of the chain. For the completely ordered all-*trans* state (the ground state), $S_{1}^{A} = 1$. The distance-dependence of the interaction potential is assumed to be separable and taken to be of the Salem type for (infinitely) long cylindrical molecules

$$V_{mn}^{A} \sim r_{mn}^{-5} \sim (A_{m} A_{n})^{5/4}. \quad (4)$$

It becomes obvious at this point that the Pink model is not a lattice model in a strict sense, but rather a topological structure which assures that each chain interacts with six neighbors corresponding to a close–packed arrangement. The chains are simply assigned different amounts of space (area) depending on their conformational state.

The ten–state Pink model for a pure lipid bilayer can be extended to describe binary lipid mixtures by explicitly incorporating a mismatch term which accounts for the incompatibility of acyl chains of different hydrophobic lengths of the two species. The Hamiltonian for a binary mixture of the two lipid species A and B is written

$$\mathcal{H} = \mathcal{H}^{A} + \mathcal{H}^{B} + \mathcal{H}^{AB}, \quad (5)$$

150

where the two first terms describe the interaction between like species and the last term the interaction between different species. The composition of the mixture is given by $x_\mathrm{B} = \sum_m \langle \mathcal{L}_{im}^\mathrm{B} \rangle = 1 - x_\mathrm{A}$. The interaction between different lipid species is described by the Hamiltonian

$$
\mathcal{H}^{AB} = \frac{-J_{AB}}{2} \sum_{\langle i,j \rangle} \sum_{m,n=1}^{10} (I_m^A I_n^B \mathcal{L}_{im}^A \mathcal{L}_{jn}^B + I_m^B I_n^A \mathcal{L}_{im}^B \mathcal{L}_{jn}^A)
$$

$$
+ \frac{\Gamma_{AB}}{2} \sum_{\langle i,j \rangle} \sum_{m,n=1}^{10} (|d_{im}^A - d_{jn}^B| \mathcal{L}_{im}^A \mathcal{L}_{jn}^B + |d_{im}^B - d_{jn}^A| \mathcal{L}_{im}^B \mathcal{L}_{jn}^A). \quad (6)
$$

The first term in $\mathcal{H}^{AB}$ describes the direct van der Waals hydrophobic contact interaction between different acyl chains. The corresponding interaction constant is taken to be the geometric average $J_{AB} = \sqrt{J_A J_B}$. Γ_{AB} in the second term of Eq. (6) represents the mismatch interaction and has been shown to be 'universal' in the sense that its value does not depend on the lipid species in question.

The model presented above for binary lipid mixtures can readily be modified to account for a mixture involving cholesterol in the case where only the interaction between the sterol and the conformational degrees of freedom of the lipid-acyl chains are taken into account. The Hamiltonian for the mixture of lipid of species A and cholesterol (C) is written

$$
\mathcal{H} = \mathcal{H}^A + \mathcal{H}^C, \quad (7)
$$

where

$$
\mathcal{H}^C = \Pi A^C \sum_i \mathcal{L}_i^C - \frac{J_A}{2} \sum_{\langle i,j \rangle} \sum_{m,n=1}^{10} I_m^A I_C (\mathcal{L}_{im}^A \mathcal{L}_j^C + \mathcal{L}_i^C \mathcal{L}_{jn}^A) - \frac{J_A}{2} \sum_{\langle i,j \rangle} I_C^2 \mathcal{L}_i^C \mathcal{L}_j^C.
$$
$$
(8)
$$

Here $\mathcal{L}_i^C = 0,1$ is the occupation variables for the cholesterol molecules, A^C is the fixed molecular cross-sectional area of cholesterol, and I_C is a phenomenological interaction parameter.

The Hamiltonian for the ten–state Pink model can also be modified to describe the effects of certain drugs and insecticides (denoted foreign molecules) on lipid bilayer properties. A certain type of modification is intended to describe the effects of those types of foreign molecules that do not change the transition enthalpy of the bilayer. The adsorbed foreign molecules can in that case be considered as interstitial impurities whose available locations are the centers of the triangles formed by three neighboring lipid acyl chains. The sites available to the foreign molecules hence constitute a honeycomb lattice embedded in the triangular lattice formed by the acyl chains. The requirement that the lipid chains and the foreign molecules occupy separate lattices and hence cannot mix, results automatically in the absence of an entropy of mixing. It is assumed that the interstitial impurities interact with the three neighboring lipid acyl chains as well as with other impurities occupying neighboring interstitial sites on the honeycomb lattice. The Hamiltonian consists of terms describing the pure lipid–bilayer interactions, the lipid–foreign molecule interactions, and the interactions between neighboring foreign molecules, respectively, i.e. for the lipid species A interacting with the foreign molecule (D)

$$\mathcal{H} = \mathcal{H}^{A} + \mathcal{H}^{D}, \tag{9}$$

where

$$\mathcal{H}^{D} = \sum_{i}\sum_{m}\sum_{\ell} J_{\mathrm{AD}}^{m} I_{m}^{A} \mathcal{L}_{im} \mathcal{L}_{\ell}^{D} - \frac{J_{\mathrm{DD}}}{2} \sum_{\langle \ell,k \rangle} \mathcal{L}_{\ell}^{D} \mathcal{L}_{k}^{D}. \tag{10}$$

$\mathcal{L}_{\ell}^{D} = 0,1$ is an interstitial site occupation variable. J_{AD}^{m} and J_{DD} are interaction constants for the lipid–foreign molecule and the foreign molecule–foreign molecule interactions, respectively. The concentration, x, of foreign molecules in the membrane is not conserved since partitioning between the membrane and the aqueous phase is assumed to occur. The use of a grand canonical ensemble is therefore appropriate and x is controlled by a chemical potential, μ. The effective Hamiltonian then becomes

$$\mathcal{H}_{\mathrm{grand}} = \mathcal{H} - \mu \sum_{\ell} \mathcal{L}_{\ell}^{D}. \tag{11}$$

Several extensions of the ten–state Pink model have been proposed in order to account for lipid–protein interactions in a specific fashion which depends on the conformational states of the lipid chains [20]. Here we consider a particular model in which the lipid-protein interactions have been incorporated into the microscopic Pink model by identifying part of the interaction parameters in terms of hydrophobic matching between the hydrophobic length of the lipid chain and hydrophobic protein length. This is implemented in the spirit of the phenomenological mattress model of lipid–protein interactions [18] which is illustrated in Fig. 5a.

The lipid–protein interactions are included in the model by assuming that the hydrophobic membrane–spanning part of the protein or polypeptide molecule is a stiff, rod–like, and hydrophobically smooth object with no appreciable internal flexibility. In this way the protein is characterized geometrically only by a cross sectional area, A_{P} (or circumference ρ_{P}), and a hydrophobic length, d_{P}. The protein can occupy one or more sites of the lipid lattice depending on its actual hydrophobic circumference.

152

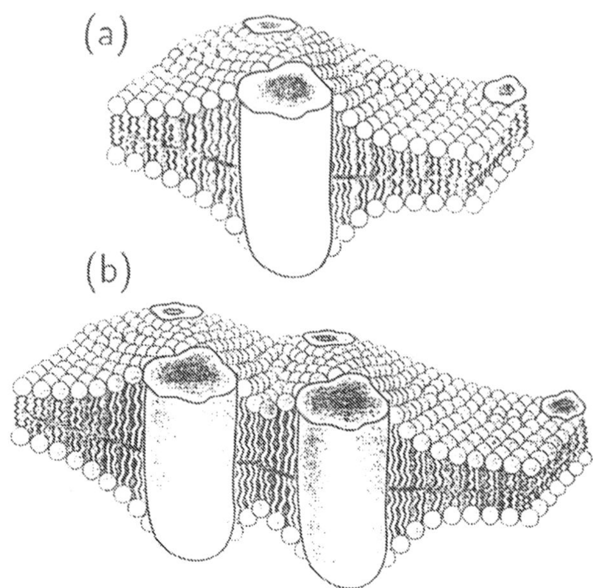

Figure 5. Schematic illustration of the hydrophobic–matching principle for lipid–protein interactions in membranes. (a): Formation of a local profile around a single protein in a lipid bilayer. The profile can be interpreted as a statistical population of acyl chains in a conformational state different from the bulk lipids, or, in the case of a binary lipid bilayer, as a local enrichment of the hydrophobically best matching lipid species at the protein interface. (b): Overlap of local profiles leads to a lipid–mediated, indirect protein–protein attraction.

The Hamiltonian of the microscopic version of the mattress model for a lipid bilayer of species A interacting with a protein (P) is now written

$$\mathcal{H} = \mathcal{H}^{A} + \mathcal{H}^{AP}, \tag{12}$$

where

$$\mathcal{H}^{AP} \;=\; \Pi A_{P} \sum_{i} L_{Pi} \tag{13}$$

$$+\frac{\Gamma_{AP}}{4}\left(\frac{\rho_{P}}{z}\right) \sum_{<i,j>} \sum_{m=1}^{10} \left(|2d_{im}^{A} - d_{P}|\mathcal{L}_{im}^{A} L_{j}^{P} + |2d_{jm}^{A} - d_{P}|\mathcal{L}_{jm}^{A} L_{i}^{P}\right)$$

$$-\frac{J_{AP}}{4}\left(\frac{\rho_{P}}{z}\right) \sum_{<i,j>} \sum_{m=1}^{10} \left(\min(2d_{im}^{A}, d_{P})\mathcal{L}_{im}^{A} L_{j}^{P} + \min(2d_{jm}^{A}, d_{P})\mathcal{L}_{jm}^{A} L_{i}^{P}\right).$$

The parameter J_{AP} is related to the direct lipid-protein van der Waals-like interaction which is associated with the interfacial hydrophobic contact of the two molecules, while the parameter Γ_{AP} is related to the hydrophobic effect. $L_{i}^{P} = 0, 1$ is the protein occupation variable. $\mathcal{L}_{im}^{A}$ and L_{i}^{P} satisfy a completeness relation at each lattice site, $\sum_{m} \mathcal{L}_{im}^{A} + L_{i}^{P} = 1$. The model in

Eq. (13) can readily be extended to binary lipid mixtures using Eqs. (5) and (6).

4.3. STRATEGY (III): THE DONIACH MODEL ON A DYNAMIC RANDOM LATTICE

A full microscopic (or first–principles) treatment of translational degrees of freedom would be ideal. Such a treatment is, however, very computationally demanding and severely limits any studies involving translational degrees of freedom in two spatial dimensions. Nielsen et al. [21] have recently proposed a set of simple models to study the phase equilibria in two–dimensional condensed systems of particles where both translational and internal degrees of freedom are present and coupled through microscopic interactions. The translational degrees of freedom of the particles are accounted for in an approximate description, in which the randomly varying spatial connectivity of the particles is represented by a random lattice whose dynamic structure is given by a triangulation of the spatial configurations. Based on this random-lattice description, two types of statistical mechanical models are then constructed. Both models are basically spin–1/2 Ising models where the spins, representing the internal degrees of freedom, are associated with hard–disk particles. Nearest–neighbor particles interact through spatially–dependent spin–spin interactions. The fluctuating number of nearest neighbors, and the possible spatial dependence of the spin–spin interactions, couple microscopically the spin degrees of freedom to the translational degrees of freedom. The first model (I) is a random–lattice Ising model with spatially–dependent nearest–neighbor spin–spin interactions. In the second model (II), the two spin states of the first model are assigned different degeneracies in the spirit of the so-called Doniach model [22]. The Doniach is identical to the multi-state Pink model described in Sec. 4.2 in the case of only two internal states of the lipid–acyl chains, a conformationally ordered state and a conformationally disordered state.

Within the random–lattice description [21], the lattice structure is dynamic (and semi–triangular): it can be seen as the result of fluidizing a regular triangular lattice through sampling over non–regular triangular lattice configurations with a fixed global topology, cf. Fig. 6a.

The global topology is here given by the Euler characteristics of the regular triangular lattice. The phase space of the translational degrees of freedom, including both the part that respects the full translational symmetry, i.e., that corresponding to fluid phases, and the part that respects the broken symmetry, i.e., that corresponding to solid phases, should therefore be well approximated by this description. The description is different from conventional off–lattice descriptions in that it only provides a

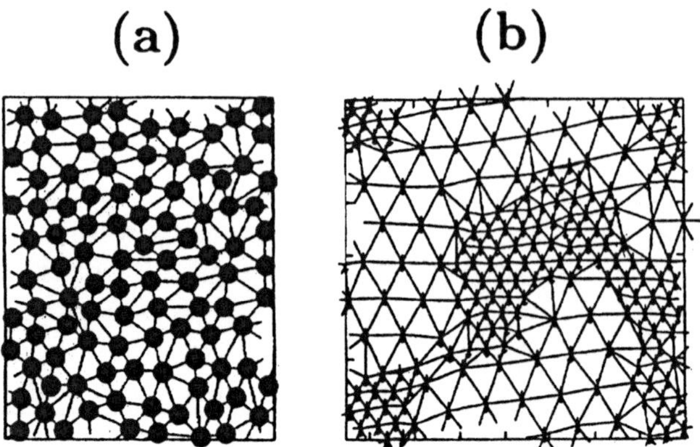

Figure 6. (a): Random lattice configuration of hard discs and tethers providing the nearest–neighbor topology. (b): Decoration of a random lattice with Ising spins. (Courtesy of Morten Nielsen.)

restricted phase space: those microscopic configurations that correspond to large density fluctuations on short length scales are effectively excluded. This approximation is sufficient for describing condensed fluid phases of systems of hard–core particles with short-range interactions. Since the definition of the model Hamiltonians along the strategy of the present section is closely tied to the simulation algorithms used, we describe the algorithm and models simultaneously. When dealing with interactions between particles in a computer simulation, the most important information required concerns the local environment of each individual particle, such as the distribution of other particles in its neighborhood and their distances to it. In conventional simulations that explicitly deal with the translational degrees of freedom, it is usually one of the most time-consuming steps to obtain and update this information from the microscopic configurations. Therefore, the algorithm proposed by Nielsen et al. [21] performs two essential tasks: 1) it generates the phase space associated with the translational degrees of freedom, and 2) it generates and retains a compact data structure that allows efficient access to structural information contained in each microscopic configuration. The data structure is based on triangulation of each spatial configuration of the particles. This triangulation is implemented as follows: an ordered configuration in which the particles are positioned on a regular triangular lattice is used as an initial state in which each site is linked to its six nearest neighbors by tethers. The lattice configuration is then represented by a network of tethers forming triangles; the term triangulation refers to this representation. The phase (or configuration) space can then be explored through a random updating (or stochastic evolution) of configurations of the lattice, which consists of three type of updates: particle moves, flips of tethers, and change of system size. The short–range repulsion between particles is modeled as a hard–core repulsion. Each particle is then considered as a hard disk of diameter d, and every site on the random lattice is occupied by such a hard–core particle. Within the first model (I), each hard core particle is associated with an Ising spin , cf. Fig. 6b. Nearest–neighbor particles are connected by tethers and interact through a distance-dependent spin–spin interaction in such a way that the tethered spins only interact if they are within a certain distance, R_0, of each other. The random–lattice Hamiltonian is then given by

$$H_{\mathrm{I}} = -J_0 \sum_{\substack{\langle i < j \rangle, \\ R_{ij} < R_0}} S_i S_j \; , \qquad (14)$$

where $\langle i < j \rangle$ again denotes a sum over all possible nearest neighbors, and R_{ij} is the distance between spins S_i and rS_j. Hence R_0 sets the range of interaction. The fact that the density of the system sets another length scale leads to a specific type of coupling between the spin and the translational degrees of freedom.

The second model (II), which is the Doniach model on a random lattice [21], is similar to model I except that one of the spin states is assigned an internal degeneracy larger than one. This model is inspired by the regular-lattice model proposed by Doniach [22] to describe the essential thermodynamic properties of lipid bilayers, in particular the main transition, that are primarily associated with the conformational degrees of freedom of lipid chains in a planar array. Doniach's original model uses two states to represent the lipid-chain conformation: one state, the ordered state (denoted $S_i = 1$), has zero internal (conformational) energy ($E_{\mathrm{O}} = 0$) and is non-degenerate ($D_{\mathrm{O}} = 1$), characteristic of the chain conformational state of lipid molecules in the gel phase. The other state, the disordered state (denoted $S_i = -1$), has a high internal energy, E_{D} (corresponding to the excitation energy associated with a conformational change), and a large degeneracy, $D_{\mathrm{D}} \gg 1$ (representing the large number of possible chain conformations which have the same value of E_{D}), characteristic of the chain conformation of lipid molecules in the fluid phase. The translational degrees of freedom of lipid chains are governed by inter–chain interactions that depend on the conformational states of the interacting chains. The random-lattice model Hamiltonian is given by

$$H_{\mathrm{II}} = H_0 + V_{\mathrm{int}}^{(1)} + V_{\mathrm{int}}^{(2)} + PA \; , \qquad (15)$$

156

where

$$H_0 = \sum_i E_D \left(\frac{1 - S_i}{2}\right) , \tag{16}$$

$$V_{\text{int}}^{(1)} = -\frac{J_0}{4} \sum_{\langle i<j \rangle} (1 + S_i)(1 + S_j) , \tag{17}$$

and

$$V_{\text{int}}^{(2)} = \frac{1}{4} \sum_{\langle i<j \rangle} V(|R_i - R_j|)(1 + S_i)(1 + S_j) . \tag{18}$$

H_0 describes the chain internal energy and $V_{\text{int}}^{(1)}$ models the chain–chain interaction, which (somewhat arbitrarily) is taken to be non–zero only if a chain and its neighbor are both in the ordered state. A is the total area of the system, and the conjugate pressure P plays the the role of a bilayer-stabilizing internal interfacial pressure. $V(R)$ in Eq. (18) is a square-well attractive potential of depth V_0 and range R_0. $V_{\text{int}}^{(1)}$ and $V_{\text{int}}^{(2)}$ together provide an approximation to the attractive intermolecular interaction between any two chains in the conformationally ordered state. The sum of the hard–core potential and the two square-well potentials, $V_{\text{int}}^{(1)}$ and $V_{\text{int}}^{(2)}$, constitutes an approximation to a standard inter-molecular potential of the Lennard–Jones type. By analogy with the Doniach lattice model, the effective interaction between any two chains in model II is taken to be zero if either one or both of the chains are in the conformationally disordered state. The square-well potential described by R_0 and V_0 controls the minimum of the potential and hence the lattice parameter of the crystalline (solid) phase. The tail of the potential extending beyond R_0 permits a possible decoupling between the two melting (or order-disorder) processes associated with the translational and conformational degrees of freedom. This model is a minimal model in the sense that it contains only the most essential physics required to model the coupling between translational and internal degrees of freedom in lipid–bilayer systems.

5. Monte Carlo simulation techniques for phase transitions

We do not intend to give a detailed account of standard Monte Carlo simulation techniques applied to lattice models of lipid bilayers. The reader is referred to Ref. [14] for details in the case of regular lattice models and Ref. [21] in the case of dynamic random lattice models. Below we provide a description of some recent techniques of data analysis that are particularly useful for the characterization of phase equilibria in lipid bilayers and lipid-protein systems when combined with Monte Carlo simulations.

These techniques are useful for lattice- as well as non-lattice simulations of membrane models [16].

The fundamental problem in numerical investigations of cooperative phenomena is associated with an unambiguous assessment of the nature of phase transitions as well as a determination of the precise location of the phase boundaries in a phase diagram. Numerical simulations share this problem with laboratory experiments. The root of the problem is that most approaches do not function on the level of the free energy but rather in terms of derivatives of the free energy, such as densities and order parameters (first derivatives) or response functions (second derivatives). In the case of one–component systems, these derivatives are seldom known with sufficient accuracy to discern between, for example, a fluctuation–dominated first–order phase transition and a continuous transition (critical point). In the case of phase coexistence in, e.g. a two–component system, there is no direct way of relating features in the response functions to the precise location of equilibrium phase boundaries. However, by use of novel techniques [23] it is possible numerically to circumvent this problem and obtain access to that part of the free energy which is necessary to locate the phase equilibria. These techniques involve calculation of distribution functions or so–called histograms of thermodynamic functions (umbrella sampling), e.g. order parameters (composition) or internal energy, thermo-dynamic re–weighting of the distribution functions in order to locate the phase transition or the phase equilibria, and then a subsequent analysis of the size dependence of the re–weighted distribution functions by means of finite-size scaling theory.

5.1. HISTOGRAM AND RE–WEIGHTING TECHNIQUES

In order to use the finite–size scaling method effectively to study phase equilibria in mixtures it is a prerequisite that a multi–dimensional distri-bution function, $\mathcal{P}(\{y_i\}_i, T, L)$, can be calculated very close to coexistence for a set of thermodynamic variables, $\{y_i\}_i$. L is the linear extension of the two–dimensional bilayer lattice which therefore consist of $N = L^2$ particles. For one–component cases, e.g. pure lipid bilayers, this set of thermodynamic variables, $\{y_i\}_i$, contains the internal energy and the order parameter, which usually is taken to be the bilayer area. For a two–component system, the set includes in addition the relative chemical potential, μ, for the two species. The distribution function for any given variable, y_j, can be obtained from the complete distribution function as follows

$$\mathcal{P}(y_j, T, L) = \sum_{i(\neq j)} \mathcal{P}(\{y_i\}_i, T, L). \qquad (19)$$

Without loss of generality we focus on the joint distribution function (or histogram), $\mathcal{P}(E, x, T, L)$, for the internal energy, E, and the composition, x. This distribution function is now, for a given system size L, calculated very accurately close to coexistence (see below for detecting coexistence). For a given set of thermodynamic parameters, (T, μ), the distribution function is defined by

$$\mathcal{P}_\mu(E, x, T, L) = \frac{n(E, x, L) \exp[-(E + \mu x)/k_B T]}{\sum_{E,x} n(E, x, L) \exp[-(E + \mu x)/k_B T]}, \tag{20}$$

where $n(E, x, L)$ is the density of states which is independent of temperature.

If $\mathcal{P}(E, x, T, L)$ is calculated with sufficient accuracy for a specific set, (T, μ), it is possible to determine the distribution function for a nearby set of parameters, (T', μ'), by employing the re–weighting method of Ferrenberg and Swendsen [24]. The method is based on standard thermodynamics and it leads to the re–weighted distribution function

$$\mathcal{P}_{\mu'}(E, x, T', L) = \frac{\mathcal{P}_\mu(E, x, T, L) \exp\left[-\left(\frac{1}{k_B T'} - \frac{1}{k_B T}\right) E - \left(\frac{\mu'}{k_B T'} - \frac{\mu}{k_B T}\right) x\right]}{\sum_{E,x} \mathcal{P}_\mu(E, x, T, L) \exp\left[-\left(\frac{1}{k_B T'} - \frac{1}{k_B T}\right) E - \left(\frac{\mu'}{k_B T'} - \frac{\mu}{k_B T}\right) x\right]}. \tag{21}$$

The two one-dimensional distribution functions can readily be derived from the combined two-dimensional distribution function, $\mathcal{P}(E, x, T, L)$, as follows

$$\mathcal{P}(E, T, L) = \sum_x \mathcal{P}(E, x, T, L) \tag{22}$$

$$\mathcal{P}(x, T, L) = \sum_E \mathcal{P}(E, x, T, L). \tag{23}$$

These distribution functions allow us to calculate average quantities, such as internal energy, order parameter, and response functions at any set of thermodynamic parameters, provided that the originally sampled distribution function is accurate enough to contain sufficient statistical information about the relevant part of phase space. Data sets very dense in, e.g., temperature can therefore be obtained, which is a prerequisite for locating the transition point and for performing a detailed finite-size scaling analysis at the transition point.

5.2. DETECTION OF PHASE EQUILIBRIA BY FINITE–SIZE SCALING

Based on the distribution functions described above, phase equilibria can be examined by a powerful method proposed by Lee and Kosterlitz [25].

This method, which is based on finite–size scaling analysis of distributions derived from computer simulations, constitutes an unambiguous technique for numerically detecting first–order transitions and phase coexistence [23]. The method involves calculation of a certain part of the free energy using the distribution functions in Eqs. (22)-(23). Free–energy–like functions, e.g. $\mathcal{F}(x, T, L)$, can be defined from these distribution functions as

$$\mathcal{F}(x, T, L) \sim - \ln \mathcal{P}(x, T, L). \tag{24}$$

The quantity $\mathcal{F}$ differs from the bulk free energy by a temperature- and L-dependent additive quantity. However, the shape of $\mathcal{F}(x, T, L)$ at fixed T and L is identical to that of the bulk free energy and furthermore $\Delta\mathcal{F}(\mu, T, L) = \mathcal{F}(x, T, L) - \mathcal{F}(x', T, L)$ is a correct measure of free-energy differences. At a first-order transition, $\mathcal{F}(x, T, L)$ has pronounced double minima corresponding to two coexisting phases at $x = x_1$ and $x = x_2$ separated by a barrier, $\Delta\mathcal{F}(\mu, T, L)$, with a maximum at $x_{\max}$ corresponding to an interface between the two phases. The height of the barrier measures the interfacial free energy between the two coexisting phases and is given by

$$\Delta\mathcal{F}(\mu, T, L) = \mathcal{F}(x_{\max}, T, L) - \mathcal{F}(x_1, T, L) = \gamma(\mu, T)L^{d-1} + \mathcal{O}(L^{d-2}), \tag{25}$$

where d is the spatial dimension of the system ($d = 2$ in the present case). $\gamma(\mu, T)$ is the interfacial free-energy density or interfacial tension. Therefore, $\Delta\mathcal{F}(\mu, T, L)$ increases monotonically with L at a first-order transition (or at phase coexistence) corresponding to a finite interfacial tension. The detection of such an increase is an unambiguous sign of a first-order transition and two-phase coexistence. In contrast, $\Delta\mathcal{F}(\mu, T, L)$ approaches a constant at a critical point, corresponding to vanishing interfacial tension in the thermodynamic limit. In the absence of a transition, $\Delta\mathcal{F}(\mu, T, L)$ tends to zero.

The iterative computational method for detecting coexistence proceeds as follows: a trial value of the chemical potential at coexistence, μ_{m}, for a given temperature is guessed or estimated from a short simulation on a very small lattice. The accurate value of μ_{m} for that system size is then determined by a long simulation at this trial value of μ and an accurate value of μ_{m} is determined by the re-weighting technique , cf. Eq. (21). If the trial value turns out to be too far from coexistence to allow a sufficiently accurate sampling of the two phases, a new trial value is estimated from the long simulation. The system size is then increased and the value of μ_{m} for the previous system size is used as trial value for the long simulation on the larger system. This rather lengthy iterative procedure assures a very accurate determination of the chemical potential at coexistence and hence leads to an accurate determination of the compositional phase diagram via the finite-size scaling analysis. The Ferrenberg-Swendsen re–weighting technique becomes more troublesome the larger the size of the system studied because the distribution functions are narrower for larger systems. The method works best for small systems which are subject to large fluctuations allowing more information about a larger part of phase space to be sampled. A particular problem is associated with the present application of the re–weighting to mixtures which involves an extra thermodynamic variable, the chemical potential. Re–weighting of data obtained for one value of the chemical potential to another value gives a larger uncertainty for the larger system simply because it is the extensive composition which is involved in the Hamiltonian and hence in the re–weighting procedure, cf. Eq. (21). This implies that a knowledge of the value of the chemical potential is required at coexistence with an uncertainty which decreases approximately as L^{-2}. Another bottleneck of a more physical nature is the time, τ, associated with the crossing of the free–energy barrier between the two phases. This time increases exponentially with the linear dimension of the two–dimensional model system, $\tau \sim \exp(\gamma L)$. In order to facilitate the re–weighting, the actual simulation time has to be much larger than τ. For a given model, this places a sharp upper bound for the system sizes which can be studied by these simulation techniques. It is possible to remedy this situation by various non-Boltzmann sampling schemes where certain functions are added to the Hamiltonian to flatten the barrier (J. Risbo, G. Besold, and O. G. Mouritsen, unpublished).

6. Phase transition and lateral structure in lipid bilayers

6.1. NATURE OF THE MAIN PHASE TRANSITION

The nature of the main transition in one-component lipid bilayer membranes is a topic of continuing controversy [16, 19]. It is usually assumed that it is of first order. There is, however, no clear–cut experimental proof of this, and the usual assumption of a first–order transition is very much due to the experimental finding of a very intense and narrow specific–heat peak with a large heat content for phospholipid bilayers at their main transition temperature, T_{m}. Nevertheless, there is no single piece of experimental observation which reveals the kind of discontinuities that are usually associated with first–order behavior. The possibility remains that the transition is at best weakly first order and possibly there is no transition at all in a strict thermodynamic sense, i.e. the bilayer transition may be close to a critical point. Whether it is on the side of the critical point corresponding to a (weakly) first–order transition or on the side where there is no phase

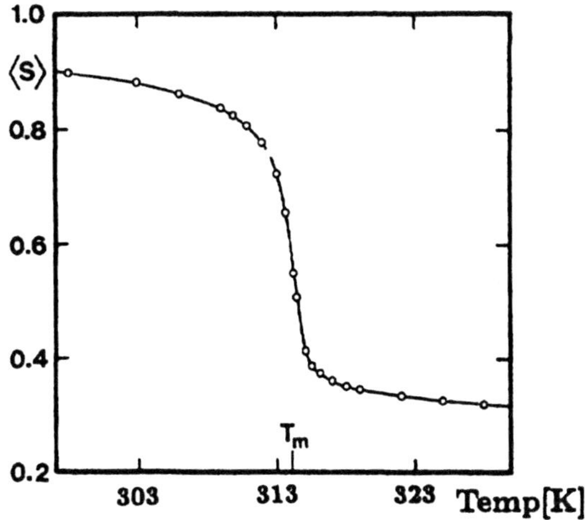

Figure 7. Average segmental acyl–chain order parameter, $\langle S \rangle = \frac{1}{2}\langle 3\cos^2\theta - 1 \rangle$ (cf. Eq. 3), as a function of temperature for $DC_{16}PC$ bilayers. $T_m = 314K$ is the transition temperature.

transition may critically depend on details of the particular system.

The current picture of a lipid bilayer at its main transition is hence one of a strongly fluctuating system [19]. The strong fluctuations are signaled on the macroscopic level by dramatic peaks in the response functions and strong variations in the order parameters. Microscopically, the fluctuating state is manifested in terms of the formation of clusters and domains leading to a heterogeneous lateral structure which is dynamically maintained.

The variation of the acyl-chain conformational order parameter, $\langle S \rangle = \frac{1}{2}\langle 3\cos^2\theta - 1 \rangle$, as determined from a Monte Carlo simulations on the multistate Pink model (cf. Sec. 4.2) of a $DC_{16}PC$ lipid bilayer is shown in Fig. 7. It is observed that the order decreases very rapidly, although in a continuous fashion, in a narrow temperature region around a temperature T_m. These numerical data are in rather close accordance with experimental measurements based on deuterium nuclear–magnetic resonance spectroscopy. Hence it appears that if the transition in $DC_{16}PC$ is of first order it is strongly smeared by thermal fluctuations. The bilayer becomes soft at the transition.

The nature of the main transition, considering only the conformational degrees of freedom of the lipid molecules, can be investigated more directly by using re–weighting and finite–size scaling techniques (cf. Sects. 5.1 and 5.2). In Fig. 8 we present the results for the free energy functional, $\mathcal{F}(A, T, L)$ (cf. Eq. 24), as a function of linear system size, L, for two dif-

162

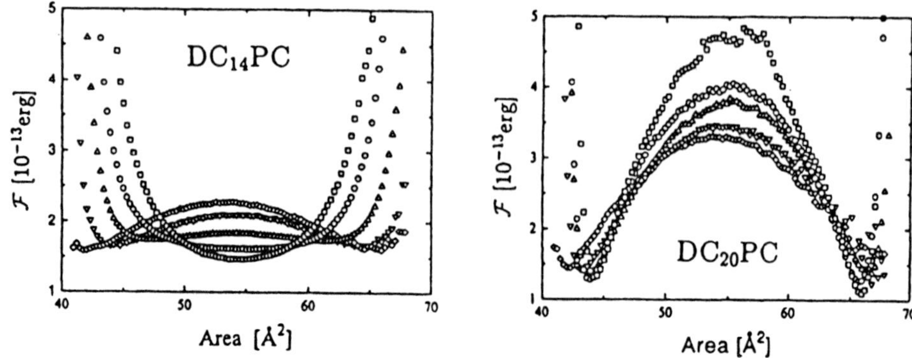

Figure 8. Free-energy functional, $\mathcal{F}(A)$, in Eq. (24) determined at the phase transition (or at the temperature where $\mathcal{F}$ is symmetric) for $DC_{14}PC$ and $DC_{20}PC$ bilayers for different system sizes, $N = L \times L$. $L = 4(\lozenge), 6(\triangledown), 10(\triangle), 15(\circ), 20(\square)$.

ferent phospholipids with different acyl-chain lengths. Here A is the lipid-bilayer area per molecule. The data are obtained from simulations on the Pink model in Eq. (1). The results refer for each system size to a temperature at which the two minima in the free energy are equally deep. It appears from this figure that for the shorter chain length there is no phase transition in the thermodynamic limit (the free–energy barrier vanishes with increasing system size) whereas for longer chains, the free–energy barrier increases signaling the occurrence of a first–order phase transition. It is found that $DC_{18}PC$ is close to being at the border line. In all cases the transitional behavior is strongly influenced by thermal density fluctuations. A macroscopic consequence of these strong fluctuations is a strong anomaly in the specific heat at the transition. As the chain length is decreased, the intensity of the specific heat is increased in the wings, in good agreement with experimental findings. Similarly, other response functions, e.g. the lateral compressibility and the elastic bending rigidity, display anomalies at the transition. Altogether, these observations suggest that the lipid bilayer softens at in the transition region. Selected simulation results for the Pink model are shown in Fig. 9.

Using the random–lattice models described in Sec. 4.3, the consequences on the phase equilibria of the interplay between the translational degrees of freedom and the internal conformational degrees of freedom of the acyl chains of lipid bilayers can be studied via the general simulation techniques described in Sec. 5 [21].

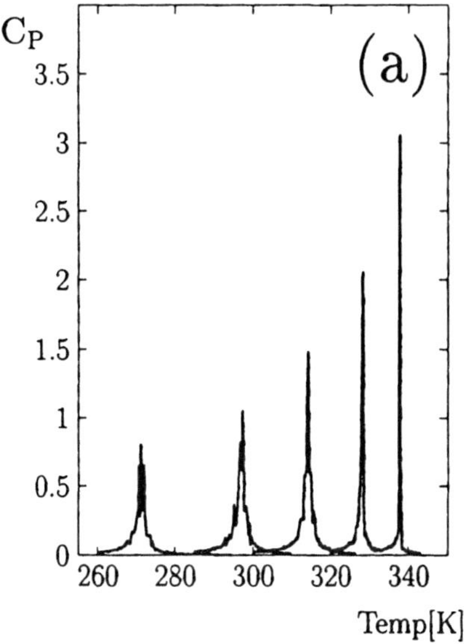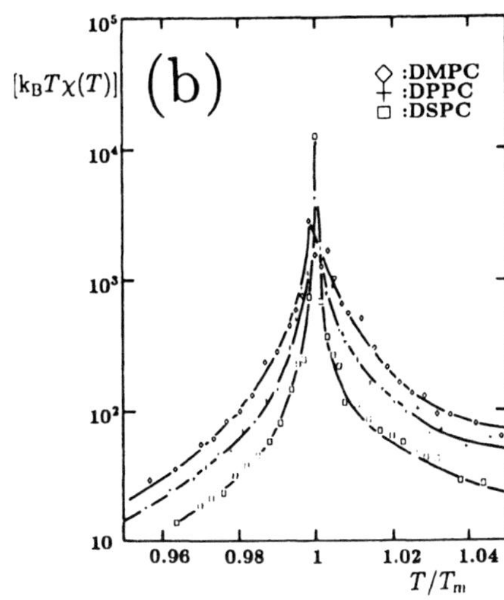

Figure 9. (a): Specific heat per lipid molecule, $C_P(T)$, for $DC_{12}PC$, $DC_{14}PC$, $DC_{16}PC$, $DC_{18}PC$, and $DC_{20}PC$ (from left to right) lipid bilayers in the transition region. (b): Lateral bilayer compressibility, $\chi(T)$, for $DC_{14}PC$, $DC_{16}PC$, and $DC_{18}PC$ bilayers in the transition region. The temperature axis is scaled with the appropriate value of the transition temperature, T_m.

Model I was found to have a rather complex phase behavior, in particular with respect to the coupling of the degrees of freedom at the macroscopic level. In Fig. 10a is shown the phase diagram for the model, given in the parameter space of the reduced pressure and the scaled temperature. Four different phases are found, solid–ordered (SO), solid–disordered (SD), liquid-ordered (LO), and liquid–disordered (LD), corresponding to the possible combinations of the labels for the two sets of degrees of freedom. In the low–pressure and high–pressure regions of the parameter space, where the first–order lattice–melting transition and the second-order (critical) spin transition are decoupled, the LO and the SD phases intervene between the SO and the LD phases. The diagram contains two special points, t_1 and t_2, which are critical end–points. Along this phase boundary, which is of first order, the translational degrees of freedom override the spin degrees of freedom and the lattice–melting transition preempts the critical spin transition, leading to a first-order singularity also in the spin order parameter.

The topology of the phase diagram for Doniach model II, cf. Fig. 10b, characterized by three phase boundaries merging into a special triple point

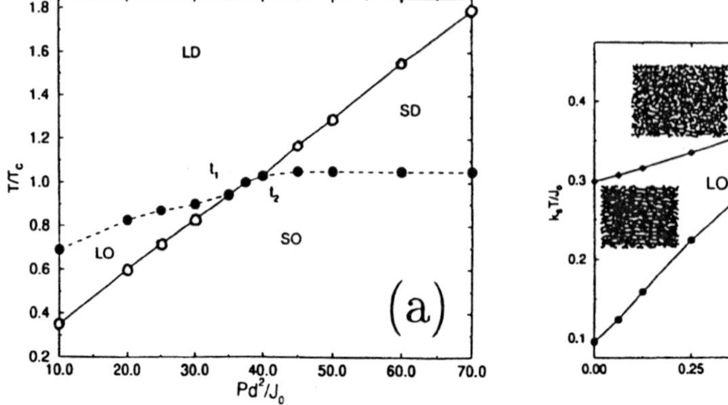

Figure 10. (a): The phase diagram for Ising model I for $R_0/d = 1.41$. The dashed phase boundary line (•) corresponds to the critical Ising–like transitions from a spin–ordered (O) to a spin–disordered (D) phase. The solid boundary line (○) corresponds to the first–order lattice–melting transition from a solid (S) phase to a liquid (L) phase. t_1 and t_2 are two critical end–points. Between the two axs critical end–points the spin order–disorder transition is coupled to the lattice melting and is of first order. (b): Phase diagram for the Doniach model on a random lattice (model II). All three phase boundaries are first-order phase boundaries. The insets show snapshots of typical micro-configurations for the three different phases labeled SO (solid-ordered), LD (liquid-disordered), and LO (liquid-ordered). Chains in the disordered state are plotted as (○) and chains in the ordered chain state as (•). The three snapshots are not given to scale. t_1 is a triple point.

t_1, resembles the low-pressure part of the topology of the phase diagram of Ising model I in Fig. 10a, with the difference that the spin (or chain conformation) order–disorder transition in this model is of first order, driven by the internal (or conformational) entropy and thus referred to as the chain-melting transition. This distinct topology indicates that, again, as in Ising model I, the thermodynamic singularity arising from the lattice melting can be either coupled or decoupled from the singularity associated with the chain–melting, depending on the values of the parameters. The three phase boundaries, all being of first order and corresponding to a lattice–melting transition, a chain–melting transition and a transition at which both melting processes take place, divide the explored region of the parameter space into three phases, the SO, LO, and LD phases. The three insets in the figure show, respectively, a characteristic microscopic configuration of each phase. In comparison with experiments the SO–LO phase line is interpreted as the so–called sub–main phase transition [26] and the LO-LD phase line as the main phase transition in long-chain phospholipid bilayers.

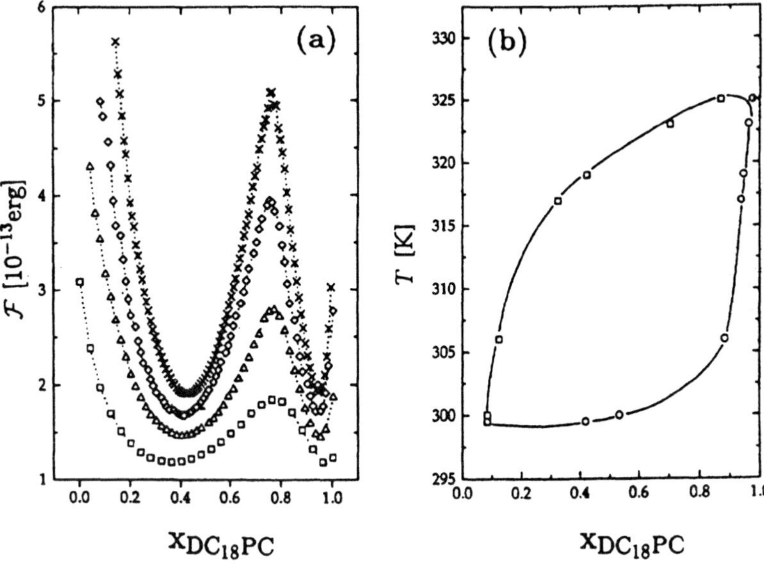

Figure 11. Monte Carlo computer-simulation data for a $DC_{14}PC$-$DC_{18}PC$ mixture. (a): Size-dependence of the free-energy functional, $\mathcal{F}(x,L)$, evaluated at coexistence at a temperature $T = 319K$. The linear sizes of the systems correspond to $L = 5$ ($\square$), 7 ($\triangle$), 9 ($\diamond$) and 11 ($\times$). (b): Phase diagram.

6.2. PHASE EQUILIBRIA IN BINARY LIPID BILAYERS

In the following we shall concentrate on phase equilibria controlled by the conformational degrees of freedom exclusively. The possible absence of a phase transition in certain one–component lipid bilayers poses some intriguing problems as to the existence of phase equilibria in lipid bilayers incorporated with other species, such as other lipids, cholesterol, or proteins. Using histogram and finite–size scaling techniques it can be shown that a first-order transition, which is absent in a one–component lipid bilayer, may be induced by mixing in small amounts of another species in such a way as to approach the critical points and the related closed coexistence regions [16]. Figure 11 shows some results from simulations on the model in Eq. (5) of binary lipid mixtures of saturated phospholipid acyl chains with different hydrophobic length in the specific case of $DC_{14}PC$-$DC_{18}PC$ mixtures. The size dependence of the free-energy functional in Eq. (24) as a function of composition, $x_{DC_{18}PC}$, is displayed in Figs. 11a. The family of distribution functions are shown at the appropriate size-dependent chemical–potential values corresponding to phase coexistence. It is seen that, as the system size is increased, there is a dramatic increase in the free-energy barrier between the two phases. This is unambiguous evidence of phase coexistence in the thermodynamic limit and that the interfacial tension tends towards a non–zero value. From data of this type the entire phase diagram can be determined very accurately, as shown in Fig. 11b. This phase diagram shows that, even though bilayers of the two pure components do not have a phase transition in a strict thermodynamic sense, a transition is induced by mixing the two components. This leads way to two critical mixing points in the phase diagram. Between these two critical points, there is a pronounced two–phase coexistence region. The phase diagram found by these techniques is rather close to experimental results. It should be remarked that for all practical experimental purposes the displacement of the critical points from the axes of the phase diagram is probably of marginal relevance, in particular for the more non–ideal mixtures.

6.3. SMALL–SCALE STRUCTURE IN LIPID BILAYERS: FLUCTUATIONS AND LIPID DOMAINS

In contrast to the trans-bilayer structure to be discussed in Sec. 7 below, the lateral bilayer structure and molecular organization is much less well characterized and its importance also much less appreciated. This is particularly the case when it comes to the small-scale structure and micro–heterogeneity in the namometer range. One reason for this is that this regime is experimentally difficult to access by direct methods [9]. Monte Carlo simulation calculations on the models described in Sec. 4.2 are ideally suited to explore the existence of such small–scale phenomena.

Simulations show that during the transitional process in the lipid bilayer (whether it be a phase transition or not) certain elements of dynamic order develop, as illustrated in Fig. 12 in the case of $DC_{16}PC$. This figure shows that the bilayer in the transition region exhibits a substantial degree of dynamic heterogeneity. Specifically for the bilayer in the conformationally–ordered gel phase, domains or clusters of chains are formed which are conformationally disordered. Conversely, in the conformationally–disordered fluid phase gel domains are formed in the fluid matrix. The domain formation is a dynamic phenomenon and could be characterized as a kind of nano–scale dynamic phase separation. Still, the overall symmetry of the mother phase is maintained, e.g. in the fluid phase the fluidity is maintained on length scales larger than the average gel domain size. While the individual lipid domains appear and disappear as they have a finite lifetime depending on their size, the average domain size is an equilibrium property. The average domain size varies strongly with temperature and it exhibits a distinct maximum at the transition. This maximum matches the maxima found in response functions, such as the specific heat and the lateral compressibility,

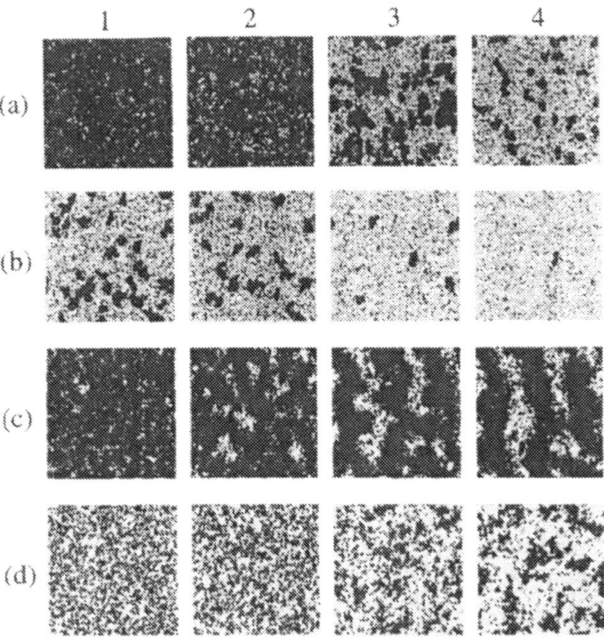

Figure 12. Top-view snapshots of typical lipid bilayer microscopic configurations as obtained from Monte-Carlo simulation on the ten-state Pink model of lipid-bilayer phase equilibria. The configurations show the bilayer heterogeneity in terms of lateral density and compositional fluctuations for pure lipid bilayers as well as certain binary mixtures consisting of DC_nPC lipids with different numbers, n, of carbons in the saturated acyl chains. (a): $DC_{16}PC$ bilayers near the phase transition at temperatures $T = 310K$ (1), 313K (2), 315K (3), and 319K (4). The transition temperature is $T_m = 314K$. Fluid domains are seen to be formed in the gel phase below T_m and gel domains in the fluid phase above T_m. (b): Microstructures at the same relative temperature, $T/T_m = 1.016$, in the fluid phase for different lipid chain lengths, $DC_{12}PC$ (1), $DC_{14}PC$ (2), $DC_{18}PC$ (3), and $DC_{20}PC$ (4). (c): Time evolution of the microstructure for a binary mixture of $DC_{12}PC$-$DC_{18}PC$ quenched form a high temperature in the fluid phase to a temperature within the gel-fluid coexistence region. (d): Compositional heterogeneity in fluid-phase binary mixtures of $DC_{16}PC$-$DC_{20}PC$ at $T = 338K$ (1), $DC_{14}PC$-$DC_{20}PC$ at $T = 338K$ (2), $DC_{12}PC$-$DC_{20}PC$ at $T = 338K$ (3), and $DC_{12}PC$-$DC_{22}PC$ at $T = 352K$ (4) at a molar fraction, $x = 0.33$, of the longer lipid. (Courtesy of dr. Kent Jørgensen.)

cf. Fig. 9. The occurrence of the domains implies that the thermal density fluctuations are correlated in space with a coherence length corresponding to the linear domain size.

The domain formation illustrated in Fig. 12a may be seen as a dynamic state of order in disorder [27]. Any membrane-associated process that takes place on time scales less than the life time of a domain (typically of the order of 10^{-4} sec) and on length scales less or comparable with the domain size will be influenced by this order-in-disorder phenomenon. The domain sizes,

168

and hence the dynamic bilayer heterogeneity, can be modified by choosing lipid species with different chain lengths, as illustrated quantitatively in Fig. 12b for the series $DC_{n_C}PC$, with $n_C = 12, 14, 18$, and 20, at the same relative temperature, T/T_m, in the fluid phase. The figure shows that the longer the acyl chain is, the smaller are the lipid domains, reflecting the fact that the short-chain lipid bilayer is closer to a critical point and sustains the strongest thermal density fluctuations.

The dynamic membrane heterogeneity and the lipid–domain formation phenomena described for one–component lipid bilayers involve gel domains. It has often been stated that such phenomena are not relevant under physiological conditions since gel domains are rarely, if ever, observed in functional biological membranes. However, dynamic lipid-domain formation in lipid-bilayer membranes need not involve gel-like domains. In fact, the liquid correlations in many-component lipid mixtures naturally lead to a local dynamic domain-like structure that is controlled by the compositional fluctuations. The compositional fluctuations are characterized by a lateral coherence length which intimately depends on the underlying phase diagram of the mixture.

For convenience we consider a series of mixtures of homologous saturated di–acyl phosphatidylcholines that only differ with respect to their acyl–chain length, cf. Fig. 4. This will make it possible, in a rather transparent fashion, to vary the degree of non–ideality of the mixture by changing the difference in acyl–chain length of the two species. Whereas the $DC_{14}PC$-$DC_{18}PC$ mixture within the temperature range of interest exhibits full miscibility within the fluid and gel phases and only has a broad gel-fluid coexistence region, the considerably more non–ideal $DC_{12}PC$-$DC_{18}PC$ mixture develops peritectic behavior with a pronounced gel–gel coexistence region. The $DC_{16}PC$-$DC_{18}PC$ system has a more ideal monotectic mixing behavior than $DC_{14}PC$-$DC_{18}PC$ and $DC_{16}PC$-$DC_{20}PC$, whereas e.g. $DC_{12}PC$-$DC_{18}PC$ and $DC_{12}PC$-$DC_{20}PC$ to be considered below also display peritectic behavior.

The description of the mixtures provided by phase diagrams is purely thermodynamic and contains no information on the local structure of the mixture. Local structure and domain–formation phenomena can be revealed from computer–simulation calculations on microscopic models that can not only reproduce phase diagrams like those in Fig. 11b, but also simultaneously provide information on fluctuations within the thermodynamic phases. In Fig. 12d are shown typical configurations of four binary mixtures, $DC_{16}PC$-$DC_{20}PC$, $DC_{14}PC$-$DC_{20}PC$, $DC_{12}PC$-$DC_{20}PC$, and $DC_{12}PC$–$DC_{22}PC$, arranged according to increased degree of non–ideality. They are all in the thermodynamic fluid phase. These configurations show that the fluid phase sustains a considerable degree of dynamic local structure. The

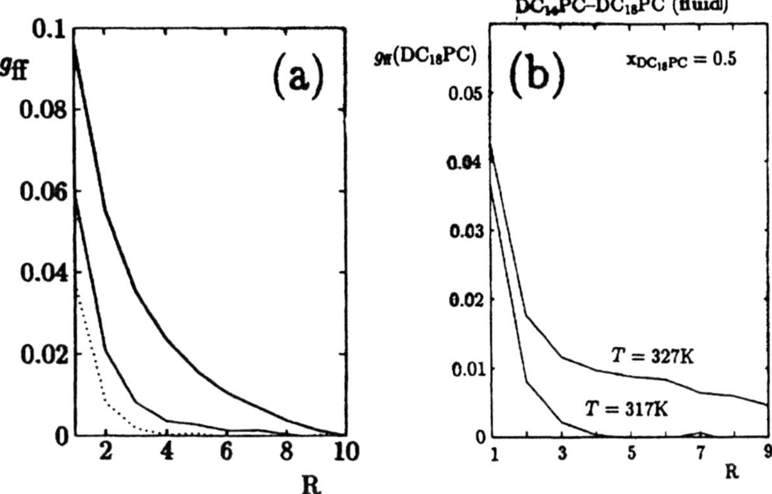

Figure 13. (a): Pair correlation function, g_ff, for fluid lipid states of the low-melting lipid in three different mixtures, $DC_{14}PC$-$DC_{18}PC$ (dotted line), $DC_{12}PC$-$DC_{18}PC$ (thin line), and $DC_{12}PC$-$DC_{20}PC$ (heavy line). Data are given at a fixed temperature, $T = 338K$, and a fixed composition, $x = 0.5$, for varying acyl-chain separation, R (in units of lattice spacings). (b): Pair correlation function, $g_\mathrm{ff}(DC_{18}PC)$, for fluid lipid states of $DC_{18}PC$ in a $DC_{14}PC$-$DC_{18}PC$ mixture at equimolar concentration shown at two different temperatures in the fluid phase, $T = 317K$ and $327K$. Results are given for varying acyl-chain separation, R (in units of lattice spacings).

lateral structure corresponds to a local compositional demixing by which chains of the same species (length) tend to cluster together. The configurations in Fig. 12d show that the larger the difference between the acyl-chain lengths, the more pronounced is the local structure.

A quantitative description of the local structure in the binary lipid mixtures is provided by the static pair correlation function, $g(R)$, which at a given concentration measures the probability that an acyl chain, in a given state of a given type of lipid, will be found at a distance, R, from the acyl chain in another given state of another given type of lipid. In the case of correlations between disordered (fluid) acyl-chain states of $DC_{n_C}PC$ this function is written $g_\mathrm{ff}(DC_{n_C}PC)(R)$ and is shown in Fig. 13a as a function of acyl-chain separation, R, for equimolar composition of three different mixtures. The data clearly show that the larger the chain-length difference between the different lipid species is, i.e. the more non-ideal the mixture is, the more long ranged is the correlation, in accordance with the visual appearance of the configurations in Fig. 13d. This effect refers to isothermal conditions and it should be kept in mind that the transition temperature of the pure lipid bilayer increases with acyl-chain length. The coherence length of the compositional fluctuations and the local structure has a strong temperature dependence, as illustrated in Fig. 13b in the case of an equimolar $DC_{14}PC$-$DC_{18}PC$ mixture at two different temperatures in the fluid phase. The figure shows that as the gel–fluid coexistence region is approached, cf. Fig. 11b, the correlation becomes more long–ranged.

In summary, the underlying phase transition in phospholipid bilayers provides a natural mechanism for lateral organization of the lipid bilayer on time- and length scales that depend on the thermodynamic conditions. It should be noted that this structural organization persists in the fluid phase. Consequently, the fluid phase, which is a disordered phase on macroscopic length scales, develops local order on the nano–scale due to the proximity of a phase transition. The possibilities for forming local order out of globally disordered lipid–bilayer phases increase when we turn to many-component systems. In these systems, lateral organization of the bilayer not only occurs in terms of static macroscopic phase separation as given by the phase diagrams. In the various one–phase regions, local order can develop in terms of lateral compositional fluctuations, e.g. in the fluid phase. This is illustrated by the computer–simulation results presented in Fig. 12d which shows how local structure persists in non-ideal binary mixtures of lipids that are miscible in the fluid phase. It is noted that the more non-ideal the mixture is the more pronounced are the compositional fluctuations. Similarly to the density fluctuations in a one–component system, the compositional fluctuations in a binary mixture correspond to lipid domains of a local composition that is very different from the global composition of the bulk phase. In physical terms, the fluid mixture may be termed a two–dimensional structured fluid. Hence, we see again that due to the underlying phase equilibria, local order and organization persist in a disordered thermodynamic phase.

So far we have only considered dynamic domain formation in equilibrium. An even larger richness in type of structural organization on a local scale arises when we also consider non–equilibrium phenomena. There are now accumulating experimental evidence that binary lipid bilayers in their gel–fluid phase coexistence region may not be fully phase separated even over time scales of hours. In the case of phase separation processes in the gel-fluid phase of phosphatidylcholine mixtures, an example of local structure formation and its evolution in time is provided by the computer-simulation results for the $DC_{12}PC$-$DC_{18}PC$ mixture shown in Fig. 12c. This figure shows how the fluid and gel domains evolve very slowly in time. Moreover, the figure indicates that not only does the mixture organize in a convoluted structure of gel and fluid domains but the structure of the fluid (mainly consisting of the low–melting lipid, $DC_{12}PC$) between the gel domains (mainly consisting of the high-melting lipid $DC_{18}PC$) is quite different from the thermodynamic fluid phase in the sense that the acyl chains of the $DC_{12}PC$ molecules are conformationally highly ordered as in their gel phase. This type of local organization is related to the phenomenon of capillary condensation.

The lifetime of the various types of lipid domains discussed in this section is difficult to assess from the theoretical model simulations, and only very little experimental data is currently available to shed light on this question. We estimate that a typical lifetime of a nano-scale domain may be as long as 10^{-4} sec in a one-component lipid bilayer and considerably longer in many-component mixtures.

6.4. MODULATION OF LIPID DOMAINS BY FOREIGN COMPOUNDS: CHOLESTEROL AND DRUGS

It is obvious that lipid bilayers that are locally ordered and dynamically organized due to fluctuations and cooperative modes controlled by the underlying phase equilibria will be sensitive to molecular agents that are active at the lipid-domain interfaces. In physical terms, the only requirement for an agent to be active in structural reorganization is that it is capable of lowering the interfacial tension of the domain interfaces. There exists a large number of molecules that interact with membranes and are capable of changing the local domain structure described in Sec. 6.3 above. Cholesterol, drugs, insecticides, as well as polypeptides and proteins couple strongly to the heterogeneity which may either be suppressed or strongly enhanced. Cholesterol as well as some drugs and insecticides act as emulsifiers and interfacially active agents by altering the lateral structure of the lipid bilayer. The interaction of proteins and polypeptides with lipids takes advantage of the lateral density and compositional fluctuations, and the coherence length of the perturbation on the lipid matrix due to the presence of proteins is directly related to the correlation length characterizing the dynamic membrane heterogeneity. We shall return to lipid-protein interactions in Sec. 8 below.

Cholesterol is a particularly prominent example that in small amounts appears to enhance the lateral density fluctuations in one-component lipid bilayers but strongly to suppress the fluctuations and the tendency for domain formation in higher concentrations [28]. In order to understand this peculiarity it is necessary to consider the full phase diagram of lipid-cholesterol mixtures. It is not possible to determine this phase diagram by the model in Eq. (7) since cholesterol's coupling to the translational degrees of freedom has to be taken into account. This is possible via the Doniach model on a random lattice, cf. Eq. (15) (Ref. [29] and unpublished work by Morten Nielsen) by introducing a second component with no internal degrees of freedom. This second component (due to the smooth sterol skeleton of the cholesterol molecule) is taken to break the lipid translational

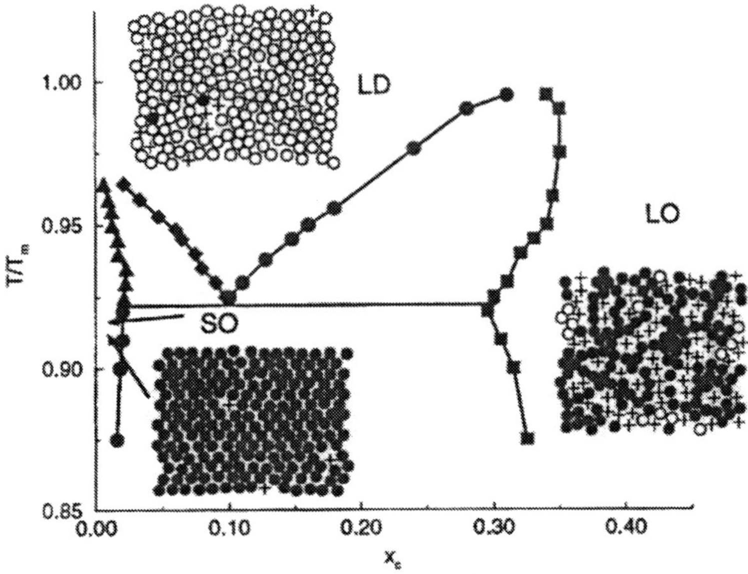

Figure 14. Lipid-cholesterol phase diagram. The inserts show typical microconfigurations for the three different phases, SO, LD, and LO (cf. Fig. 10b). Lipid molecules in ordered and disordered conformations are shown by solid and open circles, respectively. Cholesterol molecules are indicated by +. (Courtesy of Morten Nielsen.)

order and at the same time to stabilize the ordered lipid-chain conformation. Hence the effect of cholesterol is to decouple the translational and conformational degrees of freedom. Loose speaking, cholesterol can be said to renormalize the parameter V_0/J_0 in Fig. 10b and stabilize the liquid-ordered (LO) phase. The resulting phase diagram, derived by extensive use of histogram techniques, is shown in Fig. 14.

In the LO phase cholesterol strongly suppresses lipid-domain formation and lateral heterogeneity. However, at low concentrations where the cholesterol molecules predominantly influence the conformational degrees of freedom of the lipid molecules, cholesterol has the opposite effect. This can conveniently be studied by computer simulations on the simpler model in Eq. (7) [28]. The simulations show that the cholesterol molecules effectively become interfacially active and tend to promote the lipid-domain formation phenomena and consequently enhance the dynamic membrane heterogeneity. This effect is illustrated in Fig. 15 which shows the isothermal effects on a $DC_{16}PC$ lipid bilayer as cholesterol is introduced in increasing amounts up to 7.5 mol%. In this concentration range, the transitional behavior in the acyl–chain variables follows that of the narrow solid-ordered–liquid-

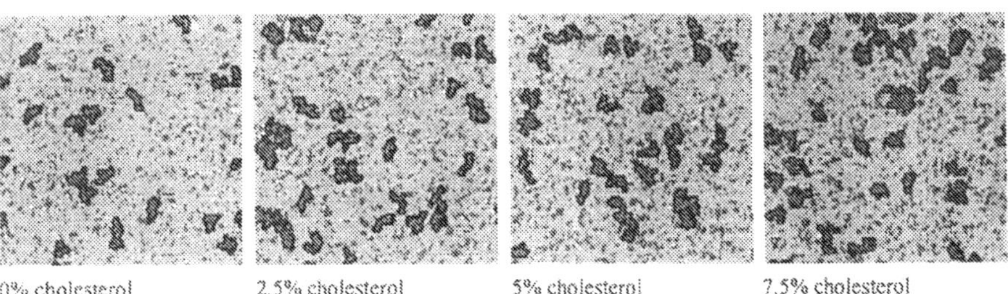

Figure 15. Typical lipid domain configurations in the fluid phase of $DC_{16}PC$-cholesterol mixtures at a temperature, $T = 319K$, in the fluid phase. Gel and fluid regions are denoted by grey and light regions, and the interfaces of the lipid domains are highlighted in black. The cholesterol molecules are denoted by $\circ$.

disordered coexistence region, cf. Fig. 14, and the three–phase line. The membrane configurations in Fig. 15 demonstrate that the domain formation is enhanced by the presence of cholesterol. At the same time cholesterol tends to accumulate at the domain interfaces. The strong enhancement of lipid fluctuations at low cholesterol contents has been confirmed by a number of experiments.

A number of chemical compounds that are used to modify membrane-associated physiological functions have a strong and often non-specific influence on the physical properties of lipid bilayers [30]. Examples include insecticides and drugs like anesthetics. It is likely that the potency of these agents is related to their capacity of altering the structure and the dynamics of the lipid bilayer, in particular the degree of order and heterogeneity. Figure 16 shows the effects of a water–soluble drug on the dynamic lipid domains in a $DC_{16}PC$ lipid bilayer in the transition region. The results are obtained from computer-simulation calculations on the model in Eqs. (9)-(11) in which the drug acts as an interstitial impurity which couples selectively to kink–like acyl–chain conformations. Since these conformations are generated predominantly at lipid-domain interfaces , the drug has a strong tendency to adsorb at the interfaces and to lower the interfacial tension. This in turn leads to a dramatic increase in the amount of interface and to a much more ramified interface structure, as seen in Fig. 16. Since the drug in this case leads to a freezing–point depression, the effect of the lipid order being destroyed by the drug is most pronounced in the gel phase. Concomitantly with the alterations in the interface morphology, there is a marked tendency for the drug molecules to accumulate at the interface. For the specific model calculation leading to Fig. 16 the local concentration

174

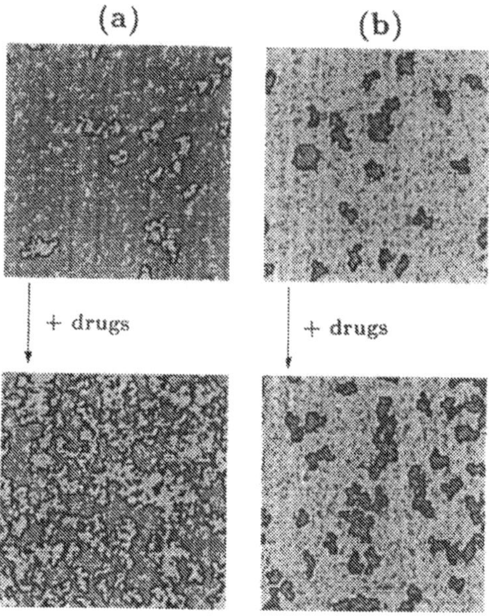

Figure 16. Effects on bilayer domain organization and order due to a water-soluble drug like halothane. Results are shown for a $DC_{16}PC$ bilayer at temperatures slightly below (a) and slightly above (b) the transition temperature of the pure lipid bilayer, $T_m = 314K$. The amount of drug dissolved in the bilayer varies significantly in the transition region. Gel and fluid regions are denoted by grey and light regions, and the interfaces of the lipid domains are highlighted in black.

of the drug in the interfaces can be several times larger than the global concentration.

7. Trans-bilayer structure

Although detailed information regarding the trans-bilayer structure can not be obtained on the basis of models of the type described in Sec. 5 above we shall for the sake of completeness briefly describe some computer-simulation results obtained from atomistic models. The trans-bilayer profile is the most well-characterized structural property of bilayers since it most easily lends itself to be monitored by e.g. X-ray and neutron-scattering techniques, magnetic resonance experiments, molecular-probe measurements, or molecular dynamics calculations. Details of the trans-bilayer structure can conveniently be obtained from molecular–dynamics calculations on atomistic models [13]. Studies of this type have revealed that the lipid bilayer in

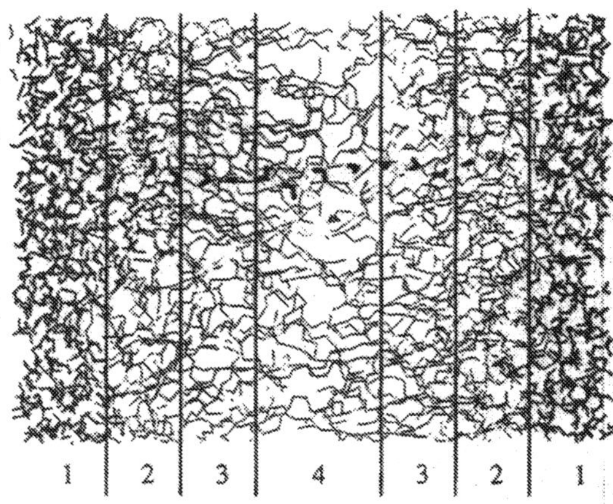

Figure 17. Trans-bilayer structure of a fluid $DC_{16}PC$ lipid bilayer as obtained from molecular dynamics calculations. The four structurally different regions are labeled (1) perturbed water, (2) hydrophilic–hydrophobic interphase involving the lipid polar–head groups, (3) soft–polymer–like region of ordered acyl–chain segments, and (4) a hydrophobic core with disordered acyl–chain segments. A possible path for the permeation of a water molecule is indicated. From Ref. [13]

its fluid phase, in contrast to the connotation of disorder implied by use of the term 'fluid', is a highly structured composite as illustrated schematically in Fig. 17.

This figure shows, in the case of a simple one-component $DC_{16}PC$ lipid bilayer, an extended structure involving (1) a layered region of perturbed (structured) water, (2) a hydrophilic–hydrophobic interphase, (3) a soft–polymer–like region of ordered acyl–chain segments, and (4) a hydrophobic core with disordered acyl-chain segments with a structure like fluid decane. The nature of this layered structure is of seminal importance for the permeability properties of the bilayer and for the way the bilayer interacts with drugs and peptides [30]. Together with experimentally determined density profiles, which determine the average positions of the different parts of the molecules, these data suggest the lipid-bilayer membrane to be a very fuzzy interface. Although the detailed nature of trans-bilayer profiles depends on the actual lipid species in question, the overall structural stratification is generic for aqueous lipid bilayers.

Since Fig. 17 is a snapshot in time it does not provide the full information about possible dynamical aspects of trans-bilayer structure. Within the time range currently accessible by full-fledged molecular dynamics calculations, it is found that the bilayer interface in the coarse of time is characterized by out-of-plane protrusion motions of the individual lipid molecules occurring on the time scale of tens of picoseconds. These motions imply that the bilayer structure, averaged over time, is fairly blurred with a substantial chemical heterogeneity in particular in layer 2. Hence the bilayer is rather rough on the molecular scale which may be important for hydration forces, peptide and protein drug transport across membranes, as well as interactions with foreign compounds, e.g. phospholipases.

8. Organization and structure of lipid bilayers with proteins

Once incorporated into lipid-bilayer membranes, amphiphilic trans-membrane polypeptides and integral membrane proteins interact with the stratified lipid bilayer structure in a complex way that both involves the liquid-crystalline degrees of freedom of the lipid molecules and the cooperative behavior of the bilayer. The presence of the proteins on the one hand modifies the lipid phase structure, the lateral molecular organization, as well as the trans-membrane profile. On the other hand the lipid structure influences the lateral organization of the proteins and sometimes their structure and functional behavior [8, 17].

One major effect of the incorporation of integral proteins and trans-membrane polypeptides into membranes is a dramatic change in the phase equilibria [17, 18, 20]. Usually, proteins are predominantly soluble in the fluid lipid-bilayer phase, and a protein-induced phase separation occurs in the gel phase. The presence of the proteins leads to structural changes in the adjacent lipid molecules as well as a modification of the local composition. One of the mechanisms proposed to relate protein-induced lipid-bilayer phase equilibria to the fundamental physical properties of the lipid-protein interfacial contact is based on a hydrophobic matching between the lipid bilayer and the hydrophobic region of the protein, cf. Fig. 5a. This hydrophobic matching concept has enjoyed considerable success in predicting phase diagrams for lipid bilayers reconstituted with specific proteins. We shall in this section describe some results obtained from computer-simulation calculations on the model in Eq. (12) which is based on the hydrophobic matching principle.

It may be anticipated that integral membrane proteins which, via a hydrophobic matching condition as illustrated in Fig. 5a, couple to the membrane lipid acyl-chain order (area density) and/or the membrane composition, are going to influence the degree of heterogeneity. Conversely, a certain degree of membrane heterogeneity will couple to the conformational state of the individual proteins as well as to the aggregational state of an ensemble of proteins.

8.1. SINGLE-PROTEIN EFFECTS

By performing computer-simulation calculations on Eq. (12) in the case of a protein of the size of bacteriorhodopsin in mixed lipid bilayers of $DC_{12}PC$–$DC_{18}PC$, it has been found [31] that a molecular sorting mechanism is operative at the protein surface. These simulations were carried out in parallel with an experiment using fluorescent-labeled lipid analogues [31]. By varying the temperature and composition across the binary lipid phase diagram, it was possible to change the effective hydrophobic length of the lipid molecules and hence the lipid-bilayer thickness. This in turn implies a change in the hydrophobic matching between the lipids and the hydrophobic surface of the protein. It was found as a consequence of this, that the lipid species, which under the conditions given best matches the protein, was enriched and selected at the lipid-protein interface. Another way of expressing this phenomenon is that the protein surrounds itself, in a statistical way, with a lipid annulus or a lipid domain predominantly composed by the lipid species which provides the best match. It should be noted that this annulus is a dynamic entity. The more pronounced the lateral bilayer heterogeneity is, i.e. the larger the lipid domains, the more pronounced is the molecular sorting. An example of a lipid-bilayer lateral profile around a protein, caused by this mechanism, is illustrated in Fig. 18 by data obtained from Monte Carlo simulations on the model in Eq.(12).

Profiles like the one shown in Fig. 18 can, at appropriate protein concentrations and profile coherence lengths, mediate the indirect protein-protein interactions to be discussed below.

Membrane proteins integrated into the lipid bilayer couple to the bilayer cooperative phenomena and modify the lipid phase equilibria. For example, at low protein concentrations, a protein-induced gel-fluid coexistence region occurs. Whereas small levels of integral membrane proteins will only have a minor effect on the thermodynamic (global) phase equilibria, it can be demonstrated by simulations in the case of bacteriorhodopsin in $DC_{14}PC$–$DC_{18}PC$ mixtures that the presence of the protein can have a marked effect on the small-scale membrane organization in terms of lipid domains and percolation structures in the gel-fluid coexistence region. At higher protein concentrations, larger protein clusters and aggregates may appear in conjunction with protein-rich phases.

8.2. LIPID–MEDIATED PROTEIN–PROTEIN INTERACTIONS: AGGREGATION AND CRYSTALLIZATION

In addition to specific short–range protein–protein interactions and direct protein–lipid interactions, a host of indirect lipid–mediated protein–protein interactions and lipid-controlled effects are operative in determining the

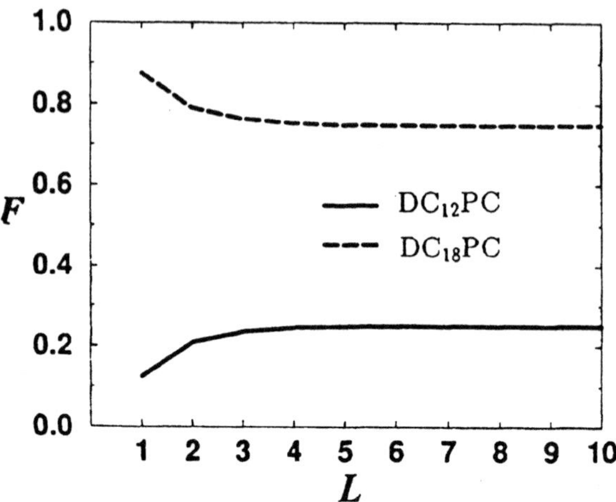

Figure 18. Compositional profile, $F(L)$, of lipid species in a 1:3 binary lipid mixture of DC$_{14}$PC (○) and DC$_{18}$PC (□) in the fluid phase measured as a function of the distance (in units of lipid acyl-chain diameters) from a large integral membrane protein that is hydrophobically best matched to the longer lipid. (Courtesy of dr. Maria M. Sperotto.)

nature of the organizational state of the lipid-protein assembly [17,18]. Whereas the specific forces usually are of enthalpic origin, many of the indirect and non-specific forces have a substantial entropic component that acts in addition to the omnipresent entropy of mixing.

One way of considering collective effects of lipid–protein interactions is based on molecular considerations. In this vein, the main idea uses the experimental observation that the proteins perturb the conformational order of neighboring lipid molecules. This idea involves details of the liquid-crystalline character of the lipid molecules. It implies that the protein couples to the lipid–acyl–chain order parameter which leads to a distinct profile of chain order around the protein, cf. Fig. 5a. An overlap of profiles from different proteins provides a mechanism for a lipid–mediated protein–protein interaction as illustrated schematically in Fig. 5b.

Another approach considers the membrane system as a continuum-mechanic elastic medium or a two–dimensional fluid and colloidal system. A fundamental characteristic of a membrane is that its surface area is much larger than the area of its cross section. Thus, it defines a quasi-two–dimensional embedding space for the proteins. This gives rise to a collective behavior that in many cases can be modeled by considering a two–dimensional system of colloids, in which the colloids model the proteins [18].

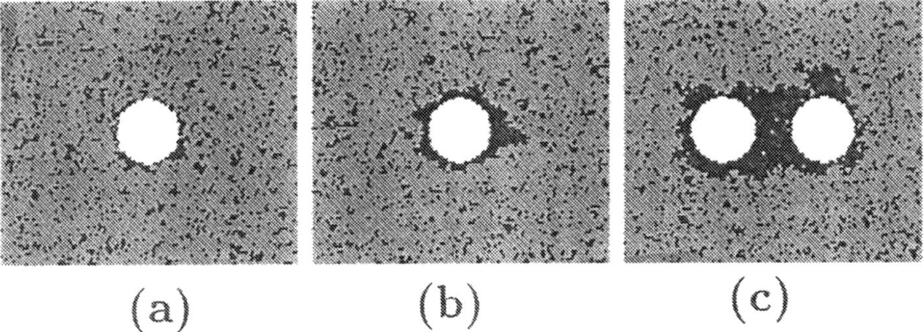

Figure 19. Computer simulated pictures of a large protein embedded in a binary lipid bilayer in which one of the lipid species exhibits preference for the protein surface. (a): Interfacial adsorption. (b): Wetting. (c): Wetting and formation of a capillary condensate between two adjacent proteins, leading to an attractive lipid-mediated joining force. (Courtesy of dr. Mads C. Sabra.)

When a membrane consists of two or more coexisting fluid (lipid) phases, the free energy associated with embedding a protein in the membrane can be lowered by surrounding it with the phase that allows for the lowest protein–lipid interaction potential. Such a preferential affinity of one of the phases to the proteins may stabilize this phase around the proteins even under conditions in which it is metastable in the protein–free membrane matrix. The preferred phase is said to wet the proteins, as illustrated in going from frame (a) to (b) in Fig. 19, where (a) is a case where only a standard molecular adsorption of the preferred species occurs at the protein surface. Due to the finite curvature of the proteins, the wetting layer in Fig. 19b only extends over a finite (non–macroscopic) distance but may well be beyond the scale of the individual lipid molecules. The possibility of a wetting layer being shared by two or more proteins (capillary condensation), as illustrated in Fig. 19c, gives rise to aggregation phenomena that reflect the interplay among several factors. These factors include the interactions that stabilize the wetting phase around the proteins, the tendency to minimize the extent of the wetting phase in the system because of its metastability, the tendency to minimize the extent of the interface between the wetting phase and the stable phases in the bulk, as well as various entropic effects. Thus, the wetting phenomenon occurs in addition to direct protein–protein (for example, van der Waals and electrostatic) and lipid–mediated interactions, predominantly in selecting the environment (membrane domains) in which they take place. Hence wetting gives rise to protein organization patterns on larger length scales.

180

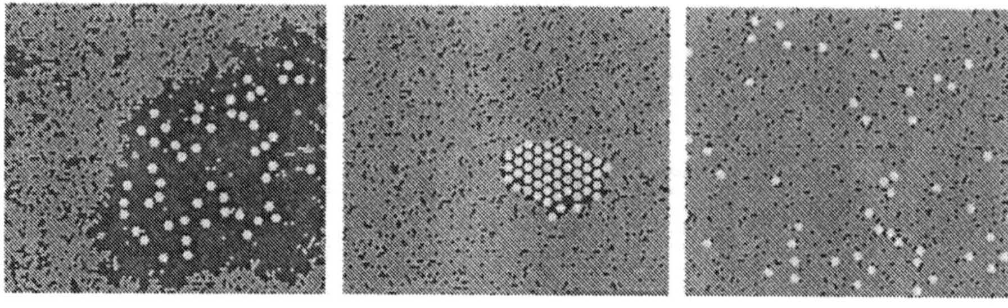

Figure 20. Computer simulated pictures of the organization of proteins in a binary lipid mixture in which one of the lipid species exhibits preference for the protein. (a): Formation of a capillary condensate of a protein rich phase of the preferred lipid. (b): Formation of a protein aggregate (crystallite). (c): A dispersion of proteins. The sequence of pictures (a)→(b)→(c) corresponds to a fixed composition and increasing temperature. (Courtesy of dr. Mads C. Sabra.)

Wetting and capillary condensation as means or protein organization in membranes, in particular protein aggregation and protein clustering, can conveniently be studied by computer–simulation calculation on specific microscopic models of lipid-protein interaction such as Eq. (12). In Fig. 20 we show results from a calculation based on the model for a binary lipid mixture of $DC_{14}PC$-$DC_{18}PC$ in Eq. (5). Wetting and capillary condensation were shown to lead to protein aggregation and to phase separation, in which one of the phases is rich in proteins, when approaching the thermodynamic conditions for a first-order fluid-fluid phase transition in the embedding membrane matrix. Results for lateral protein organization obtained from such model calculations are illustrated in Fig. 20. This figure shows, as a function of increasing temperature, the formation of (a) a capillary condensate of a protein-rich cluster/phase, (b) a densely packed protein array (two-dimensional protein crystal), and (c) a dispersion of proteins.

8.3. NON–EQUILIBRIUM ORGANIZATION OF ACTIVE PROTEINS

Most theoretical model studies as well as experimental investigations of protein organization in lipid bilayers and membranes invariably refer to thermodynamic equilibrium or at least make some attempt to achieve the conditions of equilibrium. This is naturally dictated by the fact that equilibrium thermodynamics is well–understood and its principles easily applied to phase equilibria and structure in membrane assemblies. This is a somewhat paradoxical situation, since membrane systems under physiological

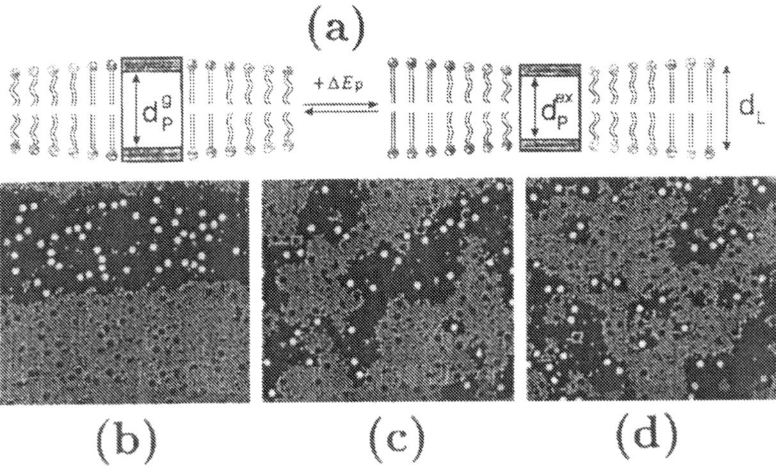

Figure 21. Computer simulated configurations of a binary bilayer incorporated with active proteins functioning according to the mechanism shown in (a). The activity is determined by a parameter Γ which denotes the strength of an external drive. (b): No drive (equilibrium gel-fluid phase separation). (c): Intermediate drive. (d): Strong drive. (Courtesy of dr. Mads C. Sabra.)

conditions operate away from thermodynamic equilibrium; there are fluxes of energy and matter through the system, and its operational state is organized in response to these fluxes and the associated dissipation of energy from the sources (proteins, enzymes, and receptors) into the lipid matrix and the aqueous compartments.

In an attempt to provide a simple setting for analyzing the consequences for lipid-protein organization under such non-equilibrium conditions, a theoretical membrane model was recently proposed and its properties in a steady-state were analyzed by computer-simulation techniques [32]. The model considers a binary lipid mixture incorporated with a trans–membrane protein, whose hydrophobic thickness could take on two different values, cf. Fig. 21a, corresponding to a ground state and an excited state, separated by an internal energy difference, ΔE_{P}. These two states, via the hydrophobic matching condition, were taken to have preference for one or the other lipid species. In equilibrium, the proteins in a given state would segregate out into that lipid phase which is enriched in the species for which the particular protein state exhibits preference. This is illustrated in Fig. 21b in the case of equal populations in the two protein states.

The system is brought out of equilibrium by driving the system externally at a rate Γ (taken on a scale relative to the diffusional motion of

182

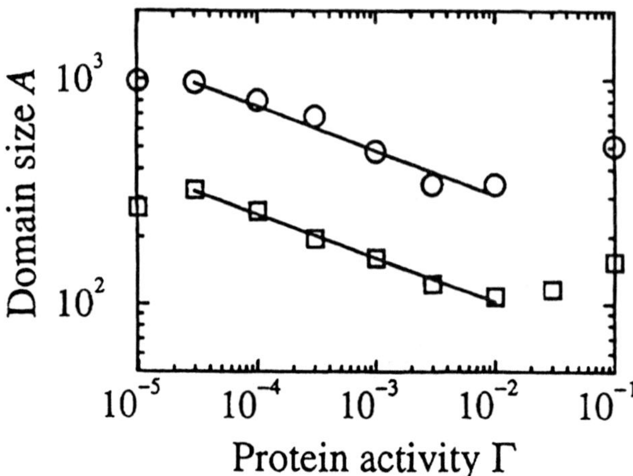

Figure 22. Average lipid domain size in binary lipid bilayer incorporated with active proteins subject to a drive of strength Γ.

the lipids). By this drive the proteins are excited from the ground state to the excited state, as schematized in Fig. 21a. De-excitation is controlled by interaction with the neighboring lipids. In steady-state, the organization of the membrane components is governed by the dissipation of the energy (provided by the external drive) into the lipid matrix and eventually into the aqueous phase that acts as a heat bath. The resulting steady-state lateral structures are shown in Fig. 21c and d. It is observed that the non-equilibrium condition introduces a new length scale, corresponding to a certain average lipid-domain area, $A(\Gamma)$, in the gel-fluid coexistence region of a binary lipid bilayer. As shown in Fig. 22, this domain size scales with the drive according to a power law, $A(\Gamma) \sim \Gamma^{-p}$. The value of p (~ 0.20) is found to be independent of temperature, protein size, as well as concentration. Hence it is believed to be a universal parameter. This type of steady-state compartmentalization of lipid-protein systems by active proteins remains to be investigated experimentally.

9. Lipid–membrane structure in relation to function

It now appears that the lipid bilayer and its structure and dynamics are much more important for membrane function that reflected in many textbooks on cell biology [2], and a substantially more refined model picture of the biological membrane is called for in order to make progress in our understanding of the relationship between membrane structure and function

[8]. We shall in this section provide two examples, one of a passive function and one of an active function, where computer simulations have been useful to unravel the possible relationship between lipid–bilayer structure and function.

9.1. PASSIVE BILAYER PERMEABILITY

One of the more dramatic consequences of strong density and compositional fluctuations in lipid bilayer membranes is the occurrence of enhanced passive trans-bilayer permeability of various species, such as ions, water, and larger amphiphilic species [33]. A general model for passive permeation of ions through lipid bilayers has been suggested which builds on the assumption that the passage predominantly takes place at interfaces between lipid domains formed in the bilayer due to dynamic lateral heterogeneity [33]. A computer simulation approach of the type described in Sec. 6.3 to the cooperative effects in lipid bilayers makes it possible to perform a quantitative estimate of the measure of heterogeneity and lipid-domain interface formation under different circumstances, e.g. near a phase transition and in the presence of other membrane components such as cholesterol, drugs, as well as polypeptides and proteins.

The bilayer configurations as presented in e.g. Fig. 12a suggest that the bilayer plane at any given time can be seen as partitioned into three regions: the bulk, the domains (clusters), and the interfaces bounding the two, corresponding to three fractional contributions to the total bilayer area, the fractional bulk area, $a_b(T)$, the fractional cluster area, $a_c(T)$, and the fractional interfacial area, $a_i(T)$, i.e. $a_b + a_c + a_i = 1$. The computer simulations using the Pink model for pure lipid bilayers, Eq. (1), show that $a_b(T)$ displays a minimum at the transition whereas both $a_c(T)$ and $a_i(T)$ have pronounced maxima. The interfacial region is found to be strongly dominated by chains in intermediate conformational states. This observation suggests that the interfaces are soft and that they act as a sink for excited acyl-chain conformations. The softness of the interfaces is of importance for prolonging the lifetime of the clusters and it is moreover crucial for the way in which other membrane components, such as cholesterol, protein, and drugs distribute themselves in the heterogeneous bilayer membrane structure.

For simplicity, we shall only describe the permeability model in the case of a pure lipid bilayer where only the temperature variation of the fractional areas is relevant. The basic assumption of the model is that the bulk and cluster regions of the bilayer each have their characteristic rates of trans-bilayer diffusion which are both considerably less than the diffusion rate in the interface. In Fig. 23 are shown some results for Na^+ permeation

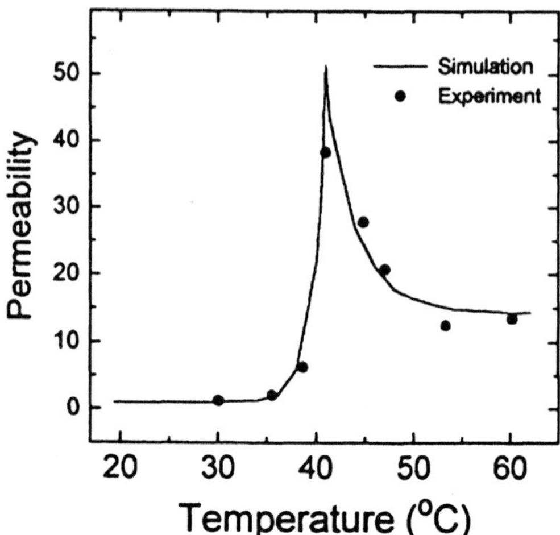

Figure 23. Relationship between passive trans-membrane relative permeability, $R(T)$, as determined by radioactive measurements of Na^+-ions (o) and computer simulations of the lipid membrane microstructure (—).

obtained from this permeability model using as an input the computer-simulation data for the temperature dependence for the different bilayer fractional areas mentioned above. It is straightforward to obtain similar predictions in the case of other ionic species. The data clearly demonstrate that there is a significant enhancement of the permeability when the acyl-chain length is decreased. The theoretically predicted passive ion perme-ability is in good agreement with experimental data for Na^+-permeation shown in Fig. 23.

9.2. ACTIVITY OF MEMBRANE-BOUND ENZYMES

Phospholipase A_2 (PLA$_2$) is an enzyme that hydrolyses phospholipids in aggregated form like lipid bilayers [34]. It has been suggested that so-called lipid-packing defects in the bilayer may promote high PLA$_2$ activity. Such packing defects may be related in a systematic way to the struc-tural microheterogeneity of the un-hydrolysed lipid bilayer as quantified from computer-simulation calculations on microscopic molecular interac-tion models of the type described in Sec. 6.3. Figure 24a shows the variation in the characteristic lag time, τ, for the typical PLA$_2$ reaction time course as a function of temperature and chain length for large unilamellar vesi-cles of DC$_{14}$PC, DC$_{16}$PC, and DC$_{18}$PC. τ is a well-defined experimental parameter that describes the rate of activation of the enzyme. The three

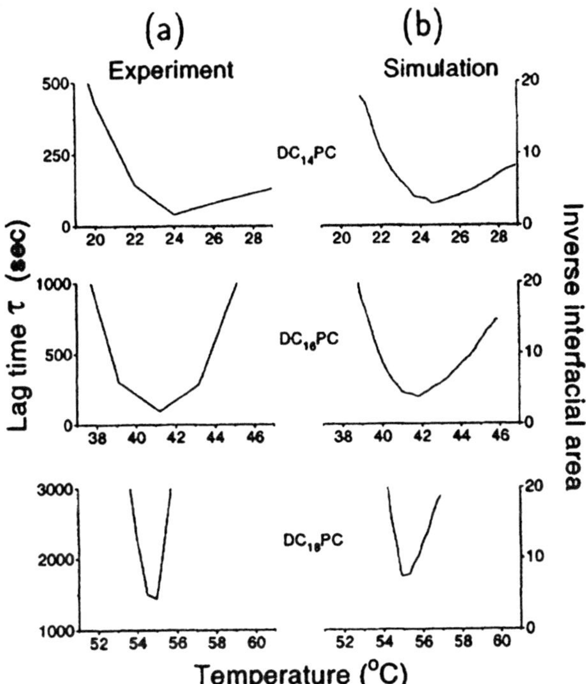

Figure 24. Relationship between phospholipase A₂ activity and lipid membrane microstructure. (a): phospholipase A₂ lag time curves. The lag time is inversely related to the activity of the enzymatic hydrolysis of unilamellar DC₁₄PC, DC₁₆PC, and DC₁₈PC liposomes. (b): Monte Carlo computer simulation of the temperature dependence of the inverse interfacial area of the heterogeneous lipid membrane structure. Results are shown for liposomes composed of DC₁₄PC, DC₁₆PC, and DC₁₈PC lipids.

curves of $\tau(T)$ each display a minimum at the respective gel-to-fluid phase transition temperature, T_{m}, which is deeper and wider the shorter the lipid chain is. This set of experimental data strongly indicates that the physical properties of the lipid bilayer substrate and the degree of microheterogeneity in the gel-to-fluid transition regions play important roles for the activity of PLA₂.

As visualized by the Monte Carlo computer simulation results in Figs. 12a and 12b, the closer to the phase transition and the shorter the lipid-chain length, the stronger the microheterogeneity. The degree of heterogeneity can readily be quantified as the amount of bilayer area occupied by the interfaces between gel and fluid lipid domains. Figure 24b shows the temperature- and lipid-chain-length dependence of the inverse interfacial area as obtained from simulations. The experimental data for $\tau(T)$

in Fig. 24a in combination with the simulated data for the inverse interfacial area in Fig. 24b demonstrate a strong semi-quantitative correlation between $\tau(T)$ and the degree of dynamic bilayer heterogeneity both with respect to temperature and lipid-chain-length variations. This correlation is valid for the width, the depth, and the asymmetry of the $\tau(T)$-function in the transition region.

The correlation between PLA_2 activity and dynamic bilayer microheterogeneity corroborates to the general hypothesis that the function of membrane-associated proteins can be modulated by collective physical properties of the lipid bilayer, specifically the dynamic bilayer heterogeneity in the nano-meter range.

Acknowledgments

Some of the work described in this paper was supported by the Danish Natural Science Research Council, the Danish Technical Research Council, as well as the Danish Medical Research Council under a center grant (Center for Drug Design and Transport). The author is an associate fellow of the Canadian Institute for Advanced Research.

References

1. Bloom, M. and Mouritsen, O.G. (1995) The evolution of membranes, in *Handbook of Biological Physics, Vol. 1.* Lipowsky, R. and Sackmann, E. (eds.), Elsevier Science B.V., Amsterdam, p. 65.
2. Mouritsen, O.G. and Andersen, O.S. (eds.) (1998) *Biol. Skr. Dan. Vid. Selsk.,* **49**, 1.
3. Bloom, M., Evans, E. and Mouritsen, O.G. (1991) *Q. Rev. Biophys.,* **24**, 293.
4. Mouritsen, O.G., Ipsen, J.H., Jørgensen, K., Sperotto, M.M., Zhang, Z., Corvera, E., Fraser, D.P. and Zuckermann, M.J. (1991) Computer simulation of phase transitions in Nature's preferred liquid crystal: the lipid bilayer membrane, in *Computer Simulation of Liquid Crystals.* Luckhurst, G. (ed.), Kluwer Academic Publishers, Amsterdam (still not published).
5. Safran, S.A. (1994) *Statistical Thermodynamics of Surfaces, Interfaces and Membranes.* Addison-Wesley Publ. Co., New York.
6. Cates, M.E. and Safran, S.A. (eds.) (1997) *Curr. Opin. Colloid. Interface Sci.,* **2**, 359.
7. Lasic, D.D. and Papahadjopoulos, D. (eds.) (1998) *Medical Applications of Liposomes.* Elsevier, Amsterdam.
8. Mouritsen, O.G. and Kinnunen, P.K.J. (1996) Role of lipid organization and dynamics for membrane functionality, in *Biological Membranes: A Molecular Perspective from Computation and Experiment.* Merz, K.M. Jr and Roux, B. (eds.), Birkhäuser, Boston, p. 463.
9. Mouritsen, O.G. and Jørgensen, K. (1997) *Curr. Opin. Struct. Biol.,* **7**, 518.
10. Merz, K.M. Jr and Roux, B. (eds.) (1996) *Biological Membranes: A Molecular Perspective from Computation to Experiment.* Birkhäuser, Boston.
11. Merz, K.M. Jr (1997) *Curr. Opin Struct. Biol.,* **7**, 511.
12. Tobias, D.J., To, K. and Klein, M.L. (1997) *Curr. Opin. Colloid. Interface Sci.,* **2**, 15.

13. Tieleman, D.P., Marrink, S.J. and Berendsen, H.J.C. (1997) *Biochim. Biophys. Acta,* **1331**, 235.
14. Mouritsen, O.G. (1990) Computer simulation of cooperative phenomena in lipid membranes, in *Molecular Description of Biological Membrane Components by Computer Aided Conformational Analysis.* Brasseur, R. (ed.), CRC Press, Boca Raton, Florida, p. 3.
15. Mouritsen, O.G., Dammann, B., Fogedby, H.C., Ipsen, J.H., Jeppesen, C., Jørgensen, K., Risbo, J., Sabra, M.C., Sperotto, M.M. and Zuckermann, M.J. (1995) *Biophys. Chem.,* **55**, 55.
16. Dammann, B., Fogedby, H.C., Ipsen, J.H., Jeppesen, C., Jørgensen, K., Mouritsen, O.G., Risbo, J., Sabra, M.C., Sperotto, M.M. and Zuckermann, M.J. (1996) Computer simulation of the thermodynamic and conformational properties of liposomes, in *Handbook of Non-medical Applications of Liposomes.* Lasic, D.D. and Barenholz, Y. (eds.), CRC Press, Boca Raton, Florida, p. 85.
17. Mouritsen, O.G. (1998) *Curr. Opin. Colloid. Int. Sci.,* **3**, 78.
18. Gil, T., Ipsen, J.H., Mouritsen, O.G., Sabra, M.C., Sperotto, M.M. and Zuckermann, M.J. (1998) *Biochim. Biophys. Acta,* **1376**, 245.
19. Mouritsen, O.G. (1991) *Chem. Phys. Lipids,* **57**, 179.
20. Mouritsen, O.G., Sperotto, M.M., Risbo, J., Zhang, Z. and Zuckermann, M.J. (1996) *Comp. Biol.,* **2**, 15.
21. Nielsen, M., Miao, L., Ipsen, J.H., Mouritsen, O.G. and Zuckermann, M.J. (1996) *Phys. Rev. E,* **54**, 6889.
22. Doniach, S. (1978) *J. Chem. Phys.,* **68**, 4912.
23. Zhang, Z., Mouritsen, O.G. and Zuckermann, M.J. (1993) *Mod. Phys. Lett. B* **7**, 217.
24. Ferrenberg, A.M. and Swendsen, R.H. (1988) *Phys. Rev. Lett.,* **61**, 2635.
25. Lee, J. and Kosterlitz, J.M. (1991) *Phys. Rev. B,* **43**, 3265.
26. Nielsen, M., Miao, L., Ipsen, J.H., Mouritsen, O.G. and Zuckermann, M.J. (1996) *Biochim. Biophys. Acta,* **1283**, 170.
27. Mouritsen, O.G. and Jørgensen, K. (1994) *Chem. Phys. Lipids,* **73**, 3.
28. Zuckermann, M.J., Ipsen, J.H. and Mouritsen, O.G. (1993) Theoretical studies of the phase behavior of lipid bilayers containing cholesterol, in *Cholesterol and Membrane Models.* Finegold, L.X. (ed.), CRC Press, Inc., Boca Raton, Florida, p. 223.
29. Zuckermann, M.J. (1998) *Biol. Skr. Dan. Vid. Selsk.,* **49**, 65.
30. Mouritsen, O.G. and Jørgensen, K. (1998). *Pharm. Res.,* **15**, 1507.
31. Dumas, F., Sperotto, M.M., Lebrun, M.-C., Tocanne, J.-F. and Mouritsen, O.G. (1997) *Biophys. J.,* **73**, 1940.
32. Sabra, M.C. and Mouritsen, O.G. (1998) *Biophys. J.,* **74**, 745.
33. Mouritsen, O.G., Jørgensen, K. and Hønger, T. (1995) Permeability of lipid bilayers near the phase transition, in *Permeability and Stability of Lipid Bilayers.* Disalvo, E.A. and Simon, S.A. (eds.), CRC Press, Boca Raton, Florida, p. 137.
34. Hønger, T., Jørgensen, K., Stokes, D., Biltonen, R.L. and Mouritsen, O.G. (1997) *Meth. Enzymol.,* **286**, 168.

FLOW PROPERTIES AND STRUCTURE OF ANISOTROPIC FLUIDS STUDIED BY NON–EQUILIBRIUM MOLECULAR DYNAMICS, AND FLOW PROPERTIES OF OTHER COMPLEX FLUIDS: POLYMERIC LIQUIDS, FERRO–FLUIDS AND MAGNETO–RHEOLOGICAL FLUIDS

SIEGFRIED HESS

Institut für Theoretische Physik,
Technische Universität Berlin, PN 7-1
Hardenbergstr. 36, D–10623 Berlin, Germany

Abstract. In the first part of this Chapter, the anisotropy of the viscosity and shear-induced structural changes of nematic liquid crystals are studied via Non-Equilibrium-Molecular-Dynamics (NEMD) simulations. The pecularities of the viscous behavior of nematics, the description of the anisotropy of the viscosity, and the physical meaning of the various viscosity coefficients associated with the names Ericksen, Leslie, and Miesowicz are presented. Methods for the computation of the viscosity by NEMD simulations of a plane Couette flow are firstly introduced for simple fluids, then the modifications and generalizations needed for the treatment of nematic liquid crystals are discussed. Results are presented for several molecular models. These are perfectly oriented, prolate and oblate, ellipsoidal particles, Gay–Berne particles, and r^{-12}–soft–spheres plus a r^{-6}–interaction with a P_2-anisotropy. The latter two types of model liquid crystals possess both nematic and smetic phases and allow the study of presmectic effects in the nematic phase. Furthermore, data are available for systems with variable degree of order: these are Gay–Berne particles, and partially stiff, partially flexible molecules composed of ten interaction sites. Shear–induced structural changes of the fluids are revealed by snapshots of configurations and by the static structure factor, presented in analogy to scattering experiments. The shear–induced transition from the smectic to the nematic phase is also analysed.

We then turn to the study of other complex fluids, in particular ferro–fluids, magneto– or electro–rheological fluids polymeric solutions and melts. In all these cases, non–newtonian effects, i.e. a pronounced dependence

189

P. Pasini and C. Zannoni (eds.), Advances in the Computer Simulations of Liquid Crystals, 189–233.

of the viscosity on the shear rate and normal stress differences are observed. Analogies between the viscosity of ferro–fluids and of discotic nematic liquid crystals as well as between that of magneto–rheological fluids and of discotics with columnar phases, in the presence of strong external fields, are pointed out. The nonlinear viscous behavior of polymeric liquids can be viewed as a self–induced anisotropy without any external orienting field. The interrelation between the rheological behavior and the structural changes in the streaming fluids is elucidated. For polymeric fluids, a comparison with neutron scattering experiments is made.

1. Introduction

The flow behavior of liquid crystals combines two properties typical for these materials: fluidity and anisotropy. O. Lehmann, who had invented the notion "liquid crystal" [1, 2], referred to the different viscosities in order to distinguish between substances which were later called "nematic", "cholesteric" and "smectic" liquid crystals. Measurements of the viscosity were already reported about a century ago by Schenck [3]. Application of a magnetic field, which is strong enough to suppress the flow orientation, renders the viscosity anisotropic. Experiments of that kind were performed by Miesowicz [4]. A phenomenological theoretical analysis of the viscous behavior and of the closely related flow alignment in nematics, was made about thirty years ago, by Ericksen and Leslie [5–9]. Since then, liquid crystals have gained great technological importance. Viscosities, together with elasticity coefficients, determine the speed with which LC–displays can be switched. Viscosity data have been reported for many substances [10]. The complete set of viscosity coefficients of nematics, however, has been measured just in a few cases [11, 12].

There have been numerous theoretical studies employing macroscopic, mesoscopic and microscopic approaches [13]. A good qualitative and reasonable quantitative understanding of the anisotropy of the viscosity and of the coefficients characterizing the flow alignment in nematic and in nematic discotic liquid crystals is obtained by the affine transformation model [14, 15]. This theory applies to perfectly oriented nematics. Following a previous study of the anisotropy of the diffusion [16], an extension of the theory to nematics with a realistic partial orientation has been put forward and tested [17]. Unified theories applicable to both the isotropic and nematic phases have been developed in the framework of irreversible thermodynamics [18–20] and by starting from (approximate) solutions of a Fokker-Planck equation for the orientational distribution function [21–28].

Non-equilibrium molecular dynamics (NEMD) computer simulations allow the computation of viscosity coefficients and provide insight into shear-induced structural changes of model fluids. For real substances, the corresponding structural information can be inferred from flow birefringence and from (light or neutron) scattering experiments. The first NEMD simulations of oriented nematics, composed of ellipsoidal particles whose interaction was obtained from a Lennard–Jones interaction by an affine transformation, were performed about 10 years ago [15]. The predictions of the affine transformation model for the anisotropy of the viscosity were confirmed. Later, the simulations were extended to nematic discotic systems [29]. The similarity between discotics and oriented ferro–fluids were pointed out [30]. In the meantime, the viscous properties of (slightly modified) Gay–Berne model systems, in the nematic phase, have been studied by NEMD simulations [31–33]. For the Gay–Berne system, the viscosity in the nematic phase has also been computed from MD simulations via the Green–Kubo relation [34]. Recently, attention was focussed on anomalies in the viscous behavior of nematics which are caused by presmectic effects. Some of the results, available for two model systems which show a nematic–smectic phase transition, have been reported in surveys covering NEMD studies of the flow behavior of simple and of selected complex fluids [35–37]. The behavior found seems to be even more complex than that reported for the real substances measured so far.

2. Description of the viscosity

2.1. GENERAL REMARKS, ISOTROPIC LIQUIDS

The viscosity characterizes the transport of linear momentum through a fluid. For a simple geometry where the flow velocity $\mathbf{v} = (v_x(y), 0, 0)$ is in the x-direction and where $\partial v_x/\partial y$ is the only nonvanishing component of the velocity gradient tensor, the (shear) viscosity is defined as the ratio of the component p_{yx} of the pressure tensor and shear rate $\Gamma = \partial v_x/\partial y$:

$$p_{yx} = -\eta \frac{\partial v_x}{\partial y}. \tag{1}$$

For a more general geometry, a tensorial description is needed. Here, cartesian components are denoted by Greek subscripts. The summation convention is used. The pressure tensor $P_{\nu\mu}$ occuring in the local conservation equation for the linear momentum $\rho dv_\mu/dt + \nabla_\nu P_{\nu\mu} = 0$, is the sum of its equilibrium value $P_{eq}\delta_{\nu\mu}$, where $\delta_{\nu\mu}$ is the unit tensor, and of the friction pressure tensor $p_{\nu\mu}$: $P_{\nu\mu} = P_{eq}\delta_{\nu\mu} + p_{\nu\mu}$. Like any second rank tensor, the latter quantity can be decomposed into its isotropic, antisymmetric and

192

symmetric traceless parts:

$$p_{\nu\mu} = p\,\delta_{\nu\mu} + \frac{1}{2}\epsilon_{\nu\mu\kappa}\,p_{\kappa}^{a} + \overline{\overline{p_{\nu\mu}}}\,, \tag{2}$$

where $p = (1/3)p_{\lambda\lambda}$ is one third of the trace of the friction pressure tensor. The antisymmetric part of the tensor involves the pseudo vector $p_{\lambda}^{a} = \epsilon_{\lambda\nu\mu}p_{\nu\mu}$, the quantity $\epsilon_{...}$ is the totally antisymmetric isotropic third rank tensor. With the help of this tensor, the cross product of two vectors $\mathbf{a}$ and $\mathbf{b}$ is given by $(\mathbf{a}\times\mathbf{b})_{\lambda} = \epsilon_{\lambda\nu\mu}a_{\nu}b_{\mu}$. The symbol $\overline{\overline{}}$ indicates the symmetric traceless part of a tensor, in particular:

$$\overline{\overline{p_{\nu\mu}}} = \frac{1}{2}(p_{\nu\mu} + p_{\mu\nu}) - \frac{1}{3}p_{\lambda\lambda}\,\delta_{\nu\mu}\,. \tag{3}$$

The velocity gradient tensor can be decomposed, in analogy to (2),

$$\nabla_{\nu}v_{\mu} = \frac{1}{3}\nabla_{\lambda}v_{\lambda}\,\delta_{\nu\mu} + \epsilon_{\nu\mu\kappa}\,\omega_{\kappa} + \gamma_{\nu\mu}\,. \tag{4}$$

The divergence $\nabla_{\lambda}v_{\lambda}$ of the velocity field vanishes for a volume conserving flow of a practically incompressible fluid. The quantity $\underline{\omega} = (1/2)\underline{\nabla}\times\mathbf{v}$ is the vorticity of the flow field . The symmetric traceless part of the velocity gradient tensor,

$$\gamma_{\nu\mu} = \overline{\overline{\nabla_{\nu}v_{\mu}}} = \frac{1}{2}(\nabla_{\nu}v_{\mu} + \nabla_{\mu}v_{\nu}) - \frac{1}{3}\nabla_{\lambda}v_{\lambda}\,\delta_{\nu\mu}\,, \tag{5}$$

is referred to as the deformation rate tensor.

The antisymmetric part of the pressure tensor is identically zero for substances composed of spherical particles without any internal rotational degree of freedom. Under most practical circumstances, in particular for stationary situations and in the absence of external fields, the antisymmetric part of the pressure tensor becomes zero for molecular fluids in the isotropic phase. As a consequence, the quantity $\mathbf{p}^{a}$ vanishes. Then, in the linear flow regime, two viscosity coefficients, viz. the above mentioned shear viscosity η and the bulk or volume viscosity η_{V}, suffice to characterize the viscous behavior of an isotropic fluid:

$$\overline{\overline{p_{\nu\mu}}} = -2\eta\,\overline{\overline{\nabla_{\nu}v_{\mu}}}\,, \quad p = -\eta_{V}\,\underline{\nabla}\cdot\mathbf{v}\,. \tag{6}$$

For an incompressible fluid, or for a volume conserving flow, one has

$$\underline{\nabla}\cdot\mathbf{v} = 0$$

Manifestations of the viscous behavior are forces acting on the walls in contact with streaming fluids, as well as friction forces and torques acting on objects moving in fluids. In addition to these phenomena, the amount of fluid, per time, which can be pushed through a pipe by a pressure difference, or the time it takes till a certain amount of fluid has flowed through a pipe, can be used to measure the viscosity. In particular, for a flow through a capillary with a rectangular cross section, thickness d, width $b >> d$, one has for the mass, per unit time, $\dot{M} = \frac{\rho}{12\eta}d^3 b\frac{\delta P}{L}$. Here ρ is the mass density, δP is the pressure difference driving the fluid through a pipe of length L. The corresponding expression for a pipe with circular cross section and diameter d is the Hagen-Poiseuille law $\dot{M} = \frac{\pi\rho}{128\eta}d^4\frac{\delta P}{L}$. The time it takes till a certain amount of fluid with mass M has flown through the capillary is $t_{flow} = M/\dot{M} \sim \eta$.

2.2. VISCOSITY OF ISOTROPIC AND LIQUID CRYSTALLINE PHASES

As a rule, in the cholesteric or smectic phase, liquid crystals are more viscous than in their isotropic phase occuring at higher temperatures. Nematic liquid crystals, in contradistinction, show a remarkable decrease of the viscosity with decreasing temperature in the vicinity of the isotropic–nematic phase transition. This was already noticed a century ago. In Fig. 1, the viscosity (in units of the viscosity of water at 0^oC, corresponding to $1.8 m\, Pa\, s$), as reported by Schenck in 1898 and 1905 [3], is displayed as function of the temperature (in centigrade). The left graph is for p-azoxyanisol (PAA) which undergoes the phase transition isotropic–nematic at $T_{ni} = 135.4^oC$. The right graph shows the data for 4,4'-bis (ethoxycarbonyl) azoxybenzene, the phase transition isotropic–smectic occurs at $T_{si} = 120^oC$. The peculiar dependence of the viscosity on the temperature of nematics will become obvious later, when the anisotropy of the viscosity and the flow alignment are discussed.

2.3. ALIGNMENT TENSOR, DIRECTOR AND ORDER PARAMETER OF NEMATICS

Liquid crystals are substances composed of nonspherical particles. The nematic phase is distinguished from the isotropic phase by the existence of a nonvanishing second rank alignment tensor , denoted by $a_{\mu\nu}$ or by $Q_{\mu\nu}$, which, apart from scalar factors, is defined by

$$a_{\mu\nu} = <\overline{u_\mu u_\nu}> . \tag{7}$$

Here $\mathbf{u}$ is a unit vector parallel to the figure axis of an effectively uniaxial particle. The brackets $< ... >$ indicate an average to be evaluated with an orientational distribution function. Phenomenologically, the alignment

194

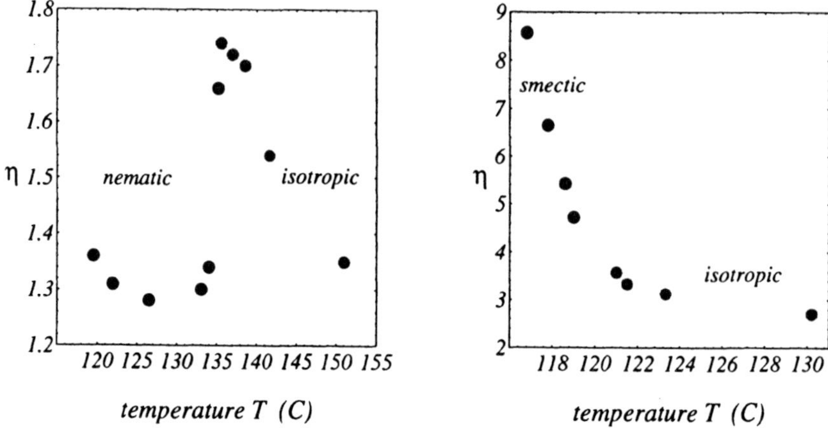

Figure 1. The viscosity, in units of the viscosity of water at $0°C$, for substances which undergo transitions from the isotropic to the nematic phase (left graph), and to a smectic phase (right graph), respectively, as functions of the temperature, in centigrade. The data were reported by Schenck. The gray circles mark the viscosity in the isotropic phase.

tensor can be introduced as a tensor proportional to the symmetric traceless part of the dielectric or the magnetic susceptibility tensor [7]. In equilibrium and outside the core of defects, the alignment of (ordinary) nematic liquid crystals is uniaxial. This means, the alignment tensor can be written as

$$< \overline{u_\mu u_\nu} >= S \, \overline{n_\mu n_\nu} \, , \quad S =< P_2(\mathbf{u} \cdot \mathbf{n}) >, \tag{8}$$

where $\mathbf{n}$, usually referred to as the "director", is a unit vector parallel to the (local) symmetry axis of the liquid crystal. The Maier-Saupe order parameter S, [38], is a measure for the degree of orientation, and it is given as an average of the second Legendre polynomial depending on the angle between the axis $\mathbf{u}$ of a molecule and the average preferential direction $\mathbf{n}$, [39]. Notice that the directions $\mathbf{n}$ and $-\mathbf{n}$ are physically equivalent. In the isotropic phase in equilibrium and in the absence of orienting external fields, one has $S = 0$. Under equivalent conditions, one finds values $S \approx 0.4$ in the nematic phase, close to the transition temperature isotropic–nematic. At lower temeratures, the order parameter increases to ≤ 0.8, before a transition to a smetic or a solid phase takes place. Perfect orientation corresponds to $S = 1$. Except at temperatures very close to the isotropic–nematic phase transition , the order parameter S is not affected by a flow. Thus, in a nematic, one treats S as constant. The director $\mathbf{n}$, however, is strongly affected by a flow due to torques caused by the velocity gradient. This situation, uniaxial alignment with constant order parameter, but variable director, is treated by the Ericksen–Leslie theory .

2.4. THE VISCOSITY COEFFICIENTS OF NEMATIC LIQUID CRYSTALS

For a nematic with the director field $\mathbf{n} = \mathbf{n}(t, \mathbf{r})$, and with corotational time derivative of the director denoted by $\mathbf{N}$:

$$N_\mu = \frac{d}{dt} n_\mu - \varepsilon_{\mu\nu\lambda}\omega_\nu n_\lambda \qquad (9)$$

the ansatz [7, 19]

$$
\begin{aligned}
-p_{\nu\mu} = \; & \alpha_1\, n_\nu n_\mu n_\lambda n_\kappa \gamma_{\lambda\kappa} + \alpha_2\, n_\nu N_\mu + \alpha_3\, n_\mu N_\nu \\
& + \alpha_4\, \gamma_{\nu\mu} + \alpha_5\, n_\nu n_\lambda \gamma_{\lambda\mu} + \alpha_6\, n_\mu n_\lambda \gamma_{\lambda\nu} \\
& + \zeta_1\, n_\lambda n_\kappa \gamma_{\lambda\kappa} \delta_{\mu\nu} + \zeta_2\, n_\nu n_\mu \nabla_\lambda v_\lambda + \zeta_3\, \nabla_\lambda v_\lambda \delta_{\mu\nu}
\end{aligned}
\qquad (10)
$$

is used. It is linear in the velocity gradient and in the corotational derivative of the director. The ansatz (10) is invariant under the replacement of $\mathbf{n}$ by $-\mathbf{n}$, as it should. The Leslie coefficients $\alpha_1, \ldots, \alpha_6$ and the coefficients ζ_1, ζ_2, ζ_3 have the dimension of a viscosity.

Decomposition of the pressure tensor and of the velocity gradient tensor into their isotropic, antisymmetric and symmetric traceless parts leads to the linear constitutive laws [19]

$$
\begin{aligned}
\overline{p_{\nu\mu}} \;&= -2\eta\, \gamma_{\nu\mu} - 2\tilde{\eta}_1\, \overline{n_\nu n_\lambda}\, \gamma_{\lambda\mu} \\
& \quad - 2\tilde{\eta}_2\, \overline{n_\nu N_\mu} - 2\tilde{\eta}_3\, \overline{n_\nu n_\mu}\, n_\lambda n_\kappa \gamma_{\lambda\kappa} \\
& \quad - \zeta_2\, \overline{n_\nu n_\mu}\, \nabla_\lambda v_\lambda \;,
\end{aligned}
\qquad (11)
$$

$$
p_\lambda^a \;= \epsilon_{\lambda\nu\mu} \left(\gamma_1\, n_\nu N_\mu + \gamma_2\, \overline{n_\nu n_\sigma}\, \gamma_{\sigma\mu} \right) \;,
\qquad (12)
$$

$$
\tfrac{1}{3} p_{\lambda\lambda} \;= -\eta_V\, \nabla_\lambda v_\lambda - \kappa\, n_\lambda n_\kappa \gamma_{\lambda\kappa} \;.
\qquad (13)
$$

The viscosity coefficients used in Eqs. (11 – 13) are related to those of Eq. (10) by

$$
\eta = \; \frac{1}{2}\left(\alpha_4 + \frac{1}{3}\left(\alpha_5 + \alpha_6 \right) \right) \;,
\qquad (14)
$$

$$
\tilde{\eta}_1 = \; \frac{1}{2}\left(\alpha_5 + \alpha_6 \right), \quad \tilde{\eta}_2 = \frac{1}{2}\left(\alpha_2 + \alpha_3 \right), \quad \tilde{\eta}_3 = \frac{1}{2}\alpha_1
\qquad (15)
$$

$$
\gamma_1 = \; \alpha_3 - \alpha_2, \quad \gamma_2 = \alpha_6 - \alpha_5 \;,
\qquad (16)
$$

$$
\eta_V = \; \frac{1}{3}\zeta_2 + \zeta_3, \quad \kappa = \zeta_1 + \frac{1}{3}\left(\alpha_1 + \alpha_5 + \alpha_6 \right) \;.
\qquad (17)
$$

The dyadic $\overline{\mathbf{n}\mathbf{n}}$ stems from the second rank alignment tensor. The coefficients linking second with first rank tensors, and second rank tensors with scalars (rank 0), obey Onsager symmetry relations . These are

$$
2\tilde{\eta}_2 = \gamma_2 \;, \quad \zeta_2 = \kappa \;.
\qquad (18)
$$

196

The first of these relations is equivalent to the Onsager-Parodi relation [6]

$$\alpha_2 + \alpha_3 = \alpha_6 - \alpha_5 . \tag{19}$$

Due to the symmetry relations, only 7 of the 9 viscosity coefficients are linearly independent. In an isotropic fluid all coefficients vanish except for the shear viscosity $\eta = \frac{1}{2}\alpha_4$ and the bulk viscosity $\eta_V = \zeta_3$. Positive entropy production requires $\eta > 0$, $\eta_V > 0$, $\gamma_1 > 0$ but $\tilde{\eta}_1$, $\tilde{\eta}_2$, $\tilde{\eta}_3$ and γ_2 may have either sign. One standard set of viscosity coefficients are the $\alpha_1, \dots, \alpha_6$ of Eq.(10). The coefficients occurring in the ansatz (11 – 13) can be preferable in connection with theoretical considerations. In experiments and in NEMD simulations, linear combinations of the "basic" viscosity coefficients are measured and calculated. Some examples are mentioned next.

In a plane Couette or a plane Poiseuille flow with the velocity in the x-direction and its gradient in the y-direction one has

$$\nabla_\nu v_\mu = \Gamma e_\nu^y e_\mu^x , \quad \omega_\mu = -\frac{1}{2}\Gamma e_\mu^z \tag{20}$$

with the shear rate

$$\Gamma = \frac{\partial v_x}{\partial y} , \tag{21}$$

$\mathbf{e}^{x,y,z}$ are unit vectors parallel to the x-,y-,z- coordinate axes.

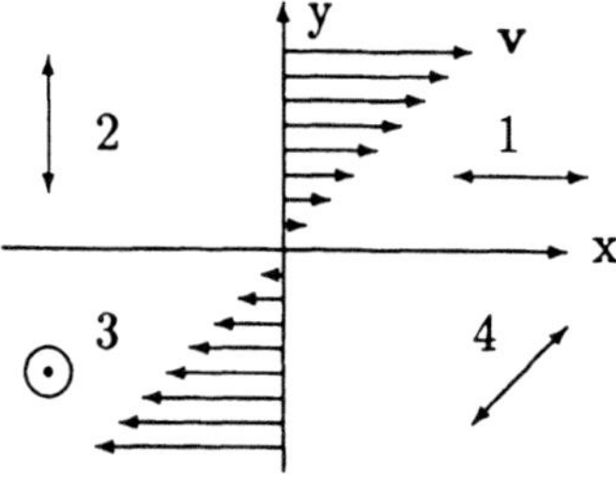

The Miesowicz orientations $1, 2, 3$ of the direc‑ tor, indicated by double arrows and by a circle with respect to the directions of the flow velocit $\mathbf{v}$ and its gradient, for the plane Couette geom‑ etry. Orientation 4 corresponds to the bisecto between directions 1 and 2.

Figure 2.   Orientation of the director for the Miesowicz viscosities η_1, η_2, η_3, and for η_4.

The Miesowicz viscosities η_i, $i = 1, 2, 3$ are defined as the ratio of the negative $yx-$component of the pressure tensor and the shear rate Γ:

$$p_{yx}^{(i)} = -\eta_i \Gamma . \tag{22}$$

The label $i = 1, 2, 3$ refers to the cases where the director $\mathbf{n}$ is parallel to the x-,y-,z-axes, respectively, cf. the following sketch for the Miesowicz di‑ rections, Fig.2. The orienting (magnetic) field has to be strong enough to overcome the flow induced orientation. A fourth coefficient η_4 analogous to

(22) with **n** parallel to the bisector between the x- and y-axes is needed to characterize the shear viscosity completely. Instead of η_4, the Helfrich viscosity coefficient $\eta_{12} = 4\eta_4 - 2(\eta_1 + \eta_2)$ is used in addition to the Miesowicz coefficients. The "rotational" viscosity γ_1 can be measured via the torque exerted on a nematic liquid crystal in the presence of a rotating magnetic field: Tsvetkov effect [39–41].

The four effective viscosities measurable in a flow experiment are related to the viscosity coefficients of Eq. (10) and Eqs. (11 – 13) by

$$
\begin{aligned}
\eta_1 &= \frac{1}{2}(\alpha_4 + \alpha_6 + \alpha_3) = \eta + \frac{1}{6}\tilde{\eta}_1 + \frac{1}{2}\tilde{\eta}_2 + \frac{1}{4}(\gamma_1 + \gamma_2)\,, \\
\eta_2 &= \frac{1}{2}(\alpha_4 + \alpha_5 - \alpha_2) = \eta + \frac{1}{6}\tilde{\eta}_1 - \frac{1}{2}\tilde{\eta}_2 + \frac{1}{4}(\gamma_1 - \gamma_2)\,, \\
\eta_3 &= \frac{1}{2}\alpha_4 = \eta - \frac{1}{3}\tilde{\eta}_1\,, \tag{23}
\end{aligned}
$$

$$
\eta_{12} = 4\eta_4 - 2(\eta_1 + \eta_2) = \alpha_1 = 2\tilde{\eta}_3\,. \tag{24}
$$

The antisymmetric part of the pressure tensor contributes to η_1 and η_2 but not to η_3 and η_{12}. Notice that

$$
\begin{aligned}
\eta_1 + \eta_2 + \eta_3 &= 3\eta + \frac{1}{2}\gamma_1, \tag{25} \\
\eta_1 - \eta_2 &= 2\tilde{\eta}_2 = \gamma_2, \tag{26} \\
\eta_1 + \eta_2 - 2\eta_3 &= \tilde{\eta}_1\,. \tag{27}
\end{aligned}
$$

As an example, viscosities measured by Kneppe, Schneider e.a. [11], for the isotropic and nematic phases of MBBA, are displayed in Fig. 3 as functions of the temperature (in $^\circ C$). The black circles, labelled by $1, 2, 3$ correspond to the Miesowicz viscosities. The gray circles, for $T > T_{ni} = 45.1^\circ C$, mark the viscosity in the isotropic phase. For $T < T_{ni}$, the gray circles indicate the computed average viscosity

$$
\eta_{av} = (\eta_1 + \eta_2 + \eta_3)/3\,. \tag{28}
$$

Clearly, there is a strong anisotropy of the viscosity in the nematic phase. The largest viscosity, that one where the director is parallel to the velocity gradient, is by a factor of almost 10 larger than the smallest one, where the director is parallel to the flow velocity. The typical order, for nematics composed of prolate particles, is

$$
\eta_1 < \eta_3 < \eta_2\,. \tag{29}
$$

Deep in the nematic phase, there is a strong increase of all viscosity coefficients with decreasing temperature. Relative viscosity coefficients, defined

198

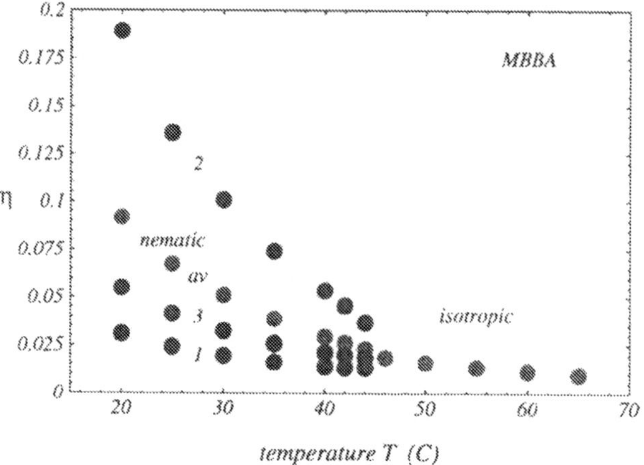

Figure 3. The viscosity in the isotropic phase and the Miesowicz viscosities in the nematic phase of MBBA as measured by Kneppe, Schneider, and coworkers.

as the ratio of the η_i to the average viscoity (28), on the other hand, approach limiting values for decreasing temperatures, as long as the system stays in the nematic phase, cf. the upper graph of Fig. 4. Such a limiting behavior is also seen for the coefficients γ_1, $-\gamma_2$ (gray), and $-\eta_{12}$, again in units of η_{av}, which are displayed in the lower graph of Fig. 4. The data for MBBA in the nematic phase, as measured in [11], are displayed as functions of the reciprocal temperature, more specifically of T_{ni}/T, where $T_{ni} = 318.3K$ is the isotropic–nematic transition temperature. The gray circles (labelled "free") in the upper graph mark the (computed) vicosity for a "free" flow in the absence of external orienting fields. The flow alignment and an expression for the relevant (effective) viscosity are discussed next. In substances with a greater nematic temperature range, like the nematic mixture N_4, the approach of limiting values of the relative viscosity coefficents, for low temperatures, is even more pronounced, cf. Fig. 9. There the relative Miesowicz viscosities , as reported in [11], are plotted as funtions of the relative temperature difference $1000K/T_{NI} - 1000K/T$, T_{NI} is the nematic-isotropic phase transition temperature.

2.5. EFFECTIVE VISCOSITY AND FLOW ALIGNMENT

For the stationary plane Couette flow , an effective viscosity η_{eff} , depending on the (arbitrary) orientation of the director, can be defined, in analogy to (22), as the ratio of $-p_{yx}$ and the shear rate. The constitutive laws (10)

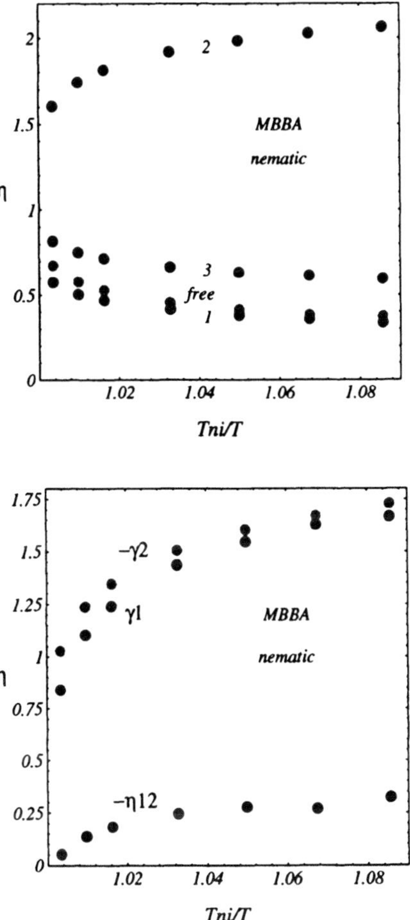

Figure 4. The Miesowicz viscosities (black circles), the viscosity for a free flow (gray), the Leslie coefficients and the Helfrich viscosity, all in units of the average viscosity η_{av}, for MBBA in the nematic phase, as functions of T_{ni}/T. The data are from measurements by Kneppe, Schneider and coworkers.

and (11, 12) lead to:

$$
\begin{aligned}
\eta_{eff} &= \frac{1}{2}\left(\alpha_4 + (\alpha_3 + \alpha_6)\, n_x^2 + (\alpha_5 - \alpha_2)\, n_y^2\right) + \alpha_1\, n_x^2 n_y^2 \\
&= \frac{1}{2}(\eta_1 + \eta_2) + \frac{1}{2}(\eta_1 - \eta_2)\,(n_x^2 - n_y^2) \\
&\quad + \left(\eta_3 - \frac{1}{2}(\eta_1 + \eta_2)\right)\, n_z^2 + \eta_{12}\, n_x^2 n_y^2. \tag{30}
\end{aligned}
$$

Here, n_x, n_y, n_z are the cartesian components of the director. For the special orientations discussed above, the viscosity coefficients $\eta_i, i = 1 - 4$, are recovered. As before, $\underline{\nabla} \cdot \mathbf{v} = 0$ has been assumed.

The antisymmetric part of the pressure tensor is associated with a torque exerted on the liquid crystal by an external field which fixes the orientation of the director. The component relevant for the Couette geometry is

$$p_z^a = p_{xy} - p_{yx} = \gamma_1 \left(n_x \frac{\partial n_y}{\partial t} - n_y \frac{\partial n_x}{\partial t} \right) + \frac{1}{2} \Gamma \left(\gamma_1 \left(n_x^2 + n_y^2 \right) + \gamma_2 \left(n_x^2 - n_y^2 \right) \right) .$$

$$(31)$$

Thus, one has, in a stationary situation, and for the Miesowicz orientations 1 and 2,

$$p_{xy}^{(1)} - p_{yx}^{(1)} = \frac{1}{2} \Gamma \left(\gamma_1 + \gamma_2 \right) , \quad p_{xy}^{(2)} - p_{yx}^{(2)} = \frac{1}{2} \Gamma \left(\gamma_1 - \gamma_2 \right) . \tag{32}$$

In contradistinction to real experiments, all components of the pressure tensor can be determined in computer simulations. Consequently, (32) can be used to compute the Leslie coefficients γ_1, γ_2 for a fluid with fixed orientation of the director.

In the absence of external fields, and far enough away from walls which might impose preferential orientations of the director, the torque balance requires that the antisymmetric part of the pressure tensor vanish. Assuming that $\mathbf{n}$ is in the xy-plane, the ansatz $n_x = \cos\varphi, n_y = \sin\varphi$ is made. Then $p_z^a == 0$, cf. (31), leads to

$$\gamma_1 \frac{\partial \varphi}{\partial t} + \frac{1}{2} \Gamma \left(\gamma_1 + \gamma_2 \cos 2\varphi \right) = 0 . \tag{33}$$

Provided that $\gamma_1 < |\gamma_2|$, a stationary solution of (33) exists. The pertaining angle $\varphi_{flow} = \varphi$, with φ from the stationary solution, is independent of the shear rate Γ. More specifically, the flow alignment angle is determined by

$$\cos 2\varphi_{flow} = -\gamma_1/\gamma_2 . \tag{34}$$

Notice that $\gamma_2 < 0$ for nematics composed of prolate particles. The angle φ_{flow} can be inferred from birefringence measurements, in particular, from the orientation of the principal axes of the dielectric tensor. Typical values for the flow alignment angle are between 20 degrees, at higher temperatures, and 5 degrees, at lower temperatures. In some substances, however, the situation $\gamma_1 > |\gamma_2|$ is realized at lower temperatures. This case, where no stationary flow alignment exists, is referred to as the tumbling regime. There, the dynamics is more intricate and it is not discussed here.

Insertion of the director into the effective viscosity (30), for the case where stationary flow alignment exists, yields the viscosity η_{free} of a nematic undergoing a "free" flow:

$$\eta_{free} = \frac{1}{2} \left(\eta_1 + \eta_2 - \gamma_1 \right) + \frac{1}{4} \eta_{12} \left(1 - (\gamma_1/\gamma_2)^2 \right) . \tag{35}$$

This expression has been used to compute the free viscosity shown in Fig. 4. Now we understand the surprising temperature dependence of the viscosity of a nematic liquid crystal in the vicinity of the isotropic–nematic phase transition, cf. the left diagram of Fig. 1. In the nematic phase, the flow alignment puts the director into an orientation close to that one where one has the smallest viscosity. Due to the large anisotropy of the viscosity, this effective viscosity is still considerably smaller than the viscosity $\eta = \eta_{av} - (1/6)\gamma_1$ cf. (25), which is approximately continuous across the phase transition.

2.6. APPLIED ORIENTING FIELD

In the presence of an external magnetic field $\mathbf{B} = B\mathbf{b}$, with $B = |\mathbf{B}|$, the torque balance reads $\mathbf{n} \times \mathbf{p}^a = \mu_0^{-1}\chi_a B^2 \mathbf{n} \times \mathbf{b}\, \mathbf{n} \cdot \mathbf{b}$. Here $\chi_a = \chi_\parallel - \chi_\perp$ is the difference between the magnetic susceptibilty for the cases where the field is parallel and perpendicular, respectively, to the director $\mathbf{n}$. When the magnetic field lies in the xy-plane and encloses the angle φ_0 with the x-axis, (33) is replaced by

$$\gamma_1 \frac{\partial\varphi}{\partial t} + \frac{1}{2}\Gamma\left(\gamma_1 + \gamma_2 \cos 2\varphi\right) + \frac{1}{2}\mu_0^{-1}\chi_a B^2 \sin 2(\varphi - \varphi_0) = 0. \qquad (36)$$

Two applications of this equation are mentioned. Firstly, consider a situation, where no flow occurs and where the field which determines the initial direction of $\mathbf{n}$ is suddenly changed to a new direction which corresponds to $\varphi_0 = 0$. Then the solution of (36) yields, for $\chi_a > 0$, $\varphi(t) = \arctan(\exp(-t/\tau_B))$ with the effective relaxation time $\tau_B = \mu_0^{-1}\chi_a B^2/\gamma_1$. The approach of the director to its new equilibrium direction can be used to measure the rotational viscosity coefficient γ_1, [39–41]; the same applies to the decay of dynamic gratings observed by optical methods, [42–44]. Secondly, consider a stationary Couette flow with a field applied in the y-direction. Then one has $\Gamma(\gamma_1 + \gamma_2 \cos 2\varphi) = \mu_0^{-1}\chi_a B^2 \sin 2\varphi$. With $c_0 = -\gamma_1/\gamma_2$ and $\tilde{B}^2 = \mu_0^{-1}\chi_a B^2/(\gamma_1\,\Gamma)$, the solution of this equation is

$$\cos 2\varphi = c(\tilde{B}) = c_0\left(1 - \tilde{B}^2\sqrt{1 - c_0^2 + c_0^2\tilde{B}^4}\right)/(1 + c_0^2\tilde{B}^4).$$

Insertion of the director into the effective viscosity (30) now yields the viscosity $\eta_{2,eff}$:

$$\eta_{2,eff}(\tilde{B}) = \frac{1}{2}(\eta_1 + \eta_2) + \frac{1}{2}(\eta_1 - \eta_2)\,c(\tilde{B}) + \frac{1}{4}\eta_{12}\left(1 - c(\tilde{B})^2\right). \qquad (37)$$

For $B = 0$, the free flow viscosity (35) is recovered. The Miesowicz viscosity η_2 is reached for $\tilde{B}^2 \gg 1$, i.e. for $\mu_0^{-1}\chi_a B^2 \gg \gamma_1\,\Gamma$, where $c \to -1$.

202

Again, $\chi_a > 0$ is assumed. A fit of the effective viscosity for intermediate values of the applied field can be used to determine the three viscosity coefficients η_1, η_2, and η_{12} from a single geometry. Preferential orientations in the vicinity of walls have not been taken into consideration here.

2.7. NORMAL PRESSURE EFFECTS

The constitutive law (11) also yields normal pressure differences and a change of the trace of the pressure tensor, linear in the shear rate Γ. For the plane Coutte geometry studied here, one has

$$p_{xx} - p_{yy} = \Gamma\, n_x n_y \left(\gamma_2 - \eta_{12} \left(n_x^2 - n_y^2 \right) \right), \tag{38}$$

$$p_{zz} - \frac{1}{2}(p_{xx} + p_{yy}) = \Gamma\, n_x n_y \left(\eta_1 + \eta_2 - 2\eta_3 + \frac{1}{2}\eta_{12} \left(1 - 3n_z^2 \right) \right), \tag{39}$$

$$\frac{1}{3}p_{\lambda\lambda} = -\Gamma\,\kappa\, n_x n_y. \tag{40}$$

All these effects vanish for the Miesowicz orientations $1, 2, 3$, but give nonzero contributions in the direction 4. In computer simulations, these relations can provide a consistency check on some of the viscosity coefficients, and (40) is needed to determine the coefficient κ. The existence of the "transverse" pressure $p_{xx} - p_{yy}$ has been been demonstrated experimentally [45].

Before NEMD simulations for the computation of the viscosity coefficients of liquid crystals are discussed, some remarks on simulations for simple fluids are in order.

3. NEMD for fluids of spherical particles

3.1. GENERAL REMARKS

Molecular dynamics (MD) computer simulations devoted to the study of nonequilibrium (NE) phenomena such as transport and relaxation processes in fluids have been developed about twenty-five years ago [46–48]. The "NEMD"-method is now well established . Reviews, conference proceedings, and some keynote references are [35–37, 49–79]. Books devoted to this topic are [80–84].

One of the first applications of NEMD was the calculation of the viscosity and the analysis of structural changes as revealed by the flow-induced distortions of the pair correlation function. Simulations intended to mimic simple fluids, which commonly are considered as prototypes of "newtonian fluids", showed a "non-newtonian", i.e. a shear–rate dependent, viscosity. Other rheological properties such as normal pressure differences, commonly observed in dense colloidal dispersions, in surfactant solutions and in polymeric liquids, have also been found in simulations of simple fluids subjected to high shear rates. In the study of the material properties of complex fluids, where the phenomenology is much richer, the computational challenges are greater than for simple fluids. Here, firstly the simulation of plane Couette flow and results for simple fluids are reviewed briefly. Then results are presented for liquid crystals (LC). Other complex fluids, in particular anisotropic fluids such as ferro-fluids, magneto- or electro-rheological (MR or ER) fluids are discussed later. Some results are also presented for polymeric liquids.

3.2. BASICS OF MOLECULAR DYNAMICS

In a molecular dynamics computer simulation for a substance composed of N spherical particles, Newton's equations of motion

$$m\,\frac{d^2}{dt^2}\,\mathbf{r}^i = \mathbf{F}^i = \sum_j \mathbf{F}^{ij} \tag{41}$$

are integrated numerically. The particles are located at positions $\mathbf{r}^i$ in a volume V. The particle density is $n = N/V$. The particle i, $(i = 1, 2, ..., N)$, feels the force $\mathbf{F}^i = \sum_j \mathbf{F}^{ij}$, which is the sum of the forces $\mathbf{F}^{ij}$ exerted by all other particles $j \neq i$ on particle i. When one wants to avoid surface effects, periodic boundary conditions and the nearest image convention are used. This means particle i either feels the force caused by particle j or by one of its images depending on which one is closest to it.

The temperature T of the system is linked with the part of the kinetic energy K which is not associated with a macroscopic motion:

$$\frac{3}{2}(N - 1)\,k_B\,T = K := \frac{1}{2}\sum_i m(\mathbf{c}^i)^2, \tag{42}$$

where k_B is Boltzmanns constant and $\mathbf{c}^i$ is the "peculiar velocity", i.e. the velocity of a particle relative to the flow velocity $\mathbf{v} = \mathbf{v}(\mathbf{r}^i)$. The center of mass of the system of N particles is constant. Thus the number of degrees of freedom, in three dimensions, is $3(N - 1)$ rather than $3N$.

In order to simulate an isothermal system the temperature has to be kept constant. The simplest version of a "thermostat" involves rescaling the peculiar velocity after each time step by the factor $(T_{wanted}/T_{measured})^{1/2}$. Other thermostats, e.g. those referred to as "Gaussian" and "Nosé-Hoover" [80], are imposed as constraints .

3.3. EXTRACTION OF DATA FROM MD SIMULATIONS

The observables of interest, such as the internal energy and the components
of the pressure or the stress tensor, can be calculated from the known positions and velocities of the particles as time averages according to the rules
of statistical physics. Typically, in a stationary state, the required data are
extracted after each 10th to each 100th time step. Similarly, more detailed
information can be obtained from the simulation, such as the velocity distribution function, the pair correlation function or the static structure factor,
which can also be measured in scattering experiments.

In ordinary molecular dynamics (MD) simulations, data are extracted
when an equilibrium state with a specified density n (or pressure) and
temperature T has been reached. In addition, dynamic quantities can be
extracted from the temporal fluctuations, e.g. transport coefficients can be
calculated from time correlation functions with the help of Green-Kubo
relations . In NEMD simulations, on the other hand, relaxation and transport phenomena are investigated more directly and in close analogy to real
experiments. Also, states far away from equilibrium are studied. In the
following, a stationary plane Couette flow is considered in more detail.

3.4. INTERACTION POTENTIAL AND SCALING

In the simulations, dimensionless or "scaled" variables are used which are
denoted by the same symbols as the physical variables when no danger of
confusion exists. For a system of particles whose forces are derived from
Lennard–Jones-(LJ)-potential Φ depending on the distance r

$$\Phi = \Phi^{LJ} = 4\Phi_0 \left(\left(\frac{r_0}{r} \right)^{12} - \left(\frac{r_0}{r} \right)^6 \right) , \qquad (43)$$

lengths and energies are presented in units of the diameter r_0 and of the
potential depth Φ_0. The units used for the particle density and for the
temperature are r_0^{-3} and $k_B^{-1} \Phi_0$. The time is scaled with the reference
time $t_0 = r_0 m^{1/2} \Phi_0^{-1/2}$, m is the mass of a particle. The pressure, the
shear rate and the viscosity of the LJ-fluid are expressed in units of $r_0^{-3}\Phi_0$,
t_0^{-1} and $r_0^{-3}\Phi_0 t_0 = r_0^{-2} m^{1/2}\Phi_0^{1/2}$. For many fluids composed of atoms or
small molecules, the specific values of r_0 and Φ_0 are available which are
needed to relate the theoretical results to the thermophysical properties of
specific substances. In the simulations, the cutoff of the interaction at a
finite distance r_c is often achieved just by putting the potential and the
force equal to zero for $r > r_c$, e.g. with $r_c = 2.5r_0$. A smoother cutoff,
where the force is (at least) continuous at r_c, however, is preferable for the
integration of the equations of motion. The "LJ-spline"-potential of Holian

and Evans [57] is an example of a modified LJ-potential with an improved cutoff.

When only the repulsive r^{-12}-part of the LJ interaction potential is taken into account, one speaks of a "soft spheres" (SS) potential. Often, this potential is written as $\Phi = \Phi_0^{SS}(\frac{r_0}{r})^{12}$ without the factor 4 of (43). When reduced SS–units with Φ_0^{SS} as energy reference are used, $T = 1/4$ in SS–units is equivalent to $T = 1$ in LJ–units. The LJ-potential cut off at its minimum $r_c\, r_0^{-1} = 2^{1/6} \approx 1.123$, which is also purely repulsive, is referred to as the WCA-potential [85].

3.5. PLANE COUETTE FLOW

For a simple shear flow in the x-direction with the gradient in the y-direction, the shear rate Γ is given by $\Gamma = \frac{\partial v_x}{\partial y}$. Such a flow can be generated either by moving boundaries or forces [46, 48], [69], or as used here, by moving image particles undergoing an ideal Couette flow with the prescribed shear rate (homogeneous shear). Let the flow be switched on at $t = 0$. Then, at time t the image particles above (below) the basic (central) box have moved in the x-direction to the right (left) by the distance $\Gamma t L$ modulo(L), where L is the length of the periodicity box in the y-direction. Of course, the periodic boundary conditions for the particles leaving and entering the basic box have to be modified (Lees-Edwards boundary conditions [47, 81]). For a system in a fluid state in equilibrium and for not too large shear rates, a linear velocity profile typical for a plane Couette flow is set up in the basic box (from which the data are extracted). At high shear rates where also plug-like flow occurs it is essential to use a velocity "profile unbiased thermostat" (PUT, [53, 71]). A shear flow can also be generated by modifying the equations of motion (SLLOD, [80–82]). For a recent review of NEMD results for rheological properties of simple and of complex fluids see [35].

3.6. PRESSURE TENSOR, VISCOSITY

Rheological properties such as the (non-newtonian) viscosity and the normal pressure differences are obtained from the cartesian components of the stress tensor $\sigma_{\mu\nu} = -p_{\mu\nu}$ or of the pressure tensor $p_{\mu\nu}$ which is the sum of "kinetic" und "potential" contributions:

$$p_{\mu\nu} = p_{\mu\nu}^{kin} + p_{\mu\nu}^{pot}, \qquad (44)$$

$$V p_{\mu\nu}^{kin} = \sum_i m_i c_\mu^i c_\nu^i, \quad V p_{\mu\nu}^{pot} = \frac{1}{2} \sum_{ij} r_\mu^{ij} F_\nu^{ij}. \qquad (45)$$

Here $\mathbf{c}^i$ is the peculiar velocity of particle i, i.e. its velocity relative to the flow velocity $\mathbf{v}(\mathbf{r}^i)$, $\mathbf{r}^{ij} = \mathbf{r}^i - \mathbf{r}^j$ is the relative position vector of particles i, j and $\mathbf{F}^{ij}$ is the force acting between them. As before, the Greek subscripts μ, ν, which assume the values $1, 2, 3$, stand for cartesian components associated with the x, y, z-directions. In the simulations, the expression for the pressure tensor given is averaged over many (10^3 to 10^6) time steps. For the present flow geometry, the (non-newtonian) viscosity η is obtained by dividing the yx(21)-component of the stress or pressure tensor by the shear rate Γ: $\eta = \sigma_{yx}/\Gamma = -p_{yx}/\Gamma$.

From the simulation, the kinetic and potential contributions to the pressure tensor and to the viscosity can be computed separately. Only the sum can be measured in a real experiment. The kinetic contribution to the viscosity dominates in dilute gases [68]. In dense fluids (liquids) the potential contribution is more important. As an example, data are presented for a

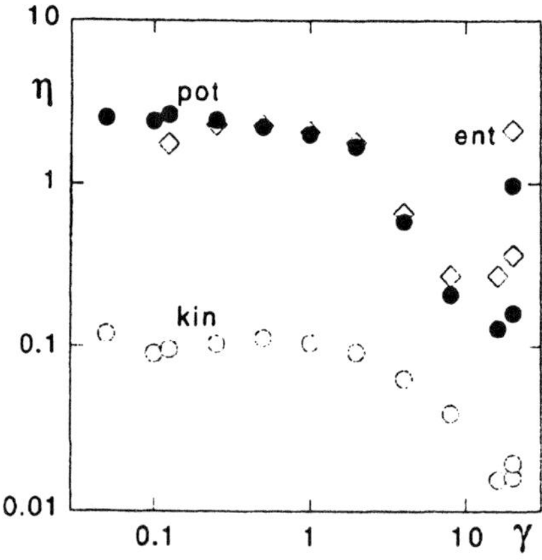

Figure 5. The kinetic (kin, open circles) and the potential (pot, closed circles) contributions to the viscosity $\eta = -p_{yx}/\Gamma$, as well as the total viscosity inferred from the entropy production (ent, diamonds) for a LJ-fluid as functions of the shear rate Γ for the density $n = 0.84$ and the temperature $T = 1$. All quantities are in LJ–units.

LJ-liquid at the density $n = 0.84 r_0^{-3}$, which corresponds to the triple point density, and the temperature $T = \Phi_0/k_B$, which is somewhat higher than the triple point temperature. The simulation was performed with $N = 512$ particles [73], and the interaction potential was cut off at $r = r_c = 2.5 r_0$. In Fig. 5 the kinetic and potential contributions to the viscosity, as obtained from the components of the pressure tensor are displayed as functions of the shear rate. The total viscosity, as it follows from the entropy production, which is proportional to $\eta\,\Gamma^2$ and is determined via the heat removed from the system by the thermostat, is also shown.

Normal stress or pressure differences, e.g. $\sigma_{xx} - \sigma_{yy} = p_{yy} - p_{xx}$, have been computed. At small shear rates one finds $-p_{yx} \sim \Gamma$ and $p_{yy} - p_{xx} \sim \Gamma^2$, as well as $p_{xx} + p_{yy} - 2p_{zz} \sim \Gamma^2$.

3.7. STRUCTURAL CHANGES

The shear rate dependence of the viscosity as displayed in the "flow curve" Fig. 5 shows four regimes: (I) The *newtonian flow* regime, where the shear viscosity η is independent of the shear rate Γ and where normal pressure differences practically vanish. In the present case, the newtonian regime corresponds to $\Gamma < 0.1$ (in LJ units). (II) A *weak shear thinning* for $0.2 < \Gamma < 2$. (III) A *strong shear thinning* for $2 < \Gamma < 20$. (IV) A *shear thickening* for $\Gamma > 20$.

These qualitative differences of the flow behavior are linked with different flow induced structural changes in the fluid. In regimes (I) and (II) these can be seen in the pair correlation function $g(\mathbf{r})$ or equivalently, in its spatial Fourier transform, the static structure factor $S(\mathbf{k})$ which determines the scattering intensity. Both quantities become anisotropic in the presence of a viscous flow. The structure factor shows distorted Debye-Scherrer rings. The Stokes-Maxwell ideas for the distortion of the structure were tested in one of the first NEMD simulations [46]. For the extension to the nonlinear flow regime cf. [86]. Some basic features of this approach will be presented below. In regime (III) a long–range partial positional ordering takes place which is apparent in real space and it is evident in snapshots [60, 61]. Of course, the long–range ordering is also seen in $g(\mathbf{r})$ and it leads to Bragg-like peaks in $S(\mathbf{k})$ [67, 68, 71, 87].

Above, the various flow regimes have been distinguished by the shear rate expressed in LJ-units. The physically relevant variable, however, is the product $\Gamma\tau$ of the shear rate and the Maxwell relaxation time τ which, in turn, is given by the small shear rate limit of the ratio η/G, i.e. of the viscosity and the (high frequency) shear modulus G. The latter quantity, which can also be extracted from the simulation, is approximately 25 for the present system and the relaxation time is $\tau \approx 0.1$ in LJ units. Thus non-newtonian flow phenomena can be observed for $\Gamma > 0.1\tau^{-1}$. In simple fluids like liquid Argon this corresponds to a shear rate which is several orders of magnitude larger than the $10^6 s^{-1}$ which can reasonably be reached in laboratory experiments. The situation is different in (dense) colloidal dispersions of spherical particles. There, considerably longer relaxation times occur and non-newtonian effects can be observed and are of importance for many applications.

208

The potential part of the pressure tensor is also determined by an integral over the pair correlation function g:

$$P_{\nu\mu}^{pot} = \frac{1}{2}n^2 \int r_\nu F_\mu\, g(\mathbf{r})\, d^3r\,. \tag{46}$$

Here $\mathbf{F}$ is the force acting between two particles. The relation (46) is the basis for the strong dependence of the viscous properties on the structure of a fluid.

3.8. GERERALIZED STOKES–MAXWELL MODEL

The structural changes in the flow regimes I and II can be treated by starting from a Kirkwood–Smoluchowski type of kinetic equation [86, 88, 89] for the pair correlation function $g = g(\mathbf{r})$:

$$\partial g/\partial t + \Gamma\, r_y\, \partial g/\partial r_x + \mathcal{D}(g) = 0\,. \tag{47}$$

Here $\mathbf{r}$ is the relative position vector between an arbitrary reference particle and any other particle in the fluid. The "damping" term $\mathcal{D}(g)$ ensures that g approches the equilibrium pair correlation function g_{eq}, which is also referred to as radial distribution function. With the relaxation time approximation

$$\mathcal{D}(g) = \tau^{-1}\left(g - g_{eq}\right), \tag{48}$$

where τ is the Maxwell relaxation time, and for a stationary situation, the kinetic equation is equivalent to

$$g = g_{eq} - \Gamma\tau\, r_y\, \partial g/\partial r_x\,. \tag{49}$$

Iteration of this equation leads to a power series in $\Gamma\tau$, the first few terms are

$$g = g_{eq} - \Gamma\tau\, r_x\, r_y\, r^{-1} g'_{eq} + (\Gamma\tau)^2 \left(r_y^2\, r^{-1} g'_{eq} + r_x^2\, r_y^2\, r^{-1}(r^{-1}g'_{eq})'\right) - .. + ... \tag{50}$$

Here the prime denotes differentiation with respect to $\dot r = |\mathbf{r}|$. Note that g_{eq} depends on r but not on the angles specifying the direction of the vector $\mathbf{r}$. The term linear in $\Gamma\tau$ of (50) corresponds to a relation suggested by Stokes and by Maxwell. The power series expansion (50) is referred to as the generalized Stokes–Maxwell model [86].

According to the linear Stokes–Maxwell relation, the function $g_+ = g_{45} - g_{135}$, i.e. the difference between the pair correlation functions g_{45} and g_{135} along the lines in the x-y plane which enclose the angles of 45° and 135° with the x-axis, is given by $-\Gamma\tau\, r g'_{eq}$. In Fig. 6, this relation is tested for a Lennard–Jones liquid at the same temperature and density as used for the data presented in Fig. 5. The shear rate $\Gamma = 0.125$ is in the linear flow regime. The relaxation time τ has been put equal to 0.14. The large gray dots mark the values of g_+ as directly extracted from the simulation, the small black dots connected by lines are the same quantity computed via the Stokes–Maxwell relation. Though the agreement is far from perfect, the essential features are described rather well.

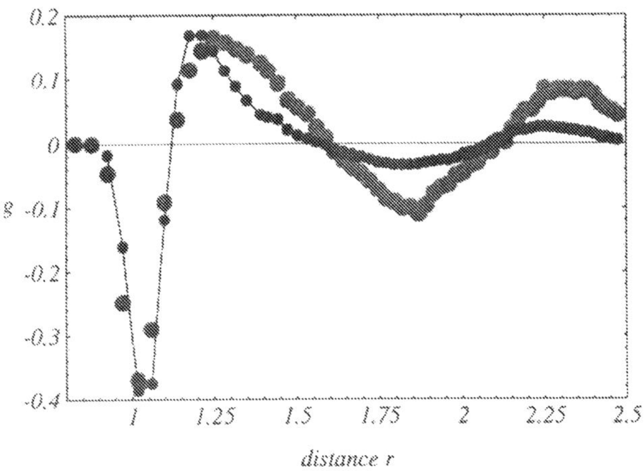

Figure 6. Test of the Stokes–Maxwell relation for a Lennard—Jones liquid in the linear flow regime. Data directly obtained in the NEMD simulation (large gray dots) are compared with those computed via the Stokes–Maxwell relation (small black dots connected by lines). The density, temperature and shear rate are $n = 0.84$, $T = 1.0$, $\Gamma = 0.125$ (in LJ units), the relaxation time $\tau = 0.14$ was used.

The Stokes–Maxwell ideas imply that the (potential contribution) to the viscosity is proportional to the relaxation time τ. More specifically, insertion of the pair correlation function (50) into (46), yields $p_{yx}^{pot} = -\eta^{pot}\,\Gamma$, in lowest order in the shear rate, with

$$\eta^{pot} = G\,\tau. \tag{51}$$

This relation has already been used above. The (high frequency) shear modulus G is given by

$$G = -\frac{1}{2}n^2 \int y\,(\partial\Phi/\partial x)\,y\,(\partial g_{eq}/\partial x)\,d^3r = \frac{1}{2}n^2 \int y^2 \partial^2\Phi/\partial x^2\,g_{eq}\,d^3r, \tag{52}$$

where x, y stand for r_x, r_y, and Φ is the binary interaction potential. Expression (52) also applies to nonspherical interaction potentials and anisotropic fluids, provided that an additional average over the orientation of a pair of particles is performed. For spherical interaction and a system in an isotropic

210

state, integration over the angles implies that y^2 and $x^2 y^2$ can be replaced by $(1/3)r^2$ and $(1/15)r^4$, respectively, and then (52) reduces to the Born–Green expression [90] $G = (1/30) \int r^{-2}(r^{-1}\Phi')' \, d^3r$.

Common fluids are composed of molecules which are modelled either as stiff nonspherical particles with angle dependent interaction or as assemblies of interaction sites, held together by extra binding forces or by constraints. Examples will be considered in the remaining parts of the Chapter.

4. Nematic liquid crystals

4.1. ELLIPSOIDS WITH PERFECT ORIENTATION

The anisotropic interaction potential $\Phi_A(\mathbf{r})$ of two (equal) ellipsoidal particles which are parallel to each other is related to the interaction Φ between spherical particles by an affine transformation of coordinates. Here $\mathbf{r}$ is the vector joining the centers of gravity of the ellipsoids. The mapping between the potentials is

$$\Phi_A(\mathbf{r}) = \Phi(\mathbf{r}^A), \tag{53}$$

where the vectors $\mathbf{r}^A$ and $\mathbf{r}$ are linked by

$$r_\mu^A r_\mu^A = r_\mu A_{\mu\nu} r_\nu, \tag{54}$$

or

$$r_\mu^A = A_{\mu\nu}^{1/2} r_\nu. \tag{55}$$

The transformation matrix $A_{\mu\nu}$ with positive eigenvalues describes the mapping of a sphere, determined by $\mathbf{r}^A \cdot \mathbf{r}^A$ in $\mathbf{r}^A$-space, onto an ellipsoid in $\mathbf{r}$-space. It is advantageous to choose the transformation matrix such that the volume of the ellipsoid is equal to that of the sphere.

For ellipsoids of revolution with axis ratio Q and their figure axis parallel to the unit vector $\mathbf{u}$, the transformation matrix and its inverse are

$$A_{\mu\nu} = Q^{2/3} \left(\delta_{\mu\nu} + (Q^{-2} - 1)u_\mu u_\nu \right), \quad A_{\mu\nu}^{-1} = Q^{-2/3} \left(\delta_{\mu\nu} + (Q^2 - 1)u_\mu u_\nu \right). \tag{56}$$

In the perfectly oriented system, the molecular axis is parallel to the director, thus $\mathbf{u} = \mathbf{n}$. The eigenvalues of $A_{\mu\nu}$ are $A_\parallel = Q^{-4/3}$ and $A_\perp = Q^{2/3}$. For spheres, one has $Q = 1$. Prolate and oblate particles (cigars and discs) correspond to $Q > 1$ and $Q < 1$, respectively. For $Q = 7/3 \approx 2.33$, the cross section through the sphere (dashed) and the corresponding ellipsoid, are displayed in the left diagram of Fig. 7. The vertical axis is parallel to the figure axis of the ellipsoid of revolution. In the simulations [15], two interaction potentials have been studied for the spherical potential of the reference fluid, viz. r^{-12}-soft spheres (SS) and Lennard–Jones (LJ) particles. The nonspherical potentials obtained by the affine transformation are

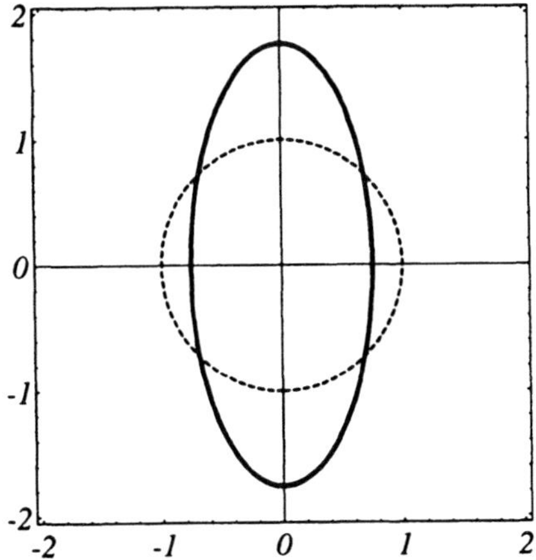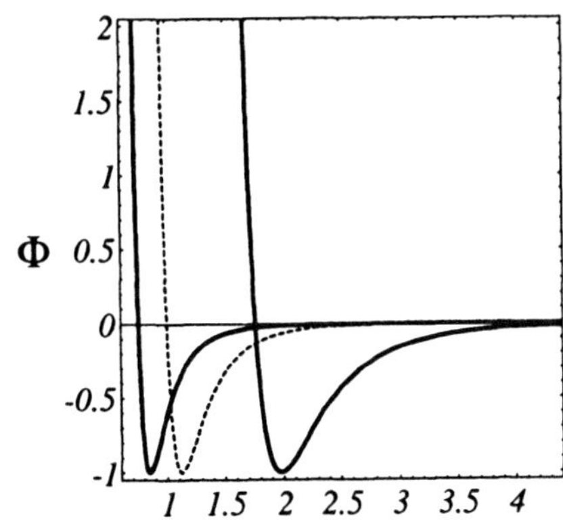

Figure 7. The mapping of a sphere (dashed) unto an ellipsoid with axes ratio $Q = 7/3$ (left diagram) and the LJE potential, as function of the distance r (right graph), for the side–side and end–end directions. For comparison, the LJ potential of spheres is also shown (dashed).

referred to as soft ellipsoids (SE) and Lennard—Jones ellipsoids (LJE). In the right graph of Fig. 7, the LJE potential is plotted as function of the distance r for the side–side ($\mathbf{r} \cdot \mathbf{u} = 0$) and the end–end ($\mathbf{r} \cdot \mathbf{u} = r$) orientations. The value $Q = 7/3$ was chosen. For comparison, the dashed curve shows the LJ potential corresponding to $Q = 1$.

All nematic viscosity coefficients, except for the bulk viscosity, have been calculated in [15], for particular state points. The prescriptions (45), which also apply to fluids of nonsherical particles, were used to compute the relevant components of the kinetic and potential contributions of the pressure tensor. The viscosity coefficients, then were obtained as discussed in the previous section. The symmetry relation (26) has been tested. Before simulation data are presented, the predictions of the affine transformation model [14] are stated. These results, calculated by a more general approach [15], could also be inferred from the Stokes–Maxwell model, as discussed above, when it is applied to the ellipsoidal particles, and when one assumes that the relaxation time is not affected by the affine transformation.

The affine transformation model relates the various viscosities of the nematic fluid to the axes ratio Q and the viscosity η_{ref} of a reference fluid of spherical particles with the same density and at the same temperature. In particular, one has

$$\eta_1^{pot} = Q^{-2}\eta_{ref}^{pot}, \quad \eta_2^{pot} = Q^2\eta_{ref}^{pot}, \quad \eta_3^{pot} = \eta_{ref}^{pot},$$

$$\eta_{12} = -(Q - Q^{-1})^2\eta_{ref}^{pot}, \quad \kappa = 0, \tag{57}$$

212

$$\gamma_1 = (Q - Q^{-1})^2 \eta_{ref}^{pot}, \quad \gamma_2 = (Q^{-2} - Q^2)\eta_{ref}^{pot}, \tag{58}$$

The bulk viscosity is not affected by the affine transformation.

The simulations were performed, for the SE fluid at the density $n = 0.6$ and the temperature $T = 0.25$, in reduced units. For the LJE liquid, the state point $n = 0.6, T = 1.15$ was chosen. The interaction was cut off at 2.5 in the $\mathbf{r}^A$-space, which means a cut off along an ellipsoidal equipotential surface in $\mathbf{r}$-space. With the computational power available over a decade ago, relatively small systems with $N = 128$ particles were used. However, it had been demonstrated before, for fluids of spherical particles, that even simulations with 108 particles yield reliable results for the viscosity. The equations of motion were integrated with a fifth–order Gear predictor–corrector method, and the temperature was kept constant by rescaling the magnitude of the velocities. In the left graph of Fig. 8, the potential contributions to

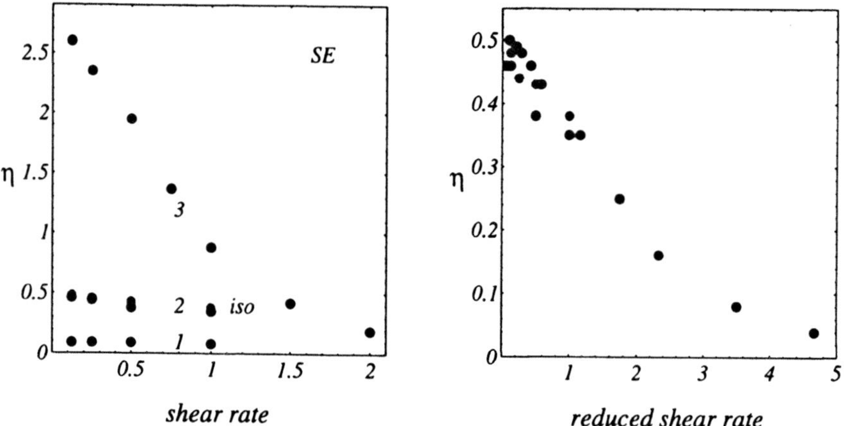

Figure 8. The potential contributions to the Miesowicz viscosities as functions of the shear rate (left) for the SE fluid, and the reduced viscosities as functions of the reduced shear rate. The gray circles mark the viscosity the reference fluid of spherical particles.

the Miesowicz viscosities of the SE fluid are displayed as functions of the shear rate, all quantities being in reduced units. The gray circles mark the viscosity of the reference fluid of spherical particles. The affine transformation also affects the velocity gradient. This this implies that all viscosities shown here should be on one master curve, when the viscosities and shear rates are scaled by factors, depending on the axes ratio Q. In particular, one has

$$\eta_1^{pot}(\Gamma) = Q^{-2}\eta_{ref}^{pot}(Q^{-1}\Gamma), \quad \eta_2^{pot}(\Gamma) = Q^2\eta_{ref}^{pot}(Q\Gamma), \quad \eta_3^{pot}(\Gamma) = \eta_{ref}^{pot}(\Gamma). \tag{59}$$

for the three Miesowicz directions. In the right graph of Fig. 8, these scaled data, e.g. $\eta^* = Q^2\eta_1^{pot}$ as function of $\Gamma^* = Q^{-1}\Gamma$ for orientation 1, etc., are presented together with the viscosity of the reference fluid η_{ref}^{pot}. Within the computational accuracy, the points indeed seem to fall onto (an unknown) master curve. As in an isotropic liquid, the kinetic contribution to the viscosity is smaller than the potential contribution. There are strong differences for the different Miesowicz orientations, however. Whereas the ratio kinetic to potential contribution is about 1/20 and 1/5 for orientations 2 and 3, it is just 1/2 for orientation 1, where one has the smallest viscosity. For the comparison with experiments, the total viscosities, the sum of potential and kinetic contributions are needed in the newtonian limit, i.e. for small shear rates.

The NEMD results, both for the SE and LJE systems, are presented in Tab. 4.1 and compared with the predictions of the affine transformation model, cf. Tab. 2. The viscosities, computed for the reference SS fluid and the LJ liquid, at the corresponding state points are $\eta_{iso} = 0.52 \pm 0.08$, and $\eta_{iso} = 0.8 \pm 0.3$, respectively.

TABLE 1. The potential contributions to the Miesowicz viscosities η_1, η_2, η_3, the Helfrich viscosity η_{12}, the Leslie viscosities γ_1, γ_2 and the coupling coefficient κ of the SE fluid at $n = 0.6, T = 0.25$, and of the LJE nematic at $n = 0.6, T = 1.15$, in SS and LJ units, respectively. In both cases, the axes ratio is $Q = 7/3$. The data are inferred from the NEMD simulations, in the limit of small shear rates.

	η_1	η_2	η_3	η_{12}	γ_1	γ_2	κ
SE	.09 ± .03	2.8 ± 0.2	.52 ± .10	-1.8 ± 0.3	1.9 ± 0.3	-2.7 ± 0.3	.0 ± .3
LJE	.15 ± .04	3.5 ± 0.3	.70 ± .10	-2.1 ± 0.4	2.2 ± 0.4	-3.5 ± 0.4	.0 ± .4

For a comparison with experimental data, viscosities in units of the average viscosity η_{av} are used. In Tab. 3, predictions of the affine transformation model and simulation results for the LJE nematic are compared

TABLE 2. Comparison of ratios of viscosity coefficients of the SE fluid and the LJE nematic with the predictions of the affine transformation model (AFT). The NEMD data are for the full viscosities, the AFT expressions involve the potential contributions only.

	η_3/η_1	η_2/η_3	η_{12}/γ_1	η_3/η_{iso}	κ
AFT	5.44	5.44	−1.00	1.00	0.00
SE	5.8 ± 0.4	5.4 ± 0.4	−0.9 ± 0.4	1.0 ± 0.2	0.0 ± 0.3
LJE	4.7 ± 0.5	5.0 ± 0.5	−1.0 ± 0.4	1.0 ± 0.2	0.0 ± 0.4

with the corresponding relative viscosities of $MBBA$ and of N_4, at specific temperatures. Good qualitative and also reasonable quantitative agreement

TABLE 3. Comparison of the predictions of the affine transformatiom model and LJE–NEMD results for relative viscosity coefficients, with the corresponding experimental data for $MBBA$ and N_4, as measured by Kneppe, Schneider, and coworkers.

	η_1/η_{av}	η_2/η_{av}	η_3/η_{av}	η_{12}/η_{av}	γ_1/η_{av}
AFT	0.08	2.46	0.45	-1.64	1.64
LJE (NEMD)	0.15 ± 0.07	2.33 ± 0.3	0.52 ± 0.1	-1.12 ± 0.4	1.40 ± 0.2
$MBBA(30^\circ C)$	0.38	1.99	0.63	-0.28	1.55
$N_4(50^\circ C)$	0.32	2.14	0.54	-0.68	1.75
$N_4((33 \pm 1)^\circ C)$	0.34	2.00	0.66	-1.21	2.00

is seen. Of course, the viscosity coefficients of real nematics with partial orientational order depend on the degree of order, e.g. expressed by the Maier–Saupe order parameter S. Motivated by the success in the comparison of MD-data for the anisotropy of the diffusion in model fluids with variable degree of orientation with a modified affine transformation model [16], similar considerations have been made for the viscosity coefficients [17]. In particular, η_{12} is reduced by a factor $S_4 =< P_4(\mathbf{u} \cdot \mathbf{n}) >$, which is smaller than $S = S_2 =< P_2(\mathbf{u} \cdot \mathbf{n}) >$.

4.2. DISCOTIC NEMATICS

The analytic results of the affine transformation model also apply to discotic nematics, where $Q < 1$. According to (57,58), the order of the Miesowcz viscosities is reversed, compared with ordinary nematics, and γ_2 becomes positive:

$$\eta_2 < \eta_3 < \eta_1, \qquad \gamma_2 > 0. \tag{60}$$

This was also confirmed in the simulations for nematic discotics [29], cf. Tab. 4.2, where NEMD results are presented for viscosities of discotic SE and LJE systems with the axes ratio $Q = 3/7$.

4.3. BIAXIAL NEMATICS

Uniaxial particles, considered so far, can also have an biaxial alignment tensor, which requires the specification of an additional direction and an additional order parameter. Fluids composed of biaxial particles, may exist in a nematic state with uniaxial or with biaxial symmetry, the latter being called biaxial nematic. Its order is characterized by four order parameters,

TABLE 4. The (potential contributions to the) Miesowicz viscosities η_1, η_2, η_3, and the Leslie viscosities γ_1, γ_2 of the discotic SE fluid at $n = 0.6, T = 0.25$, and of the LJE discotic nematic at $n = 0.6, T = 1.15$, in SS and LJ units, respectively. In both cases, the axes ratio is $Q = 3/7$. The data are inferred from the NEMD simulations, in the limit of small shear rates.

	η_1	η_2	η_3	γ_1	γ_2
SE	2.7 ± 0.3	0.10 ± 0.02	0.55 ± 0.10	1.9 ± 0.3	2.7 ± 0.3
LJE	3.1 ± 0.1	0.17 ± 0.06	0.80 ± 0.10	2.0 ± 0.1	2.9 ± 0.4

three of which are independent. The viscous behavior of this type of substances is far more complex than that of ordinary uniaxial nematics. A number of theoretical studies [91, 92], [93], and a computer simulation [94] have been made, but no experimental results are known to the author.

4.4. PRESMECTIC BEHAVIOR

For temperatures close to the transition to a smectic phase, the typical nematic order in the magnitude of the viscosity coefficients can be altered due to presmectic phenomena. As a precursor to the formation of planes, elongated particles can form disc–like clusters such that the fluid tends to behave like a nematic discotic where the viscosities η_1 and η_2 interchange their roles. An experimental example, again from [11], is shown in Fig. 10, in analogy to Fig. 9, which is for the mixture N_4, which has very broad nematic temperature range.

The substance which shows the presmetic effect is referred to as $8CBP$. Notice the increase of η_1 at the lower temperatures. The perfectly oriented ellipsoidal particles discussed above do not possess a smetic phase. Model systems where also smectic phases can occur are discussed next.

4.5. SOFT SPHERES WITH P_2-ANISOTROPY

In analogy to the ferro-fluids and magneto-rheological fluids discussed in the next section, a relatively simple model was introduced [95] where the fluid does undergo a transition from the nematic to the smectic-A phase . Incidentally, it possesses still another smectic phase (smB) in addition to the solid state. The potential is that of soft spheres and an extra anisotropic interaction of P_2 symmetry whose strength is determined by the parameter

216

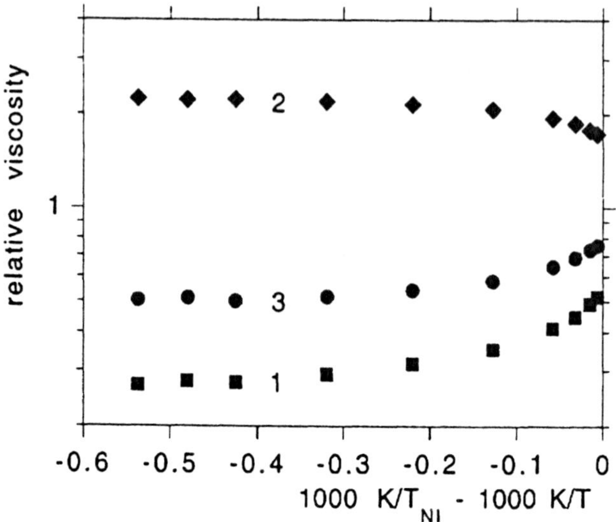

Figure 9. The relative Miesowicz viscosity coefficients $\eta_i/\eta_{av}, i = 1, 2, 3$ as functions of the relative temperature difference $1000K/T_{NI} - 1000K/T$ for the nematic mixture $N4$, T_{NI} is the nematic-isotropic phase transition temperature. Data from Kneppe, Schneider, Sharma.

Φ_{anis}:

$$\Phi = \Phi_0^{SS}(\frac{r_0}{r})^{12} + \Phi_{anis}\,(\frac{r_0}{r})^6 \left(r^{-2}(\mathbf{r}\cdot\mathbf{n})^2 - \frac{1}{3}\right) . \tag{61}$$

The unit vector $\mathbf{n}$ (director) specifies the preferential direction. A P_2 type anisotropy in the interaction, proportional to r^{-6}, has been considered by Maier and Saupe in their articles [38] on the mean field theory for the phase transition isotropic–nematic. Their P_2 fuction, however, depended on the angle between the unit vectors specifying the directions of two particles, and they disregarded the dependence of the interaction on the direction $\mathbf{r}$. For $\Phi_{anis} > 0$ this potential models elongated (prolate) particles. The shape of the particles can be charaterized by the equipotential surface $\Phi = T$. The intersection of this surface with the zx-plane, where the z-direction is parallel to the direction $\mathbf{n}$, is presented in the left graph of Fig. 11 for values $0, 1, 2, 3$ of the anisotropy parameter Φ_{anis}. Potential curves, as functions of the distance r, are displayed in the right graph of Fig. 11.

At the state point $T = 0.25, n = 0.6$, in SS-units, and with the interaction cutoff at $r = 2.5r_0$, the transition nematic–smectic-A occurs at $\Phi_{anis} \approx 2.8$ (in units of Φ_0^{SS}). The director $\mathbf{n}$ can be chosen parallel to the directions discussed above in connection with the anisotropy of the viscos-

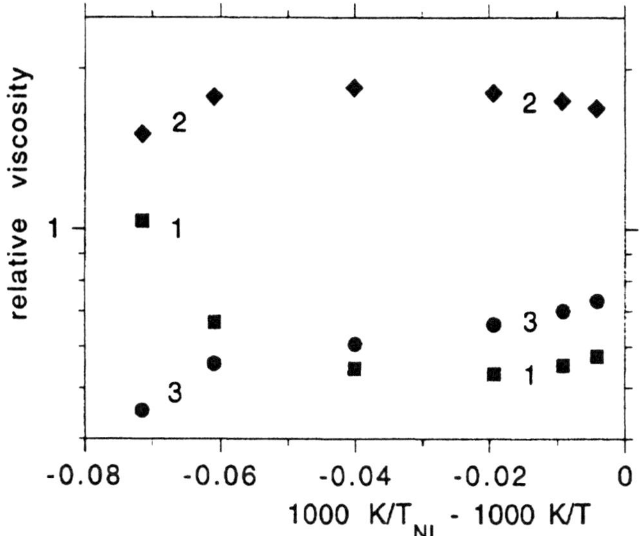

Figure 10. The relative Miesowicz viscosity coefficients η_i/η_{av}, $i = 1, 2, 3$ as functions of the relative temperature difference $1000K/T_{NI} - 1000K/T$ for the nematic $8CBP$, T_{NI} is the nematic-isotropic phase transition temperature. Data from Kneppe, Schneider, Sharma.

ity. The Miesowicz viscosities $\eta_{1,2,3}$ are displayed in Fig. 12 as functions of the anisotropy parameter Φ_{anis}. The shear rate is $\gamma = 0.1$ which, at least in the nematic phase, is in the newtonian flow regime. The typical nematic order, with $\gamma_1 < \gamma_3 < \gamma_2$, in the magnitude of the viscosities, is found for $0 < \Phi_{anis} < 1$. For $1 < \Phi_{anis} < 2.3$, presmectic effects change this behavior. In the smectic phase, for $\Phi_{anis} > 2.8$, the order of the viscosities corresponds to that of a nematic discotic system (60). The shear flow breaks the smectic layers but disc like correlated clusters remain. Of course, the shear induced structural changes can and have been analyzed with standard methods. The scenario of the crossing of the viscosity coefficients is even more complex than in experimental case shown in Fig. 10. The increase of Φ_{anis} corresponds to a decrease of the temperature. It is not known to the author whether viscous behaviour analogous to that of Fig. 12 with the sequence of the crossings of η_3 with η_2, of η_1 with η_2, and eventually of η_1 with η_3, has been observed when the temperature is lowered. But substances with such properties might well exist.

A crossing of η_1 and η_3 is also found for the perfectly oriented Gay-Berne fluid, which posseses smectic phases at temperatures below the nematic phase, cf. Fig. 14. A considerable increase of η_1, with increasing density, is

218

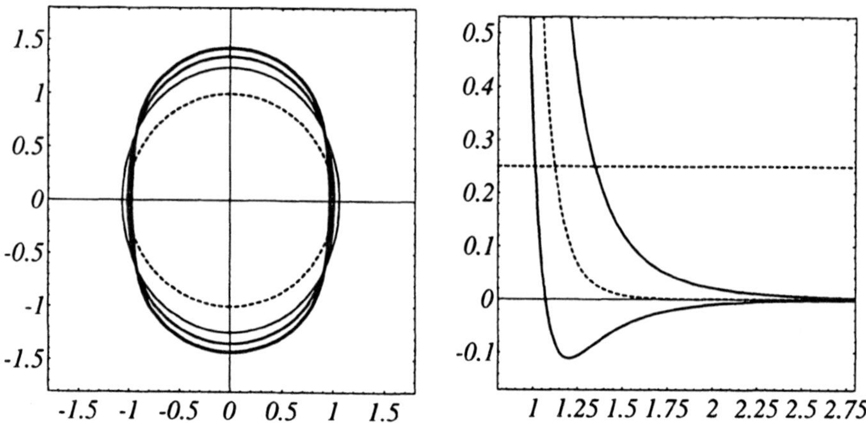

Figure 11. The left graph shows cross sections of the equipotential surface $\Phi = T = 0.25$ for $\Phi_{anis} = 1, 2, 3$ (increasing thickness) and for $\Phi_{anis} = 0$ (dashed). For $\Phi_{anis} = 2$, the potential is plotted as function of the interparticle distance r for the orientations $\mathbf{r} = \mathbf{n}r$ (purely repulsive) and $\mathbf{r}$ perpendicular to $\mathbf{n}$ (partially attractive). For comparison, the soft spheres potential is also shown (dashed).

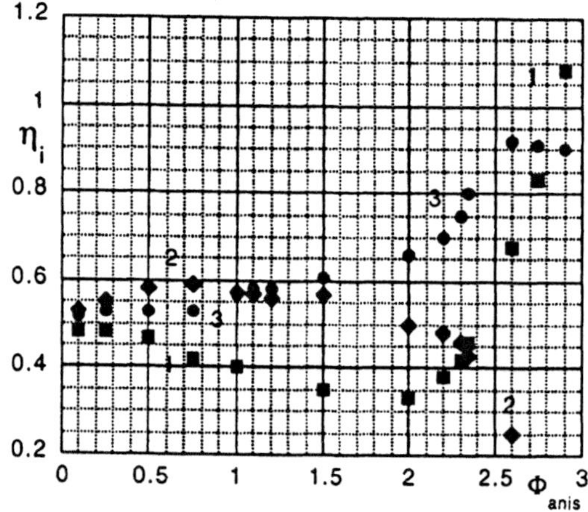

Figure 12. The Miesowicz viscosity coefficients $\eta_i, i = 1, 2, 3$ as function of the strength Φ_{anis} of anisotropic part of the interaction for $T = 0.25, n = 0.6, \gamma = 0.1$ in SS-units (C. Pereira Borgmeyer).

seen in the presmectic regime of a Gay-Berne fluid with a variable degree of alignment, cf. Fig. 15.

4.6. GAY–BERNE FLUIDS

4.6.1. *The potential*

The Gay–Berne potential [96] is a Lennard–Jones type potential where the shape of the particles and the energy depth are anisotropic. The shape is characterized by the distance $r = \sigma(\mathbf{u}_1, \mathbf{u}_2, \hat{\mathbf{r}})$, where the potential is zero. Here, the unit vectors $\mathbf{u}_1$ and $\mathbf{u}_2$ are parallel to the symmetry of the two interacting particles, $\hat{\mathbf{r}} = \mathbf{r}/r$ is the unit vector parallel to their relative position vector $\mathbf{r} = \mathbf{r}_{12} = \mathbf{r}_1 - \mathbf{r}_2$. The expression used is

$$\sigma(\mathbf{u}_1, \mathbf{u}_2, \hat{\mathbf{r}}) = r_0 \left(1 - \frac{\chi}{2r^2} \left(\frac{(\mathbf{u}_1 \cdot \mathbf{r} + \mathbf{u}_2 \cdot \mathbf{r})^2}{1 + \chi \mathbf{u}_1 \cdot \mathbf{u}_2} + \frac{(\mathbf{u}_1 \cdot \mathbf{r} - \mathbf{u}_2 \cdot \mathbf{r})^2}{1 - \chi \mathbf{u}_1 \cdot \mathbf{u}_2} \right) \right)^{-1/2}, \tag{62}$$

with a reference length r_0. The Gay–Berne potential is written as

$$\Phi^{GB} = 4\Phi_0 (\epsilon_1(\mathbf{u}_1, \mathbf{u}_2))^{\nu} (\epsilon_2(\mathbf{u}_1, \mathbf{u}_2, \hat{\mathbf{r}}))^{\mu}$$

$$\times \left(\left(\frac{r_0}{r - \sigma(\mathbf{u}_1, \mathbf{u}_2, \hat{\mathbf{r}}) + r_0} \right)^{12} - \left(\frac{r_0}{r - \sigma(\mathbf{u}_1, \mathbf{u}_2, \hat{\mathbf{r}}) + r_0} \right)^6 \right) \tag{63}$$

The functions

$$\epsilon_1(\mathbf{u}_1, \mathbf{u}_2) = \left(1 - \chi^2 \mathbf{u}_1 \cdot \mathbf{u}_2 \right)^{(-1/2)}, \tag{64}$$

and

$$\epsilon_2(\mathbf{u}_1, \mathbf{u}_2, \hat{\mathbf{r}}) = 1 - \frac{\chi'}{2r^2} \left(\frac{(\mathbf{u}_1 \cdot \mathbf{r} + \mathbf{u}_2 \cdot \mathbf{r})^2}{1 + \chi' \mathbf{u}_1 \cdot \mathbf{u}_2} + \frac{(\mathbf{u}_1 \cdot \mathbf{r} - \mathbf{u}_2 \cdot \mathbf{r})^2}{1 - \chi' \mathbf{u}_1 \cdot \mathbf{u}_2} \right), \tag{65}$$

describe the anisotropy depth of the potential minimum, Φ_0 is a reference energy. The parameters χ and χ' are related to the quantities $\kappa = r_{ee}/r_{ss}$ and $\kappa' = \epsilon_{ee}/\epsilon_{ss}$ by

$$\chi = \frac{\kappa^2 - 1}{\kappa^2 + 1}, \quad \chi' = \frac{\kappa'^{1/\mu} - 1}{\kappa'^{1/\mu} + 1}. \tag{66}$$

Here, $r_{ee} = r_0((1 + \chi)/(1 - \chi))^{1/2}$ is the end–end distance, corresponding to $\mathbf{u}_1 = \mathbf{u}_2 = \hat{\mathbf{r}}$, and $r_{ss} = r_0$ is the side–side distance, also for $\mathbf{u}_1 = \mathbf{u}_2$, but with $\hat{\mathbf{r}}$ perpendicular to $\mathbf{u}_1$, where one has $\Phi^{GB} = 0$. The depths of the energy minima in these orientations are $\epsilon_{ee} = \Phi_0 (1 - \chi^2)^{-\nu/2} (\frac{1-\chi'}{1+\chi'})^{\mu}$ and $\epsilon_{ss} = \Phi_0 (1 - \chi^2)^{-\nu/2}$. For comparison, the side–end distance and energy depth are $r_{se} = r_0(1 - \chi)^{-1/2}$ and $\epsilon_{se} = \Phi_0(1 - \chi')^{\mu}$. The effective volume associated with the biaxial ellipsoid determined by the equipotential surface $\Phi^{GB} = 0$, cf. Fig. 13, for two particles with orientations $\mathbf{u}_1$ and $\mathbf{u}_2$ is $V_{eff} = V_0(1 - \chi^2(\mathbf{u}_1 \cdot \mathbf{u}_2)^2)^{1/2}/(1 - \chi)$, where $V_0 = 4\pi r_0^3/3$ is the corresponding

220

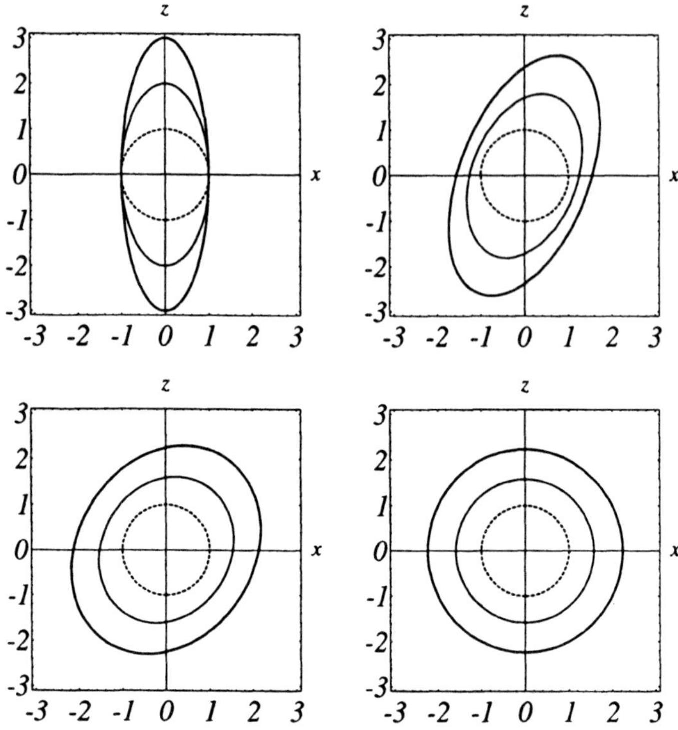

Figure 13. Cross sections through the equipotential surface $\Phi^{GB} = 0$ in the xz-plane where $\mathbf{u}_1$ is always parallel to the z-direction, $\mathbf{u}_2$ is in this plane and encloses the angles $0, 45, 75, 90°$ with the z-direction in the graphs upper left, upper right, lower left and lower right, respectively. The axis ratio κ is 2 and 3 (thicker curves) in all cases. The dashed circles mark the LJ case corresponding to $\kappa = 1$.

reference volume for spherical particles with diameter r_0. Notice that $V_0 \kappa < V_{eff} < V_0 (\kappa^2 + 1)/2$, where the limiting values are for $\mathbf{u}_1 \cdot \mathbf{u}_2 = 1$ and $\mathbf{u}_1 \cdot \mathbf{u}_2 = 0$, respectively. Thus, for $\kappa > 1$ (prolate particles), the volume $v_{eff} = V_{eff}/8$ occupied by a particle is larger than that of a particle with diameter r_0. On the other hand, r_0^{-3} is used to scale the number density of Gay–Berne particles. As a consequence, the values for the number density of a Gay–Berne system in its liquid and liquid crystalline states are considerably smaller than that of a Lennard–Jones liquid with a comparable packing fraction.

The values chosen in the simulations are $\kappa = 3$, $\kappa' = 5$, $\nu = 1$, $\mu = 2$, in accord with [96, 97]. For other choices which have also been made for these parameters, and for a modification of the potential see [98] and [99].

4.6.2. *Perfectly oriented systems*

Simulations have been performed by L. Bennett for a perfectly oriented Gay-Berne fluid. As in many studies, particles with an axis ratio 3 and a well-depth ratio for side-side and end-end configurations of 5 were chosen. In Fig. 14, the limiting values for small shear rates of the viscosities $\eta_{1,2,3}$ are shown as functions of $-1/T$, in analogy to Fig. 9 where now $1/T_{NI} = 0$. The reduced density is $n = 0.32$. The typical nematic order in the size of the coefficients is observed except for the lowest temperature where a presmetic increase of η_1 occurs.

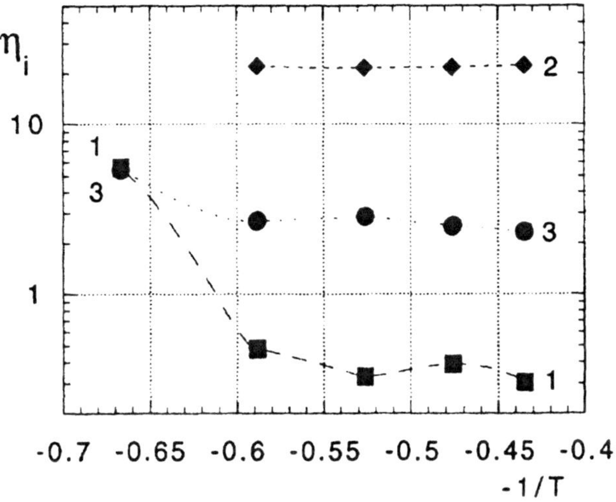

Figure 14. The Miesowicz viscosity coefficients $\eta_i, i = 1, 2, 3$ as functions of $-1/T$ for the perfectly oriented Gay-Berne fluid (L. Bennett).

4.6.3. *Variable orientation, applied magnetic field*

Recently, L. Bennett performed simulations for the Gay-Berne fluid with variable orientation. In analogy to the real experiment, an orienting field is applied. First results for the Miesowicz viscosities, cf. Fig. 15, and for the Leslie coefficients, are available. A dimensionless variable for the magnitude of the orienting magnetic field B is $B^* = (\chi_a/\mu_0)^{1/2} B$, where $\chi_a = \chi_{\parallel} - \chi_{\perp}$, characterizes the anisotropy of the magnetic susceptibility. In Fig. 15, data for $T = 0.95$, $B* = 0.9$, and for densities in the range $n = 0.18$ to 0.23 are displayed. Under equilibrium conditions, but in the presence of the orienting field, the transition nematic–smetic occurs between $n = 0.22$ and 0.23. Model fluids of short chain molecules with a stiff central part and flexible

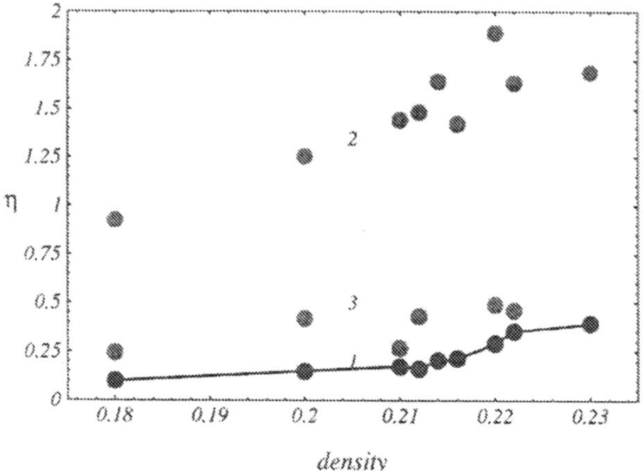

Figure 15. The Miesowicz viscosities of the Gay–Berne fluid as functions of the density. The viscosity η_1 is marked by black dots joined by straight lines, gray dots are used for η_2 and η_3, NEMD results computed by L. Bennett. In reduced units, $T = 0.95$ and $B^* = 0.9$ have been chosen for the temperature and for the magnitude of the orienting magnetic field.

ends possess broad smectic-A phase [119]. The study of their rheological properties and of the shear induced structural changes is in preparation.

5. Ferro–fluids and magneto–rheological fluids

Spherical colloidal particles with a magnetic core as occurr in ferro-fluids in the presence of an applied magnetic field have been modelled by soft spheres plus a dipole-dipole interaction (SSD) [100]:

$$\Phi = \Phi_0^{SS} \left[(\frac{r_0}{r})^{12} - \epsilon_{mag} (\frac{r_0}{r})^3 \left(r^{-2}(\mathbf{r} \cdot \mathbf{n})^2 - \frac{1}{3} \right) \right] . \tag{67}$$

The parameter $\epsilon_{mag} > 0$ is proportional to the square of the (induced) magnetic moments of the particles which are parallel to $\mathbf{n}$. The angular dependence of the nonspherical part of the interaction potential is the same as in (61), the sign of the prefactor and the r-dependence, however, are different. Pairs of particles feel a disc–like interaction since, for fixed relative kinetic energy, they can approach each other more closely in the direction parallel to $\mathbf{n}$ than in the perpendicular directions. Thus it is not surprising that ferro–fluids show an anisotropy analogous to nematic discotic liquid crystals [19, 30]. In Tab. 5, NEMD results are presented for the only state point, where all viscosities (except the bulk viscosity) have been determined. When the dipole-dipole interaction is stronger, however, chains are formed which, at higher densities, are arranged in partially ordered spa-

TABLE 5. The (potential contributions to the) Miesowicz viscosities η_1, η_2, η_3, the Helfrich viscosity η_{12}, the Leslie viscosities γ_1, γ_2 and the coupling coefficient κ of the SSD ferro–fluid at $n = 0.6, T = 0.25$, for $\epsilon_{mag} = 2.4$. The data are inferred from the NEMD simulations, in the limit of small shear rates.

	η_1	η_2	η_3	η_{12}	γ_1	γ_2	κ
SSD	$.63 \pm .05$	$.43 \pm .04$	$.50 \pm .05$	2.0 ± 0.1	$.16 \pm .02$	$.17 \pm .02$	$.6 \pm .1$

tial structures. This affects the viscous behavior in a dramatic way. An example is shown in Fig. 16 where the viscosities η_1 (magnetic field parallel to the flow velocity) and η_2 (magnetic field parallel to the gradient of the flow velocity) are plotted as functions of the anisotropy parameter ϵ_{mag}. The state point is $T = 0.25, n = 0.6$, in SS-units and the shear rate is $\gamma = 0.06$. The interaction is cut off at $r = r_c = 2.5r_0$, and $(r_0/r)^3$ in (67) is replaced by $(r_0/r)^3 - (r_0/r_c)^3 + 3(r_0/r_c)^4(r/r_0 - r_c/r_0)$ in order achieve a smoother cut off. The simulations were made with $N = 1000$ particles [101]. For $0 < \epsilon_{mag} < 3$, the discotic behavior $\eta_1 > \eta_2$ is seen in Fig. 16. For $\epsilon_{mag} > 3$, the viscosity η_2 for the field parallel to the gradient direction

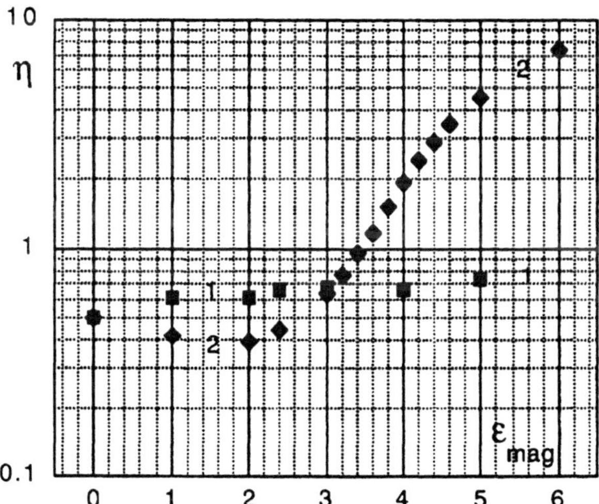

Figure 16. The viscosity coefficients η_1, η_2 as function of the strength ϵ_{mag} of the dipole-dipole interaction for $T = 0.25, n = 0.6, \gamma = 0.06$ in SS-units (T. Weider).

changes were analysed in the orientations 1 and 2, at $T = 0.25$, for number densities between $n = 0.2$ and 0.8, corresponding to packing fractions of about 0.1 to 0.8, and for the magnetic interaction strength ϵ_{mag} ranging from 0 to 8. Fig. 16 shows results for a specific case of interest. Breaking and reformation of chains and more complex structures under shear flow have been observed.

A yield stress occurs for the higher values of the dipole-dipole interaction. This is typical for the magneto–rheological (MR) fluids, which are similar to the ferro-fluids but which are composed of particles with stronger dipole-dipole interaction and usually contain a higher volume fraction of colloidal particles. Electro-rheological (ER) fluids can, with appropriate modifications, also be treated theoretically by the model. For simulations where solvent effects and hydrodynamic interactions are also taken into account, see [102].

The transition from the ferro–fluid to the magneto–rheological behavior, with increasing magnetic interaction, is analgous to what one might expect in a nematic discotic system which can undergo a transition to a columnar phase, at lower temperatures.

6. Polymeric liquids

To model a polymer melt, one starts from a simple fluid of spherical particles and introduces extra binding forces or constraints [103–105], in order to form molecular chains with a prescribed chain length of N_{ch} beads. Fixed bond angles have to be imposed when chains composed of specific chemical units, e.g. of CH_2, are treated. Longer chains can be modelled by freely joined beads where each "monomer" is a Kuhn element which stands for a few ($\approx 3-10$) chemical units. Here data are presented for polymeric liquids of the latter type.

Rheological studies for LJ-fluids where the binding was achieved by increasing the energy parameter Φ_0 for neighbors in a chain by a factor showed many features of the nonlinear flow behavior typical for polymeric melts [106, 107]. Also thermal degradation and shear induced breaking of chains were observed. The results to be presented here [109] follow from an extension of the previous simulations [108] for a system where all particles interact via the repulsive part of the LJ-potential (WCA) and an attractive FENE potential with the maximum bond length R_0 is used for the binding within the chains. More specifically, $\Phi = \Phi^{WCA} + \Phi^{FENE}$ with

$$\Phi^{WCA} = 4\Phi_0 \left[\left(\frac{r_0}{r}\right)^{12} - \left(\frac{r_0}{r}\right)^6 + \frac{1}{4} \right], \quad r \leq 2^{1/6}\, r_0, \qquad (68)$$

and $\Phi^{WCA} = 0$ for $r > 2^{1/6}\, r_0$;

$$\Phi^{FENE} = -0.5\, k^* \, \Phi_0 \frac{R_0^2}{r_0^2} ln\left[1 - \frac{r^2}{R_0^2} \right], \quad r \leq R_0, \qquad (69)$$

and $\Phi^{FENE} = \infty$ for $r > R_0$. For this potential with $R_0 = 1.5$, $k^* = 30$, $T = \Phi_0/k_B$, and $n r_0^3 = 0.85$ extensive equilibrium MD-studies have been carried out by Kremer and Grest [110]. In [108, 109] and for the data to be presented here, the same potential parameters and the same state point is used except for a slightly smaller density of $n r_0^3 = 0.84$. Molecules with chain lengths $N_{ch} = 10, 30, 60, 100, 150, 200, 300$ and 400 were studied. The systems comprised $N = 6000, 8400$ and 30000 monomers. In Fig. 17, the viscosity η is displayed as function of the shear rate γ for $N_{ch} = 30, 100, 300$.

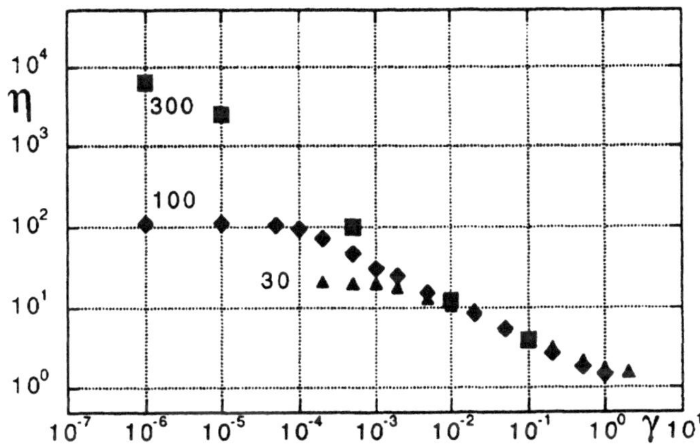

Figure 17. The viscosity of melts consisting of polymer molecules with the chain lengths 30 (triangles), 100 (diamonds), and 300 (squares) as functions of the shear rate γ. The density and the temperature are $n = 0.84$ and $T = 1$ in LJ-units (M. Kröger).

Notice that the values of η and γ, both expressed in LJ-units, span a much wider range than the data shown in Fig. 5 for a simple LJ-liquid. The newtonian limit η_{New} of the viscosity, for long molecules only reached at extremely small shear rates, is presented in Fig. 18 as a function of the chain length N_{ch}. Two regimes, referred to as the Rouse regime, where $\eta_{New} \sim N_{ch}$, and as the reptation regime, where $\eta_{New} \sim N_{ch}^{3.5}$ can be distinguished. The transition between these regimes occurs at $N_{ch} \approx 100$. This value is, as expected, about three times the entanglement length of ≈ 35 inferred from equilibrium studies [110]. A procedure to analyze and measure entanglement in MD-simulations has been developed [111].

Other rheological properties, such as the first and second normal stress differences and the viscometric functions have been computed [108, 109]. The shear–induced bond orientation, as it can be measured via the bire-

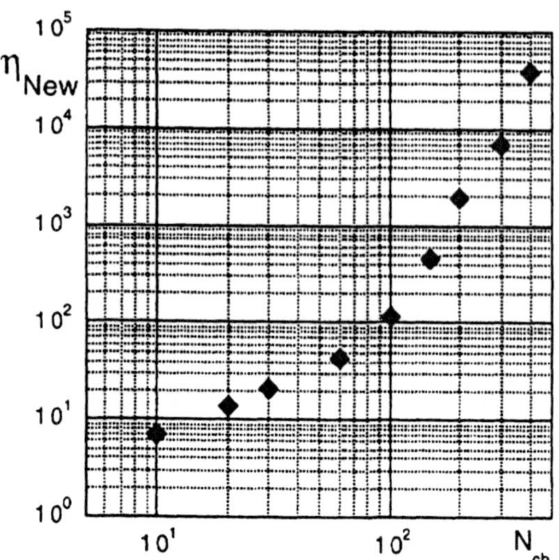

Figure 18. The newtonian viscosity η_{New} of polymer melts as function of the chain length N_{ch} (M. Kröger).

fringence, as well as the static structure factor of the whole melt or of selected chains and the shape of single polymer chains were analyzed and found to be in good agreement with experiments [112]. Other geometries, such as extensional flow of polymer melts [113] and the deformation of polymers in a glassy state, as well as systems with prescribed components of the pressure tensor [114] have also been simulated.

6.1. POLYMER SOLUTIONS

Of course, dilute polymer solutions have also been studied under nonequilibrium conditions [105, 115]. Some NEMD data for a 1 per cent solution of polymer molecules, modelled in the same fashion as the molecules in the melts discussed above, are shown in Fig. 19. The chain lengths of the polymers are 10, 30, and 60; the solvent consists of monomers. In particular, the angle χ specifying the flow orientation of the radius of gyration tensor is presented in Fig. 19 as function of the shear rate γ. The "shear resistance" m, defined by [116]

$$\tan(2\chi) = m/\beta, \quad \beta = \gamma\tau, \tag{70}$$

where τ is a relaxation time, depends on the shear rate or the "shear parameter" β, both in the simulation data of C. Aust and in the light scattering experiments of [116] for a polystyrol ($PS10$) in various solvents. For a comparison between NEMD and light scattering data, the viscosity of the solvent is used to calculate the relaxation time τ for the experiments [116]. In the simulation, τ is inferred from the reciprocal shear rate for which the orientation angle χ tends to decrease significantly from its limiting small shear shear rate value of 45°, cf. Fig. 19. The quantity $m = m(\beta)$, whose dependence on the shear rate came as a surprise some time ago, behaves similarly for the real and for the model polymer solutions, for theoretical explanations see [117].

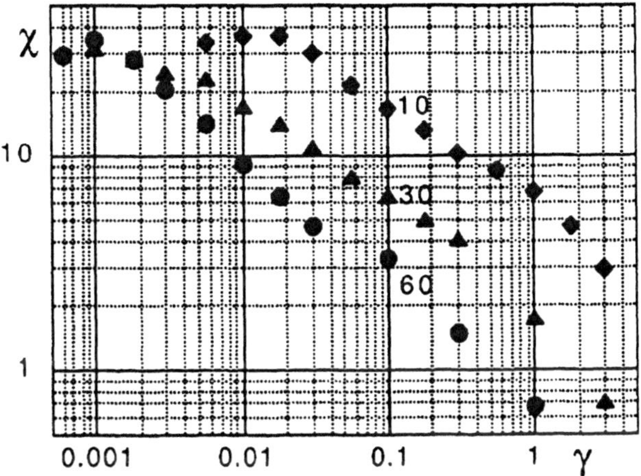

Figure 19. The flow orientation angle χ of the gyration tensor of a 1 per cent solution of polymers in a solvent of monomers as functions of the shear rate γ. The chain lengths of the polymer molecules are 10 (diamonds), 30 (triangles), and 60 (circles) (C. Aust).

Also "living polymers", i.e. chains which break and form again like the worm-like micelles in surfactant solutions, can be treated in NEMD simulations when the interaction potential is appropriately modified [118]. Furthermore, NEMD studies of the rheological behaviour are in preparation for chain molecules with stiff and flexible parts [119], as in main-chain polymeric liquid crystals.

7. Concluding remarks

Viscous properties of simple liquids, liquid crystals and of some other complex fluids have been studied by NEMD simulations. Here results have been presented for a stationary plane Couette flow. Of course, non-stationary flows and other flow geometries can and have been treated. Extensions to larger systems and longer times are needed in some rheological applications. Hybrid methods which combine molecular dynamics with hydrodynamics promise further progress. Yet the NEMD results obtained so far already provide information and insight into microscopic mechanisms of nonequilibrium phenomena similar to that available in real experiments, e.g. in light, X-ray or neutron scattering. And even more detailed information is available in molecular dynamics, e.g. snapshots showing the positions of the particles or specific contributions to physical quantities of which only the sum can be measured in the real experiment. Thus computer simulations can often yield sharper tests of the assumptions underlying conventional theories than a real experiment.

Acknowledgments

The work presented here has benefitted from conversations and collaborations with M.P. Allen, G. Ciccotti, D.J. Evans, D. Frenkel, H.J.M. Hanley, J.P. Hayter, W.G. Hoover, K. Kremer, G.R. Luckhurst, R. Pynn, S. Sarman, and C. Zannoni. Many thanks are due to my former and present coworkers F. Affouard, C. Aust, D. Baalss, L. Bennett, O. Hess, N. Herdegen, M. Kröger, W. Loose, C. Pereira Borgmeyer, R. Schramek, J. Schwarzl, H. Sollich, U. Stottut, H. Voigt, and T. Weider, who performed MD and NEMD simulations for simple and complex fluids and contributed substantially to a better understanding of the interrelations between the viscous properties and the shear induced structural changes. Furthermore, I thank H. Ehrentraut, T. Gruhn, P. Kaiser, A. Kilian, H-M. Koo, W. Muschik, R.R. Netz, M.A. Osipov, I. Pardowitz, G. Rienäcker, C. Schneggenburger, M. Schoen, A. Sonnet, H. Thurn, W. Wagner, W. Wiese for helpful discussions on theoretical aspects, as well as H.J. Eichler, G. Heppke, G. Hinrichsen, H. Hoffmann, H.-S. Kitzerow, H. Kneppe, H.-D. Koswig, H.-M. Laun, P. Lindner, R. Macdonald, F. Oestreicher, P. Pieranski, K. Praefcke, H. Rehage, G. Scherowsky, F. Schneider, J. Springer for patiently explaining experimental and chemical details to a theoretical physicist.

Financial support provided by the Deutsche Forschungsgemeinschaft (DFG) via the Sonderforschungsbereich (SFB) 335 "Anisotrope Fluide" and the Graduiertenkolleg "Polymerwerkstoffe", as well as the generous donation of computer time by the Konrad-Zuse-Zentrum für Informationstechnik (Berlin) and by the Höchstleistungsrechenzentrum der Kfa Jülich GmbH, are gratefully acknowledged.

References

1. Lehmann, O. (1890) *Z. Phys. Chem.*, **4**, 462; (1890) *Z. Phys. Chem.*, **5**, 427; (1890) *Wied. Ann. Phys.*, **40**, 401; (1890) *Wied. Ann. Phys.*, **41**, 520.
2. Kelker, H. (1973) *Mol. Cryst. Liq. Cryst.*, **21**, 1; (1988) *Mol. Cryst. Liq. Cryst.*, **165**, 1.
3. Schenck, R. (1898) *Z. Phys. Chem.*, **27**, 167; (1905) *Kristallinische Flüssigkeiten und Flüssige Kristalle.* Engelmann, Leipzig.
4. Miesowicz, M. (1935) *Nature*, **17**, 261; (1946) *Nature*, **158**, 27.
5. Ericksen, J.L. (1960) *Arch. Ratl. Mech. Anal.*, **4**, 231; (1969) *Mol. Cryst. Liq. Cryst.*, **7**, 153; Leslie, F.M. (1966) *J. Mech. Appl. Math.*, **19**, 357; (1968) *Arch. Ratl. Mech. Anal.*, **28**, 265.
6. Parodi, O. (1970) *J. Phys.*, **31**, 581.
7. de Gennes, P.G. (1974) *The Physics of Liquid Crystals.* Clarendon Press, Oxford, Chap. 3.
8. Kelker, H. and Hatz, R. (1980) *Handbook of Liquid Crystals.* Verlag Chemie, Weinheim.
9. Vertogen, G. and de Jeu, W. (1988) Thermotropic Liquid Crystals, Fundamentals, in *Chemical Physics*, **45**, Springer, Berlin.
10. Belyaev, V.V. (1989) *Russian Chem. Rev.*, **58**, 917, translated from: (1989) *Uspekhi Khimii*, **58**, 1601.
11. Kneppe, H. and Schneider, F. (1981) *Mol. Cryst. Liq. Cryst.*, **65**, 23; Kneppe, H. Schneider, F. and Sharma, N.K. (1981) *Ber. Bunsenges. Phys. Chem.*, **85**, 784; Graf, H.-H., Kneppe, H. and Schneider, F. (1992) *Mol. Phys.*, **77**, 521.
12. Beens, W. and deJeu, W. (1983) *J. Physique*, **44**, 129.
13. Muschik, W. Ehrentraut, H. Papenfuss, C. and Blenk, S. (1996) *ZAMM* , **76**, S4 129.
14. Helfrich, W. (1969) *J. Chem. Phys.*, **50**, 100; (1970) *J. Chem. Phys.* **53**, 2267.
15. Baalss, D. and Hess, S. (1986) *Phys. Rev. Lett.*, **57**, 86; (1988) *Z. Naturforsch.*, **43 a**, 662.
16. Hess, S. Frenkel, D. and Allen, M.P. (1991) *Mol. Phys.*, **74**, 765.
17. Ehrentraut, H. and Hess, S. (1995) *Phys. Rev. E*, **51**, 2203.
18. Hess, S. (1975) *Z. Naturforsch.*, **30a**, 728; 1224; (1976) *Z. Naturforsch.*, **31a**, 1507.
19. Hess, S. (1986) *J. Non-Equilib. Thermodyn.*, **11**, 175.
20. Pereira Borgmeyer C. and Hess, S. (1995) *J. Non-Equilib. Thermodyn.*, **20**, 359.
21. Hess, S. (1976) *Z. Naturforsch.*, **31a**, 1034.
22. Doi, M. (1981) *J. Polym. Sci. Polym. Phys.*, **19**, 229.
23. Maffettone, P. and Marrucci, G. (1991) *J. Non-Newtonian Fluid Mech.*, **38**, 273.
24. Maffettone, P.L. (1992) *J. Non-Newtonian Fluid Mech.*, **45**, 339.
25. Kröger, M. and Sellers, H. (1995) *J. Chem. Phys.*, **103**, 807.
26. Osipov, M. and Terentjev, E. (1989) *Z. Naturforsch.*, **44a**, 785.
27. Wu, S. and Wu, C.-S. (1990) *Phys. Rev. A*, **42**, 2219.
28. Muschik, W., Papenfuss, C. and Ehrentraut, H. (1997) *Proc. Estonian Acad. Sci. Phys. Math.*, **46**, 1/2 94.
29. Sollich, H., Baalss, D. and Hess, S. (1989) *Mol. Cryst. Liq. Cryst.*, **168**, 189.
30. Hess, S., Schwarzl, J. and Baalss, D. (1990) *J. Phys. Condens. Matter*, **2**, SA 279.
31. Sarman, S. and Evans, D.J. (1993) *J. Chem. Phys.*, **99**, 9021.
32. Sarman, S. (1997) *Physica A*, **240**, 160; (1995) *J. Chem. Phys.*, **103**, 393;

(1997) *J. Chem. Phys.*, **107**, 3144.
33. Bennett, L. and Hess, S. (1998) in *Proceedings of the 27th Freiburger Arbeitstagung Flssigkristalle*, P17.
34. Smondyrev, A.M., Loriot, G.B. and Pelcovits, R.A. (1995) *Phys. Rev. Lett.*, **75**, 2340.
35. Hess, S., Kröger, M., Loose, W., Pereira Borgmeyer, C., Schramek, R., Voigt, H. and Weider, T. (1996) Simple and Complex Fluids under Shear, in *Monte Carlo and Molecular Dynamics of Condensed Matter Systems*. Binder, K. and Ciccotti, G. (eds.), IPS Conf. Proc., **49**, Bologna p. 825.
36. Hess, S. (1996) Constraints in Molecular Dynamics, Nonequilibrium Processes in Fluids via Computer Simulations, in *Computational Physics*. Hoffmann, K.H. and Schreiber, M. (eds.), Springer, Berlin p. 268.
37. Hess, S., Aust, C., Bennett, L., Kröger, M., Pereira Borgmeyer, C. and Weider, T. (1997) *Physica A*, **240**, 126.
38. Maier, W. and Saupe, A. (1959) *Z. Naturforsch.*, **14 a**, 882; (1960) *Z. Naturforsch.*, **15 a**, 287.
39. Tsvetkov, V.N. (1939) *Acta Physicochim. U.S.S.R.*, **10**, 555.
40. Gasparoux, H. and Prost, J. (1971) *J. Phys.*, (Paris) **32**, 953; Prost, J. and Gasparoux, H. (1971) *Phys. Lett.*, **36 A**, 245.
41. Heppke, G. and Schneider, F. (1972) *Z.Naturforsch.*, **27a**, 976; (1973) *Z.Naturforsch.*, **28a**, 994.
42. Eichler, H.J. and Macdonald, R. (1991) *Phys. Rev. Lett.*, **67**, 2666; (1991) *Mol. Cryst. Liq. Cryst.*, **207**, 117.
43. Hess, O., Macdonald, R. and Eichler, H.J. (1991) *Opt. Commun.*, **82**, 526.
44. Macdonald, R. and Danlewski, H. (1994) *Mol. Cryst. Liq. Cryst.*, **251**, 145.
45. Pieranski, P. and Guyon, E. (1974) *Phys. Lett.*, **49 A**, 237.
46. Ashurst, W.T. and Hoover, W.G. (1972) *Am. Phys. Soc.*, **17**, 1196; (1972) *Phys. Rev. Lett.*, **31**, 206; (1975) *Phys. Rev.*, **A11**, 658.
47. Lees, A.W. and Edwards, S.F. (1972) *J. Phys. C*, **5**, 1921.
48. Gosling, E.M., McDonald, I.R. and Singer, K. (1973) *Mol. Phys.*, **26**, 1475.
49. Hoover, W.G. (1973) *Ann. Rev. Phys. Chem.*, **34**, 1690.
50. Hoover, W.G. and Ashurst, W.T. (1975) *Theor. Chem. Adv. and Perspectives*, **1**, 331.
51. Ciccotti, G., Jacucci, G. and McDonald, I.R. (1975) *Phys. Rev. A*, **13**, 426; (1979) *J. Stat. Phys.* **21**, 1; Trozzi, C. and Ciccotti, G. (1984) *Phys. Rev. A* , **29**, 916.
52. Heyes, D.M., Kim, J.J., Montrose, C.J. and Litovitz, T.A. (1980) *J. Chem. Phys.*, **73**, 3987.
53. Hoover, W.G. (1983) *Annu. Rev. Phys. Chem.*, **34**, 103; Evans, D.J. and Morriss, G.P. (1984) *Comp. Phys. Rep.*, **1**, 287; Evans, D.J. and Hoover, W.G. (1986) *Ann. Rev. Fluid Mech.*, **18**, 243.
54. Hanley, H.J.M. (1983) *Nonlinear Fluid Behavior*. Hanley, H.J.M. (ed.), North Holland, Amsterdam.
55. Hoover, W.G. (1983) *Physica*, **118A**, 111.
56. Evans, D.J. (1983) *Physica*, **118A**, 51.
57. Holian, B.D. and Evans, D.J. (1983) *J. Chem. Phys.*, **78**, 5157.
58. Hoover, W.G. (1984) *Physics Today*, **37**, 44.
59. Evans, D.J., Hanley, H.J. and Hess, S. (1984) *Physics Today*, **37**, 26.
60. Erpenbeck, J.J. (1984) *Phys. Rev. Lett.*, **52**, 1333.
61. Hess, S. (1985), Nonlinear behavior of dispersive media, in *J. Mécanique Théor. Appl., Numéro spécial.* Quemada, D. (ed.), p. 1; (1985) *Int. J. Thermophys.*, **6**, 657; (1985) *J.de Physique*, **46**, C3, Supplement 3, 191.
62. Ryckaert, J.P. (1985) *Mol. Phys.*, **55**, 549.
63. Ladd, A.J.C. and Hoover, W.G. (1985) *J. Stat. Phys.*, **38**, 973.
64. Heyes, D.M. (1986) *J. Chem. Soc. Faraday. II*, **82**, 1365.

65. Ciccotti, G. and Hoover, W.G. (1986) *Molecular Dynaamics Simulations of Statistical Mechanical Systems.* Ciccotti, G. and Hoover, W.G. (eds.), North Holland, Amsterdam.

66. Ciccotti, G., Frenkel, D. and McDonald, I.R. (1987) *Simulations of Liquids and Solids.* Ciccotti, G., Frenkel, D. and McDonald, I.R. (eds.) North Holland, Amsterdam.

67. Hess, S. and Loose, W. (1988) Molecular dynamics: Test of microscopic models for the material properties of matter, in *Constitutive laws and microstructure.*Axelrad, D. and Muschik, W. (eds.), Springer, Berli n, p. 92; Hess, S., Baalss, D., Hess, O., Loose, W., Schwarzl, J.F., Stottut, U. and Weider, T. (1990), in *Continuum models and discrete systems.* Maugin, G.A. (ed.), Longman, Essex, Vol. 1, p. 18.

68. Loose, W. and Hess, S. (1988) *Phys. Rev. Lett.,* **58**, 2443; (1988) *Phys. Rev. A,* **37**, 2099.

69. Hess, S. and Loose, W. (1989) *Physica A,* **162**, 138

70. Hess, O., Loose, W., Weider, T. and Hess, S. (1989) *Physica B,* **156/157**, 505.

71. Loose, W. and Hess, S. (1989) *Rheol. Acta,* **28**, 91; (1990) in *Microscopic Simulation of Complex Flows.* Mareschal, M. (ed.), (ASI series) Plenum Press; Hess, S. and Loose, W. (1989) *Physica A,* **162**, 138; (1990) *Ber. Bunsenges., Phys. Chem.,* **94**, 216.

72. Mareschal, M. (1990) *Microscopic Simulations of Complex Flow.* Mareschal, M. (ed.), Plenum, New York.

73. Hess, S. (1991) Rheology and shear induced structure of fluids in *Rheological Modelling: Thermodynamical and Statistical Approaches.* Casas-Vázques, J. and Jou, D. (eds.), Lecture Notes in Physics, **381**, Springer, Berlin, p. 51; (1988) *Physikal. Blätter,* **44**, 325.

74. Weider, T., Stottut, U., Loose, W. and Hess, S. (1991) *Physica A,* **174**, 1.

75. Mareschal, M. and Holian, B.L. (1992) *Microscopic Simulations of Complex Hydrodynamic Phenomena.* Mareschal, M. and Holian, B.L. (eds.), Plenum, New York.

76. Ciccotti, G., Pierleoni, C. and Ryckaert, J.-P. (1992) Theoretical foundation and rheological application of Non-Equilibrium Molecular Dynamics, in *Microscopic Simulations of Complex Hydrodynamic Phenomena.* Mareschal, M. and Holian, B.L. (eds.), Plenum, New York, p. 25.

77. Loose, W. and Ciccotti, G. (1992) *Phys. Rev.,* **45**, 3859.

78. Hoover, W.G. (1993) *Physica A,* **194**, 450.

79. Koplik, J. and Banavar, J.R. (1995) *Annu. Rev. Fluid Mech.,* **27**, 257.

80. Hoover, W.G. (1986) *Molecular Dynamics.* Springer, Berlin; (1991) *Computational Statistical Mechanics.* Elsevier, Amsterdam.

81. Allen, M.P. and Tildesley, D.J. (1987) *Computer Simulation of Liquids.* Clarendon, Oxford.

82. Evans, D.J. and Morris, G.P. (1990) *Statistical Mechanics of Nonequilibrium Liquids.* Academic Press, London.

83. Haberlandt, R., Fritzsche, S., Peinel, G. and Heinzinger, K. (1995) *Molekular-Dynamik.* Vieweg, Braunschweig.

84. Rapaport, D.C. (1995) *The Art of Molecular Dynamics Simulation.* Cambridge University Press, Cambridge.

85. Weeks, J.D., Chandler, D. and Andersen, H.C. (1971) *J. Chem. Phys.,* **54**, 5237.

86. Hess, S. and Hanley, H.J.M. (1983) *Phys. Lett.,* **A 98**, 35; Hanley, H.J.M., Rainwater, J.C. and Hess, S. (1987) *Phys. Rev.,* **A 36**, 1795; Rainwater, J.C., Hanley, H.J.M. and Hess, S. (1988) *Phys. Lett.,* **A 36**, 450.

87. Laun, H.M., Bung, R., Hess, S., Loose, W., Hess, O., Hahn, K., Hädicke,

232

       E., Hingmann, R., Schmidt, F. and Lindner, P. (1992) *J. Rheol.*, **36**, 743.
88. Hess, S. (1983) *Physica A*, **118**, 79.
89. Schwarzl, J.F. and Hess, S. (1986) *Phys. Rev.*, **A 33**, 4277.
90. Hess, S., Kröger, M. and Hoover, W.G. (1997) *Physica A*, **239**, 449.
91. Brand, H. and Pleiner, H. (1981) *Phys. Rev. A*, **24**, 2777;
Kini, U.D. (1984) *Mol. Cryst. Liq. Cryst.*, **108**, 71; (1984) *Mol. Cryst. Liq. Cryst.*, **112**, 265.
92. Baalss, D. (1990) *Z. Naturforsch.*, **45 a**, 7.
93. Papenfuss, C., Verhas, J. and Muschik, W. (1995) *Z. Naturforsch.*, **50 a**, 795.
94. Sarman, S. (1996) *J. Chem. Phys.*, **104**, 342; (1996) *J. Chem. Phys.*, **105**, 4211.
95. Pereira Borgmeyer, C. (1995) *Zur Theorie der Strömungsausrichtung und Anisotropie der Viskosität in Flüssigkristallen.*, Diplomarbeit, TU-Berlin.
96. Gay, J.B. and Berne, B.J. (1981) *J. Chem. Phys.*, **74**, 3316.
97. Adams, D.J., Luckhurst, G.R. and Phippen, R.W. (1987) *Mol. Phys.*, **61**, 1575;
de Miguel, E., Rull, L.F., Chalam, M.K. and Gubbins, K.E. (1991) *Mol. Phys.*, **71**, 1223; (1991) **72**, 593; (1991) **74**, 405;
de Miguel, E., Martin del Rio, E., Brown, J.T. and Allen, M.P. (1996) *J. Chem. Phys.*, **105**, 4234;
de Miguel, E. and Martin del Rio, E. (1997) *Phys. Rev. E*, **55**, 2916.
98. Zannoni, C. (1979) in *The Molecular Physics of Liquid Crystals.* Luckhurst, G.R. and Gray, G.W. (eds.), Academic Press, New York, p. 191;
Berardi, R., Emerson, A.P.J. and Zannoni, C. (1993) *J. Chem. Soc. Faraday. Trans.*, **89**, 4069.
99. Gruhn, T. and Schoen, M. (1997) *Phys. Rev. E*, **55**, 2861; (1998) *Mol. Phys.*, **93**, 681; (1998) *J. Chem. Phys.*, **108**, 9124.
100. Hess, S., Hayter, J.B. and Pynn, R. (1984) *Mol. Phys.*, **53**, 1527.
101. Weider, T. (1995) *Simulation kristalliner Strukturen unter Scherung.* Dissertation, TU-Berlin.
102. Bossis, G., Grasselli, Y., Lemaire, E., Meunier, A., Brady, J.F. and Phung, T. (1993) *Physica Scripta*, **T 49**, 89.
103. Ryckaert, J.P. and Bellemans, A. (1978) *Faraday. Disc. Chem. Soc.*, **66**, 95.
104. Berker, A., Chynoweth, S. and Michopoulos, Y. (1992) NEMD Simulations and the Rheology of Liquid Hydrocarbons, in *Microscopic Simulations of Complex Hydrodynamic Phenomena.* Mareschal, M. and Holian, B.L. (eds.), Plenum, New York, p. 75.
105. Pierleoni, C. and Ryckaert, J.P. (1993) *Phys. Rev. Lett.*, **71**, 1724; (1995) *Macromolecules*, **28**, 5087;
Ryckaert, J.P. (1996) Simulations of polymers using realistic potentials, in *Monte Carlo and Molecular Dynamics of Condensed Matter Systems.* Binder, K. and Ciccotti, G. (eds.), IPS Conf. Proc. 49, Bologna, p. 725.
106. Hess, S. (1987) *J. Non-Newtonian Fluid Mech.*, **23**, 187.
107. Kröger, M. and Hess, S. (1993) *Physica A*, **195**, 336.
108. Kröger, M., Loose, W. and Hess, S. (1993) *J. Rheol.*, **37**, 1057;
Kröger, M. (1995) *Rheologie und Struktur von Polymerschmelzen.* W&T, Berlin.
109. Kröger, M. (1995) *Appl. Rheol.*, **5**, 66.
110. Kremer, K. and Grest, G.S. (1990) *J. Chem. Phys.*, **92**, 5057.
111. Kröger, M. and Voigt, H. (1994) *Macromol. Theory Simul.*, **3**, 639.
112. Muller, R., Pesce, J.J. and Picot, C. *Macromol.*, (1993) **26**, 4356.
113. Kröger, M., Luap, C. and Muller, R. (1997) *Macromol.*, **30**, 526.
114. Voigt, H. (1996) *Struktur und Dynamik von Kettenverschlaufungen in Polymerschmelzen.* Dissertation, TU-Berlin.

115. Aust, C. (1995) *Molekulardynamik-Untersuchungen von Polymerketten in strömenden Lösungen.* Diplomarbeit, TU-Berlin.
116. Link, A., Zisenis, M., Prötzel, B. and Springer, J. (1992) *Makromol. Chem., Macromol. Symp.*, **61**, 358;
Zisenis, M. (1994) *Streulichtmessungen zum Studium der Orientierung von Hoch-Polymeren durch Scherströmung.* Dissertation, TU-Berlin.
117. Schneggenburger, C., Kröger, M. and Hess, S. (1996) *J. Non-Newt. Fluid Mech.*, **62**, 235.
118. Kröger, M. and Makhloufi, R. (1996) *Phys. Rev. E*, **53**, 2531.
119. Affouard, F., Kröger, M. and Hess, S. (1996) *Phys. Rev. E*, **54**, 5178.

SELF ATOM–ATOM EMPIRICAL POTENTIALS
FOR THE STATIC AND DYNAMIC SIMULATION
OF CONDENSED PHASES

A. GAVEZZOTTI

*Dipartimento di Chimica Strutturale e Stereochimica Inorganica,
Università di Milano
Via Venezian 21, 20133 Milano, Italy*

AND

G. FILIPPINI

*Centro CNR per lo Studio delle Relazioni tra Struttura e Reattività
Chimica
Università di Milano
Milano, Italy*

1. Introduction

The self–recognition and subsequent condensation of organic molecules into macroscopic aggregates gives rise to a rich and diverse phase behaviour. The ultimate state, at very low temperature, is ideally a pure, perfectly ordered crystal: as temperature rises, the substance may proceed to several kinds of partly disordered systems. Loss of one- or two-dimensional order produces rotator phases, plastic crystals, then liquid crystals, while total absence of order, still without molecular mobility, is proper of the glass state. After melting, liquids in the proximity of their freezing point may preserve some memory of the more ordered states, while, on increasing the temperature, molecular correlation decreases until the molecular assembly assumes the fully isotropic aspect which, in a structural and thermodynamic sense, characterizes the true liquid state.

The study of intermolecular interactions in systems made of large (molecular weight 100 to 1000 da) molecules has seen great advances in the last decades, thanks also to the detailed structural information made available by X-ray diffraction studies in crystals, through a recognition and systematization of the geometrical patterns in crystal structures. At the same time,

235

P. Pasini and C. Zannoni (eds.), Advances in the Computer Simulations of Liquid Crystals, 235–250.

236

the computer simulation of condensed phases by molecular dynamics has risen, in the last ten years or so, from the status of a theoretical chemist's dream to a relatively well established field of physicochemical research.

After a schematic description of the fundamentals of intermolecular interactions, this paper will concentrate on the computational tools which have brought the atomistic level representation of intermolecular potentials to its present state of considerable development. Since parametric empirical methods show more promise than quantum chemical ones, at least for the treatment of systems made of a large number of large molecules, the discussion will be restricted to the former.

2. Principles of intermolecular interactions

The very fact that matter exists in condensed (i.e., not gaseous) form proves the existence of forces between molecules. Intermolecular interactions arise from electromagnetic forces, indeed, for ordinary organic systems, from electric forces. In turn, these electromagnetic effects depend on the shape of the electron density clouds around atoms and molecules. We describe in this section the main components of intermolecular interaction in molecular systems [1].

2.1. LONG RANGE INTERACTIONS AMONG NON–OVERLAPPING ELECTRON DENSITIES

2.1.1. *Electrostatic potential*

There is a direct electrostatic contribution to the intermolecular potential whenever the charge distribution in the molecule is not spherically symmetric. A convenient expression for the electrostatic potential is in terms of a multipole expansion:

$$
\begin{aligned}
V(R,\theta,\phi) \;=\; & R^{-1}M + R^{-2}\left(D_x\,p_x + D_y\,p_y + D_z\,p_z\right)+ \\
& R^{-3}\left(Q_{xy}\,d_{xy} + Q_{xz}\,d_{xz} + Q_{yz}\,d_{yz} + Q_{x^2-y^2}\,d_{x^2-y^2}+ \right. \\
& \left. Q_{z^2}\,d_{z^2}\right) + ...\text{higher terms}
\end{aligned}
\tag{1}
$$

where M, D and Q are the radial parts of the monopole, dipole, and quadrupole contributions, respectively, and the angular parts are spherical harmonics (recall s, p and d atomic orbitals). When invoking a multipole expansion, the following considerations apply:
a) the number and type of non-vanishing moments varies and depends on symmetry: for example, monopole terms are non zero only for objects with net charge, dipole terms are zero for centrosymmetric molecules;
b) the truncation of the multipole expansion to its very first terms is legitimate only when the observer's distance from the molecule is much larger

than the molecular dimension - a condition seldom verified in organic condensed phases, and certainly not observed in compact organic crystals of large and flexible molecules - because

c) the representation of a molecule with just the first few moments (say, up to the dipole) is legitimate only whenever all subsequent terms in the expansion become negligible due to the higher inverse power of the intermolecular distance in the prefactor; therefore, only under strict observance of condition b) above.

Dipole–dipole interactions decay as R^{-3}, but, when the interaction is averaged over many orientations, an R^{-6} dependence is seen. For dipole–quadrupole interactions the powers are R^{-4} and R^{-8}, and for quadrupole–quadrupole interactions, R^{-5} and R^{-10}, respectively.

2.1.2. *Induction (polarization) potential*

The charge distribution of one molecule generates an electric field which modifies the charge distribution in neighbouring molecules. The purely electrostatic and the induction part of the interaction therefore influence each other in a very complex manner. An average induction potential decays as R^{-6}.

2.1.3. *Dispersion potential*

Instantaneous fluctuations in the charge distribution of one molecule produce instantaneous polarization in its charge density, and this polarization then exerts its induction power on neighbouring molecules, as described in 2.1.2. The dispersion potential can be shown to decay as a series of powers R^{-n}, with $6 < n < 10$. Inductive-dispersive interactions are always stabilizing, and the order of magnitude for an atom-atom contribution is 0.1-2.0 kJ/mole.

2.2. SHORT–RANGE INTERACTIONS AMONG WEAKLY OVERLAPPING ELECTRON DENSITIES

2.2.1. *Exchange repulsion*

There is repulsion between molecules at short range because closed-shell electrons of the same spin cannot occupy the same region of space. Phenomenologically, this intrinsically quantum mechanical effect can be described as the Pauli exclusion principle driving electrons from two approaching molecules towards mutual avoidance, thus leaving behind partially unshielded nuclei which then repel each other. Technically, a parallel can be drawn with the difference in energy between the Heitler-London singlet and triplet state in a hydrogen molecule: in the triplet state, the two unpaired electrons represent non–interacting shells, where the repulsive effect is due mostly to the exchange integral (Figure 1; hence the name). In MO lan-

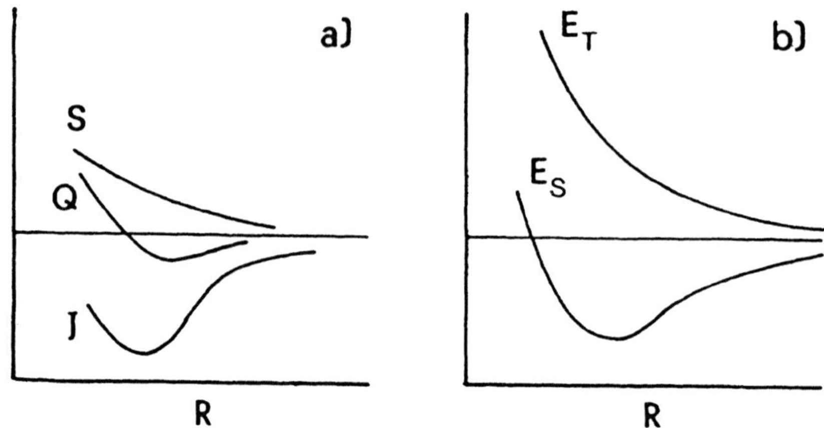

Figure 1. a) Overlap (S), coulomb (Q) and exchange (J) integrals over the hydrogen molecule as a function of interatomic distance. b) Heitler-London energy of singlet and triplet state.

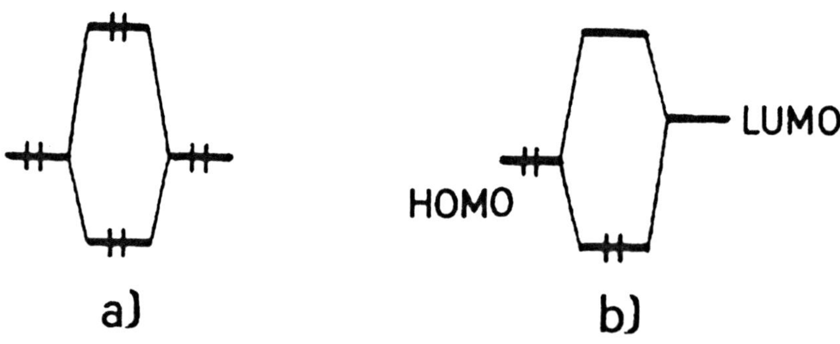

Figure 2. a) MO diagram for the interaction between two equal closed shells. b) MO diagram for a donor-acceptor interaction.

guage, this is the interaction between two filled molecular orbitals, with destabilization exceeding stabilization (Figure 2a). The exchange repulsion decays as an exponential function of distance.

2.3. ATTRACTIVE, SHORT–RANGE INTERACTIONS

2.3.1. *Charge-transfer potential*
In MO theory language, this brings together two molecules, one with a relatively high-lying filled molecular orbital, and one with a relatively low–lying empty molecular orbital, by a partial electron transfer from the former to the latter (Figure 2b). A classical example is the wide range of BEDT–TTF complexes used in the preparation of organic solids with unusual electric properties.

2.3.2. *The hydrogen bond*
There is no generally accepted definition and there is no firm theory of the hydrogen bonding, whose description ranges from an intuitive but oversimplified electrostatic picture, to several forms of three–center, many–electron interaction, especially for the very strong ones. The weak (e.g. C–H...O) "hydrogen bond" is worth 5–10 kJ/mole [2], the classical (e.g. –H...O) hydrogen bond 10–30 kJ/mole [3], and the very strong (e.g. $R-NH_3^+...N=R$) hydrogen bond 30–120 kJ/mole [4]. There is no straightforward way of deriving an approximate R–dependence law for the hydrogen bond from quantum chemical or electrostatic principles.

Table 1 shows that even when an accurate theoretical calculation of each of these effects is carried out, for small molecules, the heat of condensation is not reproduced. Much is clearly missing.

TABLE 1. Contribution to the intermolecular potential calculated by rigorous methods (kJ/mole; after ref. [1]).

Molecule	E_{coul}	E_{disp}	E_{ind}	ΔH_{subl}
water	-16.1	- 5.3	-0.9	47
ammonia	-6.2	-12.9	-0.9	29
carbon monoxide	~ 0	-1.3	~ 0	6.9

2.4. NON–ADDITIVE TERMS: MANY–BODY INTERACTIONS

The discrepancy between calculated and observed numbers in Table 1 mostly depends on the fact that calculations assume static, unperturbed electron densities, and add up the corresponding separate contributions to the total interaction energy. If the electrostatic and repulsion interactions can be safely considered as two–body interactions, it is clearly not so for induction

240

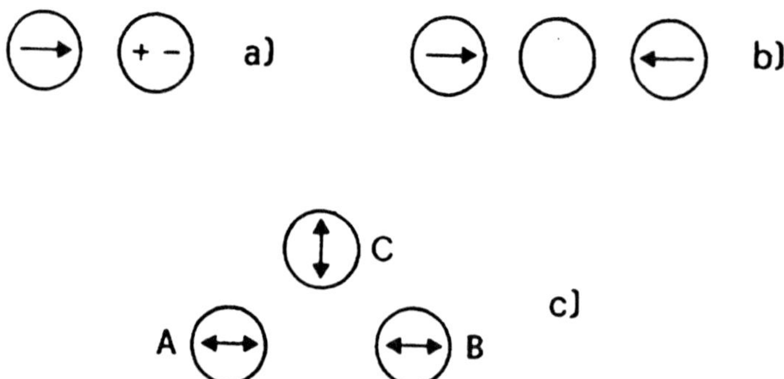

Figure 3. a) Two-body induction, b) three-body induction, cancellation; c) triple dipole term, C reduces the A–B the dispersion interaction

and dispersion terms. For example, double induction on opposite sides of a central molecule, far from adding up, causes cancellation (Figure 3a). The interaction between two dipoles is strongly influenced by the presence of a third polarizing dipole (the so–called Axilrod-Teller triple–dipole effect, Figure 3b). There is no straight-forward way of representing many-body interactions for large polyatomic molecules using empirical formulations.

3. Fitting intermolecular potentials with parametric equations

Electrostatic interactions are always present in organic media, and are often thought to be predominant. However, atom-atom electrostatic contributions are both stabilizing and destabilizing (negative and positive interaction energy, respectively), and therefore may cancel in extensive two–body summations. Induction and dispersion contributions are of relatively much smaller magnitude, but all contributions in a summation are attractive. Dispersion is present in all systems and makes a substantial contribution to the cohesive energy even of ionic bodies. It is the only attractive effect available for the cohesion of non–polar molecules. If these terms can be modeled by rather accurate formulations, many–body terms remain pretty much beyond the modeling capabilities of accurate or even semiempirical formulations. There is no hope of reproducing in computationally manageable formulas the complete physics of the interaction between large molecules, not because of intrinsic difficulties on the principles, but because the system is just too complicated. A description of interatomic potentials using the complete laws of electrostatics is therefore out of question, and there is no choice but to resort to empirical fitting.

As is always the case in such an enterprise, the value of a parametric scheme is $V = 1 - N/O$, where N is the number of parameters and O is the number of fitted observables. Therefore, V is zero if the scheme reproduces as many properties as the number of its parameters, and increases as a sharp function of parameter economy and transferability. At the same time, no economy and transferability are possible without paying the price of increasing inaccuracy. Finding the balance point, and moving in the direction of increasing value of the parametric scheme, is a process that requires a large amount of chemical sense and intuition.

The fitting of intermolecular potentials by parametric equations, for 15– to 100–atom molecules in all the most common combinations of organic chemistry, with C, H, N, O, F, Cl, S, Br atoms, has been carried out by several groups over the years (see references [5,6] for some examples and historical perspective). The fitting usually proceeded from the following principles:

a) the total interaction is represented as a summation of pairwise contributions, each of which depends solely on the distance between the interacting sites;

b) the sites from which distances are reckoned are identified with the location of atomic nuclei within the molecules (hence the name of atom–atom potentials), locations which in turn are determined in the best and easiest way using the results of X–ray diffraction experiments; no principle requires this to be so, and any molecular site ("lone pairs", even bond midpoints) could be, and has sometimes been, used as well;

c) the functional form of the interaction is determined by the requirement that the numerical evaluation of the potential and of its first two derivatives be as simple as possible; it should contain an R^{-6} term, broadly corresponding to dispersion, induction, and higher multipole interaction, an exponential or similar fastly decaying (say R^{-12}) term, corresponding to repulsion, and an R^{-1} term, broadly corresponding to electrostatic interaction between monopoles;

d) the numerical values of the parameters are to be calibrated using as a large as possible a body of experimental data, including the specific properties one wants to reproduce with the potentials. Thus, one should not use, say, virial coefficient of gases to fit potentials for liquids and crystals.

The derived parameters may be trusted to represent physical effects, or their physical significance may be neglected in exchange for practical performance. We prefer the latter attitude and hence offer the following considerations. Condition a) prevents the representation of many–body effects, which is simply not affordable for large molecules. This does not prevent such effects from being effectively incorporated in some of the extant parts of the potential. The calibration process is never carried out by allotting specific parameters to specific parts of the potential, because there is no simple means of, say, factoring an experimental heat of sublimation into dispersion, induction, electrostatic parts. Hence, the whole parameter set reproduces the whole experimental data (Table 2), and the physical significance of each parameter is lost. Conversely, this implies that electrostatic interactions may be effectively incorporated even in parametric schemes that do not include R^{-1} terms. Finally, condition b) should not be taken to imply that one atom–atom contribution represents the real potential or force between the two atoms.

TABLE 2.   Experimental quantities for the calibration of atom–atom potentials.

Property	Availability	Reliability	Usefulness
ΔH_{subl}	**	**	****
ΔH_{vap}	***	***	***
ΔH_{melt}	***	**	**
C_p	***	***	*
Lattice vibr.,ν	*	**	***
Compressibilities	*	-	****
Crystal structures	****	****	**
T_{melt}, T_{boil}	****	****	*
Excess H, S	****	***	*

This gap between parametric representations and real physics may be unpleasant, but is unavoidable, at least at the present state of affairs in intermolecular interaction. While long–term progress in the description of observables proceeds only through application of first principles, short–term progress may proceed via empirical shortcuts. Parametric approaches are in fact dead ends, but can as well effectively carry the burden of scientific progress. Of course, empirical schemes are to be discarded and renewed as new facts or new capabilities emerge.

4. Analytical forms of empirical atom–atom potentials: an example

The necessary parameters include coefficients for the repulsion, dispersion and electrostatic terms. A generalized form for the parametric expression of the non–electrostatic part of the intermolecular potential is:

$$E(R) = \epsilon \left[\frac{n+\lambda}{n+\lambda-6} - \left(\frac{R^0}{R}\right)^6 + \frac{6}{n+\lambda}\left(\frac{R^0}{R}\right)^n e^{\lambda(1-R/R^0)} \right] \qquad (2)$$

where ϵ is the well depth and R^0 the distance at the minimum. if $n = 0$ (so–called 6–exp form):

$$E(R) = A\,e^{-BR} - C\,R^{-6}; \qquad (3)$$

$$A = 6\,\epsilon\,e^{\lambda}/(\lambda-6), \quad B = \lambda/R^0, \quad C = \epsilon\,\lambda\,(R^0)^6/(\lambda-6)$$

$$\left(\frac{\partial^2 E^*}{\partial Z^2}\right)_{Z=1} = 6\,\lambda\frac{\lambda-7}{\lambda-6}$$

$$E^* = E/\epsilon; \quad Z = R/R^0.$$

If $n = 12$, $\lambda = 0$ (so–called 12-6 or Lennard–Jones form):

$$E(R) = F\,R^{-12} - G\,R^{-6}; \qquad (4)$$

$$F = \epsilon\,(R^0)^{12}, \qquad G = 2\,\epsilon\,(R^0)^6.$$

For the electrostatic part, in both intra- and intermolecular fields, atomic or site charges are derived by having them fit the molecular electrostatic potential obtained from ab initio quantum chemical calculations. In order to ensure the reliability of the molecular wavefunction, a high quality molecular orbital calculation must be carried out, and this is still problematic for very large organic molecules. If a low quality wavefunction is accepted, then the derived atomic or site charges must be regarded as no more than supplementary disposable parameters, with the further disadvantage of being non-transferable. These atom or site charges depend also on molecular conformation. Summations over R^{-1} terms are not convergent or conditionally convergent, a non-trivial technical problem in actual calculations. As a working example in the derivation of parameters , we describe here the development of a crystal force field calibrated using crystal data [7,8]. Computer simulations of condensed phases require the survey of multidimensional phase spaces, where the intermolecular potentials are to be recalculated a very large number of times. Simpler potentials and the removal of the convergence problem connected with the R^{-1} terms may save a considerable amount of computer time and, what is even more important, of human effort. Therefore, empirical parametric functions can profitably take the simple 6-exp form:

$$E_{ij} = A\,e^{-B\,R_{ij}} - C\,R_{ij}^{-6} \qquad (5)$$

244

where the double index designates any couple of atoms in different molecules, at a distance R_{ij}, A, B and C are parameters calibrated for each X...Y interaction, X or Y being C, H, N, O, S or Cl. The total packing energy is a lattice sum over i and j. Calibrating A, B and C in fact requires an estimate of ϵ, R^0 and λ.

A considerable amount of work has been devoted in the past to the development of 6-exp potentials, and reviews are available [9].The crystal properties of the elemental compounds have often been used to calibrate the potential between atoms of the same kind, relying on averaging rules for cross interactions. A key role has thus been played by N_2, O_2, Cl_2, CO_2, but such molecules, and their crystals, are radically different from the molecules and crystals of ordinary organic chemistry. Since a very large amount of structural data, and a substantial amount of thermochemical and vibrational data, are now available for organic crystals, the separate optimization of X...X, Y...Y and X...Y interactions is feasible. In particular, a careful optimization with an increase of the well depth parameter for the cross interactions simulates the missing electrostatic attraction terms. Even the peculiar features of the hydrogen bond are well reproduced by such a simple formulation.

The first problem is the calibration of equilibrium distances in the potential curves. The basic idea was to extract this information from contacts in crystal structures. Samples were retrieved from the Cambridge Structural Database (CSD), comprising data for molecular size and composition comparable to that of usual organic compounds (no molecules smaller than, say, benzene), at room-temperature, namely: hydrocarbons (C and H atoms only): 391 crystal structures (the potentials optimized for this sample have been assumed to be transferable to the hydrocarbon part of the other samples); oxahydrocarbons (C,H and O atoms): 590 crystal structures; azahydrocarbons (C,H and N atoms): 458 crystal structures; chlorohydrocarbons (C,H and Cl atoms): 103 crystal structures; sulfohydrocarbons (C,H and S atoms): 210 crystal structures; nitro compounds (C,H atoms and -NO or -NO$_2$ groups): 41 crystal structures; sulfones and sulfoxides (C, H atoms and SO or SO$_2$ groups): 53 crystal structures; aromatic nitro-amino derivatives: 8 structures; carboxylic acids: 173 structures; amides: 79 structures; alcohols: 43 structures; aromatic N-H...N compounds: 44 structures. It is easily seen that these samples cover most if not all of the organic functional groups. As usual when dealing with X–ray results, unreliable experimental H–atom positions were replaced by hydrogen atoms located by standard geometrical procedures.

The distribution of intermolecular distances is then analyzed. Calling N the number of distances in each distance interval, and M the number of molecules in each sample, the following density function is calculated for each X...Y pair:

$$D(R) = \frac{N}{M} \ \frac{1}{4\pi\,R^2\,\mathrm{d}R} \tag{6}$$

Assuming van der Waals radii as C, 1.75; H, 1.17; N, 1.5; O, 1.4 Å for C...C, O...O, N...N and N...O the sum of these radii defines an empirical limit, below which almost no contacts are observed in crystals. This rule does not always hold for mixed interactions, between atoms of different electronegativity; short contacts are found for example between carbonyl oxygen and acyl carbon atoms in carboxylic acid anhydrides; a substantial number of short C–H... O and C–H...N contacts appear. Contacts shorter than standard packing radii appear for the highly polarizable chlorine atoms.

An appropriate atom-atom potential should become repulsive only after the $D(R)$ has fallen below a certain threshold; the reason being that in non H–bonded organic crystals there are no strong attractive forces to compensate any steeply increasing repulsion at short distances. The minimum of the interatomic potential curve should be located approximately where the contact density reaches its first peak - the most frequent contacts at the shortest possible distance. Thus, the observed distributions provide a basic indication of the location of R^0 and on the beginning of the repulsive branch of the potential curve.

Next, the calibration of well depths has to be considered. Substances for which a complete X–ray crystal structure and an experimental heat of sublimation are available were collected, in order to rescale the calculated packing energy with the observed sublimation enthalpy, ΔH_{subl}.

The last problem is the calibration of the steepness parameter, λ. In principle, this requires information on the curvature at the minimum, which could come from lattice vibrational spectroscopy. Such data are however so scant that a real quantitative calibration is impossible. An universal value, $\lambda = 13.5$, was then assumed.

The result of this fitting procedure is summarized in Table 3. These potentials were developed for organic crystals, and that their performance is therefore granted only for crystals. Their transferability to other systems, like liquids and even solutions, is however promising.

5. Applications to static simulations

Static computer simulations use empirical formulations for the potential, but a fixed molecular arrangement, and neglect kinetic energy. A crystal structure is built from the information coming from the X–ray diffraction experiment, as a static array of molecules obeying crystal symmetry prescriptions. The computer program then proceeds to calculate the potential energy for any molecule within such an array, an easy task since all in-

TABLE 3. Atom–atom potentials from crystal data [7,8]: functional form $E = Ae^{-BR} - CR^{-6}$. Energies in kcal/mole, distances in Å.

Interaction	A, kcal/mole	B, Å^{-1}	C, (kcal/mole)$^{-1}$ Å^{-6}
H...H	5774	4.01	26.1
H...C	28870	4.10	113
H...N	54560	4.52	120
H...O	70610	4.82	105
H...S	64190	4.03	279
H...Cl	70020	4.09	279
C...C	54050	3.47	578
C...N	117470	3.86	667
C...O	93950	3.74	641
C...Cl	93370	3.52	923
N...N	87300	3.65	691
N...O	64190	3.86	364
O...O	46680	3.74	319
O...Cl	80855	3.63	665
S...S	259960	3.52	2571
Cl...Cl	140050	3.52	1385
HB...O (amides)	3607810	7.78	238
HB...O (acids)	6313670	8.75	205
HB...O (alcohols)	4509750	7.78	298

termolecular distances are exactly known and invariable. Such simulations, whose cost on a modern workstation is a few seconds, can provide rather accurate evaluations of heats of sublimation, of enthalpy differences between polymorphs [10], and of lattice vibrational frequencies and entropies. From these, the calculation of some thermodynamic properties of crystals, like for example the vapour pressure, is also feasible. If the potential energy is calculated along any displacement coordinate within the crystal, the corresponding activation energy barrier can be estimated; an example is molecular tumbling or lattice diffusion. More detail is to be found in [9]. When the crystal structure of a given compound is not available from X-ray diffraction experiments, it can be guessed, with moderate success, using crystal structure construction and prediction algorithms [11,12]. Normally, 10–100 crystal structures in many space groups are obtained, and no present–day force field is able to discriminate the putative from the real ones on the basis of packing energy. For all we know, this could even mean that such a discrimination is intrinsically impossible, real structures being dictated by entropy or by kinetics. In any case, the prediction exercise produces the correct crystal density and the correct sublimation enthalpy, even if they apply to a putative structure, for which these quantities do not differ significantly from the corresponding ones in real structures.

6. Dynamic simulations

In the static scheme, crystal energies are calculated as potential energy only, and hence refer in principle to 0 K. However, the X–ray results refer to average atomic positions at room T, and hence intermolecular potentials are effectively calibrated at 300 K, at least to some extent. For any disordered state, like the liquid crystalline or the liquid or glassy states, the above described static calculations are clearly inadequate, and molecular dynamics (MD) or Monte Carlo (MC) simulations must be carried out for an appropriate sampling of configurational space. Dynamic, or evolutive simulation techniques like Monte Carlo, pose then special problems:

a) the parameterization of a force field for dynamic simulations is even more problematic than for static lattice energy calculations. Since all force fields are incomplete, transferability may be a problem not only over different compounds, but over different phases of the same compound. Simulations with structural evolution may visit more frequently the repulsive branch of the potentials, poorly determined when calibrated from crystal data where molecules never climb the repulsive branch too far. In polar media, polarization is presumably one of the main driving forces in the dynamic evolution, while, as discussed before, no simple and sound method for empirically accounting for polarisation is available for large molecules;

b) the derivation of thermodynamic averages and of kinetic pathways requires a robust sampling of phase space, a real problem for large systems; besides, even on the fastest available computers, simulation times on chemically significant systems (say, 5,000 atoms) are limited to the order of nanoseconds;

c) seldom recognized, the importance of the careful design of a computational experiment increases exponentially with computing power. MD is so ingenious that its results have a tendency to be (mis)taken for truth, while they just reflect a simulation. Just as ordinary chemistry requires the design of the appropriate chemical experiment to prove an hypothesis, so computational chemistry requires the design of an appropriate computational experiment, and a sound and skeptical interpretation of its results.

7. Modern force fields and their performance

Static simulations of crystals have used, over the years, more complicated forms than just the 6-exp one. The most straightforward extension (the

Williams [13] scheme) is the 6-exp-1 one, supplementing 6-exp with a $q_i q_j / R_{ij}$ term. Even more elaborate is the distributed multipole scheme [14], which supplements the 6–exp form with multipole–multipole terms, calculated among strategically located sites over the molecule. This last scheme is more flexible and includes chemically significant anisotropies in the electron density distribution. Both schemes suffer from the drawback of requiring in principle an expensive MO calculation for each molecule. Nowadays, many widely used computer packages for the dynamic simulation of organic phases work on the basis of a generalized intra/inter molecular force field of the form:

$$U = \sum K_B (R - R_0)^2 + \sum K_\theta (\theta - \theta_0)^2 + \sum K_{UB}(S - S_0)^2 + \sum K_\chi [1 + \cos(n\chi - \delta)] + \sum K_{IMP}(\phi - \phi_0)^2 + \sum [AR^{-12} - BR^{-6} + q_i q_j R^{-1}] \tag{7}$$

where the potential terms refer, respectively, to stretching, bending, Urey–Bradley 1-3, torsional, improper dihedral, and non-bonded interactions. The latter term includes both intra- and intermolecular contributions. The problem of a consistent parametrization is at once evident, since Urey–Bradley, torsional, and intra- and intermolecular non–bonded terms are strongly coupled. In the preparation of an all–atom empirical energy function for the simulation of nucleic acids, Karplus and coworkers [15] have used experimental data as diverse as IR–Raman, crystal structures and lattice vibrations, heats of sublimation, dipole moments, backbone dihedrals and sugar puckering angles, and base-pair energies. Kollman and coworkers [16] use quantum chemical 6-31G* electron densities and potentials, fitted by point charges distributions. Jorgensen's OPLS force field [17] reproduces with great accuracy the densities and enthalpies of vaporization of organic liquids. All these large-scale force fields require a correspondingly large number of parameters; for example, the GROMOS Molecular Dynamics package [18] requires a manual running into some 1000 pages to properly illustrate its way of describing molecules. Extensive parametrization competes with the struggle to preserve the physical meaning of analytic expressions.

8. Concluding remarks

In the computer modelling of liquid crystals, global molecular potentials, like the Gay–Berne potential, partitioned into a shape–envelope and an energy part, have been succesfully applied to the simulation of many phase properties, but several attempts are now being made toward the use of atomic-level potentials. The advantage is an obvious one, in an increasing adherence to chemical detail; the disadvantages are also obvious, in the exponentially growing computer intensiveness, given the considerable size and complexity of nematogenic molecular systems.

One of the main obstacles against the use of atomic–level potentials in liquid and liquid crystal systems is a proper treatment of long–range effects. Many thermophysical properties of crystals are determined mainly by short–range interactions: the density, for example, is mainly dictated by short–range repulsion, while the bulk of the lattice energy is well reproduced also by properly calibrated R^{-6} potentials. Hydrogen bonding is also an example of essentially short–range interaction, so that the main structural determinants (the best arrangement of nearest–neighbour molecules) in a crystal can be safely modelled even without R^{-1} terms in the potential. Subtle differences between polymorphs, or such collective dynamical evolution processes as nematogenic phase transitions, may require a proper description of long–range effects, just like the reproduction of detail in liquid radial distribution curves, or of properties like viscosity or heat capacity. The problem then arises of the choice and appropriateness of the force field for each particular application. Once an empirical approach has been assumed, a safer practical attitude is perhaps to be concerned more with the validation of the adopted force field for the problem at hand, than with the adherence of the parameters or of the functional form to physical principles, which are anyway outside reach. If a sound body of experimental data for calibration were available, it could well be that simpler potentials dispensing with R^{-1} terms, similar to those which have been calibrated for organic crystals, could reproduce in a satisfactory manner some thermophysical properties of liquid crystals, and even provide a tool for at least a preliminary investigation of nematogenic phase transitions. Otherwise, as described in detail in other papers in the present proceedings, methods for an at least preliminary treatment of long–range contributions to the intermolecular potential are already being developed, at the expense of increased computing effort, with a promise of increasing accuracy.

References

1. Rigby, M., Smith, E.B., Wakeham, W.A. and Maitland, G.C. (1986) *The Forces Between Molecules*. Oxford Science Publications, Oxford.
2. Berkovitch–Yellin, Z. and Leiserowitz, L. (1984) *Acta Cryst.*, **B40**, 159.
3. Letcher, T.M. and Bricknell, B.C. (1996) *J. Chem. Eng. Data*, **41**, 166
4. Meot–Ner M. (1984) *J. Am. Chem. Soc.*, **106**, 1257
5. Kitaigorodskii, A.I and Mirskaya K.V. (1962) *Soviet Physics - Crystallography*, **6**, 408
6. Williams, D.E and Starr, T.L. (1977) *Computers & Chemistry*, **1**, 173.
7. Filippini, G. and Gavezzotti, A. (1993) *Acta Cryst.*, **B49**, 868.
8. Gavezzotti, A. and Filippini, G. (1994) *J. Phys. Chem.*, **98**, 4831.
9. Gavezzotti, A. and Filippini, G. (1997) Energetic Aspects of Crystal Packing, in

250

Theoretical Aspects and Computer Modelling and Molecular Solid State, Ch. 3. Wiley, Chichester.

10. Gavezzotti, A., and Filippini, G. (1995) *J. Am. Chem. Soc.*, **117**, 12299.

11. Gavezzotti, A. (1994) *Acc. Chem. Res.*, **27**, 309.

12. Gavezzotti, A., Filippini, G., Kroon, J., van Eijck, B.P. and Klewinghaus, P. (1997) *Chem. Eur. J.*, **3**, pp. 893–899.

13. Cox, S.R., Hsu, L.-Y. and Williams, D.E. (1981) *Acta Cryst.*, **A37**, pp. 293.

14. Price, S.L. (1996) *J. Chem. Soc. Far. Trans.*, **92**, 2997.

15. MacKerell, A.D. Jr., Wiorkiewicz–Kuczera, J. and Karplus, M. (1995) *J. Am. Chem. Soc.*, **117**, 11946.

16. Cornell, W.D., Cieplak, P., Bayly, C.I, Gould, I.R., Merz, K.M. Jr., Ferguson, D.M., Spellmeyer, D.C., Fox, T., Caldwell, J.W. and Kollman, P.A. (1995) *J. Am. Chem. Soc.*, **117**, 5179.

17. Jorgensen, W.L., Maxwell, D.S. and Tirado-Rives, J. (1996) *J. Am. Chem. Soc.*, **118**, 11225.

18. van Gunsteren, W.F., Billeter, S.R., Eising, A.A., Hunenberger, P.H., Kruger, P., Mark, A.E., Scott, W.R.P. and Tironi, I.G. (1996) *Biomolecular Simulation: The GROMOS96 Manual and User Guide.* BIOMOS b.v., Zürich–Groningen.

ATOMISTIC MODELLING OF LIQUID CRYSTAL PHASES

M.R. WILSON, M.J. COOK AND C. M^{C}BRIDE
*Department of Chemistry, University of Durham,
South Road, Durham, DH1 3LE, United Kingdom.*

1. Introduction

The last few years have seen a large number of simulation studies of liquid crystal systems. Two general methods have been popular. In the first approach, simplified molecular potentials have been used. These seek to approximate the gross features of mesogenic interactions (e.g molecular shape anisotropy, molecular dipole moments etc.) by single site potentials [1–6]. Such models have been highly successful in predicting the formation of a number of liquid crystal phases, and allowing the development of techniques for the calculation of key bulk properties, such as transport coefficients [7] and elastic constants [8]. In a second simulation approach, atomistic models have been used [9–26]. These seek to represent molecular structure as faithfully as possible by representing nonbonded interactions at the atomic level and accurately representing internal molecular motion. This approach aims to be able to accurately predict mesophase behaviour for *real systems* based only on a knowledge of molecular structure. The drawback of this method is that it is computationally expensive, and large quantities of computer time must be expended in order to compute the numerous pair interactions in such systems.

Atomistic simulations of liquid crystals have recently been reviewed in some detail [27,28]. This article seeks to point to four key areas where new developments have occurred in the past two years,

- the development of hybrid atomistic models (section 2),
- nano-second timescale simulations of atomistic models (section 3),
- the development of improved force fields for liquid crystals (section 4),
- the calculation of helical twisting power for chiral dopant materials (section 5).

251

P. Pasini and C. Zannoni (eds.), Advances in the Computer Simulations of Liquid Crystals, 251–262.
© 2000 *Kluwer Academic Publishers. Printed in the Netherlands.*

2. Hybrid Gay-Berne/Lennard-Jones models

2.1. INTRODUCTION

An interesting variant of atomistic simulations arises when rigid sections of the model mesogen are replaced by simpler potentials. In one recent example [20] the biphenyl cores of n-pentylcyanobiphenyl (5CB) were replaced by large spherical sites at the centre of each ring; thus providing large savings in computer time over united atom models of the same molecules. However, a more general hybrid model arises when parts of a molecule are replaced by anisotropic potentials (such as the Gay-Berne model) [29]. At Durham, we have developed the GBMOL [30] simulation program to carry out molecular dynamics simulations of such systems. The potential used in GBMOL is given by

$$E = \sum_{\text{bonds}} E_{\text{bond}} + \sum_{\text{angles}} E_{\text{angle}} + \sum_{\text{dihedrals}} E_{\text{dihedral}} \qquad (1)$$

$$+ \sum_{\langle ij \rangle} E_{\text{LJ}} + E_{\text{el}} + \sum_{\langle ij \rangle} E_{\text{LJ-GB}} + \sum_{\langle ij \rangle} E_{\text{GB-GB}};$$

where

$$E_{\text{bond}} = k_{\text{bond}} \left(l - l_0 \right)^2, \qquad (2)$$

$$E_{\text{angle}} = k_{\text{angle}} \left(\theta - \theta_0 \right)^2, \qquad (3)$$

$$E_{\text{dihedral}} = \sum_n c_n \cos^n \phi \qquad (4)$$

or

$$E_{\text{dihedral}} = \sum_n \frac{v_1}{2} \left(1 + \cos \phi \right) + \frac{v_2}{2} \left(1 - \cos 2\phi \right) + \frac{v_3}{2} \left(1 + \cos 3\phi \right). \qquad (5)$$

In equation 1, E_{LJ} is the Lennard-Jones pair potential; E_{el} is an electrostatic pair potential; $E_{\text{GB-GB}}$ is the Gay-Berne potential; $E_{\text{LJ-GB}}$ is the pair potential for the interaction between a Lennard-Jones site and a Gay-Berne site; and $\langle ij \rangle$ represents a sum over all pairs of sites. In equations 2-5, k_{bond} and k_{angle}, are force constants; l and l_0 are respectively the actual and natural lengths of a bond; θ and θ_0 are respectively the actual and natural bond angles; c_n, v_1, v_2, v_3 are dihedral angle coefficients; and ϕ is a dihedral angle. In GBMOL, the forces on the centres of the Gay-Berne particles are handled in the same way as the forces on atomic sites. The equations of motion are solved using an extension of the leap-frog algorithm [31], that integrates the translational motion of the atomic and Gay-Berne sites, and the motion about the centres of mass of the Gay-Berne sites separately. Additional angle bending terms are included in the angle sum of equation

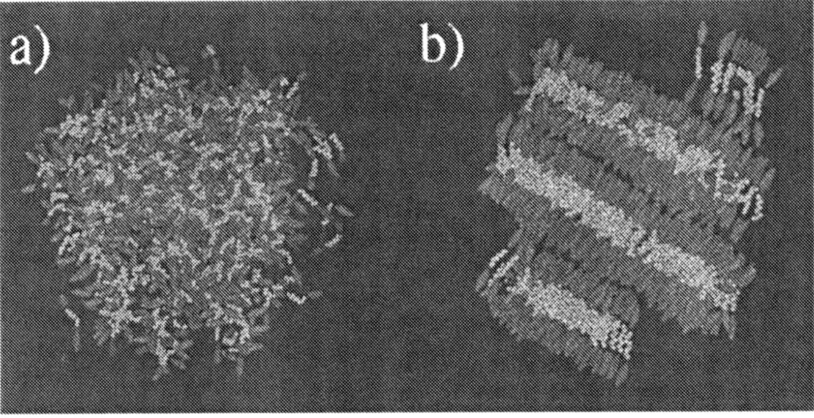

Figure 1. Snapshots showing the structure of a) isotropic phase, b) smectic-A phase for 512 liquid crystal dimer molecules composed of two Gay-Berne particles linked via an eight-atom flexible alkyl chain.

1 to prevent free rotation of the Gay-Berne particles about their centres of mass.

2.2. SIMULATION RESULTS

A number of models have already been simulated using the GBMOL potential [29, 32, 33]. Wilson [29] has examined a model for a liquid crystal dimer consisting of two Gay-Berne mesogens linked by an eight-atom flexible alkyl chain. At zero pressure this system exhibits isotropic (figure 1a), smectic-A (figure 1b) and smectic-B phases.

By comparison with the Gay-Berne potential at zero pressure [34], the flexible chains clearly stabilise the presence of a smectic-A phase. However, hybrid systems of this form seem to have a strong propensity to separate into Gay-Berne rich and Lennard-Jones rich domains (this is clearly seen in the pretransitional region of the iso tropic phase) and so nematic phases have not been observed with this system. The transition from the isotropic to the smectic-A phase occurs spontaneously on cooling, but requires simulations in excess of 6 ns. This is partly due to slow molecular reorientation, and also partly due to a change in molecular shape which accompanies the change in phase. Molecules become more elongated in the smectic phases compared to the isotropic phase, due to the quenching out of unfavourable *gauche* conformations in which the Gay-Berne units lie at an angle to the director field. Changes in conformation also lead to a small increase in the spacing of smectic layers which occurs as temperature is decreased. It is noted that molecules remain flexible even in the smectic-A phase and are readily able to

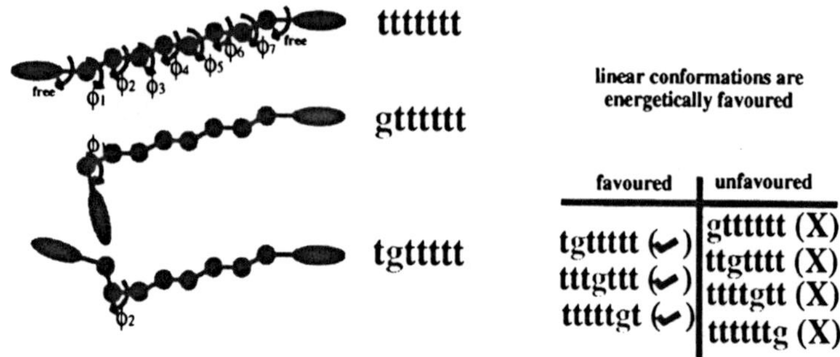

Figure 2. Schematic diagram showing three conformations for a Gay-Berne dimer system. Favoured gauche conformations are indicated.

undergo conformational changes. The dimer system exhibits a classic *odd-even effect* in bond order parameters and in dihedral angle distributions [29]. This occurs because of the preference for certain conformations in which the Gay-Berne mesogens remain orientated along (or close to) the director. This is shown in figure 2 where the even dihedral (tgttttt) clearly fulfils this criterion better than the odd dihedral (gttttt). As expected, the *all-trans* chain conformation is most favoured, and approximately 47% of molecules have this conformation in the smectic-A phase at 300 K, compared to 19% in the isotropic liquid at 330 K. Effective torsional angle potentials can be derived from dihedral angle distributions. These show that in moving from the isotropic to the smectic-A phase the energy gap between *trans* and *gauche* conformers ΔE_{gt} increases by approximately 2 kJ/mol for odd dihedrals and 0.5 kJ/mol for even dihedrals.

The computational speed of the hybrid model, compared to fully atomistic studies, allows transition temperatures to be pinned down to approximately 30 K. This is extremely encouraging with regard to future molecular engineering applications for simulations of this type.

McBride and Wilson have looked at a Gay-Berne particle with two alkyl chains (C_3 and C_7) at opposite ends of the molecule in the NVE ensemble at two separate densities [32]. At a density of approximately 752 kg m^{-3} this system shows only isotropic liquid and smectic-B phases. However, at a higher density of 970.4 kg m^{-3} a nematic phase is also seen. As with the Gay-Berne potential itself, increases in density tend to promote nematic behaviour. This is caused by the competition between orientational entropy and translational entropy in the system.

As the volume of the system is reduced, excluded volume effects become more important and the reduction in translational entropy (caused by molecules at an angle to each other excluding each other from a large volume of space) starts to strongly favour parallel alignment of molecules.

As with the dimer molecules, C_3-GB-C_7 molecules exhibit an *odd-even effect* in bond order parameters (see table 1). However, the influence of the *odd-even effect* falls off as the bonds get progressively further away from the molecular core.

TABLE 1. Bond order parameters for the C_7 chain in the C_3-GB-C_7 system of McBride *et al.* [32] at 300 K.

Bond	Order parameter
Gay-Berne axis	0.73
$C_1 - C_2$	0.05
$C_2 - C_3$	0.58
$C_3 - C_4$	0.04
$C_4 - C_5$	0.51
$C_5 - C_6$	0.04
$C_6 - C_7$	0.39

Lyulin *et al.* [33] have used GBMOL to carry out the first large scale simulations of a main chain liquid crystalline polymer (LCP). Sixty four polymer chains are considered in the simulation, with each chain consisting of ten monomer units. Each monomer consists of Gay-Berne particles linked by a flexible alkyl chain of length m. Four polymers are considered with spacer lengths $m = 5, 6, 7, 8$. Spontaneous orientational ordering occurs on cooling the polymer from the isotropic melt. However, this is a very slow process for these highly viscous materials and occurs over long timescales (in the order of 14 ns). As with experimental systems, an *odd-even* dependency of thermodynamic properties with the length of flexible spacer is observed. As with many Gay-Berne systems, the change in volume at the phase transition is very high compared to experiment. This is a major problem when making comparisons between idealised models and *real* systems. Large volume changes also occur with hard-particle models, and to date this problem has not been properly addressed by simulators.

3. Fully atomistic simulations

Fully atomistic studies of liquid crystals are becoming increasingly popular [9–26] Many early simulations suffered from a lack of available computer time. This restricted simulations to under 500 ps. However, as computers have increased in speed, it has become increasingly possible to extend such simulations into the nanosecond regime. McBride et al. [26] have

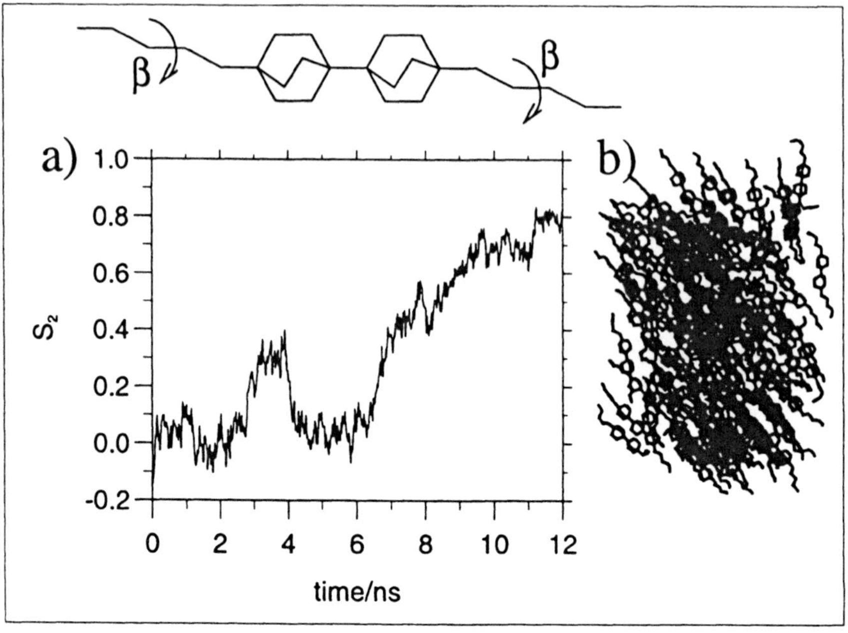

Figure 3.   a) Growth of orientational order for the molecule 4,4-'di-n-pentyl-bibicyclo-[2.2.2]octane (5,5-BBCO) at 300 K. b) A snapshot from the nematic phase of 5,5-BBCO at 250 K.

recently studied a united atom model for the molecule 4,4-'di-n-pentyl-bibicyclo[2.2.2]octane (5,5-BBCO). This is the first fully atomistic study to show the growth of an orientationally ordered phase directly from a fully isotropic liquid (see figure 3). Growth in orientational order was demonstrated by two independent quench runs starting from the isotropic phase and dropping the temperature below the isotropic–nematic phase transition.

As with the hybrid Gay-Berne/Lennard-Jones model, effective torsional angles are influenced by the change in structure of the bulk fluid surrounding a molecule that occurs when the system changes phase. McBride *et al.* also measure a change in ΔE_{gt} of 2 kJ/mol for the dihedral β (marked in figure 3). However, in this case the figure of 2 kJ/mol corrsponds to a *reduction* (rather than an increase) in the *gauche/trans* energy gap. This is because the *tgt* chain configuration becomes strongly selected by the nematic mean field for this molecule: when both chains have this conformation, they are able to lie along a similar axis to the core.

4. Improved force fields for liquid crystals

4.1. NONBONDED INTERACTIONS

For bulk simulations, the nonbonded interactions are undoubtably the most important force field components. A wide-range of parameter sets have been developed in recent years, including many all-atom [35–39] as well as united atom [40–42] force fields. Nonbonded parameters can be found to vary widely between individual parameterisations. However, a consensus is beginning to emerge that many of the force fields that were parameterised for single molecule systems, or for crystal structures, are not particularly good for modelling fluid phases.

For the simulation of liquids, it seems that the approach advocated by Jorgensen in the development of the OPLS (united-atom) and OPLS-AA (all-atom) force fields yields the best results. This involves a series of simulations for simple liquids in which the nonbonded parameters are systematically altered to produce the best fit with experimental liquid phase densities and heats of vaporisation. For example, the OPLS-AA force field makes reference to calculations on alkanes, alkenes, alcohols, ethers, acetals, thiols, sulfides, disulfides, aldehydes, ketones and amides in the parameterisation of nonbonded terms. It seems likely that nonbonded parameters developed in this way are going to be the most useful for bulk simulations of liquid crystal phases.

One serious omission from many force fields relates to the modelling of molecules which contain aromatic rings. Lennard-Jones interactions alone predict a minimum energy configuration corresponding to the stacking of the two phenyl rings, one above the other in a D_{6h} configuration. However, *parallel displaced* (C_{2h}) and *T-shaped* (C_{2v}) configurations are lower in energy [43]. In order to correctly describe the ring-ring interaction energy, a quadrupolar interaction is required in addition to Lennard-Jones sites. This can be achieved either by the use of point quadrupoles in the centre of the rings. Or by creating a quadrupole by using partial charges on individual atoms.

Polarisation effects are largely ignored in force fields. These are unlikely to be of major significance for many thermotropic mesogens. However, polarisation effects are important in many aqueous systems (including lyotropic phases) but are currently neglected in most models of water.

4.2. DIHEDRAL ANGLE PARAMETERS

Dihedral angle potentials have a major influence on molecular structure. In turn, the overall shape of a molecule has a major impact on phase behaviour and transition temperatures. Consequently, it is vital to use accurate dihedral angle data in the molecular force field. Unfortunately, very little accurate experimental data exists for these potentials.

Ab initio calculations on mesogenic fragments provide an alternative path to accurate torsional data [44]. In such calculations it is vital to include the effects of electron correlation, because this gives rise to intramolecular van der Waal's interactions. In consequence traditional quantum chemical SCF calculations are not sufficient, and calculations must be undertaken at (*at least*) the MP2 level, or by using density functional methods. The latter area of particular interest for liquid crystal fragments because, in terms of computer time, they scale much better to large molecules than MP2 calculations.

Recently, results have been reported for torsional potentials in a number of liquid crystal fragments [44]; including the ring-chain torsional potential for n-pentylcyclohexane, n-pentylbenzene and methoxybenzene; the inter-ring torsion in biphenyl, 2-fluorobiphenyl, 2,2'-difluorobiphenyl, 4-cyanobiphenyl, 5CB, phenylcyclohexane and (4-cyanophenyl)cyclohexane. The calculations also confirm that torsional potentials are largely transferable between molecules for many fragments, and can therefore be used safely in generalised molecular force fields.

5. Calculations of helical twisting power

5.1. INTRODUCTION

The helical twisting power (HTP) determines the pitch of the chiral nematic phase produced when a nematic liquid crystal phase is doped with a low concentration of chiral solute molecules. Molecules with large HTP values have major applications in electro-optic displays and in optical data processing. In this section of the paper we present a new Monte Carlo technique to calculate HTP values.

5.2. THEORY

When a nematic is doped with a chiral solute a linear relationship holds between the equilibrium wavenumber of the twist k_0 and the concentration of the chiral dopant ρ, ($\rho = N/V$)

$$k_0 = 4\pi\rho\beta \tag{6}$$

where the constant of proportionality, β, is defined as the helical twisting power.

When the chiral solute is added to a uniformly twisted nematic of pitch $\lambda = 2\pi/k$, the contribution to the elastic free energy ΔF is given by

$$\Delta F = \frac{1}{2} V K_2 \left(k - k_0\right)^2 = \frac{1}{2} V K_2 \left(k - 4\pi\rho\beta\right), \tag{7}$$

where K_2 is the Frank elastic constant for twist deformations. For a pair of chiral enantiomers at low concentrations $(N_+/V,\ N_-/V)$, equation 7 becomes [45]

$$\Delta F = \frac{1}{2} V K_2 \left(k - \frac{4\pi\beta}{V} \left[N_+ - N_-\right]\right)^2; \tag{8}$$

and in the limit N_+, $N_- \to 0$, the difference in the chemical potential between the enantiomers is given by

$$\Delta\mu = \mu_- - \mu_+ = \frac{\partial F}{\partial N_-} - \frac{\partial F}{\partial N_+} = 8\pi\beta K_2 k. \tag{9}$$

Consequently, a calculation of K_2 and $\Delta\mu$ is sufficient to determine β. Allen [45] has already used equation 9 to determine the helical twisting power for a chiral dimer consisting of two hard ellipsoids in surface contact. In this case, an expression for $\Delta\mu$ can be derived from the background pair-correlation function for monomer ellipsoids. For atomistic models of chiral dopants a different approach to the calculation of $\Delta\mu$ is required (section 5.3).

5.3. STATISTICAL PERTURBATION THEORY

Using *Statistical Perturbation Theory*, the free energy difference ΔF_{BA} between two systems A and B can be calculated by summing up all the individual contributions to the free energy calculated during the slow *hypothetical mutation* of A into B. If A and B are described by the Hamiltonians, $\mathcal{H}_A$ and $\mathcal{H}_B$, then

$$\Delta F_{BA} = \sum_{i=1}^{n} \Delta F_{\lambda_i \to \lambda_{i+1}} \tag{10}$$

where

$$\Delta F_{\lambda_i \to \lambda_{i+1}} = -k_B T \ln \left\langle \exp\left[-\left(E_{\lambda_{i+1}} - E_{\lambda_i}\right)\right]\right\rangle_{\lambda_i} \tag{11}$$

and

$$E_{\lambda_i} = \lambda_i \mathcal{H}_A + \left(1 - \lambda_i\right) \mathcal{H}_B. \tag{12}$$

In equation 11, $\langle\rangle_{\lambda_i}$ denotes an ensemble average corresponding to a point on the path defined by $\lambda = \lambda_i$, and the free energy change is carried out in a total of n steps, with $0 \leq \lambda_i \leq 1$.

260

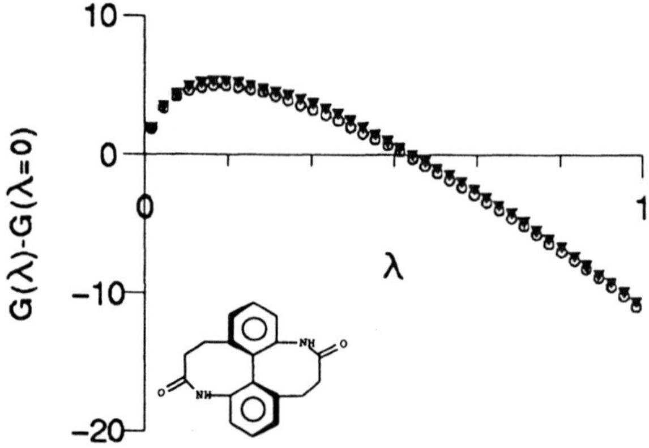

Figure 4. Free energy change for the growth of enantiomers into a twisted nematic Gay-Berne solvent. $\lambda = 0$ corresponds to no solvent present. $\lambda = 1$ corresponds to a single molecule of the enantiomer shown, fully immersed in the solvent. Squares – (+)-enantiomer, circles – (–)-enantiomers.

For a pair of enantiomers, ΔF_{BA} corresponds to the change in free energy when one enantiomer is mutated into its mirror image in the presence of a twisted nematic solvent. In recent work [46], we have considered the situation where the solvent in question is composed of Gay-Berne particles [47] and the chiral dopants are represented atomistically by the MM2 potential [48]. In this case, a series of Monte Carlo calculations can be undertaken to grow each enantiomer into the Gay-Berne solvent, and to calculate the free energy changes for these processes. The Monte Carlo calculations require the use of twisted periodic boundary conditions [49] to simulate a twisted nematic phase of Gay-Berne particles; and employ a double-wide sampling technique [50] to calculate free energy differences at each step in the perturbation. Typically, 40 perturbation steps are needed to grow a single enantiomer into a pure solvent composed of 504 Gay-Berne particles. We have employed the form of the Gay-Berne potential used in reference [34]. The twist (and other) elastic constants for this variant of the model have been calculated at a number of state points in reference [8]. Full details of the simulation method will be given elsewhere [46].

5.4. SIMULATION RESULTS

Figure 4 shows a typical set of results for a pair of enantiomers of a chiral dopant molecule. The overall reduction in the free energy for both curves arises from solvation effects. However, it is the difference in the curves at $\lambda = 1$ that provides a value of $\Delta F_{BA} =$ that can be used to calculate β in equation 9. For the material shown in figure 4, $\Delta F_{BA} = 0.425$ kJmol^{-1}, yielding a value of $\beta/\mu m^{-1} = +69$, compared to the experimental value of $\beta/\mu m^{-1} = +58$. In all, we have considered 5 chiral dopants [46]. In each case we correctly predict the sign of β and the magnitude of β to within a mean error, compared to experimental values, of approximately 15%. The one drawback of the method is that the computed free energy differences are extremely small; and therefore require long Monte Carlo simulations at a large number of perturbation steps.

6. Conclusions

Atomistic simulations of liquid crystals remain expensive, but are becoming accessible with modern advances in computer power. Currently, simulations are restricted to a few hundred molecules at the most, and lengthy simulations are required to see liquid crystal behaviour. Growth of a liquid crystal phase from the isotropic phase requires between 6-10 ns for low molecular weight materials, and longer for polymeric systems.

Hybrid models, involving Lennard-Jones and Gay-Berne sites have provided interesting data on main chain polymer systems (see section 2) and have proved extremely useful in predicting helical twisting powers (see section 5). However, current *state-of-the-art* fully atomistic calculations are still some distance short from being used in a predictive sense to forecast transition temperatures and phase sequences.

Future developments are likely to concentrate on improved force fields (section 4) and on parallel computing techniques. These will allow far more quantitative predictions to be made in this area.

References

1. Allen, M.P. and Wilson, M.R. (1989) *J. Computer-Aided Molec. Design*, **3**, 335.
2. Frenkel, D. and Mulder, B.M. (1984) *Phys. Rev. Lett.*, **52**, 287
3. Frenkel, D. and Mulder, B.M. (1985) *Molec. Phys.*, **55**, 1171.
4. Frenkel, D. and Eppenga, R. (1982) *Phys. Rev. Lett.*, **49**, 1089.
5. Frenkel, D. (1988) *J. Phys. Chem.*, **92**, 3280.
6. McGrother, S.C., Williamson, D.C. and Jackson, G. (1996) *J. Chem. Phys.*, **104**, 6755.
7. Allen, M.P. (1990) *Phys. Rev. Lett.*, **65**, 2881.
8. Allen, M.P., Warren, M.A., Wilson, M.R., Sauron, A. and Smith, W. (1996) *J. Chem. Phys.*, **105**, 2850.
9. Wilson, M.R. and Allen, M.P. (1991) *Molec. Cryst. Liq. Cryst.*, **198**, 465.
10. Wilson, M.R. and Allen, M.P. (1992) *Liq. Cryst.*, **12**, 157.
11. Wilson, M.R. (1996) *J. Molec. Liq.*, **68**, 23.
12. Komolkin, A.V., Molchanov, Yu.V. and Yakutseni, P.P. (1989) *Liq. Cryst.*, **6**, 39.
13. Picken, S.J., van Gunsteren, W.F., van Duijnen, P.Th. and de Jeu, W.H. (1989) *Liq. Cryst.*, **6**, 357.
14. Jung, B. and Schürmann, B.L. (1990) *Molec. Cryst. Liq. Cryst.*, **185**, 141.
15. Ono, I. and Kondo, S. (1991) *Molec. Cryst. Liq. Cryst.*, **8**, 69.
16. Ono, I. and Kondo, S. (1992) *Bull. Chem. Soc. Jpn.*, **65**, 1057.

17. Ono, I. and Kondo, S. (1993) *Bull. Chem. Soc. Jpn.*, **66**, 633.
18. Krömer, G., Paschek, D. and Geiger, A. (1993) *Ber. Bunsenges Phys. Chem.*, **97**, 1188.
19. Huth, J., Mosell, T., Nicklas, K., Sariban, A. and Brickmann, J. (1994) *J. Phys. Chem.*, **98**, 768.
20. Cross, C.W. and Fung, B. (1994) *J. Chem. Phys.*, **101**, 6839.
21. Yoneya, M. and Berendsen, H.J.C. (1994) *J. Phys. Soc. Jpn.*, **63**, 1025.
22. Kolmolkin, A.V., Laaksonen, A. and Maliniak, A. (1994) *J. Chem. Phys.*, **101**, 4103.
23. Yakovenko, S.Y., Muravski, A.A., Kromer, G., and Geiger, A. (1995) *Molec. Phys.*, **86**, 1099.
24. Yakovenko, S.Y., Kromer, G., and Geiger, A. (1996) *Molec. Phys.*, **275**, 91.
25. Hauptmann, S., Mosell, T., Reiling, S. and Brickmann, J. (1996) *Chem. Phys.*, **208**, 57.
26. McBride, C., Wilson, M.R. and Howard, J.A.K. (1998) *Molec. Phys.*, **93**, 955.
27. Wilson, M.R. (1998) Molecular modelling, in *Handbook of Liquid Crystals.* Demus, D., Goodby, J., Gray, G.W., Spiess, H.-W. and Vill, V. (eds.), volume 1, chapter III.3, Wiley-VCH, Weinheim.
28. Wilson, M.R. Atomistic simulations of liquid crystals, in *Structure and Bonding. Liquid Crystals.* Mingos, M. (ed.), Springer-Verlag, Heidelberg, in press.
29. Wilson, M.R. (1997) *J. Chem. Phys.*, **107**, 8654.
30. GBMOL (1996) *A replicated data molecular dynamics program to simulate combinations of Gay-Berne and Lennard-Jones sites.* Author: Wilson, M.R. University of Durham.
31. Allen, M.P. and Tildesley, D.J. (1987) *Computer Simulation of Liquids.* Chapter 3. Oxford University Press, Oxford.
32. McBride, C. and Wilson, M.R. to be published.
33. Lyulin, A.V., Al Barwani, M.S., Allen, M.P., Wilson, M.R., Neelov, I. and Allsopp, N.K. (1998) *Macromolecules*, **31**, 4626.
34. de Miguel, E., Rull, L.F., Chalam, M.K. and Gubbins, K.E. (1991) *Molec. Phys.*, **74**, 405.
35. Cornell, W.D., Cieplak, P., Bayly, C.I., Gould, I.R., Merz Jr., K.M., Ferguson, D.M., Spellmeyer, D.C., Fox, T., Caldwell, J.W. and Kollman, P.A. (1995) *J. Am. Chem. Soc.*, **117**, 5179.
36. Allinger, N.L., Yuh, Y.H.and Lii, J. (1989) *J. Am. Chem. Soc.*, **111**, 8551.
37. Jorgensen, W.L., Maxwell, D.S. and Tirado-Rives, J. (1996) *J. Am. Chem. Soc.*, **118**, 11225.
38. Jorgensen, W.L. and McDonald, N.A. (1998) *Theochem - J. Molec. Structure*, **424**, 145.
39. Halgren, T.A. (1996) *J. Comput. Chem.*, **17**, 490.
40. Weiner, S.J., Kollman, P.A., Case, D.A., Singh, U.C., Ghio, C., Alagona, G., Profeta, S. and Weiner, P. (1984) *J. Am. Chem. Soc.*, **106**, 765.
41. Karplus, M. and Gelin, B.R. (1979) *Biochemistry*, **18**, 1256.
42. Brooks, B.R., Bruccoleri, R.E., Olafson, B.D., States, D.J., Swaminathan, S. and Karplus, M. (1983) *J. Comput. Chem.*, **4**, 187.
43. Smith, G.D. and Jaffe, R.L. (1996) *J. Phys. Chem.*, **100**, 9624.
44. Adam, C.J., Clark, S.J., Wilson, M.R., Ackland, G.J. and Crain, J. (1998) *Molec. Phys.*, **93**, 947.
45. Allen, M.P. (1993) *Phys. Rev. E*, **47**, 4611.
46. Cook, M.J. and Wilson, M.R. to be published.
47. Gay, J.G. and Berne, B.J. (1981) *J. Chem. Phys.*, **74**, 3316.
48. Allinger, N.L. (1977) *J. Am. Chem. Soc.*, **99**, 8127.
49. Allen, M.P. and Masters, A. (1993) *Molec. Phys.*, **79**, 277.
50. Denti, T.Z.M., Beutler, T.C., van Gunsteren, W.F. and Diederich, F.J. (1996) *J. Phys. Chem.*, **100**, 4256.

ATOMISTIC SIMULATION AND MODELING OF SMECTIC LIQUID CRYSTALS

MATTHEW A. GLASER
*Ferroelectric Liquid Crystal Materials Research Center, and
Condensed Matter Laboratory, Department of Physics
University of Colorado
Boulder, CO 80309, USA*

Abstract. This chapter reviews recent progress in the atomistic simulation and modeling of smectic liquid crystals. Despite formidable technical challenges, atomistic simulation is coming into its own as a powerful tool for doing cutting-edge liquid crystal science, due to the convergence of unprecedented computer power, recent algorithmic developments, and the availability of high-accuracy *ab initio* methods for the creation of molecular models from first-principles. Particular emphasis is placed on the model-building process, as this is the foundation upon which 'realistic' modeling of organic materials stands or falls. Advanced methodologies for molecular simulation are also described, including the particle-mesh Ewald method for rapid evaluation of long-range interactions in periodic systems, the r-RESPA family of multiple-timestep molecular dynamics integrators, and a collective Monte Carlo method, hybrid Monte Carlo. We conclude with several illustrative examples from our recent work on smectic liquid crystals.

1. Introduction

A defining characteristic of liquid crystalline materials is an exquisitely sensitive dependence of macroscopic properties on molecular structure, with a relatively limited number of structural motifs giving rise to a rich and varied palette of liquid crystal (LC) phases and materials properties. This is one of the attributes that make LCs so appealing to the materials scientist — details of chemical structure represent a set of exceedingly sensitive 'knobs' with which the materials designer can control the properties of LC materials. The word 'control' is used in a rather loose sense here, of course

263

P. Pasini and C. Zannoni (eds.), Advances in the Computer Simulations of Liquid Crystals, 263–331.

— it is only rarely that one is able to predict LC properties *a priori*, and in most cases one is surprised (either pleasantly or unpleasantly) by the behavior of any newly synthesized material. This is one of the things that makes LC science mysterious and exciting.

The physics underlying this sensitive dependence of bulk properties on chemical detail is clear. The phase behavior and materials properties of LCs are governed by the interplay of a variety of subtle energetic and entropic effects, including electrostatic, dispersion, induction, and excluded volume interactions, and positional, orientational, and conformational entropy. These energetic and entropic effects are modulated by intramolecular chemical contrast, the presence of chemically dissimilar functional groups within the same molecule. The interplay of competing interactions leads to a situation in which alternate thermodynamic states are quite close in free energy. As a consequence, minor changes in chemical structure can shift the free energy balance between competing phases, dramatically changing the phase behavior of a material. For the same reason, modest changes in chemical structure can dramatically alter local molecular organization and correlations in a given LC phase, with correspondingly dramatic changes in the mechanical, electrical, optical, and transport properties of the material. What makes the theoretical treatment of LC materials particularly difficult is that the relatively 'weak' electrostatic, dispersion, and induction interactions cannot simply be treated as perturbations on the 'dominant' excluded volume interactions, as such weak interactions lead to completely new types of order. Gaining a basic understanding of the molecular basis for the self-organization of organic materials represents one of the outstanding intellectual challenges in soft condensed matter physics, and constitutes one of the primary goals of computer simulation of LCs.

There are two fundamental obstacles to the quantitative modeling of LC materials: the first, and most serious, is that highly accurate models of molecular structure and interactions are required, a consequence of the sensitivity of LC properties to details of molecular structure. As sufficiently accurate models are lacking in most cases, this constitutes a severe limitation. A second obstacle arises from the complexity (large number of degrees of freedom and complex interaction potentials) of LC materials, and the broad range of characteristic time and length scales that need to be accessed. To minimize finite-size effects, the simulated sample volume must be at least as large as a correlation volume, which can be quite large in LC phases, even far from phase transitions. Moreover, the duration of the simulation should be long enough to average over all the relevant characteristic timescales in the system. As shown in Table 1, relevant microscopic timescales in LCs range over many orders of magnitude. As a result, simulation of LC materials is computationally intensive. These two obstacles are related, as the computational cost of simulating complex organic materials limits the rate at which improved molecular models can be developed and tested. The first limitation is the most serious, however, as it precludes quantitative studies even in cases where adequate computational resources are available.

TABLE 1. Characteristic microscopic timescales in liquid crystals.

Process	Characteristic time (s)
Intramolecular vibration	$10^{-14} - 10^{-12}$
Conformational transition	$10^{-11} - 10^{-8}$
Reorientation of short molecular axes	$10^{-10} - 10^{-9}$
Lateral translational diffusion by molecular diameter	$10^{-10} - 10^{-8}$
Longitudinal translational diffusion by molecular length	$10^{-8} - 10^{-6}$ (nematic)
	$> 10^{-6}$ (smectic)
Reorientation of long molecular axis	$> 10^{-7}$
Decay of metastable state	$10^{-7} - \infty$

In the face of these challenges, there are two contrasting and complementary strategies. One useful approach is to study *idealized* models of increasing complexity, adding distinct physical effects (e.g. specific features of molecular structure and interactions) sequentially and studying their influence on bulk behavior in a systematic way. This is the approach followed in the majority of computer simulations of LCs. The simulation of idealized models achieves a significant reduction in computational cost through reduction in the number of degrees of freedom and simplification of the interaction potentials. Unfortunately, the cost (both in man-hours and CPU time) of systematically exploring the properties and phase behavior of a model system grows exponentially with the number of parameters in the model (i.e. with its complexity), so that the systematic investigation of the properties of models approaching chemical complexity is infeasible. Nevertheless, theoretical and simulation studies of idealized models represent our best source of information about structure-property relationships at present, although it is difficult to extrapolate from idealized models to real materials.

An alternative approach is to study models in which the molecular structure and interactions are represented with a high degree of chemical realism (leaving aside for the moment the question of whether this is really possible). As this generally involves representing molecular structure with something approaching atomic detail, such models are referred to as *atomistic* or *atomic-detail* models. By focusing on a specific point in parameter space (that corresponding to physical reality), atomistic modeling avoids the necessity of exploring a large parameter space, and gives direct access to the microphysics of real LC systems, allowing one to rapidly develop an intuitive understanding of the microscopic basis for observed LC properties. In general, atomistic modeling is the method of choice when answers to highly chemically specific questions are sought. The drawbacks of this approach are obvious: first and foremost is the need for a 'sufficiently realistic' model (one in which all important physical effects are represented with reasonable accuracy) — the creation and verification of atomistic models invariably represents a significant challenge, one which is in practice seldom met. Moreover, atomistic simulation is quite computationally intensive relative to simulation of idealized models, which fact limits the range of problems that can be addressed via atomistic simulation.

Another apparent shortcoming of atomistic simulation is that all important physical effects are included simultaneously, making it difficult to isolate the specific causes of a given phenomenon. However, we have at our disposal another strategy that can be termed 'perturbing reality': starting from the 'physical' point in parameter space, we are free to perturb our model system, e.g. by adding, removing, or modifying a specific interaction, changing the degree of molecular flexibility, adiabatically 'mutating' a molecule into a distinct chemical species, etc. In this way we can, at least in principle, isolate effects due to specific molecular-level features, and can, for example, bring thermodynamic integration to bear in order to compute the relative free energies of distinct molecular models and/or distinct states of order. This is not a strategy that has been exploited in atomistic simulations of LCs to date, although analogous approaches have been applied to biomolecular systems (see, for example, [1]).

It should be clear from the foregoing discussion that idealized models and atomistic models play complementary roles in the microscopic theory of LCs. (We note in passing that the distinction between 'idealized' and 'realistic' models is not sharp — even the most sophisticated atomistic model is, in reality, a drastically simplified representation of a much more complex quantum mechanical system, and is thus an idealized model.) Owing to the technical difficulty of atomistic simulation, however, there have been only a relatively limited number of atomistic simulation studies of thermotropic LCs [2–23]. The quality of these studies varies widely, and whether or not specific studies meet the criterion of 'chemical realism' is debatable. The first reported atomistic simulation of a thermotropic LC [2] was of total duration 60 ps, far too short a time to equilibrate even a small sample, as is obvious from Table 1. The pioneering studies of Wilson and Allen [6, 7] were probably the first atomistic simulations of sufficient duration to have a chance of reaching thermodynamic equilibrium. Subsequent work has focused on the development of improved molecular models and on the simulation of larger systems over greater spans of time.

This article is a report on the current state of the art in atomistic simulation of LCs, illustrated by examples from our work on the modeling of smectic LCs. It is our opinion that the molecular model-building process is absolutely critical to this enterprise, for the reasons outlined above. Thus, Sec. 2 is devoted to a discussion of practical strategies for creation of accurate molecular models for LC materials and of the remaining technical challenges to the development of high-accuracy interaction potentials for LC simulation. An equally important development is the recent appearance of simulation methods that make highly accurate atomistic simulations of relatively large systems and long time spans possible. These methods are described in Sec. 3.

One of the most important features of state-of-the-art methodology for molecular simulation is (nearly) linear scaling of computational effort with system size, which makes atomistic simulation of relatively large systems possible. With currently available computer hardware, we estimate that it is currently feasible to simulate LC systems consisting of ~ 100 LC molecules for time spans up to ~ 100 ns, or ~ 1000 molecules for ~ 10 ns, with full atomic detail and with most important physical effects included (although the most extensive atomistic simulations reported to date are about an order of magnitude smaller in scale than this). Although such simulations are far from routine, our experience suggests that atomistic simulation is on the verge of coming into its own as a tool for delving into some of the fundamental mysteries of LC science, and that we are on the threshold of a revolution in organic materials design and development, in which simulation and modeling will catalyze the creation of novel materials by providing a means to decipher the chemical language of self-assembly in complex organic materials such as LCs. The reasons for our optimism will become clear when we describe recent applications of atomistic simulation to smectic LC materials, in Sec. 4.

Clearly, atomistic simulation is not the method of choice for every problem. For example, systematic studies of the phase behavior of atomistic LC models similar to those that have been undertaken for idealized models (see, for example [24]) are out of reach at present, because it is not technically feasible to directly compare the relative free energies of all possible phases under given thermodynamic conditions. In particular, the problem of determining which of the astronomical number of possible crystal structures is the most stable is a seemingly insurmountable one for complex organic materials, as is the problem of determining whether specific LC phases are stable relative to the crystal phase. While it is feasible to determine which of two specific LC states is the most stable under given thermodynamic

268

conditions, the possibility that a third state (e.g. the crystalline state) is more stable than *either* LC state is ever-present, and although limited studies of the relative stability of competing LC states can provide a great deal of insight into the origins of LC self-organization, it is generally true that one can at present only study LC *states* (which could be metastable) rather than *phases* via atomistic simulation.

Similarly, large-scale atomistic simulation is not a practical calculational method for routine pre-synthesis design and evaluation of LC materials. Large-scale atomistic simulation is simply too expensive to allow effective navigation of the immense chemical 'phase space' of potential LC materials (we conservatively estimate, for example, that there are $\sim 10^{10}$ potential LC compounds with the simplest linear, two-ring, tail-core-tail structure). On the other hand, the sensitive dependence of LC properties on chemical structure would seem to require chemical specific (atomic-detail) models. Given these constraints, one is led naturally to the idea of applying mean-field theory to reduce the computational scale of LC modeling while retaining, in an average way, the effect of intermolecular interactions on molecular organization. In the simplest case, one can model a single molecule in a mean-field potential that implicitly takes into account condensed phase effects on molecular ordering. Single-molecule mean-field theory enables the rapid evaluation of LC properties for a variety of materials, but is limited to properties that can be calculated to a good approximation from the single-molecule distribution function. In general, it is useful to consider a *modeling hierarchy*, depicted schematically in Fig. 1, including methods ranging from computationally inexpensive, single-molecule mean-field approximations to expensive (but exact) many-molecule simulations, as well as intermediate-scale pair and cluster approximations. Other approximate techniques, such as liquid-state theory and density functional theory, can be regarded as intermediate-scale approaches. In general, methods of decreasing scale require increasing theoretical and/or experimental input. A great advantage of the modeling hierarchy construct is the possibility of using larger-scale (and more expensive) calculational methods to test and refine more approximate (and less expensive) smaller-scale approaches. An example of this sort of 'bootstrapping' strategy is given in Sec. 4.4.

Single-molecule mean-field theory is clearly computationally expedient, and has the further significant advantage that intermolecular interactions need not be considered explicitly, being implicitly contained in the mean-field potential. The price to be paid here is that experimental input is required to determine the form and parametrization of the mean-field potential, so that such mean-field methods are *semi-empirical* (of course, large-scale atomistic simulation is usually semi-empirical as well, requiring experimental input for the determination of intermolecular interaction

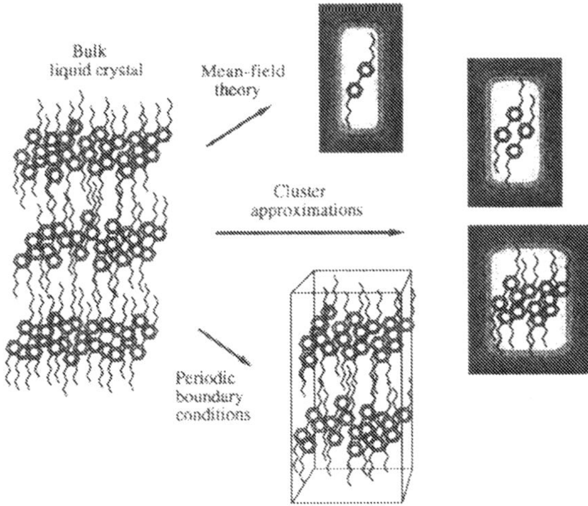

Figure 1. The modeling hierarchy.

parameters, but the reliance on experiment in mean-field models is typically greater). Single-molecule mean-field theory provides a framework for the creation of semi-empirical calculational tools for the computer-aided design of LC materials. Such semi-empirical models are only useful for exploratory design of new materials (i.e. only have predictive power) if they are in some sense *transferable*, in other words if mean-field parametrizations developed to fit the properties of a known class of materials can be applied to the modeling of novel materials. The underlying physical assumption here is one of microscopic universality: the general features of molecular organization in a particular LC phase are assumed to depend only weakly on chemical structure. In Sec. 4.3 we illustrate the mean-field approach to materials modeling through application to the prediction of ferroelectric polarization density in ferroelectric LCs.

What we currently lack are viable intermediate-scale approaches capable of systematically improving upon or extending single-molecule mean-field theory, but which are still significantly cheaper than large-scale simulation. Although some preliminary forays into this area have been attempted [25, 26], this is clearly one of the areas where opportunities to significantly extend present capabilities for modeling LC materials exist.

2. Model building

The success of the enterprise of atomistic simulation and modeling of LCs depends critically on the quality of the model for molecular structure and interactions used as input. In fact, the whole notion of 'realistic' molecular modeling of LCs only makes sense if interaction potentials that to some degree approach the ideal of chemical realism are available. It is therefore of the utmost importance to carefully consider the process of model building — even if one does not develop one's own interaction potentials, one must be cognizant of the procedures used to derive the interaction potential chosen for a particular application, specifically whether it is appropriate, sufficiently accurate, and optimal for a given problem. In this sense, the most important part of any simulation is what happens before the simulation starts.

A wide variety of interaction potentials ('force fields') for organic materials have been developed over the past three decades (for recent reviews, see [27–37]). As a general rule, the most successful of these have been developed for the modeling of specific classes of materials. Unfortunately, accurate, validated interaction potentials are not available for all (or even most) families of organic materials. Moreover, it is seldom the case that a given interaction potential is optimally matched to a particular problem of interest. These two facts, taken together, argue heavily in favor of developing interaction potentials as needed for specific applications. In this section, we sketch a pragmatic approach to the 'on-demand' derivation of interaction potentials for atomistic simulation from *ab initio* and empirical data. A more detailed account can be found elsewhere [38]. Our strategy is far from unique, and is admittedly crude — it merely provides a concrete example of the process of model building, and illustrates the technical issues that must be addressed in the context of atomistic simulation of soft matter. Our approach is similar to (but less sophisticated than) those of the Jorgensen [30, 31] and Kollman [27, 32, 36] groups.

The creation of a classical interaction potential for molecular modeling corresponds to mapping the quantum mechanical (Born-Oppenheimer) surface of a large system of electronic and nuclear degrees of freedom onto a relatively simple classical energy expression. In principle, one could dispense with this mapping, directly simulating the underlying quantum many-body system. The Car-Parrinello method represents such a first-principles approach, based on the simultaneous dynamical evolution of both nuclear and electronic degrees of freedom, with an effective Hamiltonian derived from density functional theory [39–41]. Unfortunately, this approach is exceedingly computationally intensive, so that accessing the time and length scales required to explore the statistical physics of LCs is prohibitively expensive with current computing hardware. Another drawback of the method is its reliance on density functional theory, which introduces an uncontrolled approximation (the choice of an exchange-correlation functional) into the effective interaction potential. Moreover, it is an article of faith that the statistical mechanics of organic materials such as LCs can be well represented by mapping the underlying quantum mechanical problem onto an effective classical Hamiltonian, in which the energetics of the electronic degrees of freedom are subsumed into an effective classical interaction potential for the nuclear degrees of freedom. In this case, one can still take a first-principles approach to the explicit construction of such an effective classical Hamiltonian, and, to the extent that this can be done, realize the ideal of a first-principles approach to LC materials modeling, for a tiny fraction of the computational expense of *ab initio* molecular dynamics.

In practice, it is almost never possible to construct an effective classical interaction potential for organic materials starting from first principles. A basic problem is that of energy scales: the interactions relevant to the self-organization of soft matter are exceedingly weak on the electronic energy scale characteristic of chemical bonding, and thus are quite difficult to quantify using traditional quantum chemical methods. Further difficulties arise from the complexity of the effective quantum mechanical interactions, which include Coulomb, dispersion, hydrogen-bonding, and short-range repulsive interactions, as well as many-body induction and dispersion interactions [42, 43]. The many-body interactions are particularly difficult to deal with, and are typically absorbed into effective two-body site-site interaction terms. In general, a highly simplified form for the interactions is assumed, so that the mapping onto a classical model involves a number of simultaneous and poorly controlled approximations. For instance, site-interaction models are almost universally employed, with a simple functional form for two-body dispersion-repulsion interactions and for valence (bond stretch, bond-angle bend, etc.) interactions, and with the molecular charge distribution represented by point charges and/or multipoles. A further ubiquitous but relatively untested assumption is that of *transferability*: interaction parameters derived for small organic molecules are assumed to be transferable to similar (often larger) molecules. Finally, there are some effects (e.g. quantum zero-point motion) which cannot be fully treated in a classical model. This makes it impossible to simultaneously fit vibrational spectra and vibrational amplitudes, for instance.

Thus, although a truly first-principles approach to force-field development is highly desirable, systematic procedures for effecting the explicit mapping of quantum mechanical interactions onto classical interaction potentials do not exist at present, although this is an area of intense and ongoing effort. As a consequence, almost all extant force fields are to some degree empirical, relying to a greater or lesser degree on experimental input to fix some or all parameters. As typical force fields for organic materials may contain hundreds of parameters, however, the use of first-principles information about the interaction energy of organic molecules is extremely valuable for the routine development of interaction potentials for molecular simulation. Moreover, the use of *ab initio* input enables the modeling of chemically novel materials for which reliable experimental data is lacking. Thus, the most common approach at present utilizes a combination of *ab initio* and empirical data. In the remainder of this section we describe the model-building strategy we currently employ for LC materials, as a concrete illustration of the model-building process.

A given molecular model is defined by the *form* and *parametrization* of its interaction potential. The first step in constructing a molecular model is to choose a level of chemical detail. In this regard, there are two common choices: *explicit-hydrogen* models, in which all hydrogen atoms are represented by distinct interaction sites, and *implicit-hydrogen* models, in which some or all hydrogens are combined with the atoms to which they are bonded to form effective interaction sites. Reducing the number of degrees of freedom via an effective-atom approximation leads to a large reduction in computational expense, but may result in some loss of chemical realism (of course, there is no guarantee that even an explicit-hydrogen model is chemically realistic). Further coarse-graining is possible; for example, phenyl rings can be represented as spherically symmetric interaction sites [44], or the entire LC core can be replaced by a single anisotropic interaction site [45]. As such models stray rather far from what we understand as 'atomic-detail', however, we will not consider them further here. We merely note that the line between 'realistic' and 'idealized' molecular models is not sharp, and that varying degrees of chemical detail can be useful, depending on the application.

For our work on LC materials, we adopt a hybrid representation of molecular structure, in which sp^3-hybridized CH_n groups are treated as spherically symmetric interactions sites, while sp^2-hybridized hydrogens (e.g. in phenyl rings) are represented explicitly. This choice is based on the empirical observation that implicit-hydrogen models are able to reproduce many of the thermophysical properties of liquid alkanes reasonably accurately (see, for example, [46, 47]), while implicit-hydrogen models for benzene perform poorly [48–50]. Experience also suggests that implicit-hydrogen models for sp^3-hybridized CH_n groups are not adequate for modeling crystal phase properties [51]. Thus, such hybrid models are expected to be primarily useful for modeling fluid (e.g. LC) phases.

The next step in model building is to choose a form for the interaction potential. Again, the choice is strongly conditioned by the purpose for which the model is intended. Generally, force fields range from relatively simple *thermophysical accuracy* ('Type I') force fields, intended to reproduce the thermophysical properties of materials, to relatively complex, *spectroscopic accuracy* ('Type II') force fields, designed to reproduce molecular vibrational spectra [28]. Since we are mainly concerned with reproducing the thermophysical and structural properties of LCs, we choose a relatively simple form of interaction potential, namely $U(\mathbf{r}^N) = U_{\text{str}} + U_{\text{bend}} + U_{\text{tors}} + U_{\text{inv}} + U_{\text{vdw}} + U_{\text{coul}}$, where

$$U_{\text{str}} = \sum_{\substack{\text{bonds} \\ ij}} \frac{1}{2} k_r (r_{ij} - r_{\text{eq}})^2 \tag{1}$$

$$U_{\text{bend}} = \sum_{\substack{\text{angles} \\ ijk}} \frac{1}{2} k_\theta (\theta_{ijk} - \theta_{\text{eq}})^2 \tag{2}$$

$$U_{\text{tors}} = \sum_{\substack{\text{dihedrals} \\ ijkl}} \sum_{n=0}^{12} c_{n\phi} \cos^n \phi_{ijkl} \tag{3}$$

$$U_{\text{inv}} = \sum_{\substack{\text{umbrellas} \\ ijkl}} k_\psi (\cos \psi_{\text{eq}} - \cos \psi_{ijkl})^m \qquad m = \left\{ \begin{array}{ll} 1, & \psi_{\text{eq}} = 0 \\ 2, & \psi_{\text{eq}} \neq 0 \end{array} \right. \tag{4}$$

$$U_{\text{vdw}} = {\sum_{i<j}}' 4\epsilon_{ij} \left[\left(\frac{\sigma_{ij}}{r_{ij}} \right)^{12} - \left(\frac{\sigma_{ij}}{r_{ij}} \right)^6 \right] \tag{5}$$

$$U_{\text{coul}} = {\sum_{i<j}}' \frac{q_i q_j}{r_{ij}}. \tag{6}$$

The primes on the sums in Eqs. 5 and 6 indicate that 1-2, 1-3 and 1-4 intramolecular nonbonded interactions are excluded from the sums. The internal coordinates r_{ij}, θ_{ijk}, ϕ_{ijkl}, and ψ_{ijkl} are defined by

$$r_{ij} = |\mathbf{r}_{ij}| = |\mathbf{r}_j - \mathbf{r}_i| \tag{7}$$

$$\theta_{ijk} = \cos^{-1} \left[-\frac{\mathbf{r}_{ij} \cdot \mathbf{r}_{jk}}{r_{ij} \, r_{jk}} \right] \tag{8}$$

$$\phi_{ijkl} = \cos^{-1} \left[\frac{(\mathbf{r}_{ij} \times \mathbf{r}_{jk}) \cdot (\mathbf{r}_{jk} \times \mathbf{r}_{kl})}{|\mathbf{r}_{ij} \times \mathbf{r}_{jk}| \, |\mathbf{r}_{jk} \times \mathbf{r}_{kl}|} \right] \tag{9}$$

$$\psi_{ijkl} = \sin^{-1} \left[\frac{\mathbf{r}_{ij} \cdot (\mathbf{r}_{ik} \times \mathbf{r}_{il})}{r_{ij} \, |\mathbf{r}_{ik} \times \mathbf{r}_{il}|} \right]. \tag{10}$$

ψ_{ijkl} measures the angle between $\mathbf{r}_{ij}$ and the plane defined by $\mathbf{r}_{ik}$ and $\mathbf{r}_{il}$ for three-coordinated atoms i. To retain a symmetric form for the inversion potential, the total inversion potential for an atom i is taken to be the average of three umbrella torsion terms, corresponding to the three possible choices of the special bond $\mathbf{r}_{ij}$.

As the form of the interaction potential implies, induction interactions are not included explicitly, but are instead absorbed into effective dispersion parameters ϵ_{ij} and σ_{ij}. The dispersion parameters also effectively account for many-body dispersion interactions as well as artifacts arising from the interaction-site approximation. In a sense, therefore, all the 'trash' gets swept into the dispersion parameters, which must therefore be determined empirically by fitting simulated properties to experimental data. This approach, while crude, has been shown to work well for molecular fluids under thermodynamic conditions close to those for which the empirical determination of dispersion parameters is carried out. Our interaction potential explicitly includes long-range electrostatic interactions, so the interaction parameters are not cutoff-dependent. This appears to be the most physically motivated choice, one which has been shown to be essential to the proper treatment of thermodynamic and structural properties of molecular fluids.

The remaining step in building a molecular model is fixing the interaction parameters k_r, r_{eq}, k_θ, θ_{eq}, $c_{n\phi}$, k_ψ, ψ_{eq}, ϵ_{ij}, σ_{ij}, and q_i. As parameters need to be provided for every distinct combination of atom types, we need to determine a large number (perhaps hundreds) of interaction parameters. As this statement implies, the initial step in parametrizing a classical interaction potential is to specify atom types. Generally speaking, we define as many atom types as are needed to account for significant variations in geometrical parameters, torsional potentials, etc., among chemically distinct compounds. As usual, what constitutes a 'significant' difference will depend on the specific application. Of course, special 'effective' atom types must be introduced for implicit-hydrogen models. In our current work, we utilize a rather limited number of atom types (see, for example, [38]).

Once the atom typing scheme is specified, we are still left with the problem of determining a large number of interaction parameters, so we need ways of effectively reducing the number of parameters. For a start, we make use of standard geometrical mean 'combining rules' to define van der Waals parameters for interactions between unlike atoms in terms of those between identical atoms, i.e. $\epsilon_{ij} = \sqrt{\epsilon_{ii}\epsilon_{jj}}$ and $\sigma_{ij} = \sqrt{\sigma_{ii}\sigma_{jj}}$. Because we are not concerned with accurately reproducing vibrational spectra, we assume generic (Dreiding II [52]) values for the bond stretch, bond angle bend, and inversion force constants k_r, k_θ, and k_ψ. These two simplifications together result in a great reduction in the number of parameters that need to be determined.

We utilize a combination of *ab initio* and empirical input to determine the remaining parameters. The geometrical parameters r_{eq} and θ_{eq} as well as the torsional parameters $c_{n\phi}$ are determined from fits to moderate-level (MP2/6-31G(d)//HF/6-31G(d) or B3LYP/6-31G(d)//HF/6-31G(d)) *ab initio* calculations [53] of optimized geometries and torsional potentials for LC substructures (see [38]). As the site charges q_i are generally considered to be less transferable than other parameters, these are determined from electrostatic potential (ESP) fits [54] to semi-empirical (AM1) electron densities for the LC molecules of interest (moderate-level *ab initio* electron density calculations for molecules as large as typical LC molecules exceed readily available computational resources, so we must resort to semi-empirical methods). As discussed above, the van der Waals parameters ϵ_{ii} and σ_{ii} are the main empirical parameters; their values are chosen to reproduce the thermophysical properties of small organic molecules in the liquid state. In our work to date, we have used van der Waals parameters from the literature, in particular OPLS (Optimized Potentials for Liquid Simulations) parameters [55] and the alkane parameters of Siepmann et al. [46], both of which are optimized for liquid-state simulations.

The success of our approach to model building can be gauged from the applications described in Sec. 4. Overall, this strategy seems to be rather successful, although there are clear avenues for improvement, particularly in the treatment of induction interactions.

3. Methodology

Once a molecular model has been defined, one needs a means for calculating the statistical (ensemble average) properties of the model. For complex molecular systems at high densities, the method of choice is molecular dynamics (MD) simulation. Although a number of collective Monte Carlo (MC) methods have been proposed for the simulation of dense molecular systems, including configurational-bias MC [56] and the hybrid MC method [57] described below, the general consensus seems to be that MD outperforms even the cleverest collective MC methods for general-purpose simulation of dense molecular systems. It is quite possible that this situation will change in the future, as the MC method allows an essentially unlimited number of variations. At present, however, MD is the workhorse for routine simulation of organic materials, and has the further advantage of giving access to dynamical properties as well as static properties.

In the remainder of this section we briefly outline state-of-the-art algorithms for MD simulation of molecular systems, including the particle-mesh Ewald (PME) method for evaluating long-range Coulomb and dispersion interactions [58–61], and the r-RESPA multiple-timestep MD integration algorithm [62, 63]. As a preamble to our discussion of the r-RESPA algorithm, we include a short digression on the stability of MD integrators, a subject that has a bearing on the effectiveness of the r-RESPA method, and provides the requisite background for our discussion of the hybrid MC method in Sec. 3.4. As both PME and r-RESPA are discussed in the lectures by Procacci, our treatment of these methods will be brief. Advanced methods for the simulation of organic materials are also discussed in the excellent recent review of Berne and Straub [64].

For evaluating statistical averages in single-molecule mean-field theory, a wider range of methods are available. For molecules with a limited number of internal (conformational) degrees of freedom, it is sometimes possible to evaluate configurational averages directly, provided that the conformational states are restricted to a discrete set, as in the rotational isomeric state (RIS) approximation [65]. More generally, either MC or MD can be used to evaluate ensemble averages, although special care must be taken in applying MD to small systems [56, 66]. In Sec. 3.4 we describe a specific collective MC method, hybrid Monte Carlo (HMC) [57], which is useful for statistical sampling in small molecular systems. HMC is a hybrid of MD and MC, in which short MD trajectories are used to generate trial MC moves. As a thorough exposition of HMC is lacking in the literature, we include a detailed derivation and discussion of the method here. While HMC seems to be an effective method for simulating small systems, it is far less effective for simulating large systems, for reasons discussed below, although we cannot rule out the possibility that a variant of the HMC method will prove to be effective for large-scale simulation.

3.1. THE PARTICLE-MESH EWALD METHOD

One of the most time-consuming parts of a molecular simulation is the evaluation of long-range electrostatic interactions. Whereas it is possible to treat long range dispersion interactions in an average way by applying a long-range correction (which is equivalent to the approximation that the pair correlation function is unity beyond some cutoff radius) [67], applying the same procedure in the case of Coulomb interactions leads to a divergent long-range 'correction', so it becomes necessary to consider long-range electrostatic interactions explicitly. Moreover, simply neglecting Coulomb interactions between pairs of particles separated by a distance greater than some cutoff r_c is not a viable option. Not only does this give results that depend on the particular choice of cutoff, but it also leads to artifactual structural and thermodynamic behavior [68–78].

In keeping with the philosophy of 'realistic' atomistic simulation, it is requisite for us to attempt to treat all of the important physical effects at a reasonable quantitative level, so we cannot simply neglect long-range Coulomb interactions. We thus require accurate yet efficient methods for evaluating such interactions explicitly. Ironically, it turns out to be significantly more difficult to do so for systems with free boundary conditions (which are finite in extent) than for systems with periodic boundary conditions (which are infinite in extent). We will focus on systems with periodic boundary conditions here, as this is the most commonly encountered situation. Fast methods for evaluating Coulomb interactions in systems with free boundary conditions or with mixed periodic and free boundary conditions (e.g. freely suspended thin films or organic monolayers on solid substrates) are discussed elsewhere [79–82]. Finally, the use of a long-range correction for dispersion (r^{-6}) interactions is suspect in systems with extended positional correlations (as in many LC systems) or systems with long-range or quasi-long-range positional order (e.g. smectics). In such cases, it is desirable to have efficient and accurate methods available to evaluate long-range dispersion interactions explicitly. As discussed in this section, the particle mesh Ewald (PME) method [58–61] provides a computationally expedient route to the evaluation of long-range Coulomb and dispersion interactions in periodic systems (for a recent review of Ewald summation techniques, see [83]).

We consider a unit cell containing N point charges q_i, periodically replicated to fill space, and with $\sum_{i=1}^{N} q_i = 0$ (overall charge neutrality). The unit cell is taken to be a parallelepiped having edge vectors $\mathbf{a}_1$, $\mathbf{a}_2$, and $\mathbf{a}_3$. A given point charge interacts with all other charges in the primitive unit cell as well as the periodic images of all charges (including itself). The total electrostatic energy per unit cell is thus

$$U_{\text{coul}}(\mathbf{r}^N) = \frac{1}{2} \sum_{\mathbf{R}} \sum_{i,j=1}^{N} {}' \frac{q_i q_j}{|\mathbf{r}_j - \mathbf{r}_i + \mathbf{R}|}. \tag{11}$$

Here, $\mathbf{r}^N$ denotes the set of point charge coordinates $\{\mathbf{r}_i\}$, the outer sum ranges over Bravais lattice vectors $\mathbf{R} = n_1\mathbf{a}_1 + n_2\mathbf{a}_2 + n_3\mathbf{a}_3$, where n_1, n_2, and n_3 are integers, and the prime on the inner sum indicates that terms with $i = j$ and $\mathbf{R} = 0$ are omitted. Evaluation of U_{coul} presents serious difficulties, as it is a conditionally convergent series, with a value that depends on the order in which the terms are summed.

The classical way out of this difficulty is that proposed by Ewald [84], who transformed the conditionally convergent sum into two rapidly convergent sums, one in real space and the other in reciprocal space, i.e. $U_{\text{coul}} = U_{\text{dir}} + U_{\text{rec}}$, where

$$U_{\text{dir}} = \frac{1}{2} \sum_{\mathbf{R}} \sum_{i,j=1}^{N} {}' \frac{q_i q_j \, \text{erfc}(\alpha|\mathbf{r}_j - \mathbf{r}_i + \mathbf{R}|)}{|\mathbf{r}_j - \mathbf{r}_i + \mathbf{R}|}, \tag{12}$$

and

$$U_{\text{rec}} = \frac{2\pi}{V} \sum_{\mathbf{K} \neq 0} \frac{\exp(-K^2/4\alpha^2)}{K^2} |S(\mathbf{K})|^2 - \frac{\alpha}{\sqrt{\pi}} \sum_{i=1}^{N} q_i^2. \tag{13}$$

Here, erfc is the complementary error function, α is a convergence parameter, and the reciprocal-space sum ranges over all nonzero reciprocal lattice vectors $\mathbf{K} = m_1\mathbf{b}_1 + m_2\mathbf{b}_2 + m_3\mathbf{b}_3$, where the m_α are integers and the reciprocal lattice primitive vectors $\mathbf{b}_\alpha$ are defined by $\mathbf{b}_\alpha \cdot \mathbf{a}_\beta = 2\pi\delta_{\alpha\beta}$ ($\delta_{\alpha\beta}$ is the Kronecker delta function). The structure factor $S(\mathbf{K})$ is defined as

$$S(\mathbf{K}) = \sum_{i=1}^{N} q_i \exp(i\mathbf{K} \cdot \mathbf{r}_i). \tag{14}$$

Optimizing the Ewald method with respect to variation of the convergence parameter α yields a method with $O(N^{3/2})$ complexity [85, 86].

In the particle mesh Ewald (PME) method [58–61], the exponential that appears in the structure factor $S(\mathbf{K})$ in the reciprocal-space term is interpolated onto a grid in real space so that the structure factor becomes a discrete Fourier transform, which may be evaluated using fast Fourier transforms (FFTs).

First, note that we can write

$$
\begin{aligned}
S(\mathbf{K}) &= \sum_{i=1}^{N} q_i \exp(i\mathbf{K} \cdot \mathbf{r}_i) \\
&= \sum_{i=1}^{N} q_i \exp\left[2\pi i \left(\frac{m_1 u_{i1}}{M_1} + \frac{m_2 u_{i2}}{M_2} + \frac{m_3 u_{i3}}{M_3}\right)\right],
\end{aligned}
\tag{15}
$$

where the $u_{i\alpha}$ are the scaled coordinates of particle i,

$$u_{i\alpha} = \frac{M_\alpha \mathbf{b}_\alpha \cdot \mathbf{r}_i}{2\pi}, \tag{16}$$

with $0 \leq u_{i\alpha} < M_\alpha$. If the complex exponentials are now interpolated onto a grid,

$$\exp\left[2\pi i \left(\frac{m_\alpha u_{i\alpha}}{M_\alpha}\right)\right] \approx \sum_{k_\alpha=-\infty}^{\infty} W(u_{i\alpha} - k_\alpha) \exp\left[2\pi i \left(\frac{m_\alpha k_\alpha}{M_\alpha}\right)\right], \tag{17}$$

Eq. 15 can be written as a discrete Fourier transform, which may be evaluated using FFTs.

Note that computing $S(\mathbf{K})$ using FFTs requires $O(M \log M)$ work, where $M = M_1 M_2 M_3$ is the total number of grid points, and that the number of grid points M is proportional to the volume V, assuming a constant grid spacing. At a given density, V is proportional to the number of atoms N, so the computational complexity of the PME algorithm scales as $O(N \log N) \simeq O(N)$.

In Fig. 2 we compare SGI R10000 CPU times obtained using the optimized Ewald method [85, 86] with those obtained with the optimized PME method, for the same relative accuracy (10^{-4}), and for various numbers of charges N. The PME method is faster than the optimized Ewald method for systems as small as 1000 charges, and is a factor of 5 faster for systems of 5000 charges. This disparity is expected to grow rapidly with system size, since the computational cost of the Ewald method scales as $N^{3/2}$, while that of the PME scales as $N \log N$. For typical system sizes (on the order of a few thousand atoms), use of the PME method leads to a speedup in our molecular simulation code of more a factor of 2. The PME method can also be used to efficiently evaluate long-range dispersion interactions [59] in cases where this is necessary (e.g. in spatially inhomogeneous systems).

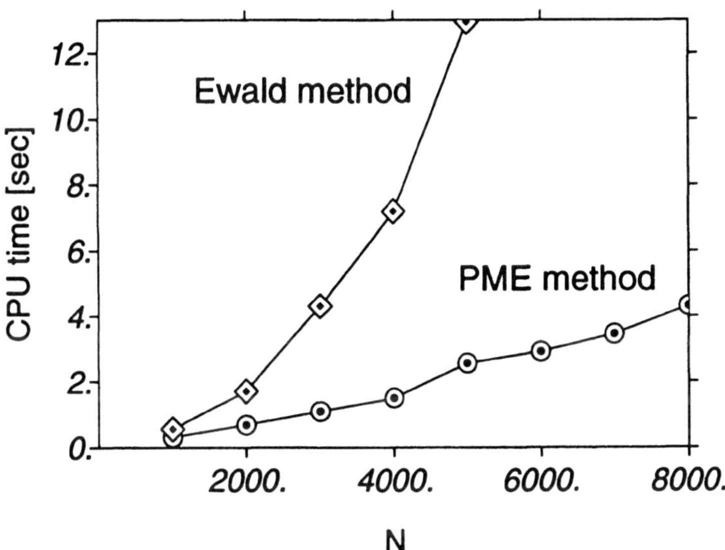

Figure 2. CPU times for the evaluation of the electrostatic energy and forces in systems of N charges, for both the PME method (circles) and the optimized Ewald method (diamonds). The times shown are for the SGI R10000 processor.

3.2. WHY IS MOLECULAR DYNAMICS STABLE?

Conventional molecular dynamics is advertised as a method for carrying out simulations in the microcanonical ensemble, that is under conditions of constant particle number, volume, and energy (constant NVE). (More precisely, MD samples the $NVEP$ ensemble, because MD also conserves the system center of mass momentum $\mathbf{P}$.) When one pauses to reflect for a moment, this seems little short of miraculous: because an MD integration represents a discretization of Hamilton's equations of motion, the energy is not exactly conserved (even though the energy is a constant of motion of the underlying Hamiltonian dynamics), and each integration step introduces a small error ΔE in the total energy. Assuming uncorrelated errors, after N_{MD} integration steps the energy should drift away from its original value by $\sim \sqrt{N_{\mathrm{MD}}}\Delta E$. In other words, the MD trajectory should be unstable. The miraculous thing is that *stable* MD algorithms exist, and that these are some of the simplest MD integrators, for example the leapfrog and Verlet-type integrators, whereas more accurate integrators such as the Gear predictor-corrector and Runge-Kutta algorithms are unstable. Obviously, the assumption of uncorrelated errors is not valid for a special class of MD integrators, for which, amazingly enough, the error in the energy appears to remain *strictly bounded* for extremely long MD simulations (eventually, of course, the finite precision of the computer makes itself known, and the accumulation of roundoff error leads to energy drift on very long time scales).

What accounts for the stability of certain MD integrators? This turns out to be a fairly subtle question, but the short answer is that the stable integrators are those that respect the symmetry properties of the under-lying Hamiltonian dynamics, specifically the volume-preserving (or area-preserving) character of the Hamiltonian flow (Liouville's theorem). In particular, this means that if we follow the time evolution of an ensemble of systems, the phase space volume will not expand or contract, but will be preserved. Such integrators are known as *symplectic* integrators, as they preserve a geometric invariant called the *symplectic form* [87]. The symplec-tic property follows if the integrator is constructed so that each integration step corresponds to a canonical transformation of phase-space coordinates. Conservation of the symplectic form is a more stringent condition than, but implies, conservation of phase space volume. However, in the following we will take *symplectic* to be synonymous with *area-preserving*. It can be shown formally that the discrete mapping defined by a symplectic inte-grator is generated by a Hamiltonian-like function (the *map Hamiltonian* [88]) that approaches the true Hamiltonian of the system as the integra-tion stepsize goes to zero [56,63,88–90]. The map Hamiltonian is a constant of the motion of the discrete dynamics, which, for sufficiently small step sizes, remains close to the constant energy surface corresponding to the system Hamiltonian for all times. In other words, the boundedness of the energy error for symplectic integrators is due to the existence of a conserved energy-like quantity (the map Hamiltonian), which remains, in a limiting sense, close to a true constant energy surface at all times.

Symplectic integrators can be explicitly constructed via an approximate factorization of the Liouville time-evolution operator [56, 62]. A *symmetric* factorization of the Liouville operator yields *time-reversible* (as well as *area-preserving*) integrators, and, in fact, the second-order (in the integration stepsize Δt) integrators derived in this way turn out to be the familiar Verlet or leapfrog integrators. Liouville operator-splitting methods can also be used to derive the r-RESPA multiple-timestep MD integrators described in Sec. 3.3, which are generalized, hierarchical leapfrog integrators. The stability of the r-RESPA family of integrators is a consequence of their symplectic structure. As we will see, the area-preserving and time-reversible nature of these integrators also makes them a suitable basis for the hybrid MC algorithms described in Sec. 3.4.

3.3. MULTIPLE-TIMESTEP MOLECULAR DYNAMICS

The idea of multiple-timestep molecular dynamics (MTSMD) is a natural one when one wishes to carry out MD simulations of systems in which the particle motions involve more than one characteristic timescale, for instance when the constituent particles are subjected to forces varying on multiple timescales, or when the system consists of several species of particles with widely disparate masses. Generally speaking, good stability (i.e. good energy conservation) requires the use of an integration timestep that is small compared to the shortest characteristic timescale for particle motion. Given this constraint, it is natural to look for ways to avoid evaluating the more slowly varying contributions to the force as frequently as the rapidly varying components.

A good example of a case in which this might be quite advantageous is provided by systems of particles interacting via long-range pairwise forces (e.g. Coulombic or gravitational forces). The force in this case can usually be split into a rapidly varying short-range part and a more slowly varying long-range part. In principle, MTSMD should make it possible to evaluate the expensive long-range part of the force less frequently than the (relatively inexpensive) short-range part, thereby realizing a large savings in computational cost. Another example relevant to the present discussion is that of a system of molecules with intramolecular degrees of freedom, in which the intramolecular force consists of a sum of contributions of varying strength, including bond stretching, bond-angle bending, and dihedral torsion terms, in addition to intra- and intermolecular nonbonded (van der Waals and Coulomb) forces. Such systems exhibit a hierarchy of timescales, ranging from the rapid bond-stretching motions to the significantly more slowly varying (and more expensive) nonbonded interactions, so MTSMD can in principle be used to great advantage.

MTSMD methods were first described by Street et al. [91], and their method was subsequently applied to a variety of systems [92–94]. However,

282

this method involves the computation of derivatives of the force, and is not time-reversible or area preserving, which leads to instabilities for long integration times, and which makes the approach unsuitable as a basis for hybrid Monte Carlo (see Sec. 3.4). Recently, a class of reversible and area-preserving (symplectic) MTSMD integrators was derived by Tuckerman et al. [62] and independently by Sexton and Weingarten [63] via an approximate factorization of the Liouville operator (see [56]). The former authors refer to their methods as *reversible reference system propagator algorithms* (r-RESPA), and we adhere to their terminology here. MD integrators of the r-RESPA family are *flexible* (permitting one to easily construct a wide variety of MTSMD schemes), *simple* (being straightforward generalizations of the familiar leapfrog or Verlet integrators), *stable* (a consequence of the their area-preserving property), and *efficient* (for molecular simulation, r-RESPA is a factor of 2-10 more efficient than single-timestep MD, depending on the application). As a consequence, the r-RESPA approach has come to enjoy increasing popularity in the molecular simulation community in recent years, as an attractive alternative to constraint dynamics.

A good discussion of the application of r-RESPA to molecular simulation can be found in the papers of Procacci and co-workers [95–100] and in the lecture notes by Procacci in this volume. These authors also discuss how the method can be combined with extended system methods to simulate NVT and NPT ensembles (see also [62, 101]). The r-RESPA algorithm is particularly powerful when combined with the PME method for evaluating long-range interactions. The two methods in combination give a large savings in computational expense for systems of more than a few thousand particles.

3.4. HYBRID MONTE CARLO

A natural question that arises in the context of molecular simulation is: can one devise Monte Carlo methods that sample configuration space as efficiently (or more efficiently) than molecular dynamics for molecular fluids at high (e.g. liquid crystalline) densities? In particular, it would be quite useful to have at our disposal MC methods for catalyzing slow processes in LCs, to avoid the high computational cost of equilibrating and obtaining adequate statistical sampling in atomic-detail MD simulations of LCs. This essentially requires a *collective* MC method, one in which many degrees of freedom are updated in each MC move. Many collective MC methods have been proposed [56,67,102–104], but none of these methods has proven to be competitive with MD for simulation of high-density molecular fluids. Another strategy is to use information about the 'natural' dynamics of a system to devise 'smart' MC methods [67]. The most recent and promising MC algorithm that utilizes this idea is Hybrid Monte Carlo [57], which is a MC method built on MD, in which short MD trajectories are used to generate trial MC moves. Such a procedure can be shown to constitute a proper MC method (i.e. satisfy the detailed balance requirement) if the MD integrator used to generate the trajectories is reversible and area-preserving. As discussed above, the Verlet/leapfrog-type algorithms satisfy this requirement, as do the r-RESPA integrators. This constitutes a major advantage of HMC for molecular simulation, as HMC can exploit the inherent efficiency of r-RESPA. Moreover, writing an HMC program involves a relatively trivial extension of a working MD program, an important consideration if one is interested in developing robust, general-purpose code for molecular simulation.

The potential advantages of HMC over MD arise from the fact that MC is inherently a much more flexible (and powerful) method than MD. One has unlimited freedom in the choice of MC moves, provided that they satisfy detailed balance, including potentially efficacious moves that bear no relation to the natural dynamics. Thus, MD-like (HMC) MC moves can be freely mixed with other types of MC moves designed, for instance, to catalyze specific slow processes in the simulated system. HMC moves can also be mixed with volume-changing MC moves to simulate an *NPT* ensemble. Moreover, as shown below, there is *no restriction* on the form of the 'guidance' Hamiltonian used to generate HMC trajectories. The guidance Hamiltonian need not be the same as the Hamiltonian of the system being simulated, so that a completely artificial dynamics can be constructed for the purpose of overcoming configuration-space bottlenecks. Similarly, the temperature characterizing the initial velocity distribution used to generate HMC trajectories (the 'sampling' temperature) need not coincide with the temperature at which the system is being simulated. Unfortunately, the acceptance rate (and hence the efficiency) of HMC tends to decrease sharply if the guidance Hamiltonian differs too much from the true Hamiltonian, or if the sampling temperature deviates substantially from the true temperature. Nevertheless, the considerable freedom that one has in constructing HMC schemes raises the hope that useful HMC-based MC methods for molecular simulation can be devised. In the remainder of this section, we discuss the HMC method and its variants in more detail, and describe some preliminary tests of the efficiency of HMC for molecular simulation.

HMC was originally developed by Duane et al. [57] in the context of lattice field theory simulation, and it has since become a method of choice for computer simulation of lattice field theories with fermionic degrees of freedom. The fermionic degrees of freedom lead to effective interactions between bosonic degrees of freedom that are highly non-local, so that local updates require calculations that are just as expensive as global updates.

Thus, methods for generating collective lattice updates are highly desirable. One strategy for accomplishing this is to introduce an artificial dynamics for the lattice degrees of freedom, and to integrate the equations of motion describing the evolution of the system in a fictitious time variable. Unfortunately, this procedure is only exact in the limit that the integration step size goes to zero, for the simple reason that the true Hamiltonian dynamics is not generated by a simulation with a finite integration timestep. As discussed in Sec. 3.2, discretization error causes the distribution function of the simulated system to deviate from the exact distribution function (generally speaking, the system behaves as though it were governed by a *map Hamiltonian* that differs slightly from the true Hamiltonian). Consequently, an extrapolation to zero step size must be carried out, which can be prohibitively expensive. The winning advantage of HMC in this context is that it is exact (free of discretization error) even for relatively large step sizes, obviating the need for an expensive limiting procedure.

It is interesting to note that the inexactness of MD is usually ignored in condensed phase simulations (i.e., MD and MC are generally considered to give the same results, even though only the latter exactly reproduces the true distribution function). Generally speaking, the accuracy of most molecular simulations is not sufficiently high that such an effect would be noticed. However, in some instances it may be important to use a formally exact method. For example, HMC has been used in recent simulations of melting in simple two-dimensional systems [105], where it was feared that discretization error could introduce artifactual shifts in the relative free energies of solid, liquid, and hexatic phases.

Before introducing HMC, we briefly review standard Metropolis Monte Carlo to highlight the similarities and differences between the two methods. In Metropolis MC, a Markov chain of configurations $\mathbf{r}^N$ (in the following, $\mathbf{r}^N$ represents the set of all configurational degrees of freedom for N particles) is generated to sample configurations according to a specified distribution

$$\rho_U(\mathbf{r}^N) = Z_U^{-1} e^{-\beta U(\mathbf{r}^N)}, \tag{18}$$

where $\beta = 1/(k_B T)$, $U(\mathbf{r}^N)$ is the potential energy of the system, and $Z_U = \int d\mathbf{r}^N e^{-\beta U(\mathbf{r}^N)}$ is the configurational partition function [67, 102]. The Markov chain generates configurations consistent with the desired distribution function $\rho_U(\mathbf{r}^N)$ if

$$\int d\mathbf{r}'^N [\rho_U(\mathbf{r}'^N)\pi(\mathbf{r}'^N \to \mathbf{r}^N) - \rho_U(\mathbf{r}^N)\pi(\mathbf{r}^N \to \mathbf{r}'^N)] = 0, \tag{19}$$

where $\pi(\mathbf{r}^N \to \mathbf{r}'^N)$ is the underlying transition probability for the Markov chain. This condition is difficult to enforce in practice, so one usually imposes the more restrictive (but still sufficient) *detailed balance* condition,

$$\rho_U(\mathbf{r}^N)\pi(\mathbf{r}^N \to \mathbf{r}'^N) = \rho_U(\mathbf{r}'^N)\pi(\mathbf{r}'^N \to \mathbf{r}^N), \tag{20}$$

which ensures that Eq. 19 is satisfied. It is usually convenient to decompose the overall transition probability π into a *proposal* probability P_P (which can be chosen to have a simple form) and an *acceptance* probability P_A (which will depend on $U(\mathbf{r}^N)$):

$$\pi(\mathbf{r}^N \to \mathbf{r}'^N) = P_P(\mathbf{r}^N \to \mathbf{r}'^N)P_A(\mathbf{r}^N \to \mathbf{r}'^N). \tag{21}$$

Detailed balance is satisfied for the (non-unique) Metropolis choice for P_A,

$$P_A(\mathbf{r}^N \to \mathbf{r}'^N) = \min\left[1, \frac{\rho_U(\mathbf{r}'^N)P_P(\mathbf{r}'^N \to \mathbf{r}^N)}{\rho_U(\mathbf{r}^N)P_P(\mathbf{r}^N \to \mathbf{r}'^N)}\right]. \tag{22}$$

If P_P is further chosen to be *symmetric*, i.e. $P_P(\mathbf{r}^N \to \mathbf{r}'^N) = P_P(\mathbf{r}'^N \to \mathbf{r}^N)$, then we recover the usual form for the Metropolis acceptance probability,

$$P_A(\mathbf{r}^N \to \mathbf{r}'^N) = \min\left[1, e^{-\beta\Delta U}\right], \tag{23}$$

where $\Delta U = U(\mathbf{r}'^N) - U(\mathbf{r}^N)$.

HMC can be regarded as a generalization of Metropolis MC in which one introduces 'auxiliary' variables $\mathbf{p}^N$ and considers a Markov chain in an expanded state space $(\mathbf{r}^N, \mathbf{p}^N)$, such that

$$\begin{aligned}
\pi(\mathbf{r}^N \to \mathbf{r}'^N) &= \int d\mathbf{p}^N d\mathbf{p}'^N P_P[(\mathbf{r}^N, \mathbf{p}^N) \to (\mathbf{r}'^N, \mathbf{p}'^N)] \\
&\quad \times P_A[(\mathbf{r}^N, \mathbf{p}^N) \to (\mathbf{r}'^N, \mathbf{p}'^N)].
\end{aligned} \tag{24}$$

In HMC, the $\mathbf{p}^N$ are taken to be a set of momenta canonically conjugate to the variables $\mathbf{r}^N$, with Hamiltonian $H(\mathbf{r}^N, \mathbf{p}^N)$, and the proposal probability is that generated by a discretized Hamiltonian flow.

A single HMC move consists of the following three steps:

1. Given the current configuration $\mathbf{r}^N$, sample new momenta $\mathbf{p}^N$ from a Maxwellian distribution,

$$\rho_K(\mathbf{p}^N) = Z_K^{-1}e^{-\beta K(\mathbf{p}^N)}, \tag{25}$$

   where

$$K(\mathbf{p}^N) = \sum_{i=1}^{N} \frac{|\mathbf{p}_i|^2}{2m_i} \tag{26}$$

   is the kinetic energy, the m_i are the particle masses, and Z_K is a normalization factor.

2. Evolve the system forward in time for N_{MD} MD timesteps of size Δt (i.e. through a time interval $t = N_{\mathrm{MD}}\Delta t$) according to a specific discretization of Hamilton's equations of motion,

$$\frac{d\mathbf{r}_i}{dt} = \frac{\partial H}{\partial \mathbf{p}_i} \tag{27}$$

and

$$\frac{d\mathbf{p}_i}{dt} = -\frac{\partial H}{\partial \mathbf{r}_i}, \tag{28}$$

to produce a new set of coordinates and momenta $(\mathbf{r}'^N, \mathbf{p}'^N)$.

3. Accept the new configuration $\mathbf{r}'^N$ with probability

$$P_A[(\mathbf{r}^N, \mathbf{p}^N) \to (\mathbf{r}'^N, \mathbf{p}'^N)] = \min\left[1, e^{-\beta \Delta H}\right], \tag{29}$$

where $\Delta H = H(\mathbf{r}'^N, \mathbf{p}'^N) - H(\mathbf{r}^N, \mathbf{p}^N)$ is the change in total energy. Otherwise, reset the coordinates to $\mathbf{r}^N$.

Note that if the equations of motion were integrated exactly, H would be a constant of motion, and the acceptance rate would be unity. Because the integration timestep is finite, the Hamiltonian is not conserved exactly, and the *discretization error* ΔH is non-zero, leading to an average acceptance probability smaller than unity. The basic parameters of the HMC algorithm are the number of integration timesteps per HMC move, N_{MD}, and the integration stepsize, Δt. As discussed in more detail below, the optimal choice of integration timestep represents a tradeoff between 'cheap' dynamics (i.e. large step size) and a reasonable acceptance rate. Notice that for systems with discontinuous potentials (e.g. hard core systems, or systems of particles with square well potentials) the HMC acceptance probability will always be unity, since the energy is conserved exactly in MD simulations of such systems, even for a finite timestep.

Provided that the MD algorithm used to integrate the equations is *time-reversible* and *area-preserving* ($|J| = |\partial(\mathbf{r}'^N, \mathbf{p}'^N)/\partial(\mathbf{r}^N, \mathbf{p}^N)| = 1$) it can be shown that the HMC procedure satisfies the detailed balance requirement, and generates configurations distributed according to a canonical distribution. We now proceed to prove this.

First, note that the proposal probability corresponding to steps 1 and 2 above is:

$$P_P[(\mathbf{r}^N, \mathbf{p}^N) \to (\mathbf{r}'^N, \mathbf{p}'^N)] = Z_K^{-1} e^{-\beta K(p)} \delta[(\mathbf{r}'^N, \mathbf{p}'^N) - G(t)(\mathbf{r}^N, \mathbf{p}^N)]. \tag{30}$$

Here, δ is the Dirac delta function, and $G(t)$ is a discrete evolution operator,

$$G(t) = G(\Delta t)^{N_{\mathrm{MD}}}, \tag{31}$$

where $G(\Delta t)$ is some (reversible and area-preserving) discretization of Hamilton's equations of motion. For example, $G(\Delta t)$ could correspond to the leapfrog integrator or the r-RESPA integrator [62]. Time reversibility in the present context means that propagation of the system forward in time through a time interval t followed by a reversal of all momenta, propagation forward in time through another time interval t, and a final reversal of momenta brings the system back to the original point $(\mathbf{r}^N, \mathbf{p}^N)$ in phase space. Symbolically, this can be written

$$R_{\mathbf{p}}G(t)R_{\mathbf{p}}G(t) = I, \tag{32}$$

where the operator $R_{\mathbf{p}}$ reverses all momenta, and I is the identity operator. Equivalently, we can write the reversibility condition as

$$G^{-1}(t) = R_{\mathbf{p}}G(t)R_{\mathbf{p}}, \tag{33}$$

where $G^{-1}(t)$ is the inverse of $G(t)$. Although reversibility is obviously a property of the true Hamiltonian dynamics, it is not a property shared by all discretizations of Hamilton's equations of motion. For example, the symplectic Euler algorithm [89, 90] is not reversible (although it is area-preserving).

With this background, let us proceed to prove that HMC satisfies detailed balance, in other words that Eq. 20 is satisfied for the HMC choice of transition probability (Eqs. 24, 29, and 30). First, note that

$$
\begin{aligned}
\rho_U(\mathbf{r}^N)\pi(\mathbf{r}^N \to \mathbf{r}'^N) &= Z^{-1}e^{-\beta U(\mathbf{r}^N)}\int d\mathbf{p}^N d\mathbf{p}'^N e^{-\beta K(\mathbf{p}^N)} \\
&\quad \times \delta[(\mathbf{r}'^N, \mathbf{p}'^N) - G(t)(\mathbf{r}^N, \mathbf{p}^N)] \\
&\quad \times \min\left[1, e^{-\beta[H(\mathbf{r}'^N, \mathbf{p}'^N) - H(\mathbf{r}^N, \mathbf{p}^N)]}\right] \\
&= Z^{-1}\int d\mathbf{p}^N d\mathbf{p}'^N \delta[(\mathbf{r}'^N, \mathbf{p}'^N) - G(t)(\mathbf{r}^N, \mathbf{p}^N)] \\
&\quad \times \min\left[e^{-\beta H(\mathbf{r}'^N, \mathbf{p}'^N)}, e^{-\beta H(\mathbf{r}^N, \mathbf{p}^N)}\right], \tag{34}
\end{aligned}
$$

where $Z = Z_U Z_K$, and where we have used the fact that $H(\mathbf{r}^N, \mathbf{p}^N) = K(\mathbf{p}^N) + U(\mathbf{r}^N)$. Similarly,

$$
\begin{aligned}
\rho_U(\mathbf{r}'^N)\pi(\mathbf{r}'^N \to \mathbf{r}^N) &= Z^{-1}e^{-\beta U(\mathbf{r}'^N)}\int d\mathbf{p}^N d\mathbf{p}'^N e^{-\beta K(\mathbf{p}'^N)} \\
&\quad \times \delta[(\mathbf{r}^N, \mathbf{p}^N) - G(t)(\mathbf{r}'^N, \mathbf{p}'^N)] \\
&\quad \times \min\left[1, e^{-\beta[H(\mathbf{r}^N, \mathbf{p}^N) - H(\mathbf{r}'^N, \mathbf{p}'^N)]}\right] \\
&= Z^{-1}\int d\mathbf{p}^N d\mathbf{p}'^N \delta[(\mathbf{r}^N, \mathbf{p}^N) - G(t)(\mathbf{r}'^N, \mathbf{p}'^N)] \\
&\quad \times \min\left[e^{-\beta H(\mathbf{r}'^N, \mathbf{p}'^N)}, e^{-\beta H(\mathbf{r}^N, \mathbf{p}^N)}\right]. \tag{35}
\end{aligned}
$$

Making the variable transformation $\mathbf{p}^N \to -\mathbf{p}^N, \mathbf{p}'^N \to -\mathbf{p}'^N$ in Eq. 35 and using the fact that the Hamiltonian is invariant under this transformation, we have

$$\rho_U(\mathbf{r}'^N)\pi(\mathbf{r}'^N \to \mathbf{r}^N) = Z^{-1} \int d\mathbf{p}^N d\mathbf{p}'^N \delta[(\mathbf{r}^N, -\mathbf{p}^N) - G(t)(\mathbf{r}'^N, -\mathbf{p}'^N)]$$
$$\times \min\left[e^{-\beta H(\mathbf{r}'^N, \mathbf{p}'^N)}, e^{-\beta H(\mathbf{r}^N, \mathbf{p}^N)}\right]. \tag{36}$$

Now consider the delta function in Eq. 34. In general, we can write

$$\delta[(\mathbf{r}'^N, \mathbf{p}'^N) - G(t)(\mathbf{r}^N, \mathbf{p}^N)] = \frac{1}{|J|}\delta[G^{-1}(t)(\mathbf{r}'^N, \mathbf{p}'^N) - (\mathbf{r}^N, \mathbf{p}^N)], \tag{37}$$

where

$$|J| = \left|\frac{\partial(\mathbf{r}'^N, \mathbf{p}'^N)}{\partial(\mathbf{r}^N, \mathbf{p}^N)}\right| \tag{38}$$

is the Jacobian determinant for the variable transformation $(\mathbf{r}^N, \mathbf{p}^N) \to (\mathbf{r}'^N, \mathbf{p}'^N)$. Eq. 37 is the counterpart, for the Dirac delta function, of the classical change of variables formula for multidimensional integrals (see, for example, [106]). If the dynamics is *area-preserving* ($|J| = 1$) and *reversible* ($G^{-1}(t) = R_\mathbf{p}G(t)R_\mathbf{p}$), then

$$\delta[(\mathbf{r}'^N, \mathbf{p}'^N) - G(t)(\mathbf{r}^N, \mathbf{p}^N)] = \delta[G^{-1}(t)(\mathbf{r}'^N, \mathbf{p}'^N) - (\mathbf{r}^N, \mathbf{p}^N)]$$
$$= \delta[R_\mathbf{p}G(t)R_\mathbf{p}(\mathbf{r}'^N, \mathbf{p}'^N) - (\mathbf{r}^N, \mathbf{p}^N)]$$
$$= \delta[R_\mathbf{p}G(t)(\mathbf{r}'^N, -\mathbf{p}'^N) - (\mathbf{r}^N, \mathbf{p}^N)]$$
$$= \delta[G(t)(\mathbf{r}'^N, -\mathbf{p}'^N) - R_\mathbf{p}^{-1}(\mathbf{r}^N, \mathbf{p}^N)]$$
$$= \delta[G(t)(\mathbf{r}'^N, -\mathbf{p}'^N) - (\mathbf{r}^N, -\mathbf{p}^N)]. \tag{39}$$

In the next-to-last step we have made the change of variables $(\mathbf{r}^N, \mathbf{p}^N) \to R_\mathbf{p}^{-1}(\mathbf{r}^N, \mathbf{p}^N)$, and in the final step have used the fact that $R_\mathbf{p}^{-1} = R_\mathbf{p}$. Substituting Eq. 39 into Eq. 34 we have

$$\rho_U(\mathbf{r}^N)\pi(\mathbf{r}^N \to \mathbf{r}'^N) = Z^{-1} \int d\mathbf{p}^N d\mathbf{p}'^N \delta[G(t)(\mathbf{r}'^N, -\mathbf{p}'^N) - (\mathbf{r}^N, -\mathbf{p}^N)]$$
$$\times \min\left[e^{-\beta H(\mathbf{r}'^N, \mathbf{p}'^N)}, e^{-\beta H(\mathbf{r}^N, \mathbf{p}^N)}\right]. \tag{40}$$

Comparing Eqs. 40 and 36, we see that

$$\rho_U(\mathbf{r}^N)\pi(\mathbf{r}^N \to \mathbf{r}'^N) = \rho_U(\mathbf{r}'^N)\pi(\mathbf{r}'^N \to \mathbf{r}^N), \tag{41}$$

i.e. the detailed balance requirement is satisfied.

This rather tedious proof immediately suggests several interesting generalizations of the basic HMC algorithm. First, notice that we only require the evolution operator $G(t)$ to be *reversible* and *area-preserving* (in fact, it appears that even the *area-preserving* property is inessential, so long as the acceptance probability is modified by the appropriate Jacobian factor). The Hamiltonian used to generate the dynamics need not be the same as the true Hamiltonian. In general, the evolution operator $G(t)$ may describe the phase-space flow generated by a *guidance* Hamiltonian $H_g(\mathbf{r}^N, \mathbf{p}^N)$ different from $H(\mathbf{r}^N, \mathbf{p}^N)$ (the dynamics need not necessarily even be Hamiltonian in nature, provided it is reversible and area-preserving).

The possibility of using a guidance Hamiltonian different from the true Hamiltonian was realized in the first paper on HMC [57], where it was shown that the acceptance probability for HMC could be optimized for lattice field theory simulations by using an H_g slightly different from H. Another appealing strategy, applicable to molecular systems for which evaluating the forces can be quite computationally expensive (e.g. due to long-range electrostatic interactions), is to use a different, less expensive energy expression to generate 'cheap' dynamics for proposing HMC moves, reserving the computationally costly evaluation of the exact energy for the acceptance decision. One can also imagine accelerating configurational sampling on rough potential energy surfaces, where the accessible region of configuration space consists of a large number of nearly equivalent local minima separated by large potential energy barriers, by using a guidance Hamiltonian corresponding to a smoother potential energy landscape, with lower barriers between disjoint regions of configuration space. The guidance Hamiltonain can even be time-dependent, provided that its time variation is symmetric over the MD trajectory.

Note also that the momentum variables are introduced simply as auxiliary variables, and need not have an initial distribution that is Maxwellian. In principle, *any* distribution $\rho_K(\mathbf{p}^N)$ can be used in place of the Maxwell distribution $Z_K^{-1} e^{-\beta K(\mathbf{p}^N)}$, provided ρ_K is invariant under the transformation $\mathbf{p}^N \to -\mathbf{p}^N$. For a general $\rho_K(\mathbf{p}^N)$, the appropriate HMC acceptance probability is

$$P_A[(\mathbf{r}^N, \mathbf{p}^N) \to (\mathbf{r}'^N, \mathbf{p}'^N)] = \min\left[1, \frac{\rho_K(\mathbf{p}'^N)}{\rho_K(\mathbf{p}^N)} e^{-\beta \Delta U}\right], \qquad (42)$$

where $\Delta U = U(\mathbf{r}'^N) - U(\mathbf{r}^N)$.

Unfortunately, if H_g differs too markedly from H then H is no longer even approximately conserved in the dynamics (H_g is the (approximately) conserved quantity), and the acceptance probability (Eq. 29), which depends on the *true* Hamiltonian H, drops sharply. One must thus be judicious in the choice of guidance Hamiltonian.

Similarly, radical deviations of the momentum sampling distribution from the Maxwell distribution appropriate to the simulated temperature are likely to give extremely poor results. For example, taking the initial momenta $\mathbf{p}^N$ to be distributed on a spherical shell (i.e. with fixed magnitudes and random directions) will always give an acceptance rate of zero, as the final momenta $\mathbf{p}'^N$ have essentially zero probability of being distributed in this way. In general, the best results will be obtained when the momentum sampling distribution differs only slightly from the 'physical' Maxwell distribution. For instance, ρ_K can be chosen to be a Maxwell distribution at to a temperature slightly higher than the simulated temperature, to speed up configurational sampling.

The foregoing discussion illustrates the flexibility of HMC relative to MD, but should not be considered to exhaust the interesting possible variations of HMC. For example, the performance of HMC can be improved by sampling the trajectory length N_{MD} from a distribution instead of working with a fixed trajectory length [107, 108], and HMC can be formulated in generalized coordinates rather than cartesian coordinates [109, 110].

Before discussing tests of HMC, we derive two useful identities. We first prove that

$$\left\langle e^{-\beta \Delta H} \right\rangle = 1. \tag{43}$$

This identity is useful for checking the correctness of HMC code, and for detecting poor sampling of configuration space. It is a straightforward consequence of the area-preserving property of the dynamics:

$$
\begin{aligned}
\left\langle e^{-\beta \Delta H} \right\rangle &= Z^{-1} \int d\mathbf{r}^N d\mathbf{p}^N e^{-\beta H(\mathbf{r}^N, \mathbf{p}^N)} e^{-\beta [H(\mathbf{r}'^N, \mathbf{p}'^N) - H(\mathbf{r}^N, \mathbf{p}^N)]} \\
&= Z^{-1} \int d\mathbf{r}^N d\mathbf{p}^N e^{-\beta H(\mathbf{r}'^N, \mathbf{p}'^N)} \\
&= Z^{-1} \int d\mathbf{r}'^N d\mathbf{p}'^N e^{-\beta H(\mathbf{r}'^N, \mathbf{p}'^N)} \\
&= 1.
\end{aligned}
$$

In the second-to-last step we have made a change of variables $(\mathbf{r}^N, \mathbf{p}^N) \to (\mathbf{r}'^N, \mathbf{p}'^N)$, using the fact that the Jacobian determinant for this transformation is unity ($|J| = 1$), and in the final step we have used the definition of Z.

For small Δt, it can be argued [111] that the discretization error (ΔH) distribution is, to a good approximation, Gaussian, so that only the first two cumulants of the distribution (the mean and the variance) are non-negligible. Moreover, Eq. 43 implies a relation between the mean and the variance, namely

$$
\begin{aligned}
\langle \Delta H \rangle &= \frac{\beta}{2} \left\langle (\Delta H - \langle \Delta H \rangle)^2 \right\rangle \\
&= \frac{\beta \sigma_{\Delta H}^2}{2}, \tag{44}
\end{aligned}
$$

where $\sigma^2_{\Delta H}$ is the variance in ΔH.

In this case, the mean acceptance probability is

$$
\begin{aligned}
\langle P_A \rangle &= \frac{1}{\sqrt{2\pi\sigma^2_{\Delta H}}} \int_{-\infty}^{\infty} dx \, \min\left[1, e^{-\beta x}\right] \exp\left[-\frac{(x - \langle \Delta H \rangle)^2}{2\sigma^2_{\Delta H}}\right] \\
&= \operatorname{erfc}\left(\frac{1}{2}\sqrt{\beta\langle\Delta H\rangle}\right).
\end{aligned}
\tag{45}
$$

This relation between $\langle P_A \rangle$ and $\langle \Delta H \rangle$ is well obeyed in practice.

We can use Eq. 45 to deduce how the performance of the HMC algorithm depends on system size. There are two essential points to grasp:

1. For MD trajectories of more than a few integration timesteps using Verlet-type integrators such as r-RESPA, $|\Delta H - \langle\Delta H\rangle| \sim \Delta t^2$. The *local* error in a single timestep scales as Δt^3, but the *global* error after a number of timesteps is *bounded* and scales as Δt^2, provided Δt is small enough that the dynamics is stable. Thus, $\langle\Delta H\rangle \sim \Delta t^4$, since it is proportional to the variance.

2. In typical cases, the average discretization error scales linearly with system size, i.e. $\langle\Delta H\rangle \sim N$. This is essentially an empirical observation.

Overall, therefore, we have $\langle\Delta H\rangle \sim N\Delta t^4$.

Generally speaking, HMC is only efficient for a relatively high average acceptance probability ($\langle P_A \rangle > 50\%$). To maintain a constant average acceptance rate for variable system size, we require $N\Delta t^4 = $ constant, or $\Delta t \sim N^{-1/4}$. Because the timestep must be made smaller as the system size increases, the computational cost $\mathcal{E}$ of HMC increases faster than linearly with system size, $\mathcal{E} \sim N^{5/4}$, whereas for MD, $\mathcal{E} \sim N$. Thus, the basic HMC method will always become less efficient than MD for some critical system size, even if it is more efficient for small systems.

Although one might imagine that HMC could be made significantly more efficient than MD by using an integration timestep Δt for generating trial HMC moves that is substantially larger than that used in a stable MD integration, it is easy to see that this will not work. The basic reason is that HMC is only efficient if each proposed HMC move results in a configuration significantly different from the original configuration. Otherwise, the diffusion through configuration space is slow (i.e. we have a random walk with very small steps) and the computational effort required to generate a statistically independent configuration is large. Thus, the total duration of the MD trajectory used to generate each HMC move, $t = N_{\mathrm{MD}}\Delta t$, must be large on the scale of the decorrelation times characteristic of the collective dynamics, which typically translates into MD trajectories of hundreds of integration timesteps. At the same time, the discretization error must be kept within reasonable bounds, or else the HMC acceptance rate will become unworkably small, leading to poor statistical sampling. These two requirements together imply that the integration timestep used in HMC cannot be significantly larger than the maximum Δt that generates stable MD trajectories.

Moreover, even if HMC is more efficient than MD for small systems, the decrease in acceptance rate with increasing system size discussed above implies that HMC will inevitably become less efficient than MD for sufficiently large systems. One way of getting around the unfortunate size scaling of HMC is to treat it as a *local* MC update, in which the positions of a fixed number of atoms $n < N$ are updated in each HMC move, with all other atomic positions fixed. If n is independent of N, the computational effort will scale linearly with system size (at the cost of some overall loss of efficiency). To our knowledge, this approach has not been tested to any significant extent.

In fact, our tests suggest that HMC is less efficient than MD even for rather small systems, as we now proceed to show for a specific molecular system. We studied a system consisting of 27 $C_{24}H_{50}$ chains in the melt, at a density of $\rho = 0.68$ g/cm^3 and a temperature of $T = 473$ K. We used the united-atom alkane model of Siepmann et al. [46], augmented by bond-stretching and bond angle bending potentials. In this model, non-bonded interactions between sites on the same molecule separated by three or fewer bonds are excluded. Lennard-Jones interactions between distinct sites are truncated smoothly between 11 and 12 Å, with a standard long-range correction applied to account for more distant interactions [67]. For both HMC and MD, a r-RESPA integrator with four distinct intervals of force evaluation is employed. An initial equilibration run of 60 ps was carried out prior to the efficiency tests described here.

The efficiency of a particular method can be gauged by calculating the *decorrelation times* for specific observables, which measure the computational effort required to generate a statistically independent sample (the relevant time here is CPU time). The most efficient algorithm is that which *minimizes* the relevant decorrelation time, allowing the maximum number of statistically independent states to be sampled for a given expenditure of computational effort. For complex molecular systems, a variety of distinct microscopic processes can be probed, each with its own characteristic timescale, so we chose to measure several distinct quantities to assess the efficiency of HMC relative to MD.

A measure of the rate of orientational diffusion is the time required for decorrelation of the molecular end-to-end vector,

$$\tau_{ee} = \int_0^\infty C_{ee}(\tau)d\tau, \tag{46}$$

where C_{ee} is the end-to-end vector autocorrelation function,

$$C_{ee}(\tau) = \frac{\langle \mathbf{r}_{ee}(\tau) \cdot \mathbf{r}_{ee}(0) \rangle}{\langle |\mathbf{r}_{ee}|^2 \rangle}, \tag{47}$$

and $\mathbf{r}_{ee}$ is the molecular end-to-end vector. A measure of the rate of change of molecular shape ('shape' diffusion) is the squared radius of gyration decorrelation time,

$$\tau_{\mathrm{srg}} = \int_0^\infty C_{\mathrm{srg}}(\tau) d\tau, \tag{48}$$

where C_{srg} is the squared radius of gyration autocorrelation function,

$$C_{\mathrm{srg}}(\tau) = \frac{\langle R_g^2(\tau) R_g^2(0) \rangle - \langle R_g^2 \rangle^2}{\langle R_g^4 \rangle - \langle R_g^2 \rangle^2}. \tag{49}$$

Here,

$$R_g^2 = \frac{1}{M} \sum_{i=1}^{n} m_i (\mathbf{r}_i - \mathbf{R}_{\mathrm{cm}})^2 \tag{50}$$

is the squared molecular radius of gyration, M is the molecular mass, m_i is the mass of site i, $\mathbf{R}_{\mathrm{cm}}$ is the center of mass position of the molecule, and the sum ranges over all n sites in the molecule (in the present case, $n = 24$). We can also measure the rate at which distinct molecular conformations are sampled. A measure of conformational diffusion is the isomerization time,

$$\tau_{\mathrm{iso}} = \int_0^\infty C_{\mathrm{iso}}(\tau) d\tau, \tag{51}$$

where C_{iso} is the rotational isomer autocorrelation function,

$$C_{\mathrm{iso}}(\tau) = \frac{1}{n-3} \sum_{j=1}^{n-3} \frac{\langle W_j(\tau) W_j(0) \rangle - \langle W_j \rangle^2}{\langle W_j^2 \rangle - \langle W_j \rangle^2}, \tag{52}$$

and W_j is an integer variable indicating the rotational isomeric state of the jth rotatable bond,

$$W_j = \begin{cases} 1 & \text{if bond is in } \textit{trans} \text{ configuration} \\ 0 & \text{if bond is in } \textit{gauche} \pm \text{ configuration} \end{cases} \tag{53}$$

Finally, the rate of translational diffusion can be gauged by the effective diffusion constant for chains,

$$D_{\mathrm{eff}} = \lim_{t \to \infty} \frac{\langle |\mathbf{R}_{\mathrm{cm}}(t) - \mathbf{R}_{\mathrm{cm}}(0)|^2 \rangle}{6t}. \tag{54}$$

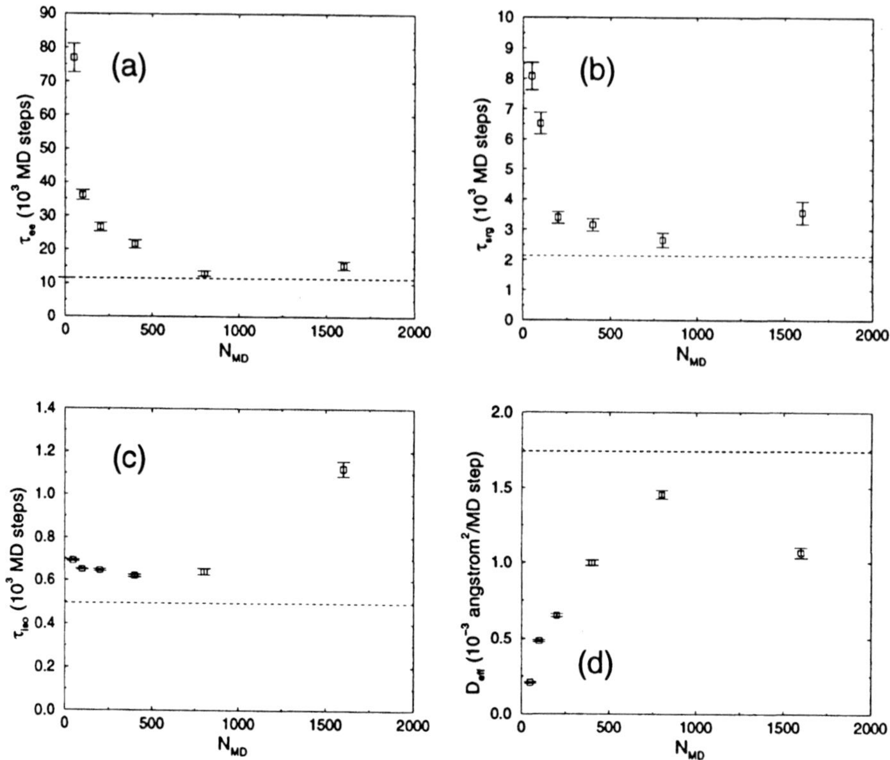

Figure 3. Comparison of the relative efficiency of HMC (squares) and MD (dashed line) as a function of HMC trajectory length N_{MD} for $\Delta t = 0.0075$ ps. Several measures of configurational sampling efficiency are shown: τ_{ee} (a), τ_{srg} (b), τ_{iso} (c), and D_{eff} (d). The unit of computational effort is one MD integration timestep. An integration stepsize of $\Delta t = 0.0075$ ps was used in the MD simulation.

The most efficient algorithm from the standpoint of translational diffusion is that which *maximizes* the effective diffusion constant, measured in terms of mean-squared displacement as a function of CPU time. All of the measures of sampling efficiency defined above are averaged over all molecules in the system.

In Fig. 3 we have compared the relative efficiency of HMC and MD as a function of HMC trajectory length N_{MD}, using the four measures of sampling efficiency described in the previous paragraph. An r-RESPA integrator with four distinct rates of force evaluation and a maximum integration stepsize of $\Delta t = 0.0075$ ps was used for both MD and HMC [112]. The unit of computational effort is an MD timestep, as the cost of a single integration timestep is essentially the same for both MD and HMC. By all measures, HMC is less efficient than MD for all trajectory lengths. As

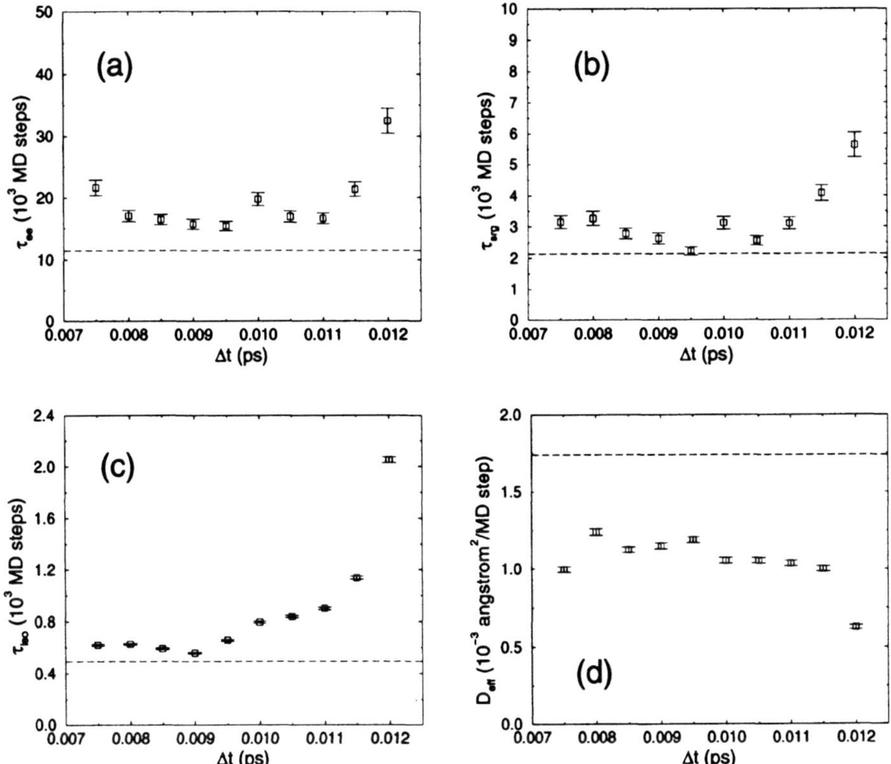

Figure 4. Comparison of the relative efficiency of HMC (squares) and MD (dashed line) as a function of HMC integration timestep Δt for $N_{\mathrm{MD}} = 400$. Several measures of configurational sampling efficiency are shown: τ_{ee} (a), τ_{srg} (b), τ_{iso} (c), and D_{eff} (d). The unit of computational effort is one MD integration timestep. An integration stepsize of $\Delta t = 0.0075$ ps was used in the MD simulation.

discussed above, the algorithm is expected to be quite inefficient for small N_{MD}, and the efficiency is almost independent of N_{MD} for large N_{MD}. In Fig. 4 we have examined the effect of varying the HMC integration stepsize Δt on sampling efficiency for $N_{\mathrm{MD}} = 400$. Increasing the integration stepsize does not significantly improve the performance of the HMC algorithm, and in fact degrades its performance for large enough Δt. Again, this is essentially as expected from the general considerations outlined above, as the MD integrator becomes unstable for large Δt, leading to poor HMC performance. Thus, even for this relatively small system, HMC is less efficient than MD. The relative performance of HMC will be even worse for larger systems.

We cannot rule out the possibility that some variant of HMC might be found that performs well relative to MD, but at present it appears that HMC is not a competitive large-scale simulation method relative to MD. We have, on the other hand, found HMC to be useful for importance sampling in small systems, for example in the single-molecule mean-field theory calculations described in Sec. 4.3, and it appears that in this case HMC is competitive with other methods.

4. Applications

4.1. MICROSCOPIC STRUCTURE OF A PARTIAL BILAYER SMECTIC

The defining characteristic of a smectic liquid crystal is a one-dimensional modulation of the molecular number (or electron, mass, etc.) density, which is characterized by the period of the modulation, the smectic layer spacing d. Exactly which physical effects conspire to determine the layer spacing is a subtle question. In simple 'monolayer' SmA LCs, the smectic layer spacing is approximately equal to the fully extended molecular length l. Even in this case, however, d has a nontrivial temperature dependence, generally decreasing with increasing temperature. In other cases, d can differ significantly from l, for instance in *bilayer* smectics (e.g. the orthogonal SmA_2 phase), for which $d \sim 2l$, or in *partial bilayer* smectics (e.g. the orthogonal SmA_d phase), for which $l < d < 2l$. Bilayer or partial bilayer smectic phases are typically exhibited by systems composed of molecules having either hydrogen-bonding or strongly polar terminal groups, in which short-ranged dimeric or antiparallel association plays a critical role in determining supermolecular structure.

The prototypical partial bilayer smectic is 4-octyl-4′-cyanobiphenyl (8CB), which forms a SmA_d phase with layer spacing $d = 31.432$ Å at room temperature [113], approximately 1.4 times the fully extended molecular length of 22.1 Å, and in which the layer spacing *increases* with increasing temperature, in contrast to most monolayer SmA materials. The chemical structure and phase sequence of 8CB are shown in Fig. 5. Despite the fact that SmA_d 8CB is in no way a typical smectic, it is undoubtedly the most studied smectic material, owing to its wide availability, chemical stability, and room-temperature smectic phase range. A key feature of the molecular structure of 8CB is the terminal cyano group, which possesses a large dipole moment ($\mu = 4.05$ D). The formation of a partial bilayer smectic phase in 8CB is thought to reflect short-range antiparallel association of molecules due to strong dipole-dipole interactions between terminal dipoles and/or dipole-induced dipole interactions between terminal dipoles and polarizable cores. In the conventional 'cartoon' view of the microscopic organization of SmA_d 8CB [114–116], each smectic layer consists of two strongly interdigitated polar sublayers with oppositely oriented terminal dipoles, as depicted schematically in Fig. 6.

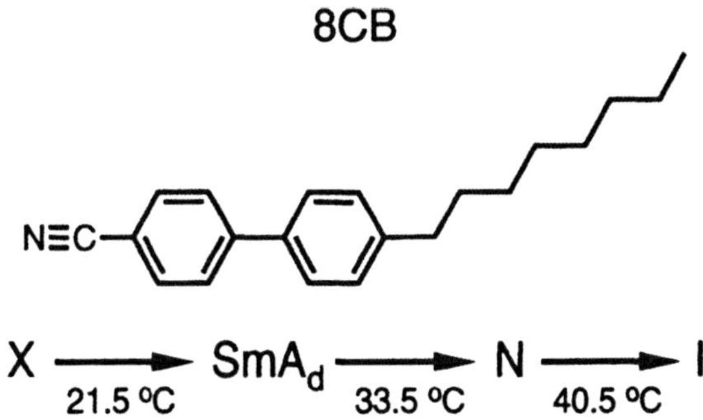

Figure 5. Chemical structure and phase sequence of 8CB.

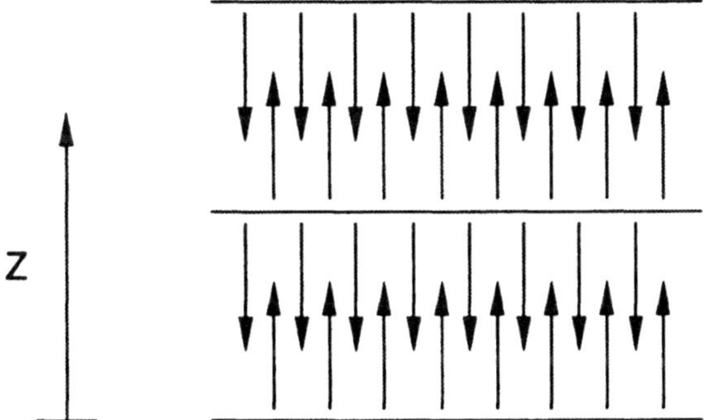

Figure 6. Schematic representation of the microscopic structure of 8CB, in which each smectic layer consists of two interdigitated polar sublayers. 8CB molecules are depicted as arrows, with the terminal cyano group corresponding to the head of the arrow.

In an effort to move our conceptual understanding of the microscopic structure of partial bilayer smectics beyond the cartoon stage, we have initiated a series of large-scale atomistic simulation studies of 8CB (a more detailed account of this work will appear elsewhere [117]). This is part of a larger simulation project aimed at elucidating the microscopic organization of cyanobiphenyls, the most intensively studied chemical family of LCs. We were particularly interested in learning whether we could model 8CB at a sufficient level of physical realism to reproduce the experimental layer spacing. This represents a sensitive test of the interaction potential employed in the simulation, and constitutes a zero-order requirement for quantitative atomistic simulation of smectics. Moreover, there exists a wealth of other experimental data on 8CB, making this an excellent system for testing and refining 'realistic' simulation methodology.

Intra- and intermolecular interaction potentials were derived using a combination of *ab initio* and empirical information, according to the procedure described in Sec. 2. It is worth noting that in this simulation, as in others described in this article, we neglect induction interactions, which may be important in this system, as the 8CB molecule possesses both a large permanent dipole and a highly polarizable core. Due to this neglect of induction effects as well as the other approximations of convenience that have been made, it is not at all guaranteed that results in consonance with experiment will be obtained.

We carried out an *NPT* MD simulation [118] of a system containing 100 8CB molecules at $P = 1$ atm and $T = 290$ K, utilizing a highly optimized multiple-timestep MD (r-RESPA) scheme [119]. Long-range electrostatic interactions were evaluated to high accuracy using the particle-mesh Ewald method. Methylene (CH_2) and methyl (CH_3) groups were treated as effective atoms (i.e. single spherically symmetric interaction sites), but full atomic detail is employed otherwise. In particular, hydrogens attached to sp^2-hybridized carbons in the phenyl rings were included explicitly. However, we set the hydrogen mass equal to that of carbon in this case, to avoid introducing a distinct level of multiple-timestep MD for fast intramolecular vibrational modes involving hydrogen. This has no effect on the equilibrium distribution, but does modify the dynamics. Specifically, the use of 'heavy' hydrogens modifies the intramolecular vibrational spectrum (which we don't expect to be accurately reproduced by our simple force field anyway) and, more importantly, slightly modifies the overall timescale for the dynamics.

A bilayer (SmA_2-like) initial condition, shown in Fig. 7(a), was used, with perfect polar ordering of 8CB molecules within each half of the bilayer. As can also be seen from this figure, a rather unusual placement of smectic layers within the monoclinic unit cell is employed, in which each bilayer connects to the adjacent bilayer across the periodic boundaries. Thus, in reality there is only a single continuous smectic bilayer present in the system. This was done to promote homogeneity, i.e. to avoid situations in which the ordering varies from layer to layer, and to minimize finite-size effects arising from the effective constraint of a fixed number of molecules per layer. In this situation, a monoclinic unit cell must be employed if ordinary periodic boundary conditions are used (i.e. if one wishes the cell edge vectors to be Bravais lattice vectors). Alternatively, one could employ an orthorhombic unit cell, at the price of introducing shifted periodic boundary conditions.

The final configuration from the simulation (after 8.06 ns) is shown in Fig. 7(b). Although smectic layering is still apparent, the organization of the layers is quite complex and fairly disordered, bearing essentially no resemblance to the cartoon of Fig. 6. The basic question here is: how faithful a representation of the microscopic organization of the SmA_d phase of 8CB is this? To begin to answer this question, we must determine whether the properties of the simulated system are consistent with experimental measurements.

Before doing so, however, we need to determine whether the simulated system is equilibrated. In Fig. 8 we have plotted the time evolution of several instantaneous thermodynamic and structural properties. These include the mass density ρ, the potential energy U, and the eigenvalues of the instantaneous ordering tensor $\mathbf{Q}$. The ordering tensor is defined as

$$Q_{\alpha\beta} = \frac{1}{N} \sum_{i=1}^{N} \left(\frac{3}{2} \hat{u}_{i\alpha} \hat{u}_{i\beta} - \frac{1}{2} \delta_{\alpha\beta} \right), \tag{55}$$

where $\hat{u}_i$ is a unit vector coincident with the instantaneous long axis of molecule i (defined as the minimum-moment principal axis of the molecular inertia tensor), $\alpha, \beta = x, y, z$, and the sum ranges over all N molecules. The largest eigenvalue of the ordering tensor defines the nematic order parameter S, and the corresponding eigenvector defines the nematic director $\hat{n}$. All quantities appear to have reached a steady state by the end of the simulation, aside from a slight decrease in the nematic order parameter over the last 3 ns of the simulation. The layer spacing d, shown in Fig. 9, shows a more significant long-term variation, although it is approximately constant for the last 2 ns of the simulation. We thus calculate the average properties of the system for the last 2 ns of the simulation for comparison with experiment. Before discussing the comparison with experiment, however, we note that there is a sobering message here, namely that proper equilibration of the SmA_d phase of 8CB by MD requires in excess of 6 ns (corresponding to about a month of CPU time on a fast workstation). As we will show below, the situation for other systems is even worse.

The average layer spacing for the last 2 ns of the simulation is $\langle d \rangle = 31.9$ Å, in remarkable agreement with the measured layer spacing of 31.432 Å at $T = 24$ °C [113]. The average simulated mass density, $\langle \rho \rangle = 1.018$ g/cm^3, is also quite close to the experimental density of 1.027 g/cm^3 at $T = 28$ °C [114]. Moreover, the average nematic order parameter for the LC core, defined as the largest eigenvalue of the average core ordering tensor $\langle \mathbf{Q}_c \rangle$ (computed from Eq. 55, with $\hat{u}_i$ taken to be the long axis of the biphenyl core), is $S_c = 0.78$, in very good agreement with Raman and optical dichroism measurements, which yield $S_c = 0.77$ at $T = 28$ °C [120].

We have also compared our simulation results with experimental measurements that probe the orientational distribution of specific functional groups of 8CB, namely deuteron magnetic resonance (DMR) measurements of nematic order parameters for C-D bonds in the alkyl tails of specifically deuterated 8CB [121]. We find that the simulated order parameters are uniformly somewhat larger than the experimental ones, although the qualitative trend as a function of carbon position is well reproduced. The discrepancy between simulation and experiment may be due to the small

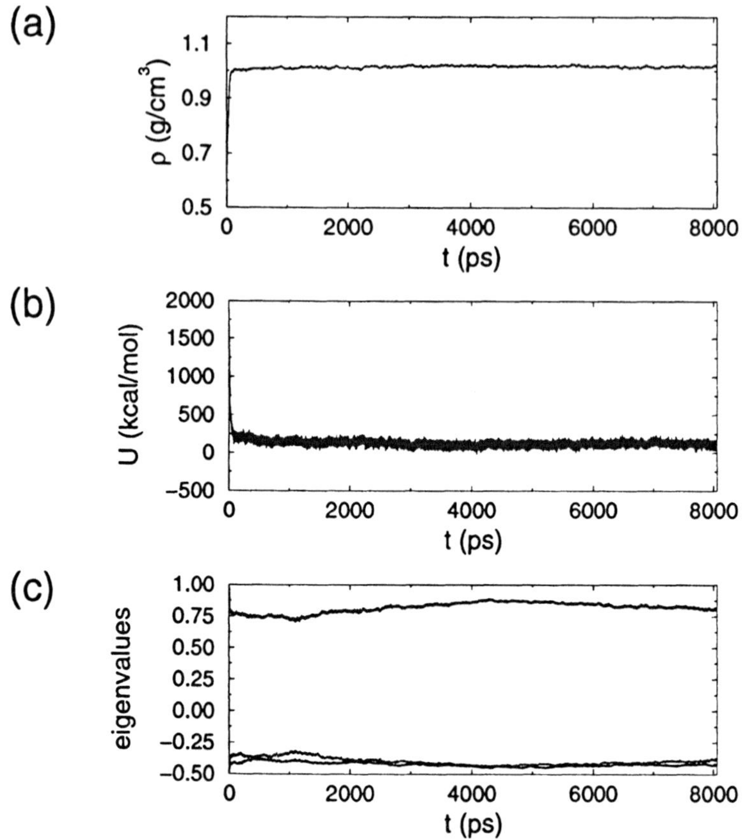

Figure 8. Time evolution of mass density ρ (a), potential energy U (b), and eigenvalues of the ordering tensor $\mathbf{Q}$ (c), over the duration of an 8.06 ns simulation of 8CB.

size of the simulated sample (finite size effects will tend to increase the degree of ordering), but more likely reflects inaccuracies in the molecular model.

This favorable overall level of agreement between simulation and experiment indicates that the essential physical effects are included in our model for 8CB. In particular, these results suggest that induction effects (which are *not* included explicitly in this simulation) are of secondary importance in determining the structure of SmA_d 8CB. Given that this seems to be a reasonable model for SmA_d 8CB, we can proceed to analyze its structure in more detail. Of particular interest are the structure of the smectic layers and the short-range correlations between 8CB molecules.

To characterize the structure of the smectic layers in more detail, we have computed mass and number density profiles along the layer normal

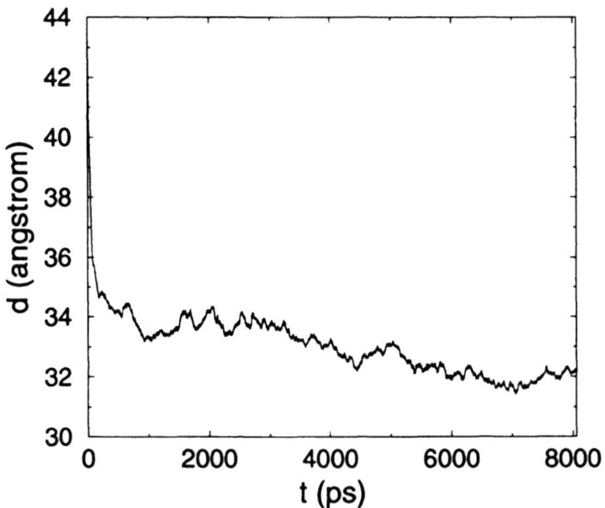

Figure 9. Time evolution of smectic layer spacing d over the duration of an 8.06 ns simulation of 8CB.

$\hat{z}$. These are shown in Fig. 10. The overall mass density profile, shown in Fig. 10(a), is approximately sinusoidal. We would expect the electron density profile to be quite similar, and so it is no surprise that only a single Bragg peak is observed in the x-ray diffraction pattern of 8CB. The molecular center of mass number density profile, shown in Fig. 10(b), is roughly in accord with the notion of a bilayer: molecules with their terminal dipoles directed in the $-Z$ direction (dot-dashed curve) are largely confined to the upper half of the layer, while molecules with terminal dipoles pointing in the $+Z$ direction (dashed curve) reside mainly in the lower half of the layer. However, there is a significant probability of finding a molecule having a given polarity in the 'wrong' half of the layer. The number density profile for cyano groups, shown in Fig. 10(c), indicates that the cyano groups are reasonably well localized in the middle of the layer, with a slight offset of the distribution of cyano groups with the C$\equiv$N bond pointing in the $-Z$ direction (dot-dashed curve) relative to $+Z$-directed cyano groups.

Overall, the density profiles in Fig. 10 reveal a much more complex layer structure than might have been expected. This complex structure reflects the intermolecular correlations present in this system. To shed light on the role of dipole-dipole interactions in determining the structure of the SmA$_d$ phase of 8CB, we measured positional and orientational correlation functions for terminal cyano groups. In Fig. 11 we show the pair distribution function for cyano groups, $g(r)$, and the corresponding polar orientational

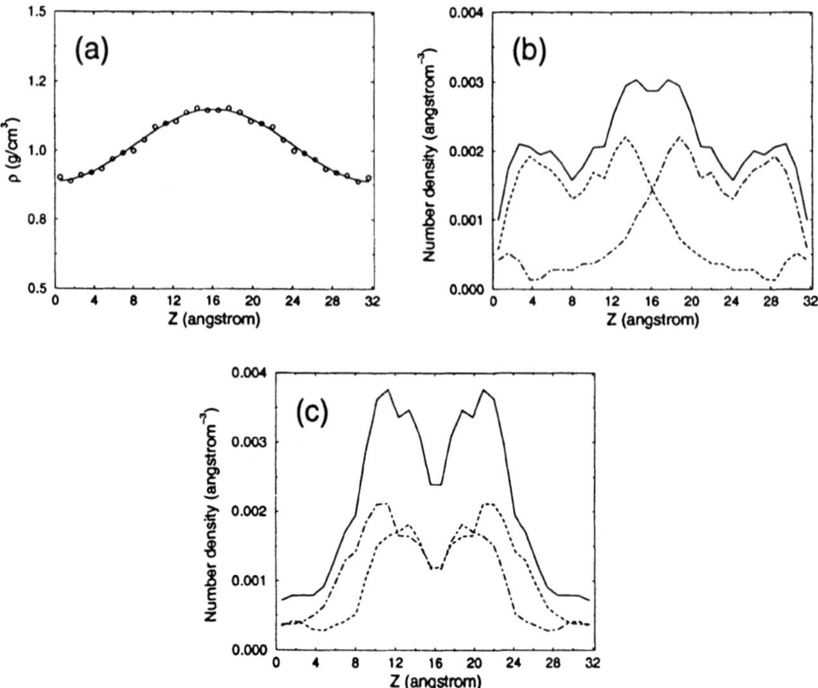

Figure 10. Mass density profile (a), molecular center of mass number density profile (b), and cyano group number density profile (c) for SmA$_d$ 8CB. Also shown in (b) and (c) are the partial number density profiles for molecules or cyano groups with C≡N bonds oriented in the $+Z$ direction (dashed curves) and $-Z$ direction (dot-dashed curves).

correlation function,

$$g_1(r) = \frac{\langle P_1(\hat{\mathbf{u}}_{CN}(r) \cdot \hat{\mathbf{u}}_{CN}(0)\rangle}{g(r)}, \qquad (56)$$

where $\hat{\mathbf{u}}_{CN}$ is a unit vector directed along the C≡N bond, and $P_1(x) = x$ is the first Legendre polynomial. $g(r)$ (Fig. 11(a)) exhibits a peak at $r = 3.7$ Å, but is otherwise essentially featureless. The prominent negative feature in $g_1(r)$ (Fig. 11(a)) reveals that cyano groups in the first coordination shell surrounding a given cyano are, with a high probability, antiparallel to the central cyano group ($g_1(r)$ reaches a value smaller than -0.8, whereas the maximum negative value is -1.0).

A more detailed picture emerges when we consider the two-dimensional correlation functions $g(r_\parallel, r_\perp)$ and $g_1(r_\parallel, r_\perp)$, where $r_\parallel = \mathbf{r} \cdot \hat{\mathbf{u}}_{CN}$ and $r_\perp = |\mathbf{r} - r_\parallel \hat{\mathbf{u}}_{CN}|$. These correlation functions are shown in Fig. 12. The reference cyano group is situated at the origin, with the C≡N bond oriented in the $+r_\parallel$ direction. Notice first that there is a 'correlation hole' for -18Å $<$

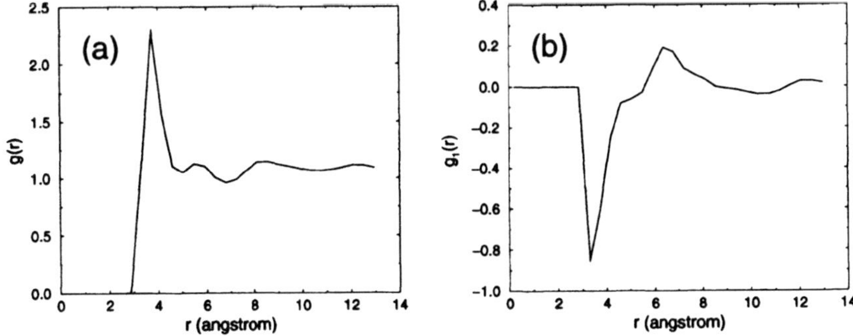

Figure 11. Pair correlation function $g(r)$ (a) and polar orientational correlation function $g_1(r)$ (b) for terminal cyano groups in the SmA$_d$ phase of 8CB. The prominent negative feature in $g_1(r)$ near $r = 3.5$ Å reveals strong antiparallel correlation of cyano groups in the first coordination shell.

$r_\parallel < 4$Å and $r_\perp < 3$Å, that roughly defines the excluded volume of the central molecule. The other prominent feature of the data is a ridge of high probability near $r_\perp = 4$Å extending from $r_\parallel = -15$Å to $r_\parallel = 3$Å. Along this ridge are three peaks, the most prominent of which is at $r_\parallel \approx 0.5$Å, $r_\perp \approx 3.5$Å. This peak reaches a height of 6, corresponding to a density of C$\equiv$N groups six times their average density. Examination of $g_1(r_\parallel, r_\perp)$ (Fig. 12(b)) reveals that the groups associated with this peak have a strong tendency to be antiparallel to the central C$\equiv$N group. Thus, the most significant correlations between cyano groups can be traced to their strong antiparallel association at short distances. The other structure evident in Fig. 12(b) is quite complex, and much more difficult to interpret, but it is clear that the significant correlations do not extend much beyond the first coordination shell.

We feel that this study of 8CB is quite promising, demonstrating that it is possible to model smectics at a quantitative level, provided that the most important physical effects are incorporated in the simulation. The results are particularly encouraging given that 8CB is a 'difficult' smectic, with a complex layer structure. The existence of a reasonable model for 8CB makes it possible to undertake more ambitious projects, such as the study of photoactive solutes in a SmA$_d$ 8CB host, described in the next section.

4.2. PHOTO-CONTROLLED NANOPHASE SEGREGATION

Liquid crystals are exotic solvents, in that they impose their state of order on solutes. The orientational ordering of non-spherical solutes in nematic solvents is well-studied (see, for example [122]). Smectic solvents are particularly interesting, as they impose one-dimensional positional order as well

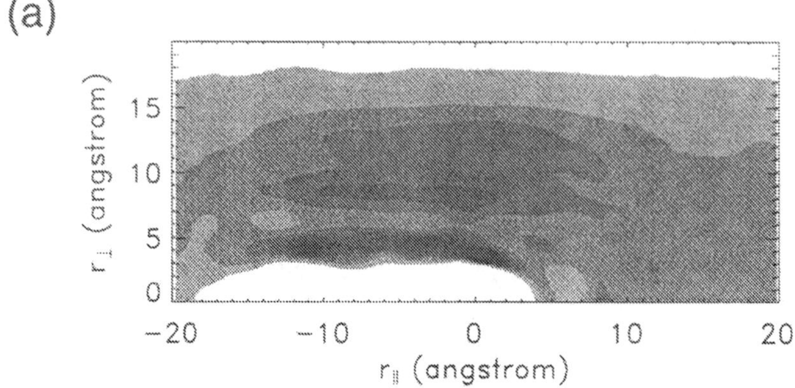

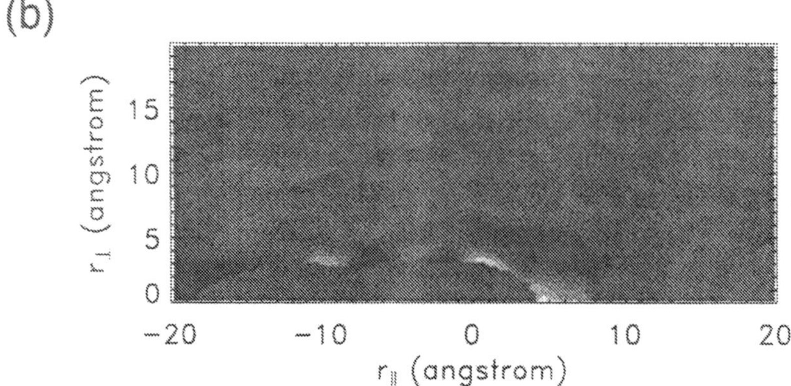

Figure 12. Two-dimensional pair correlation function $g(r_\parallel, r_\perp)$ (a) and polar orientational correlation function $g_1(r_\parallel, r_\perp)$ (b) for terminal cyano groups in the SmA$_d$ phase of 8CB.

as orientational order on solutes. This positional ordering can be regarded as a sort of nanophase segregation. Symmetry considerations dictate that there are just two distinct possibilities: solutes in a smectic host are either concentrated between smectic layers (interlamellar segregation) or within smectic layers (intralamellar segregation). These two situations are illustrated schematically in Fig. 13. Interlamellar segregation tends to be the rule for aliphatic solutes such as hexane, leading to the creation of 'organic lyotropics', in which the smectic layer spacing increases strongly with solute concentration [123]. By contrast, solutes that are chemical similar to LC core groups (e.g. those containing aromatic rings) tend to be exhibit intralamellar segregation, being incorporated into the smectic layers [123, 124]. Nanoscale concentration of solutes in smectic hosts leads to dramatically enhanced polymerization rates for photopolymerizable monomers in smectic hosts [124].

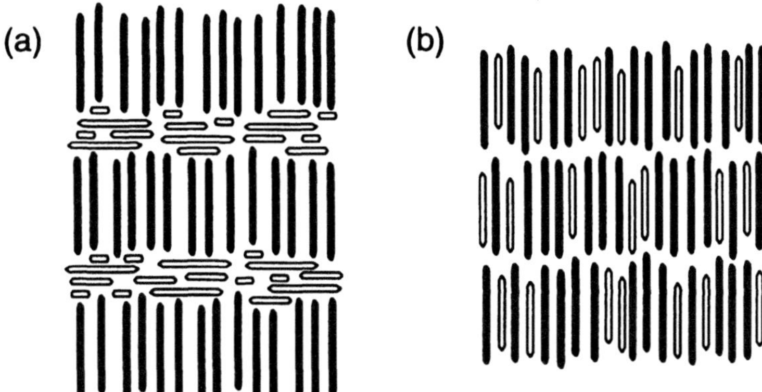

Figure 13. Schematic depiction of the nanophase segregation of non-spherical solute molecules (unfilled rods) in a smectic host (filled rods). Symmetry dictates that there are only two possible situations: interlamellar segregregation (a), and intralamellar segregation (b).

A further consequence of the symmetry of a smectic phase is that the orientational and conformational distributions of solutes in smectic solvents vary with their position along the layer normal, i.e. relative to the plane of the smectic layers. Conversely, the positional distribution of a solute in a smectic host depends on its conformational state (as does its orientational distribution). An immediate consequence of this is that *photoactive* solutes in smectic hosts that undergo a conformational change upon photoisomerization have a positional distribution that depends upon photochemical state (provided that the photochemical transformation does not destroy the smectic ordering of the host, or lead to macroscopic phase separation).

We have recently carried out a large-scale atomistic simulation study of photoactive solutes in smectic LCs, in order to characterize changes in positional and orientational distribution due to changes in photochemical state, as well as changes in the structure of the host [125]. This study was motivated by the recent observation of a novel photomechanical effect in mixtures of a photolabile solute (*p,p'*-diheptylazobenzene, or 7AB) with SmA 8CB, involving a reversible increase in the layer spacing upon *trans–cis* photoisomerization of 7AB [113,126–128]. The chemical structure of 7AB as well as the conformational change induced by UV irradiation is shown in Fig. 14. Our simulations show that the observed increase in smectic layer spacing is due to photoactive molecules being driven from within the smectic layers to the interlayer region upon *trans–cis* photoisomerization. Stating it somewhat differently, our results demonstrate effective photo-control of the positional distribution of 7AB in SmA 8CB, with the *trans–cis* photoisomerization of 7AB being accompanied by a change from *intralamellar* segregation to *interlamellar* segregation. Photocontrolled nanophase segregation of chiral photoactive dopants in SmC hosts offers an attractive route to the photomodulation of spontaneous polarization in ferroelectric LCs (FLCs), a prerequisite to the use of FLCs as optical data storage materials.

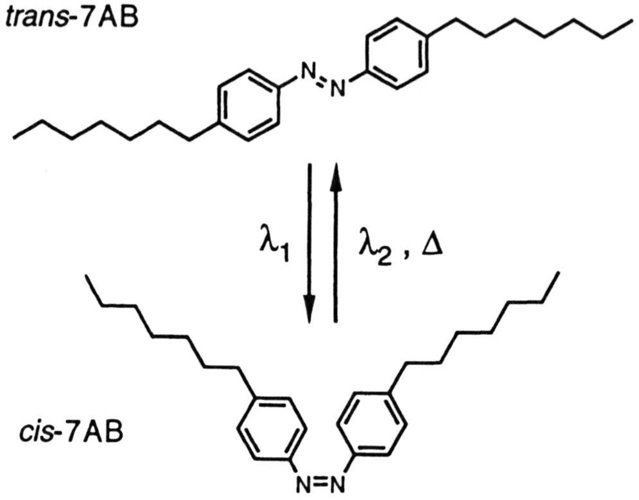

Figure 14.   Chemical structures of *trans-* and *cis*-7AB. In the absence of illumination, *trans*-7AB is ∼ 50 kJ/mol more stable than *cis*-7AB, but can be converted to the *cis* form by electronic excitation under UV irradiation. The reverse reaction proceeds thermally or upon illumination with longer-wavelength light.

We carried out *NPT* MD simulations of systems containing 100 8CB molecules and 12 7AB molecules at $P = 1$ atm and $T = 290$ K, with periodic boundary conditions. The composition of the simulated 8CB/7AB mixtures (10.7 mol% 7AB) is close to that studied experimentally. Two simulations were carried out: one with all 7AB molecules in the *trans* conformation (which we refer to as the *trans*-7AB simulation) and one with all 7AB molecules in the *cis* conformation (the *cis*-7AB simulation). The initial configurations for the two simulations were made as similar as possible. In both cases, a SmA_2-like initial configuration was used for 8CB, and 7AB molecules were placed within the smectic layers, with their long axes oriented along the layer normal. As in the simulation of pure 8CB described in Sec. 4.1, the smectic layers were placed within the monoclinic cell in such a way that the systems effectively consist of only a single bilayer. The initial configurations for the *trans*-7AB and *cis*-7AB simulations are shown in Figs. 15(a) 16(a), respectively. Molecular models and interaction potentials were derived according to the procedure described above, and the simulation methodology was essentially identical to that used in the simulation of pure 8CB (see Sec. 4.1).

The central result of this simulation study is apparent from Figs. 15 and 16: while *trans*-7AB remains largely incorporated into the smectic layers, *cis*-7AB is expelled from the smectic layers, and segregates between smectic layers. This difference in positional ordering is reflected in a difference in unit cell dimensions — the unit cell for the *trans*-7AB simulation is relatively 'fat', reflecting the increase in in-layer area caused by incorporation of *trans*-7AB molecules into the smectic layers, while the unit cell for the *cis*-7AB simulation is relatively 'skinny', reflecting the increase in smectic layer spacing caused by interlamellar segregation of *cis*-7AB.

This behavior can be seen more quantitatively from Fig. 17, which shows the smectic layer spacing for the two simulations as a function of simulation time. The *cis*-7AB simulation converges to a significantly larger equilibrium value than does the *trans*-7AB simulation. The average layer spacing for the last 2 ns of the *cis*-7AB simulation is $d_{\text{cis}} = 35.3$ Å, while that for the last 2 ns of the *trans*-7AB simulation is $d_{\text{trans}} = 32.6$ Å, giving a layer spacing increase upon *trans*–*cis* photoisomerization of $\Delta d_{\text{sim}} = d_{\text{cis}} - d_{\text{trans}} = 2.7$ Å. By contrast, the experimentally measured layer spacing under UV illumination is $d_{\text{UV}} = 30.274$ Å, while that in the absence of UV illumination is $d_{\text{vis}} = 30.119$ Å, giving $\Delta d_{\text{exp}} = d_{\text{UV}} - d_{\text{vis}} = 0.155$ Å. At first glance, it would appear that the results of our simulation are in extremely poor agreement with experiment. However, DMR measurements on partially deuterated 7AB reveal that only $\sim 6\%$ of the 7AB molecules are converted to the *cis* form under UV irradiation [129], whereas our simulations effectively assume 100% conversion. If the experimental results are extrapolated to 100% conversion, assuming a linear effect, we find $\Delta d = 2.6$ Å, in extremely good accord with the photomechanical effect estimated from our simulation. It thus appears that the microscopic mechanism for the observed photomechanical effect, namely photo-controlled nanophase segregation, is a plausible one, despite the fact that the absolute values of d differ significantly from the experimental ones. The discrepancy between simulation and experiment in the present case can perhaps be attributed to inaccuracies in the intermolecular interaction potential for the 7AB-8CB mixture.

The expulsion of *cis*-7AB from the smectic layers (interlamellar segregation) is clear from an examination of the mass density profiles for 8CB and 7AB along the layer normal direction, shown in Fig. 18. The 7AB mass density profile for the *trans*-7AB simulation exhibits a weak maximum near the center of the 8CB layer (Fig. 18(a)), whereas in the *cis*-7AB simulation the 7AB mass density profile has a deep minimum near the center of the smectic layer (Fig. 18(b)).

308

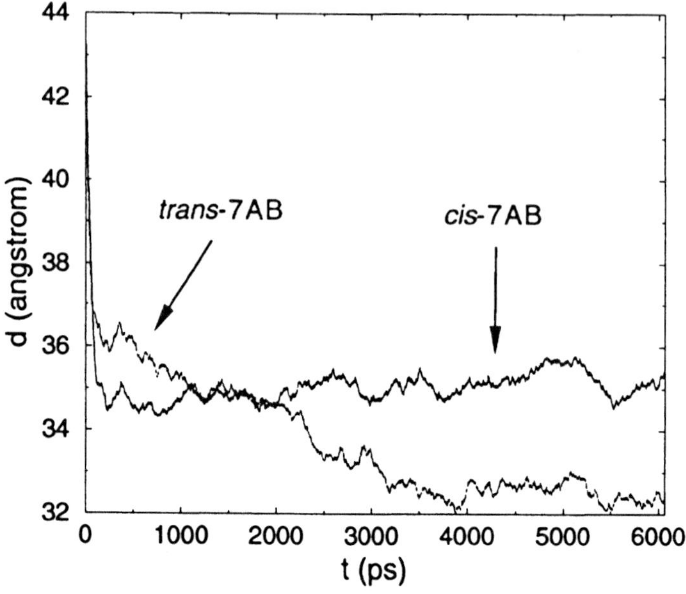

Figure 17. Smectic layer spacing *d* vs. time for the *trans*-7AB and *cis*-7AB simulations. The two simulations converge to significantly different values of *d*, reflecting a significantly different positional distribution of 7AB molecules.

The possibility of controlling the nanophase segregation of photolabile solutes in smectic hosts has interesting implications for the design of photoresponsive LC materials. Indeed, LCs are unique photoactive media, in that their fluidity and optical anisotropy offer the possibility of collective amplification of weak photochemical effects. For example, photocontrolled nanophase segregation suggests a mechanism for the photomodulation of ferroelectric polarization in ferroelectric LCs [130–133]. More generally, the phenomenon of photo-controlled nanophase segregation has intriguing implications for the nanoscale engineering and manipulation of organic materials, and may have relevance to the understanding of biological photosystems, in which photosensitive molecules are imbedded in a fluid lamellar environment.

This study clearly illustrates the power of large-scale atomistic simulation as an exploratory tool. Even in this case, however, the limitations of the method are apparent. For instance, although a pretty clear idea of the microscopic mechanism for the observed photomechanical behavior of 8CB/7AB mixtures emerges from this study, the thermodynamic driving forces for the expulsion of *cis*-7AB molecules from the smectic layers

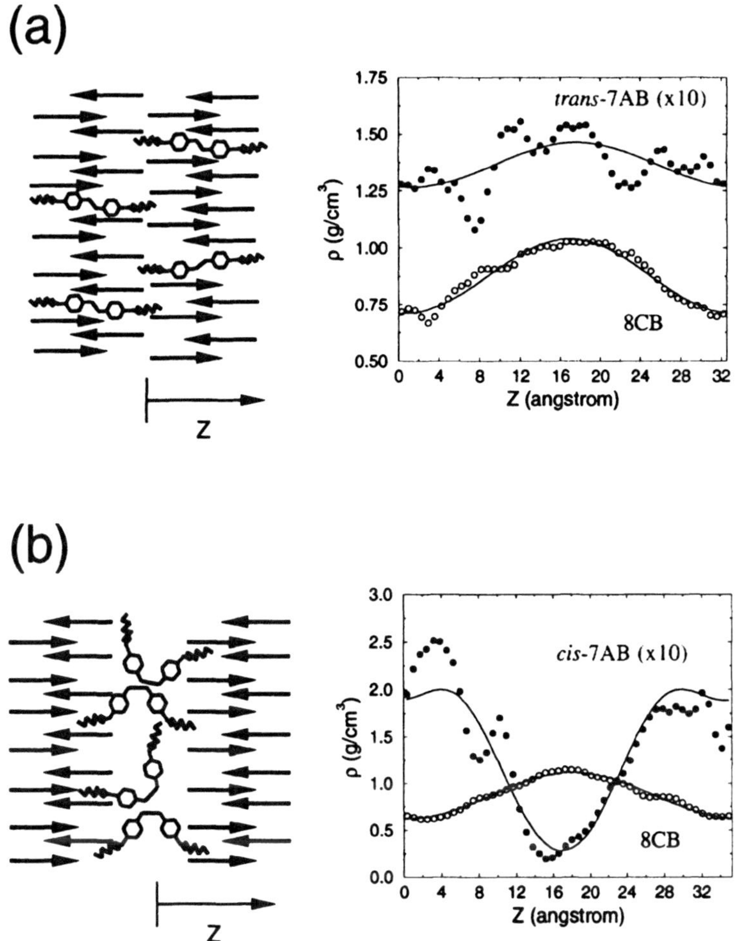

Figure 18. Mass density profiles for 7AB (filled circles) and 8CB (open circles) for the *trans*-7AB (a) and *cis*-7AB (b) simulations. The 7AB densities have been multiplied by a factor of 10 for clarity. As indicated schematically at left, *trans*-7AB is preferentially incorporated into the 8CB layers (with 8CB molecules represented by arrows), while *cis*-7AB molecules are concentrated between smectic layers.

are unclear. Determining which specific free energy contributions govern nanophase segregation of solutes in smectic LCs will require at least an order of magnitude more computational effort than the present study.

4.3. COMPUTER-AIDED DESIGN OF FERROELECTRIC LIQUID CRYSTALS

As discussed in Sec. 4.2, liquid crystals are exotic solvents, which impose their state of order on solutes. Of particular interest to us are SmC LCs, which possess polar orientational order (in a sense discussed in more detail below), and which, as a consequence, impose polar orientational order on solute molecules. For solute molecules of sufficiently low symmetry, namely *chiral* molecules, this polar orientational ordering gives rise to a spontaneous ferroelectric polarization. This, in our view, is the basic microscopic mechanism for ferroelectricity in the majority of SmC* LCs (SmC materials containing chiral components). The same basic mechanism also applies to one-component chiral SmC* materials, where we now consider a chiral 'solute' molecule immersed in a SmC 'solvent' of identical molecules. The application of a guest-host paradigm to both one-component and multicomponent SmC* materials turns out to be a useful approach, as it suggests a simple single-molecule mean-field approach to modeling ferroelectric polarization, in which the polar orientational order imposed on a single 'solute' molecule by a SmC 'solvent' is treated via an effective mean-field potential, or potential of mean force. In this section, we show that such a mean-field approach provides a computationally expedient route to the calculation of ferroelectric polarization density. The success of this approach suggests that the application of mean-field theory is a promising general strategy for the computer-aided design of LC materials.

Any discussion of the origins of ferroelectricity in the SmC* phase inevitably starts with a discussion of the macroscopic symmetries of this phase. First, consider the *achiral* SmC phase. The point group symmetries of the SmC phase include π (C_2) rotation about any axis normal to the tilt plane (the plane defined by the layer normal $\hat{z}$ and the nematic director $\hat{n}$) and lying within the midplane of a smectic layer or within a plane midway between two adjacent smectic layers, and mirror reflection (σ_v) in any plane parallel to the tilt plane. The remaining point symmetry operation, namely inversion (I), can be composed from the other two elements. Because the macroscopic properties of a phase are invariant under its point group symmetry operations (Neumann's principle), ferroelectric polarization is precluded in the achiral SmC phase, since any vector quantity changes sign under inversion. In a SmC phase composed of or containing *chiral* molecules (a SmC* phase), on the other hand, the only point symmetry operation present is C_2 rotation about an axis normal to the tilt plane, as mirror reflection and inversion change the handedness of the constituent molecules, and thus do not leave the structure invariant. As first pointed out by Meyer [134], C_2 symmetry precludes a ferroelectric polarization in the tilt plane, but does not prohibit (rather, implies) a spontaneous ferroelectric polarization density along the C_2 axis (the normal to the tilt plane) [135].

Similar symmetry arguments can be applied to understand the origins of ferroelectricity at a molecular level. The essential point to grasp here is that the single-molecule distribution function $f_1(\mathbf{r}^n)$ exhibits the macroscopic symmetries of the phase. In other words, a given molecular configuration (i.e. a given molecular position, orientation, and conformation) and any configuration obtained from it by one of the symmetry operations of the phase are present with equal probability — $f_1(\mathbf{r}^n)$ is invariant under the transformation of molecular coordinates implied by any macroscopic symmetry operation.

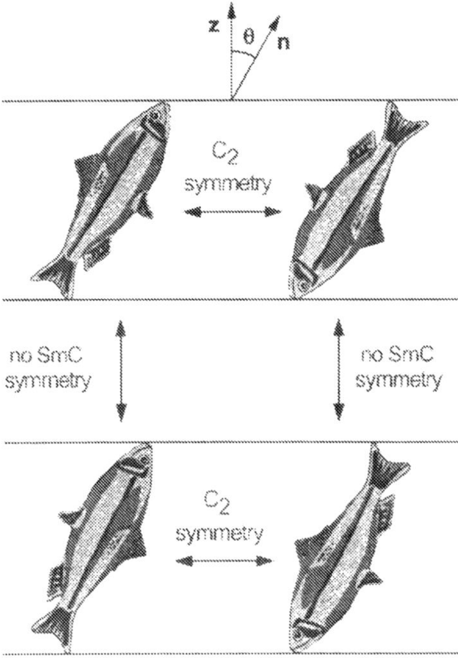

Figure 19. This figure illustrates polar orientational ordering in the achiral SmC phase, with a low-symmetry (but non-chiral) molecule represented as a fish having a distinct head and tail and distinct front and back. The macroscopic symmetries of the SmC phase imply that the two configurations at the top, which are related by a π (C_2) rotation about the normal to the tilt plane, are present with equal probability. Similarly, the two configurations shown at the bottom are present with equal probability. By contrast, the two configurations on the left, which are related by a π rotation about $\hat{\mathbf{n}}$, are in general present with *different* probabilities, because π rotation about $\hat{\mathbf{n}}$ is not a macroscopic symmetry of the SmC phase. This dissymmetry represents a sort of *polar orientational ordering*, in that a molecule-fixed coordinate system will, on average, orient in a polar way with respect to the laboratory coordinate frame (we more often see the right eye of the fish than the left eye, for instance).

The implications of this for the SmC phase are illustrated in Fig. 19. We consider an idealized molecule (here represented by a fish) with a head that is distinguishable from its tail and a belly that is distinguishable from its back, but which is left–right symmetric (achiral). For the moment we consider only rigid molecules having no internal degrees of freedom. The macroscopic symmetries of the SmC phase imply that the two configurations at the top of Fig. 19, which are related by a C_2 rotation, are present with equal probability. Similarly, the two configurations shown on the bottom of Fig. 19 are present with equal probability. By contrast, the two configurations on the left side of Fig. 19, which are related by a π rotation about $\hat{n}$, are in general present with *different* probabilities, because π rotation about $\hat{n}$ is not a macroscopic symmetry of the SmC phase. In other words, our fish prefers to have either its ventral fin or its dorsal fin near the center of a smectic layer — symmetry doesn't tell us which orientation is favored, but does tell us that there will be a preference for one or the other state. This dissymmetry represents a sort of *polar orientational ordering*, in that a molecule-fixed coordinate system will, on average, orient in a polar way with respect to the laboratory coordinate frame. If the molecule-fixed coordinate system is defined by orthogonal unit vectors $\hat{u}$, $\hat{v}$, and $\hat{w}$, with $\hat{v}$ normal to the mirror symmetry plane of the molecule, and the laboratory frame is defined by orthogonal unit vectors $\hat{x}$, $\hat{y}$, and $\hat{z}$, where $\hat{z}$ is the layer normal and $\hat{y}$ is the normal to the tilt plane (the $x - z$ plane), then the polar order parameter $\langle \hat{v} \cdot \hat{y} \rangle$ has a nonzero value, as configurations with $\hat{v}$ pointing in the $+y$ direction (top configurations in Fig. 19) occur with a different probability than configurations with $\hat{v}$ pointing in the $-y$ direction (bottom configurations in Fig. 19). Because the molecules are mirror symmetric, however, there can be no component of electric dipole moment normal to the mirror plane, and so this polar orientational ordering does not manifest itself in a ferroelectric polarization. If our molecule is made *chiral*, however, for instance by decorating it with an electric dipole along $\hat{v}$, then polar orientational ordering does manifest itself in a ferroelectric polarization density P in the y-direction, because $\langle \hat{v} \cdot \hat{y} \rangle \neq 0$. Even if the chirality is completely passive, having no effect on the orientational distribution of the molecules, the polar orientational order *already present* in the *achiral* SmC phase gives rise to a nonzero P. In this picture, the chirality manifests itself at the single-molecule level, simply contributing a net dipole moment transverse to the (approximate) mirror symmetry plane of the molecule. This represents a minimal microscopic model for ferroelectricity in SmC* materials, and is simply a transcription of the Meyer symmetry argument into molecular terms. Polar anisotropy in the orientational distribution about the molecular long axis is the key feature of the single-molecule distribution function f_1 in this context.

So far we have considered only rigid molecules. Real LC molecules are flexible, however, and in the LC phases an astronomical number of distinct molecular conformations are present. For this reason, it is nonsense to speak about the symmetry of a LC molecule — the vast majority of molecular conformations have no symmetries whatsoever (and are therefore chiral). Under these circumstances, it makes sense only to talk about *statistical* symmetry, in other words the symmetry of the statistical ensemble of all molecular conformations, orientations, and positions, as embodied in the single-molecule distribution function f_1, which, as before, possesses all of the symmetries of the phase. In the achiral SmC phase, in particular, mirror symmetry implies that a given (chiral) molecular conformation and its mirror image are present with precisely equal probabilities. Thus, although a given chiral configuration makes a contribution to the ferroelectric polarization by virtue of the polar orientational order characteristic of the phase, as outlined above, its mirror image, present with equal probability, makes a contribution to the polarization of equal magnitude and opposite sign, so that the net polarization is exactly zero. In the chiral SmC* phase, by contrast, mirror-image conformers are no longer present with equal probabilities (in fact, conformers of the 'wrong' handedness are not present at all), and so a nonzero ferroelectric polarization will be present. For flexible molecules, the observed polarization P depends on the conformational distribution as well as the anisotropy of the orientational distribution about the long molecular axis.

The foregoing symmetry arguments give insight into the microscopic origins of ferroelectric polarization, but provide little guidance in the *quantitative* calculation of P. Calculation of the sign and magnitude of P requires specific assumptions about the form of the single-molecule orientational-conformational distribution function f_1. To this end we have developed a simple mean-field approach to the calculation of P, the *Boulder model*, which utilizes empirical knowledge about SmC materials to create a specific model for f_1, and hence P. The key assumption in the Boulder model is the so-called *zig-zag model*, the assumption that molecular cores are more tilted with respect to the layer normal than are molecular tails, implying a specific form of anisotropy in the orientational distribution about the molecular long axis. The zig-zag model was originally proposed to account for the discrepancy between tilt angles measured optically and those deduced from x-ray layer spacing measurements [136], and gives a consistent account of the behavior of most SmC materials for which layer spacing data exists. Thus, zig-zag ordering seems to be a quite general feature of the microscopic organization of SmC LCs.

The other key ingredient of the Boulder model is a mean-field approximation. Zig-zag ordering is considered to be imposed on a given molecule in the SmC phase by a mean-field potential, or 'binding site', that represents the net effect of its interactions with its neighbors. The Boulder model is illustrated schematically in Fig. 20, which shows a LC molecule in a specific, zig-zag shaped, conformation immersed in a mean-field potential having the symmetry of a bent cylinder. In one orientation, the molecule fits well into the zig-zag shaped binding site, and has relatively low energy, whereas in the other orientation (obtained from the first by a π rotation about $\hat{n}$) the molecule fits poorly into the binding site, and has a relatively high energy. The configuration on the left thus occurs with higher probability than that on the right. In other words, the zig-zag shaped binding site produces polar anisotropy in the orientational distribution about $\hat{n}$. If the molecule is chiral, with a net dipole moment normal to the plane of the zig-zag, then the polar orientational ordering imposed by the binding site produces a nonzero contribution to P. The total ferroelectric polarization can be calculated as an average over all molecular conformations, appropriately weighted by the probability of each conformation, which depends on the form of the mean-field potential as well as the intramolecular potential energy.

An implicit assumption of the Boulder model is that the form of the mean-field potential does not depend on whether the phase is chiral or not. The local environment of a single molecule is assumed to be achiral, so that chirality only enters at the single-molecule level. With this assumption, the Boulder model applies equally well to neat SmC* materials and to chiral dopants in achiral SmC hosts.

In order to turn the Boulder model into a calculational method, we need to specify the form of the mean-field potential. The mean-field potential is *defined* by the relation

$$f_1(\mathbf{r}^n) = Z^{-1} \exp\{-\beta[U_{\mathrm{mf}}(\mathbf{r}^n) + U_{\mathrm{int}}(\mathbf{r}^n)]\}, \tag{57}$$

where $\mathbf{r}^n$ denotes the set of atomic coordinates for a single molecule, $U_{\mathrm{int}}(\mathbf{r}^n)$ is the intramolecular potential energy for an isolated molecule, and $U_{\mathrm{mf}}(\mathbf{r}^n)$ is the mean-field potential, which accounts for the effect of intermolecular interactions on f_1. If the exact form of f_1 were known, Eq. 57 would constitute an *exact* definition of U_{mf}. In practice, we have only limited information about f_1 even in the most favorable case. Experimental measurements such as NMR and FTIR, for example, give low-order moments of the orientational distributions for specific functional groups, which serve to constrain f_1, but do not uniquely determine it. Given the absence of detailed information about f_1, the only reasonable approach is to adopt a *minimal* model that incorporates the essential physics we wish to include, but which depends on a small enough number of parameters to permit a (nearly) unique parametrization on the basis of experimental data.

In our work to date, we have adopted a simple but flexible form for U_{mf},

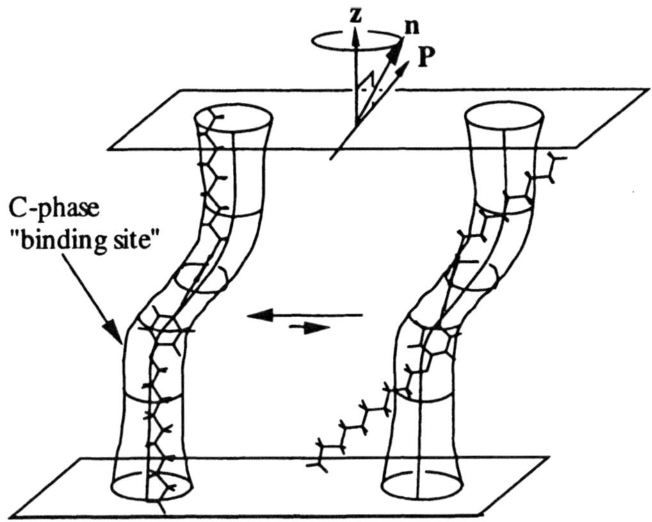

Figure 20. Schematic depiction of the orientational ordering of a molecule in the zig-zag shaped Boulder model 'binding site', or mean field potential. In one orientation, the molecule fits well into the zig-zag shaped binding site (left), and has relatively low energy, whereas in the other orientation (obtained from the first by a π rotation about $\hat{n}$) the molecule fits poorly into the binding site (right), and has a relatively high energy. The configuration on the left thus occurs with higher probability than that on the right. If the molecule is chiral, with a net dipole moment normal to the plane of the zig-zag shaped molecule, then the polar orientational ordering imposed by the binding site produces a nonzero contribution to P.

namely

$$U_{\mathrm{mf}}(\mathbf{r}^n) = -\sum_{i=1}^{N_t} u_0 \left|\mathbf{v}_i\right| P_2\left(\hat{\mathbf{v}}_i \cdot \hat{\mathbf{n}}_t\right) - u_0 \left|\mathbf{v}_c\right| P_2\left(\hat{\mathbf{v}}_c \cdot \hat{\mathbf{n}}_c\right) - \alpha \left|\mathbf{v}_{ee}\right|. \qquad (58)$$

U_{mf} is essentially a sum of orientational potentials acting on specific rigid segments of the molecule [137]. The sum ranges over all N_t bond vectors $\mathbf{v}_i$ between heavy atoms in both LC tails. $\mathbf{v}_c$ is the end-to-end vector of the LC core (defined as the minimal contiguous ring-containing portion of the molecule), and $\mathbf{v}_{ee}$ is the end-to-end vector of the molecule as a whole. $\hat{\mathbf{n}}_t$, the tail 'director', specifies the preferred orientation of tail segments, and $\hat{\mathbf{n}}_c$, the core 'director', specifies the preferred orientation of the core. P_2 is the second Legendre polynomial, $P_2(x) = (3x^2 - 1)/2$. The last term in Eq. 58 is included to suppress 'folded' or 'compact' configurations, which otherwise occur with relatively high probability in the model, particularly when attractive intramolecular nonbonded interactions are included in U_{int}. In our more recent work, we have used purely repulsive Weeks-Chandler-Andersen (WCA) intramolecular nonbonded interactions [138], which in part (but not entirely) obviates the need for this term.

The form of Eq. 58 is quite similar to that of mean-field models used successfully to model NMR measurements on nematic and SmA LCs [139–141]. What is new is the inclusion of distinct orienting directions for the tails and the core, permitting the imposition of zig-zag ordering on a single molecule, and yielding a model appropriate to the SmC phase. The model depends on only three parameters, namely the energy scale of the orientational potential, u_0, the relative orientation of core and tail 'directors', $\theta_{\rm rel} = \cos^{-1}(\hat{\bf n}_{\rm c} \cdot \hat{\bf n}_{\rm t})$, and the elongational energy constant, α. Once these parameters are specified, the average of any quantity $A({\bf r}^n)$ that depends only on the atomic coordinates of a single molecule can be calculated as a simple average over all molecular conformations and orientations, weighted by $f_1({\bf r}^n)$:

$$\langle A \rangle = \int d{\bf r}^n f_1({\bf r}^n) A({\bf r}^n). \tag{59}$$

The observables that may be calculated from $f_1({\bf r}^n)$ include the ferroelectric polarization P, orientational order parameters of specific functional groups probed by NMR and FTIR spectroscopy, and, to a good approximation, linear and nonlinear optical properties.

Because the parameters in our model must be determined from fits to experimental data, this approach is inherently *semi-empirical*. Single-molecule mean-field theory can be employed to interpret experimental data, e.g. to provide an interpretation of a few order parameters measured via polarized IR spectroscopy in terms of the single-molecule distribution function. A more interesting prospect is to use mean-field theory as the basis for a semi-empirical approach to molecular design. The intramolecular potential energy $U_{\rm int}$ is derived from *ab initio* quantum chemistry calculations, as described in Sec. 2, while the parametrization of $U_{\rm mf}$ is empirical, i.e. established from fits to NMR, FTIR, or other experimental measurements. The 'raw' output of such a theory is the single-molecule distribution function, which provides an interpretation of the experiments used to fix the parameters. Furthermore, if a 'universal' or 'transferable' parametrization that applies to a broad class or family of materials can be found, then such a theory can be used to predict the properties of as yet unsynthesized materials. In the specific case of SmC LCs, this assumption of 'microscopic universality', i.e. the assumption of a universal 'binding site' for all SmC materials, is the key to the creation of a predictive model for ferroelectric polarization in SmC* LCs.

The calculation of statistical averages requires a means of evaluating the high-dimensional integral in Eq. 59. In most of our work to date we have used the hybrid Monte Carlo method discussed in Sec. 3.4 to evaluate statistical averages via importance sampling. HMC has proven to be quite efficient at sampling molecular conformations and orientations, and, as the

HMC method is based on MD, the calculations can be carried out with the same r-RESPA-based MD code used to perform the MD simulations described elsewhere in this article. Another approach is to work within the rotational isomeric state (RIS) approximation, in which the dihedral angle for each rotatable bond is restricted to a discrete set of values, corresponding to local minima in the rotational potential. In this case, the average in Eq. 59 involves a weighted sum over all discrete molecular conformations and an integral over molecular orientations. For molecules that are not too large, it is possible to explicitly enumerate all conformations and evaluate orientational averages numerically. This approach requires far less CPU time than importance sampling for molecules of modest size, and is free of statistical error. However, it involves an additional approximation (the RIS approximation) and cannot be applied to large LC molecules (the number of molecular conformations increases exponentially with the number of rotatable bonds).

As an initial test of the Boulder model, we have calculated the sign and magnitude of P for 28 FLC materials, to see if the model can be calibrated/parametrized to reproduce the experimentally measured P [142]. The procedure used to parametrize the model is described in our earlier paper [142]. This set of SmC^* materials included 14 neat SmC^* materials and 14 mixtures of a chiral dopant with an achiral SmC host. In the latter case, the measured polarization was extrapolated to 100% concentration of chiral component. A plot of measured vs. calculated polarization is shown in Fig. 21. The Boulder model reproduces the correct sign of polarization in every case except one, and generally reproduces the magnitude of polarization at a semiquantitative level. Moreover, the model appears to work equally well for both neat SmC^* materials and mixtures, providing strong support for our unified treatment of SmC^* materials. The data also exhibits several outliers, however, points which lie more than 4 standard deviations off of the line $P_{exp} = P_{calc}$. We will have more to say about these outliers in a moment. Before doing so, however, we point out the latent possibilities suggested by these results. The model seems to be capable of reliably distinguishing between high-P and low-P materials. As a result, the 'calibrated' model should be useful as a tool for pre-synthesis screening of FLC materials for high polarization.

At the moment, however, we are reluctant to use the Boulder model to guide synthetic work until we better understand the limitations of the model (e.g. the outliers in Fig. 21). Although use of improved molecular models, charge distributions, etc., can modify the results somewhat, our more recent work (described in Sec. 4.4) suggests that the discrepancies between theory and experiment reflect a fundamental shortcoming of the mean-field model itself. Specifically, it appears that the model tends to fail

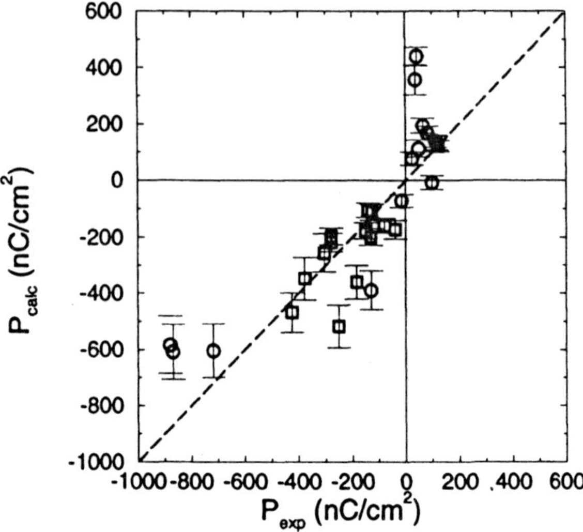

Figure 21. Experimental (P_{exp}) vs. calculated (P_{calc}) ferroelectric polarization densities for a series of 28 FLC materials. Data is shown both for pure materials (squares) and mixtures (circles). The correct sign of P is reproduced in every case except one, and the overall correlation between theory and experiment is quite encouraging, indicating that the Boulder model can be used as a reliable semi-quantitative predictor of P.

when the chiral center and associated dipoles are close to the core (on the first two sites in the chiral tail). Similarly, we have encountered difficulties in fitting FTIR data using the 'minimal' model outlined above, and in modeling the temperature dependence of P for some FLC materials.

4.4. MICROSCOPIC ORGANIZATION OF THE SMECTIC C PHASE

The foregoing discussion of the Boulder model for the microscopic origins of ferroelectric polarization density in SmC* liquid crystals suggests that this model provides a sound conceptual basis for understanding the molecular basis for ferroelectric polarization and for formulating predictive computational strategies. However, the very success of this model raises a number of questions, such as:

- How universal are the basic assumptions that go into the Boulder model, in particular the assumption that cores are more tilted than tails in the SmC phase (the zig-zag hypothesis), and the assumption of an *achiral* 'binding site' (weak chiral expression in the local supermolecular organization of the SmC* phase)?

 – What are the limitations of the Boulder model? We have already encountered cases in which the Boulder model predictions are in poor numerical agreement with experiment. What is the physical reason for the failure of the model in these cases, and can the model be systematically improved to deal with these cases?

We can also ask a number of more fundamental questions regarding the nature of the SmC phase, including:

 – Why are molecular cores are more tilted than tails, in general?
 – More fundamentally, why do tilted smectic phases occur at all? What is the microscopic mechanism for molecular tilt?
 – One can distinguish between intralayer tilt coupling and interlayer tilt coupling. What effective interactions are responsible for communicating tilt direction from layer to layer? Why is *anticlinic* ordering favored in certain materials (e.g. antiferroelectrics), whereas most SmC materials exhibit synclinic ordering?

Finally, there are a number of materials properties of relevance to FLC device physics in addition to the ferroelectric polarization density P (see, for example [143], such as the optical tilt angle θ, the smectic layer spacing d, and the rotational viscosity η_ϕ, which depend critically on intermolecular correlations, and are thus inaccessible to single-molecule mean field theory.

An obvious route to testing the assumptions underlying the Boulder model is to carry out large-scale (many-molecule) simulations of prototypical SmC and SmC* materials, making direct comparisons between results from large-scale simulation and mean-field models. Such simulations can additionally lead to the development of improved predictive mean-field models for ferroelectric polarization, by giving insight into the causes of failure of the Boulder model in specific cases. Finally, exploratory simulations of this sort can ultimately aid in understanding the microscopic origins of molecular tilt and zig-zag ordering, and in the understanding of the molecular basis of a variety of materials properties.

As a first step in this direction, we have carried out an extensive simulation study of a 'generic' achiral phenylbenzoate-based SmC mesogen, 4,n-hexyloxyphenyl-4,n'-decyloxybenzoate (HOPDOB) (a more detailed account of this work will appear elsewhere [144]). The molecular structure and phase sequence of HOPDOB are shown in Fig. 22. We carried out an *NPT* MD simulation of 90 HOPDOB molecules at $P = 1$ atm and $T = 351$ K, a state point in the SmA phase, close to the SmA-SmC transition. However, as shown below, our model system exhibits SmC ordering at this temperature and pressure, and so the SmA–SmC transition is apparently at a somewhat higher temperature in the model than in experiment. As for the systems described in Secs. 4.1 and 4.2, periodic boundary conditions were used, with smectic layers arranged within a monoclinic box such that the

320

system in effect contains only a single continuous smectic layer, although the long dimension of the box is equivalent to three smectic layers. This boundary condition tends to ensure a uniform tilt direction throughout the sample, and is similar to that used for the simulations of 8CB and 7AB–8CB mixtures described above. Note that anticlinic ordering (in which the tilt direction alternates from layer to layer) is effectively precluded by this choice of unit cell, as it would require the creation of at least two domain walls. To simulate anticlinic states requires at least two distinct layers. A SmA-like initial condition was used, with molecular long axes initially oriented along the layer normal. The total duration of the simulation was 9.56 ns, rather long by the standards of atomistic simulation (corresponding to roughly 2 months of CPU time on an R10000-based SGI workstation).

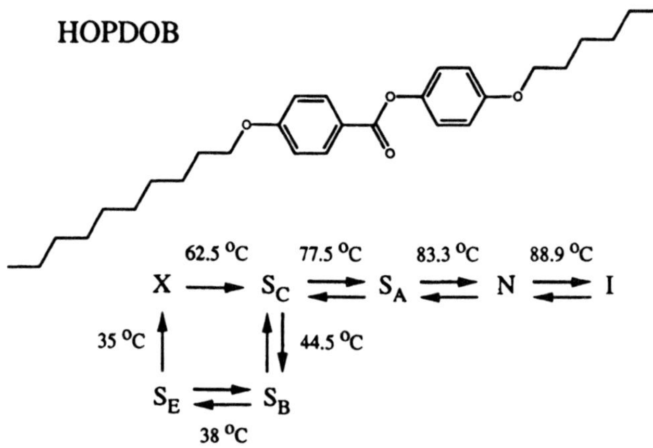

Figure 22.   Chemical structure and phase sequence of HOPDOB.

The initial and final configurations of the simulation are shown in Fig. 23. Evidently, reasonably uniform SmC ordering has developed by the end of the simulation, with the average molecular director tilted substantially with respect to the layer normal. In fact, well-developed SmC ordering only appears after $\sim$ 6 ns. Moreover, even after almost 10 ns the system is probably not completely equilibrated, as can be seen from Fig. 24, which shows the time evolution of mass density, potential energy, and smectic layer spacing over the duration of the run. Although the mass density and potential energy appear to approach a steady state within $\sim$ 1 ns, the smectic layer spacing has not achieved a steady state even after $\sim$ 10 ns. This result underlines the difficulty of equilibrating smectic LCs, particularly tilted smectics, and gives some indication of the level of computational expense required to make a meaningful measurement of the smectic layer spacing via atomistic simulation. Despite the fact that the layer spacing has probably not equilibrated, it is still possible to investigate the properties of the SmC state that develops after $\sim$ 6 ns. In the following, we present averages over the final 2 ns of the simulation.

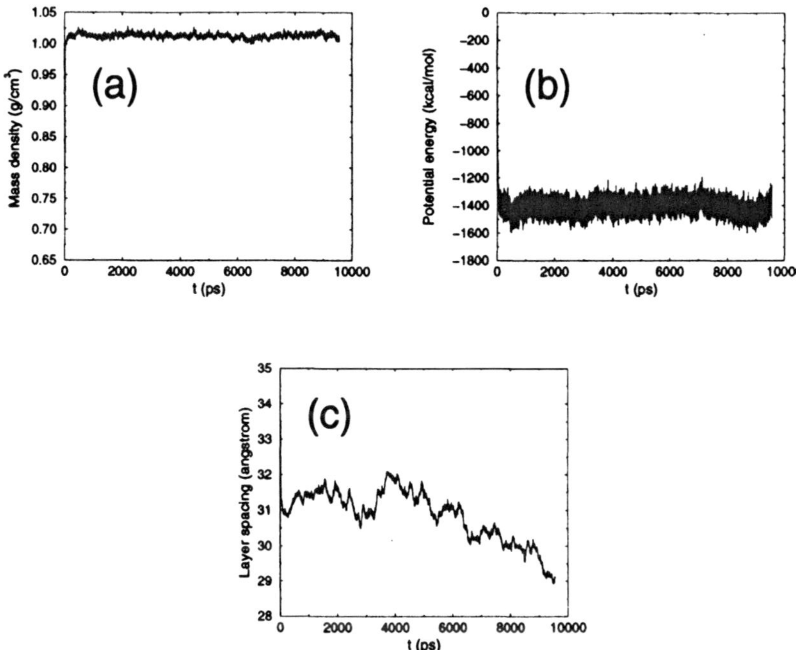

Figure 24. Time evolution of mass density (a), potential energy (b), and smectic layer spacing (c). The smectic layer spacing exhibits significant variation even after 9.56 ns, indicating that the system is not fully equilibrated.

As is apparent from Fig. 23, the smectic layers remain quite well-defined after 9.56 ns. The in-layer positional and bond-orientational correlations are consistent with SmC ordering, with liquid-like (short-range) positional correlations and some evidence of tilt-induced long-range bond-orientational order, as expected for the SmC phase [145].

One of the first questions we would like to answer is: what is the average tilt angle in the SmC state, and are molecular cores really more tilted than tails? To answer these questions, we computed the average tilt angles of the core and of each of the tails. As described above, this was done by calculating the average ordering tensor $\mathbf{Q}$ for each of the molecular segments individually, and then calculating the angle between the maximum-eigenvalue principal axis of $\mathbf{Q}$ and the layer normal. The instantaneous orientation of the segment in question is defined by the segment end-to-end vector. We find that the core, with a tilt angle of 25.2°, is indeed more tilted than

322

the tails, which have average tilt angles of 20.1° (decyloxy tail) and 21.5° (hexyloxy tail). In short, our simulations support the zig-zag hypothesis.

Having a SmC sample at our disposal gives us an opportunity to make a direct comparison between the single-molecule distribution function from large-scale simulation and that derived from a simple mean-field model. This represents an extremely powerful way of testing and refining mean-field models for LCs. In particular, this approach enables the direct comparison of the single-molecule distribution function obtained from simple mean-field models with that obtained from large-scale simulation. The mean-field potential used to model the single-molecule distribution function for SmC HOPDOB is nearly the same as that used in the Boulder model calculation of FLC polarization (Eq. 58). As before, statistical averages over molecular conformations and orientations were calculated using the hybrid Monte Carlo method.

To compare large-scale simulation and mean-field theory, we have calculated various orientational order parameters of interest. Of particular relevance to the microscopic origins of ferroelectric polarization are order parameters that measure the degree and nature of polar orientational ordering in the SmC phase. We have probed the polar orientational order as a function of position in the two alkoxy tails of HOPDOB via the polar order parameters

$$\langle v_{iy} \rangle = \langle \hat{\mathbf{v}}_i \cdot \hat{\mathbf{y}} \rangle, \tag{60}$$

where $\mathbf{v}_i = \mathbf{b}_i \times \mathbf{b}_{i+1}$ is the normal to an adjacent pair of bond vectors in one of the alkoxy tails, and $\hat{\mathbf{y}}$ is a unit vector normal to the tilt plane defined by the layer normal $\hat{\mathbf{z}}$ and the director $\hat{\mathbf{n}}$. Roughly speaking, the magnitude of $\langle v_{iy} \rangle$ measures how effectively a chiral substituent attached to site i is aligned along the polar axis, and its sign is related to the resulting sign of P.

In Fig. 25 we compare polar order parameters calculated from mean-field theory with those obtained from large-scale simulation, plotted as a function of atom position in the two alkoxy tails of HOPDOB. The parameters in the mean-field model have been chosen to optimize agreement between mean-field theory and experiment with respect to $\langle v_{iy} \rangle$ and various other orientational order parameters. For sites two or more bonds distant from the core ($i \geq 3$), there is good agreement between mean-field theory and large-scale simulation, both in the magnitude and sign of $\langle v_{iy} \rangle$. In particular, this implies that mean-field theory is likely to correctly reproduce the odd-even alternation in the sign of P that is commonly observed experimentally when the position of the stereocenter is moved along the tail [146]. However, there are large discrepancies between mean-field theory and large-scale simulation for the first two sites in both tails, with mean-field theory giving $\langle v_{iy} \rangle \approx 0$ and large-scale simulation yielding a relatively large $|\langle v_{iy} \rangle|$ for $i = 1, 2$. Note also that the even-odd alternation of sign observed for $i \geq 3$ does not hold for the first two sites in each tail in the large-scale simulation (sites 2 and 3 have the same sign of $\langle v_{iy} \rangle$ rather than opposite signs).

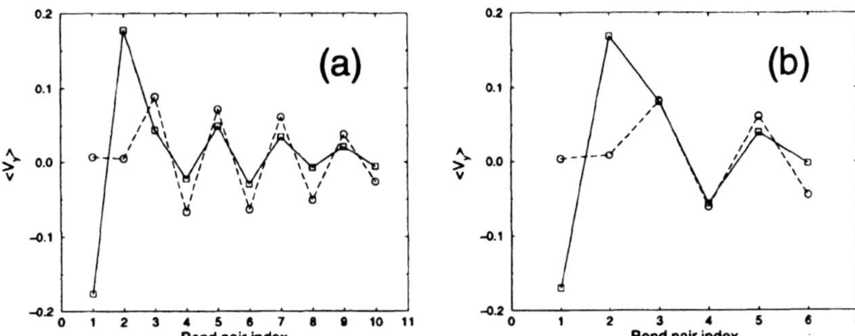

Figure 25. Polar order parameters $\langle v_{iy} \rangle$ for various sites i in the decyloxy (a) and hexyloxy (b) tails of HOPDOB, from mean-field theory (circles) and large-scale simulation (squares). Significant discrepancies are seen for the first two sites in each tail.

The discrepancy between mean-field theory and many-molecule simulation has its origin in the fact that the orientational distributions of groups in the core about its long axis are nearly isotropic in the mean-field model, but exhibit significant anisotropy in the large-scale simulation. Presumably, this anisotropy arises from the excluded volume, dispersion, and electrostatic interactions between cores (molecular 'packing' effects). The intramolecular coupling of tails and core implies that any orientational anisotropy in the orientational distribution of the core is communicated to the first two sites in each tail, but gives way to more generic 'zig-zag' type ordering for sites more distant from the core. The implications of this behavior are clear — for compounds having chiral substituents on sites distant from the core, the Boulder model is likely to give reliable predictions, while for compounds with chiral groups close to the core, Boulder model calculations of P can be expected to be significantly in error. In fact, the two outliers in Fig. 21 correspond to compounds with chiral substituents immediately adjacent to the core. However, in cases in which the core–tail coupling is sufficiently strong (e.g. the nitro-alkoxy series of FLCs [147]) the orientational distribution of at least a portion of the core is determined by zig-zag ordering, and in this case the Boulder model calculations appear to be reliable even for chiral centers close to the core.

This study provides a nice illustration of how large-scale simulation can be used to test and develop approximate theories, in particular shedding light on how the Boulder model might be improved. In general, we feel that that combination of large-scale simulation and theory is a powerful strat-

324

egy that we plan to continue to exploit to develop calculational strategies appropriate for a variety of problems in LC science.

5. Conclusion

Atomistic simulation, at its best, represents a powerful theoretical tool for probing the basic physics of soft condensed matter. Moreover, we are now at a point in time where the convergence of unprecedented computing power, high-accuracy electronic structure methods, and advanced simulation methodology make it possible to model complex organic materials from first principles. As a result, molecular simulation is now coming into its own as a tools for the exploratory investigation of organic materials, and the first-principles design and discovery of new materials and new materials paradigms is no longer an unrealizable vision. The near future should see atomistic simulation evolving from a strictly exploratory technique into a systematic, quantitative tool for exploring the molecular-scale physics of liquid crystals. In particular, we need to begin applying a standard of rigor to atomistic simulations comparable to that exemplified by the best simulations of idealized models. It will not be easy to realize this goal, but it is well worth the effort, as atomistic simulation holds the potential to give the materials designer true control over the self-organization and resulting properties of liquid crystals and other organic materials. The real challenge here is to have the imagination and creativity to take full advantage of this unprecedented opportunity.

Acknowledgements

A number of people made important contributions to the work described in this chapter, including Yves Lansac, Edgardo Garcia, Titus Weider, Valeriy Ginzburg, Noel Clark, Dave Walba, Daan Frenkel, Demetri Photinos, and Oleg Lavrentovich. Special thanks is due to Tom Darden for generously sharing his particle-mesh Ewald code with us, and to Dave Coleman for his help in producing the molecular graphics. This work was supported by NSF Materials Research Science and Engineering Center Grant DMR-9809555, AFOSR MURI Grant F49620-97-1-0014, NASA Grant NAG3-1846, and a grant of computer time from the Kent State Liquid Crystal Institute.

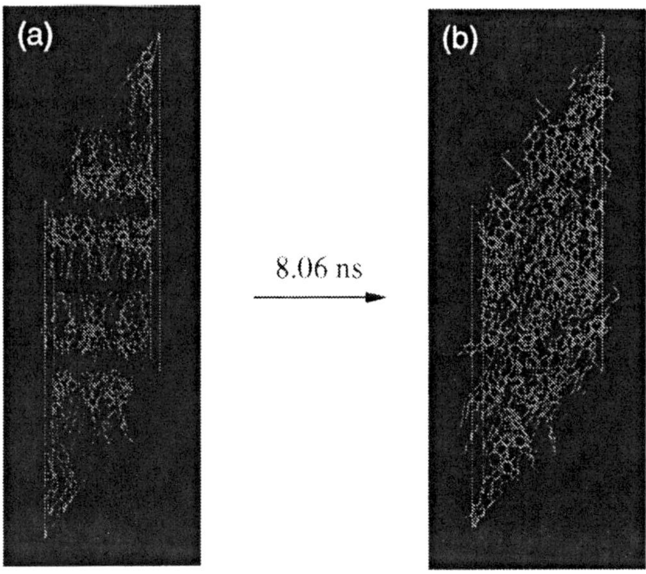

Figure 7. Initial (a) and final (b) configurations of an 8.06 ns simulation of 8CB in the SmA_d phase.

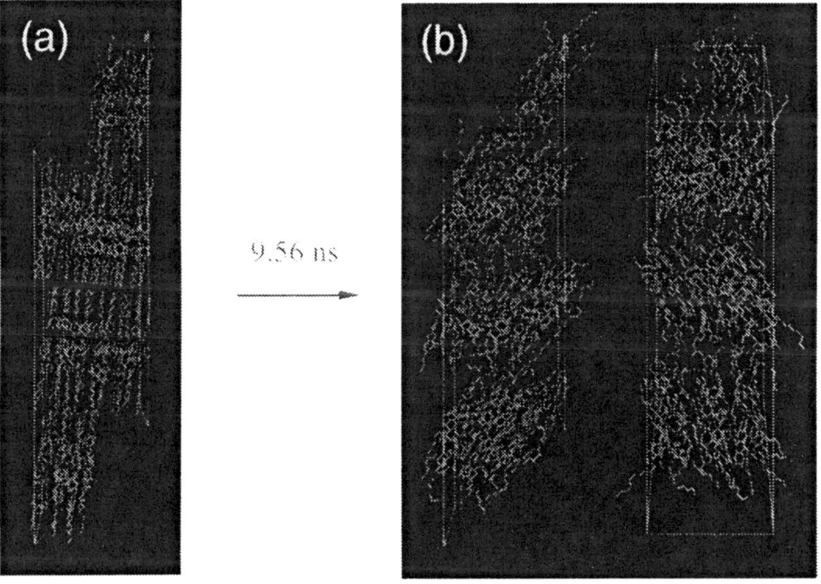

Figure 23. Initial (left) and final (right) configurations from a 9.56 ns simulation of HOPDOB.

326

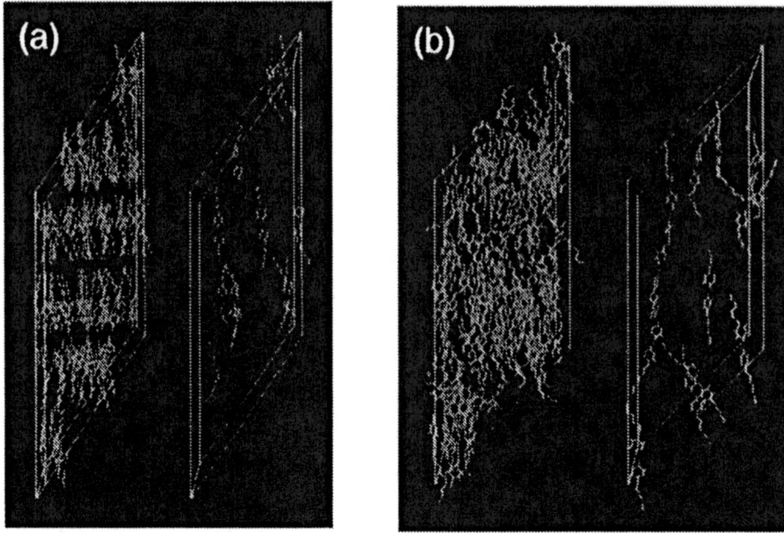

Figure 15. Initial (a) and final (b) configurations of the 6.06 ns *trans*-7AB simulation. For clarity, 8CB and 7AB molecules are displayed separately, in the left and right panels of each image, respectively. After 6.06 ns, *trans*-7AB is largely incorporated into the smectic layers.

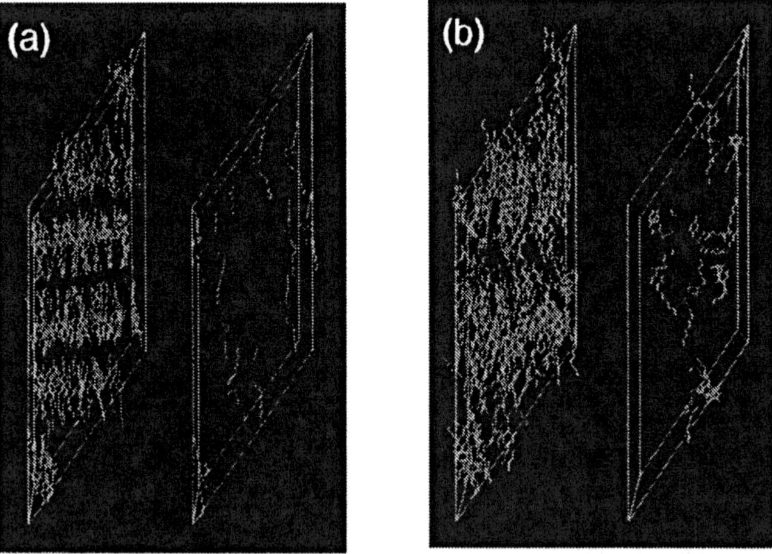

Figure 16. Initial (a) and final (b) configurations of the 6.06 ns *cis*-7AB simulation. For clarity, 8CB and 7AB molecules are displayed separately, in the left and right panels of each image, respectively. After 6.06 ns, *cis*-7AB is segregated between the smectic layers.

References

1. Straatsma, T.P., Zacharias, M., and McCammon, J.A. (1993) Free energy difference calculations in biomolecular systems, in *Computer Simulation of Biomolecular Systems: Theoretical and Experimental Applications, Vol. 2.* van Gunsteren, W.F., Weiner, P.K., and Wilkinson, A. J. (eds.), ESCOM, Leiden, p. 349.
2. Picken, S.J., van Gunsteren, W.F., van Duijnen, P.Th., and de Jeu, W.H. (1989) *Liq. Cryst.*, **6**, 357.
3. Komolkin, A.V., Molchanov, Yu.V., and Yakutseni, P.P. (1989) *Liq. Cryst.*, **6**, 39.
4. Jung, B., and Schürmann, B. L., (1990) *Mol. Cryst. Liq. Cryst.*, **185**, 141.
5. Ono, I., and Kondo, S. (1991) *Mol. Cryst. Liq. Cryst. Letters*, **8**, 69.
6. Wilson M.R., and Allen, M.P. (1991) *Mol. Cryst. Liq. Cryst.*, **198**, 465.
7. Wilson M.R., and Allen, M.P. (1992) *Liq. Cryst.*, **12**, 157.
8. Maliniak, A. (1992) *J. Chem. Phys.*, **96**, 2306.
9. Krömer, G., Paschek, D., and Geiger, A. (1993) *Ber. Bunsen-Ges. Phys. Chem.*, **97**, 1188.
10. Huth, J., Mosell, T., Nicklas, K., Sariban, A., and Brickmann, J. (1994) *J. Phys. Chem.*, **98**, 7685.
11. Yakovenko, S.Ye., Minko, A.A., Krömer, G., and Geiger, A. (1994) *Liq. Cryst.*, **17**, 127.
12. Glaser, M.A., Malzbender, R., Clark, N.A., and Walba, D.M. (1994) *J. Phys.:Condens. Matter*, **6**, A261.
13. Komolkin, A.V., Laaksonen, A., and Maliniak, A. (1994) *J. Chem. Phys.*, **101**, 4103.
14. Glaser, M.A., Malzbender, R., Clark, N.A., and Walba, D.M. (1995) *Mol. Sim.*, **14**, 343.
15. Yakovenko, S.Y., Muravski, A.A., Krömer, G., and Geiger, A. (1995) *Mol. Phys.*, **86**, 1099.
16. Patnaik, S.S., Plimpton, S.J., Pachter, R., and Adams, W.W. (1995) *Liq. Cryst.*, **19**, 213.
17. Komolkin A.V. and Maliniak, A. (1995) *Mol. Phys.*, **84**, 1227.
18. Wilson, M.R. (1996) *J. Mol. Liq.*, **68**, 23.
19. Sandstroem, D., Komolkin A.V., and Maliniak, A. (1996) *J. Chem. Phys.*, **104**, 9620.
20. Sandstroem, D., Komolkin A.V., and Maliniak, A. (1997) *J. Chem. Phys.*, **106**, 7438.
21. Yakovenko S.Y. and Muravski, A.A. (1998) *Liq. Cryst.*, **24**, 657.
22. McBride, C., Wilson, M.R., and Howard, J.A.K. (1998) *Mol. Phys.*, **93**, 955.
23. Paschek, D., Yakovenko, S.Y., Muravski, A.A., and Geiger, A. (1998) *Ferroelectrics*, **212**, 45.
24. Bolhuis, P. and Frenkel, D. (1997) *J. Chem. Phys.*, **106**, 666.
25. Zannoni, C. (1986) *J. Chem. Phys.*, **84**, 424.
26. de Loof, H., Harvey, S.C., Segrest, J.P. and Pastor, R.W. (1991) *Biochemistry*, **30**, 2099.
27. Cornell, W.D., Howard, A.E., and Kollman, P. (1991) *Curr. Opin. Struct. Biol.*, **1**, 201.
28. Hagler A.T. and Ewig, C.S., (1994) *Comput. Phys. Commun.*, **84**, 131.
29. Halgren, T.A. (1995) *Curr. Opin. Struct. Biol.*, **5**, 205.
30. Kaminski G. and Jorgensen, W.L. (1996) *J. Phys. Chem.*, **100**, 18010.
31. Jorgensen, W.L., Maxwell, D.S., and Tirado–Rives, J. (1996) *J. Am. Chem. Soc.*, **118**, 11225.
32. Kollman, P.A. (1996) *Acc. Chem. Res.*, **29**, 461.
33. J. P. Bowen and L. Guyan (1997) New vistas in molecular mechanics, in *Practical Applications of Computer-Aided Drug Design.* Charifson, P.S. (ed.), Dekker, New York, p. 495.
34. Moult, J. (1997) *Curr. Opin. Struct. Biol.*, **7**, 194.

35. Hünenberger, P.H. and van Gunsteren, W.F. (1997) Empirical classical interaction functions for molecular simulation, in *Computer Simulation of Biomolecular Systems: Theoretical and Experimental Applications, Vol. 3.* van Gunsteren, W.F., Weiner, P.K., and Wilkinson, A.J. (eds.), Kluwer Academic Publishers, Dordrecht, p. 3.
36. Kollman, P., Dixon, Cornell, R.W., Fox, T., Chipot, C. and Pohorille, A. (1997) The development/application of a 'minimalist' organic/biochemical molecular mechanics force field using a combination of *ab initio* calculations and experimental data, in *Computer Simulation of Biomolecular Systems: Theoretical and Experimental Applications, Vol. 3.* van Gunsteren, W.F., Weiner, P.K., and Wilkinson, A.J. (eds.), Kluwer Academic Publishers, Dordrecht, p. 83.
37. MacKerell, A.D. Jr, Bashford, D., Bellott, M., Dunbrack, R.L., Evanseck, J.D., Field, M.J., Fischer, S., Gao, J., Guo, J., Ha, S., Joseph–McCarthy, D., Kuchnir, L., Kuczera, K., Lau, F.T.K., Mattos C., Michnick, S., Ngo, T., Nguyen, D.T., Prodhom, B., Reiher, W.E., III, Roux, B., Schlenkrich, M., Smith, J.C., Stote, R., Straub, J., Watanabe, M., Wiorkiewicz-Kuczera, J., Yin, D., and Karplus M. (1998) *J. Phys. Chem. B*, **102**, 3586.
38. Glaser, M.A., Clark, N.A., Garcia, E., and Walba, D.M. (1997) *Spectrochimica Acta A*, **53**, 1325.
39. Gillan, M.J. (1991) Calculating the properties of materials from scratch, in *Computer Simulation in Materials Science.* Meyer M. and Pontikis V. (eds.), Kluwer Academic Publishers, Dordrecht, p. 257.
40. Galli G. and Parrinello, M. (1991) Ab-initio molecular dynamics: principles and practical implementation, in *Computer Simulation in Materials Science.* Meyer M. and Pontikis V. (eds.), Kluwer Academic Publishers, Dordrecht, p. 283.
41. Galli, G. and Pasquarello, A. (1993) First-principles molecular dynamics, in *Computer Simulation in Chemical Physics.* Allen, M.P. and Tildesley, D. (eds.), Kluwer Academic Publishers, Dordrecht, p. 261.
42. Stone, A.J. (1996) *The Theory of Intermolecular Forces.* Oxford University Press, New York.
43. Sprik, M. (1993) M. Effective pair potentials and beyond, in *Computer Simulation in Chemical Physics.* Allen, M.P. and Tildesley, D. (eds.), Kluwer Academic Publishers, Dordrecht, p. 211.
44. Cross, C.W., Fung, B.M. (1995) *Mol. Cryst. Liq. Cryst. Sci. Technol. A*, **262**, 1795.
45. Wilson, M.R. (1997) *J. Chem. Phys.*, **107**, 8654.
46. Siepmann, J.I., Karaborni, S., and Smit, B. (1993) *Nature*, **365**, 330.
47. Paul, W., Yoon, D.Y., and Smith, G.D. (1995) *J. Chem. Phys.*, **103**, 1702.
48. Shi, X. and Bartell, L.S. (1988) *J. Phys. Chem.*, **92**, 5667.
49. Bartell, L.S., Sharkey, L.R., and Shi, X. (1988) *J. Am. Chem. Soc.*, **110**, 7006.
50. Nagy, J., Smith, V.H., and Weaver, D.F. (1995) *J. Phys. Chem.*, **99**, 13868.
51. Pant, P.V.K., Han, J., Smith, G.D., and Boyd, R.H. (1993) *J. Chem. Phys.*, **99**, 597.
52. Mayo, S.L., Olafson, B. D., and Goddard, W.A., III (1990) *J. Phys. Chem.*, **94**, 8897.
53. Frisch, M.J., Trucks, G.W., Schlegel, H.B., Gill, P.M.W., Johnson, B.G., Robb, M.A., Cheeseman, J.R, Keith, T., Petersson, G.A., Montgomery, J.A., Raghavachari, K., Al-Laham, M.A., Zakrzewski, V.G., Ortiz, J.V., Foresman, J.B., Cioslowski, J., Stefanov, B.B., Nanayakkara, A., Challacombe, M., Peng, C.Y., Ayala, P.Y., Chen, W., Wong, M.W., Andres, J.L., Replogle, E.S., Gomperts, R., Martin, R.L., Fox, D.J., Binkley, J.S., Defrees, D.J., Baker, J., Stewart, J.P., Head–Gordon, M., Gonzalez, C., and J.A. Pople, J.A. (1995) *Gaussian 94, Revision D.4.* Gaussian, Inc., Pittsburgh PA.
54. Breneman, C.M. and Wiberg, K.B., (1990) *J. Comp. Chem.*, **11**, 361.
55. Jorgensen, W.L., Madura, J.D., and Swenson, C.J. (1984) *J. Am. Chem. Soc.*, **106**, 6638; Briggs, J.M., Matsui, T., and Jorgensen, W.L. (1990) *J. Comput. Chem.*, **11**, 958; Jorgensen, W.L., Laird, E.R., Nguyen, T.B., and Tirado–Rives, J. (1993) *J. Comput.*

Chem., **14**, 206; Briggs, J.M., Nguyen, T.B., and Jorgensen, W.L. (1991) *J. Phys. Chem.*, **95**, 3315.

56. Frenkel, D. and Smit, B. (1996) *Understanding Molecular Simulation.* Academic Press, San Diego.

57. Duane, S., Kennedy, A.D., Pendleton, B.J., and Roweth, D. (1987) *Phys. Lett. B,* **195**, 216.

58. Darden, T., York, D., and Pedersen, L. (1993) *J. Chem. Phys.*, **98**, 10089.

59. Essmann, U., Perera, L., Berkowitz, M.L., Darden, T.A., Lee, H., and Pedersen, L.G. (1995) *J. Chem. Phys.*, **103**, 8577.

60. Lee, H., Darden, T.A., and Pedersen, L.G. (1995) *J. Chem. Phys.*, **102**, 3830.

61. Pedersen, L.G. (1995) *J. Chem. Phys.*, **103**, 3668.

62. Tuckermann, M., Berne, B.J., and Martyna, G.J. (1992) *J. Chem. Phys.*, **97**, 1990.

63. Sexton, J.C. and Weingarten, D.H (1992) *Nucl. Phys. B,* **380**, 665.

64. Berne, B.J. and Straub, J.E. (1997) *Curr. Opin. Struct. Biol.*, **7**, 181.

65. Flory, P.J. *Statistical Mechanics of Chain Molecules.* Interscience, New York, 1969.

66. Kusnezov, D., Bulgac, A., and Bauer, W. (1990) *Ann. Phys.*, **204**, 155.

67. Allen, M.P. and Tildesley, D.J. (1987) *Computer Simulation of Liquids.* Clarendon Press, Oxford.

68. Smith, P.E. and Pettitt, B.M. (1991) *J. Chem. Phys.*, **95**, 8430.

69. Bader, J.S. and Chandler, D. (1992) *J. Phys. Chem.*, **96**, 6423.

70. Schreiber, H. and Steinhauser, O. (1992) *Chem. Phys.*, **168**, 75.

71. Alper, H.E., Bassolino, D., and Stouch, T.R. (1993) *J. Chem. Phys.*, **98**, 9798.

72. York, D.M., Darden, T.A., and Pedersen, L.G. (1993) *J. Chem. Phys.*, **99**, 8345.

73. Smith, P.E. and van Gunsteren, W.F. (1993) Methods for the evaluation of long range electrostatic forces in computer simulations of molecular systems, in *Computer Simulation of Biomolecular Systems: Theoretical and Experimental Applications, Vol. 2.* van Gunsteren, W.F., Weiner, P.K., and Wilkinson, A.J. (eds.), ESCOM, Leiden, p. 182.

74. Buono, G.S.D., Cohen, T.S., and Rossky, P.J. (1994) *J. Mol. Liq.*, **60**, 221.

75. Roberts, J.E. and Schnitker, J. (1994) *J. Chem. Phys.*, **101**, 5024.

76. Perera, L., Essmann, U., and Berkowitz, M.L. (1995) *J. Chem. Phys.*, **102**, 450.

77. Tironi, I.G., Sperb, R., Smith, P.E., and van Gunsteren, W.F. (1995) *J. Chem. Phys.*, **102**, 5451.

78. Feller, S.E., Pastor, R.W., Rojnuckarin, A., Bogusz, S., and Brooks, B.R. (1996) *J. Phys. Chem.*, **100**, 17011.

79. Tildesley, D.J. (1993) The Molecular Dynamics Method, in *Computer Simulation in Chemical Physics.* Allen, M.P. and Tildesley, D.J. (eds.), Kluwer Academic Publishers, Dordrecht, p. 23.

80. Clark, A.T., Madden, T.J., and Warren, P.B. (1996) *Mol. Phys.*, **87**, 1063.

81. Clark, A.T., Madden, T.J., and Warren, P.B. (1997) *Mol. Phys.*, **92**, 947.

82. Wasserman, E., Rustad, J.R., Felmy, A.R., Hay, B.P., and Halley, J.W. (1997) *Surf. Sci.*, **385**, 217.

83. Toukmaji A.Y. and Board, J.A. Jr. (1996) *Comput. Phys. Commun.*, **95**, 73.

84. Ewald, P.P. (1921) *Ann. Physik.*, **64**, 253.

85. Perram, J.W., Petersen, H.G., and de Leeuw, S.W. (1988) *Mol. Phys.*, **65**, 875.

86. Fincham, D. (1994) *Mol. Sim.*, **13**, 1.

87. Arnold, V.I. (1978) *Mathematical Methods of Classical Mechanics.* Springer-Verlag, New York.

88. Bruce, B., (1995) *Construction and Properties of Symplectic Integrators for Hamiltonian Systems.* Ph.D. thesis, University of Colorado, Boulder.

89. Yoshida, H. (1992) Symplectic integrators for Hamiltonian systems: basic theory, in *Chaos, Resonance and Collective Dynamical Phenomena in the Solar System.* Ferraz-Mello S. (ed.), IAU, the Netherlands, p. 27.

90. Yoshida, H. (1993) *Celest. Mech.*, **56**, 27.

91. Streett, W.B., Tildesley, D.J., and Saville, G. (1978) *Mol. Phys.*, **35**, 639.

92. Nicolas, J.J., Gubbins, K.E., Streett, W.B., and Tildesley, D.J. (1979) *Mol. Phys.*, **37**, 1429.

93. Swindoll R.D. and Haile, J.M. (1984) *J. Comput. Phys.*, **53**, 289.

94. Teleman, O. and Jönsson, B. (1986) *J. Comput. Chem.*, **7**, 58.

95. Procacci, P. and Berne, B.J. (1994) *Mol. Phys.*, **83**, 255.

96. Procacci, P. and Berne, B.J. (1994) *J. Chem. Phys.*, **101**, 2421.

97. Procacci, P., Darden, T.A., and Marchi, M. (1996) *J. Phys. Chem.*, **100**, 10464.

98. Procacci, P. and Marchi, M. (1996) *J. Chem. Phys.*, **104**, 3003.

99. Procacci, P., Darden, T.A., Paci, E., and Marchi, M. (1997) *J. Comput. Chem.*, **18**, 1848.

100. Procacci, P., Marchi, M. and Martyna, G.J. (1998) *J. Chem. Phys.*, **108**, 8799.

101. Martyna, G.J., Tuckerman, M.E., Tobias, D.J. and Klein, M.L. (1996) *Mol. Phys.*, **87**, 1117.

102. Tildesley, D.J. (1993), The Monte Carlo method, in *Computer Simulation in Chemical Physics*. Allen, M. P. and Tildesley D.J. (eds.), Kluwer Academic Publishers, Dordrecht, p. 1.

103. Frenkel, D. (1993), in *Advanced Monte Carlo techniques* in *Computer Simulation in Chemical Physics*, Allen, M.P. and Tildesley D.J. (eds.), Kluwer Academic Publishers, Dordrecht, p. 93.

104. Vlugt, T.J.H., Martin, M.G., Smit, B., Siepmann, I.J. and Krishna, R. (1998) *Mol. Phys.*, **94**, 727.

105. Bagchi, K., Andersen, H.C. and Swope, W. (1996) *Phys. Rev. Lett.*, **76**, 255.

106. Saichev, A.I. and Woyczyński, W.A. (1997) *Distributions in the Physical and Engineering Sciences*, Birkhäuser, Boston.

107. Mackenzie, P.B. (1989) *Phys. Lett. B*, **226**, 369.

108. Kennedy, A.D. and Pendleton, B. (1991) *Nucl. Phys. B (Proc. Suppl.)*, **20**, 118.

109. Forrest, B.M. and Suter, U.W. (1994) *Mol. Phys.*, **82**, 393.

110. Forrest, B.M. and Suter, U.W. (1994) *J. Chem. Phys.*, **101**, 2616.

111. Gupta, S., Irbäck, A., Karsch, F. and Petersson, B. (1990) *Phys. Lett. B*, **242**, 437.

112. Our r-RESPA MD integrator uses four distinct intervals for force evaluation, corresponding to bond stretching (0.625 fs), bond angle bending (1.25 fs), dihedral torsion (2.5 fs), and nonbonded interactions (7.5 fs).

113. Krentsel, T.A., Lavrentovich, O.D. and Kumar, S. (1997) *Mol. Cryst. Liq. Cryst.*, **304**, 463.

114. Leadbetter, A.J., Durrant, J.L.A. and Rugman, M. (1977) *Mol. Cryst. Liq. Cryst. Lett.*, **34**, 231.

115. Leadbetter, A.J., Richardson, R.M. and Colling, C.N. (1975) *J. de Physique*, **36**, C1-37.

116. Lydon, J.E. and Coakley, C.J. (1975) *J. de Physique*, **36**, C1-45.

117. Lanšac, Y., Glaser, M.A. and Clark, N.A. (1999), in preparation.

118. Our *NPT* ensemble simulations employed the weak-coupling algorithm of Berendsen, H.J.C., Postma, J.P.M., van Gunsteren, W.F., Dinola, A. and Haak, J.R. (1984) *J. Chem. Phys.*, **81**, 3684.

119. To deal effectively with the broad range of characteristic timescales associated with various force contributions, we use a r-RESPA MD algorithm with five distinct intervals for force evaluation, corresponding to bond stretching (0.417 fs), bond angle bending (0.833 fs), dihedral torsion (1.667 fs), short range nonbonded (5 fs), and long range electrostatic interactions (10 fs).

120. Kobinata, S., Kobayashi, T., Yoshida, H., Chandani, A.D.L. and Maeda, S. (1986) *J. Mol. Struct.*, **146**, 373.

121. Boden, N., Clark, L.D., Bushby, R.J., Emsley, J.W., Luckhurst, G.R. and Stockley, C.P. (1981) *Mol. Phys.*, **42**, 565.

122. Barnhoorn, J.B.S., de Lange, C.A. and Burnell, E.E. (1993) *Liq. Cryst.*, **13**, 319.

123. Rieker, T.P. (1995) *Liq. Cryst.*, **19**, 497.

124. Guymon, C.A., Hoggan, E.N., Clark, N.A., Rieker, T.P., Walba, D.M., Bowman,

C.N. (1997) *Science*, **275**, 57.

125. Lansac, Y., Glaser, M.A., Clark, N.A. and Lavrentovich, O.D. (1999), *Nature*, **398**, 54.

126. Folks, W.R., Reznikov, Yu.A. Chen, L., Khizhnyak, A.I. and Lavrentovich, O.D. (1995) *Mol. Cryst. Liq. Cryst.*, **261**, 259.

127. Folks, W.R., Reznikov, Yu.A., Yarmolenko, S.N. and Lavrentovich, O.D. (1997) *Mol. Cryst. Liq. Cryst.*, **292**, 183.

128. Folks, W.R., Keast, S., Krentsel, T.A., Zalar, B., Zeng, H., Reznikov, Yu.A., Neubert, M., Kumar, S., Finotello, D. and Lavrentovich, O.D. (1998) *Mol. Cryst. Liq. Cryst.*, **320**, 77.

129. Zalar, B. and Finotello, D. unpublished results.

130. Ikeda, T., Sasaki, T., Ichimura, K. (1993) *Nature*, **361**, 428.

131. Sasaki, T., Ikeda, T., Ichimura, K. (1994) *J. Am. Chem. Soc.*, **116**, 625.

132. Sasaki, T., Ikeda, (1995) *J. Phys. Chem.*, **99**, 13002; *ibid.*, **99**, 13008; *ibid.*, **99**, 13013.

133. Coles, H.J., Walton, H.G., Guillon, D., Poetti, G. (1993) *Liq. Cryst.*, **15**, 551.

134. Meyer, R.B., Liébert, L., Strzelecki, L. and Keller, P. (1975) *J. Phys. (Paris) Lett.*, **36**, 69.

135. C_2 rotation is only a *local* symmetry, as bulk SmC* materials exhibit a helical precession of tilt direction around the layer normal on longer length scales, and so are more properly described as *helielectric*. However, confining a sufficiently thin SmC* between solid substrates can unwind the helix, producing a ferroelectric thin film (the surface-stabilized ferroelectric liquid crystal (SSFLC) geometry). See Clark, N.A. and Lagerwall, S.T. (1980) *Appl. Phys. Lett.*, **36**, 899.

136. Bartolino, R., Doucet, J. and Durand, G. (1978) *Ann. Phys.*, **3**, 389.

137. The form of mean-field potential used in our original study of ferroelectric polarization [142] was slightly more complicated.

138. Weeks, J.D., Chandler, D. and Andersen, H.C. (1971) *J. Chem. Phys.*, **54**, 5237.

139. Marcelja, S. (1974) *J. Chem. Phys.*, **60**, 3599.

140. Photinos, D.J., Samulski, E.T. and Toriumi, H. (1991) *Mol. Cryst. Liq. Cryst.*, **204**, 161.

141. Luckhurst, G.R. (1994) *Mol. Phys.*, **82**, 1063.

142. Glaser, M.A., Ginzburg, V.V., Clark, N.A., Garcia, E., Walba, D.M. and Malzbender, R. (1995) *Mol. Phys. Rep.*, **10**, 26.

143. Skarp, K. and Handschy, M.A. (1988) *Mol. Cryst. Liq. Cryst.*, **165**, 439.

144. Glaser, M.A., Lansac, Y., Fung, B.M., Fernsler, J. and Clark, N.A. (1999), in preparation.

145. Pershan, P.S. (1988) *Structure of Liquid Crystal Phases*, World Scientific, Singapore.

146. Goodby, J.W. (1991) Properties and structures of ferroelectric liquid crystals, in *Ferroelectric Liquid Crystals: Principles, Properties, and Applications*, Goodby, J.W. (ed.), Gordon and Breach, Philadelphia, p. 99.

147. Walba, D.M., Blanca Ros, M., Clark, N.A., Shao, R., Robinson, M.G., Liu, J.-Y., Johnson, K.M. and Doroski, D. (1991) *J. Am. Chem. Soc.*, **113**, 5471.

MULTIPLE TIME STEPS ALGORITHMS FOR THE ATOMISTIC SIMULATIONS OF COMPLEX MOLECULAR SYSTEMS

PIERO PROCACCI

Department of Chemistry, University of Florence, Via G. Capponi 9, Firenze I-50121 Italy

AND

MASSIMO MARCHI

Section de Biophysique des Protéines et des Membranes, DBCM, DSV, CEA, Centre d'Études, Saclay, 91191 Gif-sur-Yvette Cedex, FRANCE

1. Introduction

In this Chapter we shall illustrate some state of the art techniques for the molecular dynamics (MD) simulation of atomistic models of complex molecular systems. In atomistic models the coordinates of all atomic nuclei including hydrogens are treated explicitly and interactions between distant atoms are represented by a pairwise additive dispersive-repulsive potential and a Coulomb contribution due to the atomic charges. Furthermore, nearby atoms interact through special two body, three body and four body functions representing the valence bonds, bending and torsional interaction energies surfaces.

The validity of such an approach as well as the reliability of the various potential models proposed in the literature [1–4] is not the object of the present Chapter. For reading on this topic, we refer to the extensive and ever growing literature [2, 5–7]. Here, we want only to stress the general concept that atomistic simulations usually have more predictive power than simplified models, but are also very expensive with respect to the latter from a computational standpoint. This predictive power stems from the fact that, in principle, simulations at the atomistic level do not introduce any uncontrolled approximation besides the obvious assumptions inherent in the definition of the potential model and do not assume any *a priori* knowledge of the system, except of course its chemical composi-

P. Pasini and C. Zannoni (eds.), Advances in the Computer Simulations of Liquid Crystals, 333–387.
© 2000 *Kluwer Academic Publishers. Printed in the Netherlands.*

334

tion and topology. Therefore, the failure in predicting specific properties
of the system for an atomistic simulation is due only to the inadequacy
of the adopted interaction potential. We may define this statement as the
brute force postulate. In practice, however, in order to reduce the computational burden, severe and essentially uncontrolled approximations such as
neglect of long range interactions, suppression of degrees of freedom, du-
bious thermostatting or constant pressure schemes are often undertaken.
These approximations, however, lessen the predictive power of the atomistic approach and incorrect results may follow due to the inadequacy in
the potential model, baseless approximations or combinations of the two.
Also, due to their cost, the predictive capability of atomistic level simulations might often only be on paper, since in practice only a very limited
phase space region can be accessed in an affordable way, thereby providing
only biased and unreliable statistics for determining the macroscopic and
microscopic behavior of the system. It is therefore of paramount importance in atomistic simulations to use computational techniques that do not
introduced uncontrolled approximations and at the same time are efficient.

As stated above, simulations of complex systems at the atomistic level,
unlike simplified models, have the advantage of representing with realistic detail the full flexibility of the system and the potential energy surface
according to which the atoms move. Both these physically significant ingre-
dients of the atomistic approach unfortunately pose severe computational
problems: on one hand the inclusion of full flexibility necessarily implies the
selection of small step size thereby reducing in a MD simulation the sampling power of the phase space. On the other hand, especially the evaluation
of inter-particle long range interactions is an extremely expensive task using conventional methods, its computational cost scaling typically like N^2
(with N being the number of particles) quickly exceeding any reasonable
limit. In this Chapter we shall devote our attention to the methods, within
the framework of classical MD simulations, that partially overcome the difficulties related to the presence of flexibility and long range interactions
when simulating complex systems at the atomistic level.

Complex systems experience different kind of motions with different
time scales: intramolecular vibrations have a period not exceeding few hundreds of femtoseconds while reorientational motions and conformational
changes have much longer time scales ranging from few picoseconds to
hundreds of nanoseconds. In the intra-molecular dynamics one may also
distinguish between fast stretching involving hydrogens with period smaller
than 10 fs, stretching between heavy atoms and bending involving hydrogens with double period or more, slow bending and torsional movements
and so on. In a similar manner in the diffusional regime many different
contributions can be identified.

In a standard integration of Newtonian equations, all these motions, irrespectively of their time scales, are advanced using *the same* time step whose size is inversely proportional to the frequency of the *fastest* degree of freedom present in the system, therefore on the order of the femtosecond or even less. This constraint on the step size severely limits the accessible simulation time.

One common way to alleviate the problem of the small step size is to freeze some supposedly irrelevant and fast degrees of freedom in the system. This procedure relies on the the so-called SHAKE algorithm [8–10] that implements holonomic constraints while advancing the *Cartesian* coordinates. Typically, bonds and/or bending are kept rigid thus removing most of the high frequency density of states and allowing a moderate increase of the step size. The SHAKE procedure changes the input potential and therefore the output density of the states. Freezing degrees of freedom, therefore, requires in principle an *a priori* knowledge of the dynamical behavior of the system. SHAKE is in fact fully justified when the suppressed degrees of freedom do not mix with the "relevant" degrees of freedom. This might be "almost" true for fast stretching involving hydrogens which approximately defines an independent subspace of internal coordinates in almost all complex molecular systems [11] but may turn to be wrong in certain cases even for fast stretching between heavy atoms. In any case the degree of mixing of the various degrees of freedom of a complex system is not known *a priori* and should be on the contrary considered one of the targets of atomistic simulations. The SHAKE algorithm allows only a moderate increase of the step size while introducing, if used without caution, essentially uncontrolled approximations. In other words indiscriminate use of constraints violates the brute-force postulate.

A more fruitful approach to the multiple time scale problem is to devise a more efficient "multiple time step" integration of the equation of motion. Multiple time step integration in MD is a relatively old idea [12–17] but only in recent times, due to the work of Tuckerman, Martyna and Berne and coworkers [18–23] is finding widespread application. These authors introduced a very effective formalism for devising multilevel integrators based on the symmetric factorization of the Liouvillean classical time propagator. The multiple time step approach allows integration of all degrees of freedom at an affordable computational cost. In the simulation of complex systems, for a well tuned multilevel integrator, the speed up can be sensibly larger than that obtained imposing bond constraints. Besides its efficiency, the multiple time steps approach has the advantage of not introducing any *a priori* assumption that may modify part of the density of the state of the system.

The Liouvillean approach to multiple time steps integrator lends itself to the straightforward, albeit tedious, application to extended Lagrangian systems for the simulation in the canonical and isobaric ensembles: once the equations of motions are known, the Liouvillean and hence the scheme, is *de facto* available. Many efficient multilevel schemes for constant pressure or constant temperature simulation are available in the literature [23–25].

As already outlined, long range interactions are the other stumbling block in the atomistic MD simulation of complex systems. The problem is particularly acute in biopolymers where the presence of distributed net charges makes the local potential oscillate wildly while summing, e.g. onto spherical non neutral shells. The conditionally convergent nature of the electrostatic energy series for a periodic system such as the MD box in periodic boundary conditions (PBC) makes any straightforward truncation method essentially unreliable [10, 26, 27].

The reaction field [28, 29] is in principle a physically appealing method that correctly accounts for long range effects and requires only limited computational effort. The technique assumes explicit electrostatic interactions within a cutoff sphere surrounded by a dielectric medium which exerts back in the sphere a "polarization" or reaction field. The dielectric medium has a dielectric constant that matches that of the inner sphere. The technique has been proved to give results identical to those obtained with the *exact* Ewald method in Monte Carlo simulation of dipolar spherocylinders where the dielectric constant that enters in the reaction field is updated periodically according to the value found in the sphere. The reaction field method does however suffer of two major disadvantages that strongly limits its use in MD simulations of complex systems at the atomistic level: during time integration the system may experience instabilities related to the circulating dielectric constant of the reaction field and to the jumps into the dielectric of molecules in the sphere with explicit interactions. The other problem, maybe more serious, is that again the method requires an *a priori* knowledge of the system, that is the dielectric constant. In pure phases this might not be a great problem but in inhomogeneous systems such as solvated protein, the knowledge of the dielectric constant might be not easily available. Even with the circulating technique, an initial unreliable guess of the unknown dielectric constant, can strongly affect the dielectric behavior of the system and in turn its dynamical and structural state.

The electrostatic series can be computed in principle *exactly* using the Ewald re-summation technique [30, 31]. The Ewald method rewrites the electrostatic sum for the periodic system in terms of two absolutely convergent series, one in the direct lattice and the other in reciprocal lattice. This method, in its standard implementation, is extremely CPU demanding and scales like N^2 with N being the number of charges with the unfortunate consequence that even moderately large size simulations of inhomogeneous biological systems are not within its reach. The rigorous Ewald method, which does not suffers of none of the inconveniences experienced by the reaction field approach, has however regained resurgent interest very recently after publication by Darden, York and Pedersen [32] of the Particle Mesh technique and later on by Essmann, Darden *et al.* [33] of the variant termed Smooth Particle Mesh Ewald (SPME). SPME is based on older idea idea of Hockney [34] and is essentially an interpolation technique with a charge smearing onto a regular grid and evaluation via fast Fourier Transform (FFT) of the interpolated reciprocal lattice energy sums. The performances of this techniques, both in accuracy and efficiency, are astonishing. Most important, the computational cost scales like $N \log N$, that is essentially *linearly* for any practical application. Other algorithm like the Fast Multipole Method (FMM) [35–38] scales exactly like N, even better than SPME. However FMM has a very large prefactor and the break even point with SPME is on the order of several hundred thousand of particles, that is, as up to now, beyond any reasonable practical limit.

The combination of the multiple time step algorithm and of the SPME [11] makes the simulation of large size complex molecular systems such as biopolymers, polar mesogenic molecules, organic molecules in solution etc., extremely efficient and therefore affordable even for long time spans. Further, this technique do not involve any uncontrolled approximation[1] and is perfectly consistent with standard PBC.

The present Chapter is organized as follow: In section 2 we first discuss some fundamental properties that numerically computed trajectories must satisfies and the illustrate a technique based on the Liouville formalism which allows to construct stepwise integrators with those properties built-in. We next discuss the derivation of the multiple time step integrators in the microcanonical ensemble for complex molecular systems with different and overlapping time scales. In section II we generalized the methods discussed in the previous section to non microcanonical ensembles. We illustrate in details the multiple time step scheme for the most general extended Lagrangian in MD simulation that is the Parrinello-Rahman-Nosé Lagrangian for the isothermal isobaric ensemble with anisotropic stress tensor. In the last section we briefly review the smooth particle mesh Ewald method and describe the combination of the latter with multiple time step schemes. A general protocol for the multiple time step integration to be adopted in complex systems is then discussed and illustrated with practical examples.

---

[1]Of course SPME *is* itself an approximation of the true electrostatic energy. This approximation is however totally under control since the energy can be determined to any given accuracy and the effect of finite accuracy can be easily controlled on any computed property of the system. The approximation *is not* uncontrolled.

338

2. Symplectic and reversible integrators

In an Hamiltonian problem, the symplectic condition and microscopic reversibility are inherent properties of the true time trajectories which, in turn, are the exact solution of Hamilton's equation. A stepwise integration defines a t-flow mapping which may or may not retain these properties. Non symplectic and/or non reversible integrators are generally believed [39–42] to be less stable in the long-time integration of Hamiltonian systems. In this section we shall illustrate the concept of reversible and symplectic mapping in relation to the numerical integration of the equations of motion.

2.1. CANONICAL TRANSFORMATION AND SYMPLECTIC CONDITIONS

Given a system with n generalized coordinates q, n conjugated momenta p and Hamiltonian H, the corresponding Hamilton's equations of motion are

$$\dot{q}_i = \frac{\partial H}{\partial p_i}$$

$$\dot{p}_i = -\frac{\partial H}{\partial q_i} \qquad i = 1, 2, ...n \tag{1}$$

These equations can be written in a more compact form by defining a column matrix with $2n$ elements such that

$$\mathbf{x} = \begin{pmatrix} \mathbf{q} \\ \mathbf{p} \end{pmatrix}. \tag{2}$$

In this notation the Hamilton's equations (1) can be compactly written as

$$\dot{\mathbf{x}} = \mathbf{J}\frac{\partial H}{\partial \mathbf{x}} \qquad \mathbf{J} = \begin{pmatrix} 0 & 1 \\ -1 & 0 \end{pmatrix} \tag{3}$$

where $\mathbf{J}$ is a $2n \times 2n$ matrix, $\mathbf{1}$ is an $n \times n$ identity matrix and $\mathbf{0}$ is a $n \times n$ matrix of zeroes. Eq. (3) is the so-called symplectic notation for the Hamilton's equations.[2]

Using the same notation we now may define a transformation of variables from $\mathbf{x} \equiv \{q,p\}$ to $\mathbf{y} \equiv \{Q,P\}$ as

$$\mathbf{y} = \mathbf{y}(\mathbf{x}) \tag{4}$$

For a restricted canonical transformation [43, 44] we know that the function $H(\mathbf{x})$ expressed in the new coordinates $\mathbf{y}$ serves as the Hamiltonian function

---

[2]Symplectic means "intertwined" in Greek and refers to the interlaced role of coordinate and momenta in Hamilton's equations.

for the new coordinates $\mathbf{y}$, that is the Hamilton's equations of motion in the $\mathbf{y}$ basis have exactly the same form as in Eq. (3):

$$\dot{\mathbf{y}} = \mathbf{J}\frac{\partial H}{\partial \mathbf{y}} \tag{5}$$

If we now take the time derivative of Eq. (4), use the chain rule relating $\mathbf{x}$ and $\mathbf{y}$ derivatives and use Eq. (5), we arrive at

$$\dot{\mathbf{y}} = \mathbf{MJM}^t \frac{\partial H}{\partial \mathbf{y}}. \tag{6}$$

Here $\mathbf{M}$ is the Jacobian matrix with elements

$$M_{ij} = \partial y_i/\partial x_i, \tag{7}$$

and $\mathbf{M}^t$ is its transpose. By comparing Eqs. (5) and (6), we arrive at the conclusion that a transformation is canonical if, and only if, the Jacobian matrix $\mathbf{M}$ of the transformation Eq. 4 satisfies the condition

$$\mathbf{MJM}^t = \mathbf{J}. \tag{8}$$

Eq. (8) is known as the symplectic condition for canonical transformations and represents an effective tool to test whether a generic transformation is canonical. Canonical transformations play a key role in Hamiltonian dynamics. It may also be proved that there exist a canonical transformation, parametrically dependent on time, changing the coordinates and momenta $\mathbf{y}(t) \equiv \{p, q\}$ at any given time t to their initial values $\mathbf{x} \equiv \{p_0, q_0\}$ [43]. Obtaining such a transformation is of course equivalent to finding the solution for Hamilton's equations of motion. [3] The transformation ϕ

$$\mathbf{z}(t) = \phi(t, \mathbf{z}(0)) \tag{9}$$

from the initial values $\{p_0 q_0\} \equiv \mathbf{z}(0)$ to the generalized coordinates $\{P, Q\} \equiv \mathbf{z}(t)$ at a generic time t defines the t-flow mapping of the systems and, being canonical, its Jacobian matrix obeys the symplectic condition (8). An important consequence of the symplectic condition, is the invariance under canonical (or symplectic) transformations of many properties of the phase space. These invariant properties are known as "Poincare invariants" or canonical invariants. For example transformations or t-flow's mapping

---

[3]This concept may be effectively understood using the generating function approach [39, 43] for canonical transformation applied to the particular case of an infinitesimal canonical transformation (ict) dependent on a "trajectory" parameter α. If the generator is the Hamiltonian itself and the parameter is time t, then the canonical transformation made up of a succession of infinitesimal ict's is the solution of Hamilton's equation [43].

340

obeying Eq. (8) preserve the phase space volume. This is easy to see, since the infinitesimal volume elements in the $\mathbf{y}$ and $\mathbf{x}$ bases are related by

$$dy = |\det \mathbf{M}| dx \tag{10}$$

where $|\det \mathbf{M}|$ is the Jacobian of the transformation. Taking the determinant of the symplectic condition Eq. (8) we see that $|\det \mathbf{M}| = 1$ and therefore

$$dy = dx. \tag{11}$$

For a canonical or symplectic t-flow mapping this means that the phase total space volume is invariant and therefore Liouville theorem is automatically satisfied.

A stepwise numerical integration scheme defines a Δt-flow mapping or equivalently a coordinates transformation, that is

$$\begin{aligned} Q(\Delta t) &= Q(q(0), p(0), \Delta t) \\ P(\Delta t) &= P(q(0), p(0), \Delta t) \end{aligned} \qquad \mathbf{y}(\Delta t) = \ \mathbf{y}(\mathbf{x}(0)). \tag{12}$$

We have seen that exact solution of the Hamilton equations has t-flow mapping satisfying the symplectic conditions (8). If the Jacobian matrix of the transformation (12) satisfies the symplectic condition then the integrator is termed to be symplectic. The resulting integrator, therefore, exhibits properties identical to those of the exact solution, in particular it satisfies Eq. (11). Symplectic algorithms have also been proved to be robust, i.e resistant to time step increase, and generate stable long time trajectory, i.e. they do not show drifts of the total energy. Popular MD algorithms like Verlet, leap frog and velocity Verlet are all symplectic and their robustness is now understood to be due in part to this property. [18, 21, 41, 42]

2.2. LIOUVILLE FORMALISM: A TOOL FOR BUILDING SYMPLECTIC AND REVERSIBLE INTEGRATORS

In the previous paragraphs we have seen that it is highly beneficial for an integrator to be symplectic. We may now wonder if there exists a general way for obtaining symplectic and possibly, reversible integrators from "first principles". To this end, we start by noting that for any property which depends on time implicitly through $p, q \equiv \mathbf{x}$ we have

$$\begin{aligned} \frac{dA(p,q)}{dt} &= \sum_{q,p} \left(\dot{q} \frac{\partial A}{\partial q} + \dot{p} \frac{\partial A}{\partial p} \right) = \sum_{q,p} \left(\frac{\partial H}{\partial p} \frac{\partial A}{\partial q} - \frac{\partial H}{\partial q} \frac{\partial A}{\partial p} \right) \\ &= iLA \tag{13} \end{aligned}$$

where the sum is extended to all n degrees of freedom in the system. L is the Liouvillean operator defined by

$$iL \equiv \sum_{q,p}\left(\dot{q}\frac{\partial}{\partial q}+\dot{p}\frac{\partial}{\partial p}\right)=\sum_{q,p}\left(\frac{\partial H}{\partial p}\frac{\partial}{\partial q}-\frac{\partial H}{\partial q}\frac{\partial}{\partial p}\right). \tag{14}$$

Eq. (13) can be integrated to yield

$$A(t) = e^{iLt}A(0). \tag{15}$$

If A is the state vector itself we can use Eq. (15) to integrate Hamilton's equations:

$$\begin{bmatrix} q(t) \\ p(t) \end{bmatrix} = e^{iLt}\begin{bmatrix} q(0) \\ p(0) \end{bmatrix}. \tag{16}$$

The above equation is a formal solution of Hamilton's equations of motion. The exponential operator e^{iLt} times the state vector defines the t-flow of the Hamiltonian system which brings the system phase space point from the initial state q_0, p_0 to the state $p(t), q(t)$ at a later time t. We already know that this transformation obeys Eq. (8). We may also note that the adjoint of the exponential operator corresponds to the inverse, that is e^{iLt} is unitary. This implies that the trajectory is exactly time reversible. In order to build our integrator, we now define the *discrete time propagator* $e^{iL\Delta t}$ as

$$e^{iLt} = \left[e^{iLt/n}\right]^n; \qquad \Delta t = t/n \tag{17}$$

$$e^{iL\Delta t} = e^{\sum_{q,p}\left(\dot{q}\frac{\partial}{\partial q}+\dot{p}\frac{\partial}{\partial p}\right)\Delta t}. \tag{18}$$

In principle, to evaluate the action of $e^{iL\Delta t}$ on the state vector p, q one should know the derivatives of all orders of the potential V. This can be easily seen by Taylor expanding the discrete time propagator $e^{iL\Delta t}$ and noting that the operator $\dot{q}\partial/\partial q$ does not commute with $-\partial V/\partial q(\partial/\partial p)$ when the coordinates and momenta refer to *same* degree of freedom. We seek therefore approximate expressions of the discrete time propagator that retain both the symplectic and the reversibility property. For any two linear operators A, B the Trotter formula [45] holds:

$$e^{(A+B)t} = \lim_{n\to\infty}(e^{At/n}e^{Bt/n})^n \tag{19}$$

We recognize that the propagator Eq. (18) has the same structure as the left hand side of Eq. (19); hence, using Eq. (19), we may write for Δt sufficiently small

$$e^{iL\Delta t} = e^{\left(\dot{q}\frac{\partial}{\partial q}+\dot{p}\frac{\partial}{\partial p}\right)\Delta t} \simeq e^{\dot{q}\frac{\partial}{\partial q}\Delta t}e^{\dot{p}\frac{\partial}{\partial p}\Delta t}+O(\Delta t^2) \quad . \tag{20}$$

342

Where, for simplicity of discussion, we have omitted the sum over q and p in the exponential. Eq. (20) is exact in the limit that $\Delta t \to 0$ and is first order for finite step size. Using Eq. (8) it is easy to show that the t-flow defined in Eq. (20) is symplectic, being the product of two successive symplectic transformations. Unfortunately, the propagtor Eq. (20) is not unitary and therefore the corresponding algorithm is not time reversible. Again the non unitarity is due to the fact that the two factorized exponential operators are non commuting. We can overcome this problem by halving the time step and using the approximant:

$$e^{(A+B)t} \simeq e^{At/2}e^{Bt/2}e^{Bt/2}e^{At/2} = e^{At/2}e^{Bt}e^{At/2}. \tag{21}$$

The resulting propagator is clearly unitary, therefore time reversible, and is also correct to the second order [46]. Thus, requiring that the product of the exponential operator be unitary, automatically leads to more accurate approximations of the true discrete time propagator [46, 47]. Applying the same argument to the propagator (18) we have

$$e^{iL\Delta t} = e^{\left(\dot{q}\frac{\partial}{\partial q}+\dot{p}\frac{\partial}{\partial p}\right)\Delta t} \simeq e^{\dot{p}\frac{\partial}{\partial p}\Delta t/2}e^{\dot{q}\frac{\partial}{\partial q}\Delta t}e^{\dot{p}\frac{\partial}{\partial p}\Delta t/2} + O(\Delta t^3). \tag{22}$$

The action of an exponential operator $e^{(a\partial/\partial x)}$ on a generic function $f(x)$ trivially corresponds to the Taylor expansion of $f(x)$ around the point x at the point $x + a$, that is

$$e^{a\partial/\partial x}f(x) = f(x + a). \tag{23}$$

Using Eq. (23), the time reversible and symplectic integration algorithm can now be derived by acting with our Hermitian operator Eq. (22) onto the state vector at $t = 0$ to produce updated coordinate and momenta at a later time Δt. The resulting algorithm is completely equivalent to the well known velocity Verlet:

$$\begin{aligned} p(\Delta t/2) &= p(0) + F(0)\Delta t/2 \\ q(\Delta t) &= q(0) + \left(\frac{p(\Delta t/2)}{m}\right)\Delta t \\ p(\Delta t) &= p(\Delta t/2) + F(\Delta t)\Delta t/2. \end{aligned} \tag{24}$$

We first notice that each of the three transformations obeys the symplectic condition Eq. (8) and has a Jacobian determinant equal to one. The product of the three transformation is also symplectic and, thus, phase volume preserving. Finally, since the discrete time propagator (22) is unitary, the algorithm is time reversible.

One may wonder what it is obtained if the operators $\dot{q}\partial/\partial q$ and $-\partial V/\partial q(\partial/\partial p)$ are exchanged in the definition of the discrete time propagator (22). If we do so, the new integrator is

$$
\begin{aligned}
q(\Delta t/2) &= q(0) + \frac{p(0)}{m}\Delta t/2 \\
p(\Delta t) &= p(0) + F[q(\Delta t/2)]\Delta t \\
q(\Delta t) &= q(\Delta t/2) + \frac{p(\Delta t)}{m}\Delta t/2.
\end{aligned}
\tag{25}
$$

This algorithm has been proved to be equivalent to the so-called Leap-frog algorithm [48]. Tuckerman *et al.* [18] called this algorithm *position Verlet* which is certainly a more appropriate name in the light of the exchanged role of positions and velocities with respect to the velocity Verlet. Also, Eq. (21) clearly shows that the position Verlet is essentially identical to the Velocity Verlet. A shift of a time origin by $\Delta t/2$ of either Eq. (25) or Eq. (24) would actually make both integrator perfectly equivalent. However, as pointed out in Ref. [19], half time steps are not formally defined, being the right hand side of Eq. (21) an approximation of the discrete time propagator for the *full* step Δt. Velocity Verlet and Position Verlet, therefore, do not generate numerically identical trajectories although of course the trajectories are similar. We conclude this section by saying that is indeed noticeable that using the same Liouville formalism different long-time known schemes can be derived. The Liouville approach represent therefore a unifying treatment for understanding the properties and relationships between stepwise integrators.

2.3. POTENTIAL SUBDIVISION AND MULTIPLE TIME STEPS INTEGRATORS FOR NVE SIMULATIONS

The ideas developed in the preceding sections can be used to build multiple time step integrators. Multiple time step integration is based on the concept of reference system. Let us now assume that the system potential V be subdivided in n terms such that

$$
V = V_0 + V_1 + ... + V_n.
\tag{26}
$$

Additionally, we suppose that the corresponding average values of the square modulus of the forces $F_k = |\partial V_k/\partial \mathbf{x}|$ and of their time derivatives $\dot{F}_k = |d/dt(\partial V_k/\partial \mathbf{x})|$ satisfy the following condition:

$$
\begin{aligned}
F_0^2 &>> F_1^2 >> .. >> F_n^2 \\
\dot{F}_0^2 &>> \dot{F}_1^2 >> .. >> \dot{F}_n^2.
\end{aligned}
\tag{27}
$$

These equations express the situation where different time scales of the system correspond to different pieces of the potential. Thus, the Hamiltonian of the k-th *reference system* is defined as

$$H = T + V_0 + ..V_k, \tag{28}$$

with a perturbation given by:

$$P = V_{k+1} + V_{k+2}.. + ..V_n. \tag{29}$$

For a general subdivision of the kind given in Eq. (26) there exist n reference *nested* systems. In the general case of a flexible molecular systems, we have fast degrees of freedom which are governed by the stretching, bending and torsional potentials and by slow intermolecular motions driven by the intermolecular potential. As we shall discuss with greater detail in section 4, in real systems there is no clearcut condition between intra and intermolecular motions since their time scales may well overlap in many cases. The conditions Eq. (27) are, hence, never fully met for any of all possible potential subdivisions.

Given a potential subdivision Eq. (26), we now show how a multi-step scheme can be built with the methods described in section 2.2. For the sake of simplicity, we subdivide the interaction potential of a molecular system into two components only: One intra molecular, V_0, generating mostly "fast" motions and the other intermolecular, V_1, driving slower degrees of freedom. Generalization of the forthcoming discussion to a n-fold subdivision, Eq. (26), is then straightforward.

For the 2-fold inter/intra subdivision, the system with Hamiltonian $H = T + V_0$ is called the intra-molecular reference system whereas V_1 is the intermolecular perturbation to the reference system. Correspondingly, the Liouvillean may be split as

$$\begin{aligned} iL_0 &= \dot{q}\frac{\partial}{\partial q} - \frac{\partial V_0}{\partial q}\frac{\partial}{\partial p} \\ iL_1 &= -\frac{\partial V_1}{\partial q}\frac{\partial}{\partial p}. \end{aligned} \tag{30}$$

Here L_0 is the Liouvillean of the 0-th reference system with Hamiltonian $T + V_0$, while L_1 is a perturbation Liouvillean. Let us now suppose now that Δt_1 is a good time discretization for the time evolution of the perturbation, that is for the slowly varying intermolecular potential. The discrete $e^{iL\Delta t_1} \equiv e^{(iL_0 + iL_1)\Delta t_1}$ time propagator can be factorized as

$$\begin{aligned} e^{iL\Delta t_1} &= e^{iL_1\Delta t_1/2}(e^{iL_0\Delta t_1/n})^n e^{iL_1\Delta t_1/2} \\ &= e^{iL_1\Delta t_1/2}(e^{iL_0\Delta t_0})^n e^{iL_1\Delta t_1/2}, \end{aligned} \tag{31}$$

where we have used Eq. (21) and we have defined

$$\Delta t_0 = \frac{\Delta t_1}{n} \tag{32}$$

as the time step for the "fast" reference system with Hamiltonian $T + V_0$. The propagator (31) is unitary and hence time reversible. The external propagators depending on the Liouvillean L_1 acting on the state vectors define a symplectic mapping, as it can be easily proved by using Eq. (8). The full factorized propagator is therefore symplectic as long as the inner propagator is symplectic. The Liouvillean $iL_0 \equiv \dot{q}\partial/\partial q - \partial V_0/\partial q\,\partial/\partial p$ can be factorized according to the Verlet symplectic and reversible breakup described in the preceding section, but with an Hamiltonian $T + V_0$. Inserting the result into Eq. (31) and using the definition (30), the resulting double time step propagator is then

$$e^{iL\Delta t_1} = e^{\frac{-\partial V_1}{\delta q}\frac{\partial}{\delta p}\Delta t_1/2}\left(e^{\frac{-\partial V_0}{\delta q}\frac{\partial}{\delta p}\Delta t_0/2}e^{\dot{q}\frac{\partial}{\delta q}\Delta t_0}e^{\frac{-\partial V_0}{\delta q}\frac{\partial}{\delta p}\Delta t_0/2}\right)^n e^{\frac{-\partial V_1}{\delta q}\frac{\partial}{\delta p}\Delta t_1/2} \tag{33}$$

This propagator is unfolded straightforwardly using the rule (23) generating the following symplectic and reversible integrator from step $t = 0$ to $t = \Delta t_1$:

$$
\begin{aligned}
p\left(\tfrac{\Delta t_1}{2}\right) &= p(0) + F_1(0)\tfrac{\Delta t_1}{2} \\
\text{DO} \quad & \text{i=1,n} \\
p\left(\tfrac{\Delta t_1}{2} + i\tfrac{\Delta t_0}{2}\right) &= p\left(\tfrac{\Delta t_1}{2} + [i-1]\tfrac{\Delta t_0}{2}\right) + F_0\left([i-1]\tfrac{\Delta t_0}{2}\right)\tfrac{\Delta t_0}{2} \\
q\left(i\Delta t_0\right) &= q\left([i-1]\Delta t_0\right) + p\left(\tfrac{\Delta t_1}{2} + i\tfrac{\Delta t_0}{2}\right)\tfrac{\Delta t_0}{m} \\
p\left(\tfrac{\Delta t_1}{2} + i\Delta t_0\right) &= p\left(\tfrac{\Delta t_1}{2} + i\tfrac{\Delta t_0}{2}\right) + F_0\left(i\Delta t_0\right)\tfrac{\Delta t_0}{2} \\
\text{ENDDO} \quad & \\
p\left(\Delta t_1\right) &= p'\left(\tfrac{\Delta t_1}{2}\right) + F_1(n\Delta t_0)\tfrac{\Delta t_1}{2}
\end{aligned}
\tag{34}
$$

Note that the slowly varying forces F_1 are felt only at the beginning and the end of the macro-step[4] Δt_1. In the inner n steps loop the system moves only according to the Hamiltonian of the reference system $H = T + V_0$. When using the potential breakup, the inner reference system is rigorously

---

[4]When the large step size at which the intermittent impulses are computed matches the period of natural oscillations in the system, one can detect instabilities of the numerical integration due to resonance effects. Resonances occurs for pathological systems such as fast harmonic oscillators in presence of strong, albeit slowly varying, forces [41] and can be cured easily by tuning the time steps in the multilevel integration. However, for large and complex molecules it is unlikely that an artificial resonance could sustain for any length of time [41]

conservative and the total energy of the reference system (i.e. $T + V_0 + \ldots + V_k$) is conserved during the P micro-steps.[5]

The integration algorithm given an arbitrary subdivision of the interaction potential is now straightforward. For the general subdivision (26) the corresponding Liouvillean split is

$$iL_0 = \dot{q}\frac{\partial}{\partial q} - \frac{\partial V_i}{\partial q}\frac{\partial}{\partial p}; \quad iL_1 = -\frac{\partial V_1}{\partial q}\frac{\partial}{\partial p} \ldots; \quad iL_n = -\frac{\partial V_k}{\partial q}\frac{\partial}{\partial p}. \tag{35}$$

We write the discrete time operator for the Liouville operator $iL = L_0 + \ldots L_n$ and use repeatedly the Hermitian approximant and Trotter formula to get a hierarchy of *nested* reference systems propagator, viz.

$$e^{i(\sum_{i=0}^{n} L_i)\Delta t_n} = e^{iL_n\frac{\Delta t_n}{2}}\left(e^{i(\sum_{i=0}^{n-1} L_i)\Delta t_{n-1}}\right)^{P_{n-1}} e^{iL_n\frac{\Delta t_n}{2}} \tag{36}$$

$$\Delta t_n = \Delta t_{n-1}P_{n-1}$$

$$e^{i(\sum_{i=0}^{n-1} L_i)\Delta t_{n-1}} = e^{iL_{n-1}\frac{\Delta t_{n-1}}{2}}\left(e^{i(\sum_{i=0}^{n-2} L_i)\Delta t_{n-2}}\right)^{P_{n-2}} e^{iL_{n-1}\frac{\Delta t_{n-1}}{2}} \tag{37}$$

$$\Delta t_{n-1} = \Delta t_{n-2}P_{n-2}$$

$$\ldots$$

$$e^{i(L_0+L_1+L_2)\Delta t_2} = e^{iL_2\frac{\Delta t_2}{2}}\left(e^{i(L_1+L_0)\Delta t_1}\right)^{P_1} e^{iL_2\frac{\Delta t_2}{2}} \tag{38}$$

$$\Delta t_2 = \Delta t_1 P_1$$

$$e^{i(L_0+L_1)\Delta t_1} = e^{iL_1\frac{\Delta t_1}{2}}\left(e^{iL_0\Delta t_0}\right)^{P_0} e^{iL_1\frac{\Delta t_1}{2}} \tag{39}$$

$$\Delta t_1 = \Delta t_0 P_0$$

where Δt_i is the generic integration time steps selected according to the time scale of the i-th force F_i. We now substitute Eq. (39) into Eq. (38) and so on climbing the whole hierarchy until Eq. (36). The resulting multiple time steps symplectic and reversible propagator is then

$$e^{iL\Delta t_n} = e^{F_n\frac{\partial}{\partial p}\Delta t_n}\left(e^{F_{n-1}\frac{\partial}{\partial p}\frac{\Delta t_{n-1}}{2}} .. \right.$$

$$\left. .. \left(e^{F_0\frac{\partial}{\partial p}\frac{\Delta t_0}{2}} e^{\dot{q}\frac{\partial}{\partial q}\Delta t_0} e^{F_0\frac{\partial}{\partial p}\frac{\Delta t_0}{2}}\right)^{P_0} .. e^{F_{n-1}\frac{\partial}{\partial p}\frac{\Delta t_{n-1}}{2}}\right)^{P_{n-1}} e^{F_n\frac{\partial}{\partial p}\frac{\Delta t_n}{2}} \tag{40}$$

[5]In the original *force* breakup [17, 18], the energy is not generally conserved during the unperturbed motion of the inner reference systems but only at the end of the full macrostep. Force breakup and potential breakup have been proved to produce identical trajectories [21]. With respect to the force the breakup, implementation of the potential breakup is slightly more complicated when dealing with intermolecular potential separation, but the energy conservation requirement in any defined reference system makes the debugging process easier.

The integration algorithm that can be derived from the above propagator was first proposed by Tuckerman, Martyna and Berne and called r–RESPA, reversible reference system propagation algorithm [18]

2.4. CONSTRAINTS AND R–RESPA

The r-RESPA approach makes unnecessary to resort to the SHAKE procedure [8, 9] to freeze some fast degrees of freedom. However the SHAKE and r-RESPA algorithms are not mutually exclusive and sometimes it might be convenient to freeze some degrees of freedom while simultaneously using a multi-step integration for all other freely evolving degrees of freedom. Since r-RESPA consists in a series of nested velocity Verlet like algorithms, the constraint technique RATTLE [49] used in the past for single time step velocity Verlet integrator can be straightforwardly applied. In RATTLE both the constraint conditions on the coordinates and their time derivatives must be satisfied. The resulting coordinate constraints is upheld by a SHAKE iterative procedure which corrects the positions exactly as in a standard Verlet integration, while a similar iterative procedure is applied to the velocities at the half time step.

In a multi time step integration, whenever velocities are updated, using part of the overall forces (e.g. the intermolecular forces), they must also be corrected for the corresponding constraints forces with a call to RAT-TLE. This combined RATTLE-r-RESPA procedure has been described for the first time by Tuckerman and Parrinello [50] in the framework of the Car-Parrinello simulation method. To illustrate the combined RATTLE-r-RESPA technique in a multi-step integration, we assume a separation of the potential into two components deriving from intramolecular and intermolecular interactions. In addition, some of the covalent bonds are supposed rigid, i.e.

$$d_a = d_a^{(0)} \tag{41}$$
$$\dot{d}_a = 0 \tag{42}$$

equation where a runs over all constrained bonds and $d_a^{(0)}$ are constants. In the double time integration (34), velocities are updated four times, i.e. two times in the inner loop and two times in the outer loop. To satisfy (41), SHAKE must be called to correct the position in the inner loop. To satisfy (42), RATTLE must be called twice, once in the inner loop and the second

348

time in the outer loop according to the following scheme

$$
\begin{aligned}
p' \left(\tfrac{\Delta t_1}{2}\right) &= p(0) + F_1(0)\tfrac{\Delta t_1}{2} \\
p \ (\Delta t_1) &= RATTLE_p \left\{ p'\left(\tfrac{\Delta t_1}{2}\right) \right\} \\
\text{DO} \quad & i=1,n \\
& p'\left(\tfrac{\Delta t_1}{2} + i\tfrac{\Delta t_0}{2}\right) = p\left(\tfrac{\Delta t_1}{2} + [i-1]\tfrac{\Delta t_0}{2}\right) + F_0\left([i-1]\tfrac{\Delta t_0}{2}\right)\tfrac{\Delta t_0}{2} \\
& p\left(\tfrac{\Delta t_1}{2} + i\tfrac{\Delta t_0}{2}\right) = RATTLE_p \left\{ p'\left(\tfrac{\Delta t_1}{2} + i\tfrac{\Delta t_0}{2}\right) \right\} \\
& q'\left(i\Delta t_0\right) = q\left([i-1]\Delta t_0\right) + p\left(\tfrac{\Delta t_1}{2} + i\tfrac{\Delta t_0}{2}\right)\tfrac{\Delta t_0}{m} \\
& q\left(i\Delta t_0\right) = RATTLE_q \left\{ q'\left(i\Delta t_0\right) \right\} \\
& p\left(\tfrac{\Delta t_1}{2} + i\Delta t_0\right) = p\left(\tfrac{\Delta t_1}{2} + i\tfrac{\Delta t_0}{2}\right) + F_0\left(i\Delta t_0\right)\tfrac{\Delta t_0}{2} \\
\text{ENDDO} & \\
p \ (\Delta t_1) &= p'(\tfrac{\Delta t_1}{2}) + F_1(n\Delta t_0)\tfrac{\Delta t_1}{2}.
\end{aligned}
\tag{43}
$$

Where $RATTLE_p$ and $RATTLE_q$ represent the constraint procedure on velocity and coordinates, respectively.

2.5. APPLICATIONS

As a first simple example we apply the double time integrator (34) to the NVE simulation of flexible nitrogen at 100 K.

The overall interaction potential is given by

$$
V = V_{intra} + V_{inter}
$$

Where V_{inter} is the intermolecular potential described by a Lennard-Jones model between all nitrogen atoms on different molecules [51]. V_{intra} is instead the intramolecular stretching potential holding together the two nitrogen atoms of each given molecule. We use here a simple harmonic spring depending on the molecular bond length r_m, namely:

$$
V_i = \frac{1}{2}\sum_m k(r - r_0)^2,
$$

with r_0 and r the equilibrium and instantaneous distance between the nitrogen atoms, and k the force constant tuned to reproduce the experimental gas-phase stretching frequency [52]. As a measure of the accuracy of the numerical integration we use the adimensional energy conservation ratio [11, 21, 22, 53]

$$
R = \frac{<E^2> - <E>^2}{<K^2> - <K>^2}
\tag{44}
$$

TABLE 1. Energy conservation ratio R for various integrators (see text). The last three entries refer to a velocity Verlet with bond constraints. $< V_i >$ and $< V_m >$ are the average value of the intra-molecular and intermolecular energies (in KJ/mole), respectively. CPU is given in seconds per picoseconds of simulation and Δt in fs. Single time step velocity Verlet with $\Delta t = 4.5$ fs is unstable.

Δt	n	R	CPU	$< V_i >$	$< V_m >$
0.3	**1**	**0.005**	**119**	**0.1912**	**-4.75**
0.6	1	0.018	62	0.1937	-4.75
1.5	1	0.121	26	0.2142	-4.75
4.5	1	-	-	-	-
0.6	2	0.004	59	0.1912	-4.75
1.5	5	0.004	28	0.1912	-4.75
3.0	10	0.005	18	0.1912	-4.75
4.5	15	0.006	15	0.1912	-4.75
6.0	20	0.008	12	0.1912	-4.74
9.0	30	0.012	10	0.1911	-4.74
3.0	-	0.001	14	-	-4.74
6.0	-	0.004	8	-	-4.75
9.0	-	0.008	6	-	-4.74

where E and K are the total and kinetic energy of the system, respectively. In table 1 we show the energy conservation ratio R and CPU timings on a IBM-43P/160MH/RS6000 obtained for flexible nitrogen at 100 K with the r-RESPA integrator as a function of n and Δt_1 in Eq. (34) and also for single time step integrators. Results of integrators for rigid nitrogen using SHAKE are also shown for comparison. The data in Table 1 refer to a 3.0 ps run without velocity rescaling. They were obtained starting all runs from coordinates corresponding to the experimental $Pa3$ structure [54, 55] of solid nitrogen and from velocities taken randomly according to the Boltzmann distribution at 100 K. The entry in bold refers to the "exact" result, obtained with a single time step integrator with a very small step size of 0.3 fs. Note that R increases quadratically with the time step for single time step integrators whereas r-RESPA is remarkably resistant to outer time step size increase. For example r-RESPA with $\Delta t_1 = 9.0 fs$ and $P = 30$ (i.e. $\Delta t_0 = 0.3$ fs) yields better accuracy on energy conservation than single time step Velocity Verlet with $\Delta t = 0.6 fs$ does, while being more than

350

six times faster. Moreover, r-RESPA integrates all degrees of freedom of the systems and is almost as efficient as Velocity Verlet with constraints on bonds. It is also worth pointing out that energy averages for all r-RESPA integrators is equal to the exact value, while at single time step even a moderate step size increase results in sensibly different averages intra-molecular energies. As a more complex example we now study a cluster of eight

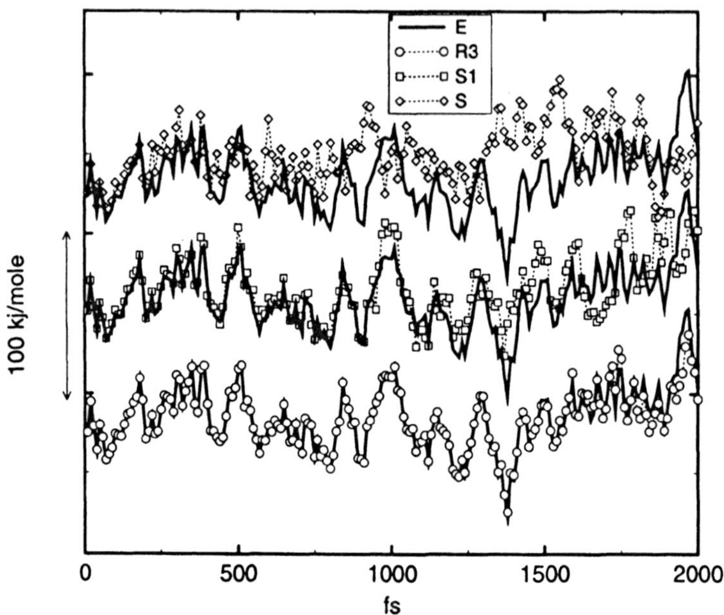

Figure 1. Time record of the torsional potential energy at about 300 K for a cluster of eight molecules of $C_{24}H_{50}$ obtained using three integrators: solid line integrator E; circles integrator R3; squares integrator S1; diamonds integrator S (see text)

single chain alkanes $C_{24}H_{50}$. In this case the potential contains stretching, bending and torsional contributions plus the intermolecular Van-der-Waals interactions between non bonded atoms. The parameter are chosen according to the AMBER protocol [2] by assigning the carbon and hydrogen atoms to the AMBER types ct and hc, respectively. For various dynamical and structural properties we compare three integrators, namely a triple time step r-RESPA (R3) single time step integrator with bond constraints on $X - H$ (S1) and a single time step integrator with all bonds kept rigid (S). These three integrators are tested, starting from the same phase space point, against a single time step integrator (E) with a very small time step generating the "exact" trajectory. In Fig. 1 we show the time record of the torsional potential energy. The R3 integrator generates a trajectory practically coincident with the "exact" trajectory for as long as 1.5 ps. The single time step with rigid $X - H$ bonds also produces a fairly accurate trajectory, whereas the trajectory generated by S quickly drifts away from the exact time record. In Fig. 2 we show the power spectrum of the velocity autocorrelation function obtained with R3, S1 and S. The spectra are compared to the exact spectrum computed using the trajectories generated by the accurate integrator E. We see that R3 and S1 generates the same spectral profile within statistical error. In contrast, especially in the region above 800 wavenumbers, S generates a spectrum which differs appreciably from the exact one. This does not mean, of course, that S is unreliable for the "relevant" torsional degrees of freedom. Simply, we cannot *a priori* exclude that keeping all bonds rigid will not have an impact on the equilibrium structure of the alkanes molecules and on torsional dynamics. Actually, in the present case, as long as torsional motions are concerned all three integrators produce essentially identical results. In 20 picoseconds of simulation, R3 S1 and S predicted 60, 61, 60 torsional jumps, respectively, against the 59 jumps obtained with the exact integrator E. According to prescription of Ref. [56], in order to avoid period doubling, we compute the power spectrum of torsional motion form the autocorrelation function of the vector product of two normalized vector perpendicular to the dihedral planes. Rare events such as torsional jumps produce large amplitudes long time scale oscillations in the time autocorrelation function and therefore their contribution overwhelms the spectrum which appears as a single broaden peak around zero frequency. For this reason all torsions that did undergo a barrier crossing were discarded in the computation of the power spectrum. The power spectrum of the torsional motions is identical for all integrators within statistical error when evaluated over 20 ps of simulations.

From these results it can be concluded that S1 and R3 are very likely to produce essentially the same dynamics for all "relevant" degrees of freedom. We are forced to state that also the integrator S appears to accurately predict the structure and overall dynamics of the torsional degrees of freedom at least for the 20 ps time span of this specific system.[6] Since torsions *are*

---

[6] For example, our conclusions on the effect of SHAKE onto torsional motions for highly flexible systems differs form the results published by Watanabe and Karplus [53] for another flexible system. i.e. met-enkephalin *in vacuo*. They compared SHAKE on $X - H$ against full flexibility and found that the power spectrum of torsional degrees of freedom differs significantly. For met-enkephalin their spectrum, evaluated on a 10 ps time span, shows a single strong peak at 10 or 40 wavenumbers, with and without constraints, respectively. The different behavior of the constrained and totally flexible system might be ascribed in their case to the the specificity of the system and/or the potential, although this seems unlikely [22]. In their study, on the other hand, we must

352

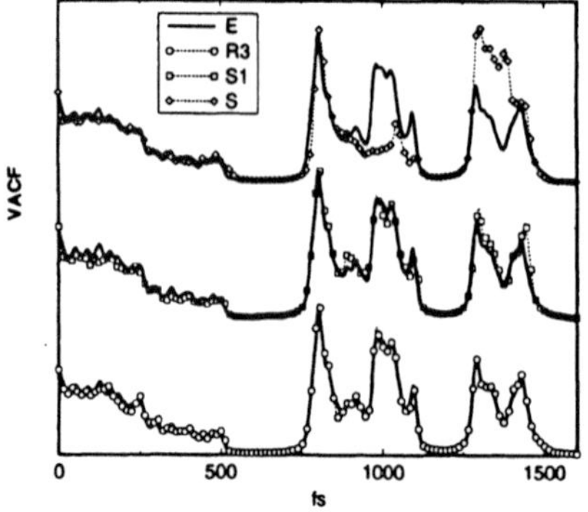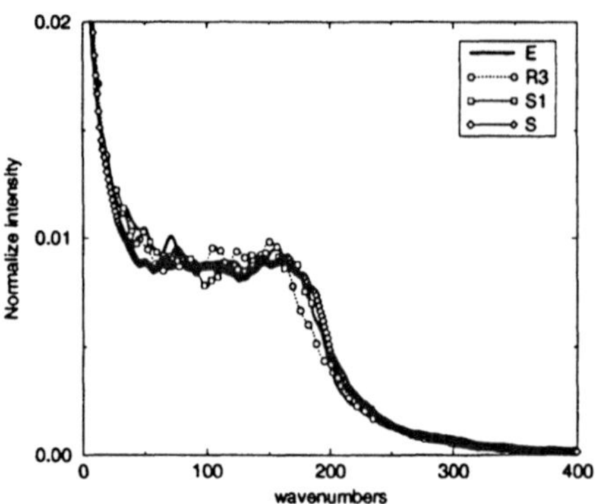

Figure 2. Power spectra of the velocity autocorrelation function (left) and of the torsional internal coordinates (right) at 300 K for a cluster of 8 $C_{24}H_{50}$ molecules calculated with integrators E, R3, S1 and S (see text) starting from the same phase space point

not normal coordinates and couple to higher frequency internal coordinates such as bending and stretching, the ability of the efficient S integrator of correctly predicting low frequency dynamics and structural properties *cannot* be assumed *a priori* and must be, in principle, verified for each specific case. We also do not know how the individual eigenvectors are affected by the integrators and, although the overall density for S and S1 appears to be the same, there might be considerable changes in the torsional dynamics. R3 does not require any assumption, is accurate everywhere in the spectrum (see Fig. 2) and is as efficient as S. For these reasons R3, or a multi-step version of the equally accurate S1, must be the natural choice for the simulation of complex systems using *all-atoms* models

3. Multiple time steps algorithm for the isothermal-isobaric ensemble

The integrators developed in the previous section generates dynamics in the microcanonical ensemble where total energy, number of particles and volume are conserved. The derivation based on the Liouvillean and the corresponding propagator, however lends itself to a straightforward generalization to non microcanonical ensembles. Simulations of this kind are based on the concept of extended system and generate trajectories that sample the phase space according to a target distribution function. The ex-

remark the unusual shape of the spectral torsional profile with virtually no frequencies above 100 wavenumbers and with strong peaks suspiciously close the minimum detectable frequency according to their spectral resolution.

tended system method is reviewed in many excellent textbooks and papers [10, 23, 57–62] to which we refer for a complete and detailed description. Here it suffices to say that the technique relies on the clever definition of a modified or *extended* Lagrangian which includes extra degrees of freedom related to the intensive properties (e.g. pressure or temperature) one wishes to sample with a well defined distribution function. The dynamics of the extended system is generated in the *microcanonical* ensemble with the true n degrees of freedom and, additionally, the extra degrees of freedom related to the macroscopic thermodynamic variables. With an appropriate choice, the equations of motion of the extended system will produce trajectories in the *extended* phase space generating the desired equilibrium distribution function upon integration over the extra (extended) variables. There are several extended system techniques corresponding to various ensembles, e.g. constant pressure in the NPH ensemble simulation with isotropic [63] and anisotropic [64] stress, constant temperature simulation [65] in the NVT ensemble and isothermal–isobaric simulation [66] in the NPT ensemble. As we shall see, the dynamic of the *real* system generated by the extended system method is never Hamiltonian. Hence, symplecticness is no longer an inherent property of the equations of motion. Nonetheless, the Liouvillean formalism developed in the preceding section, turns out to be very useful for the derivation of multiple time step reversible integrators for a general isothermal–isobaric ensemble with anisotropic stress, or NPT[7]. This extended system is the most general among all non microcanonical simulations: The NPT, NPH the NVT and even NVE ensemble may be derived from this Lagrangian by imposing special constraints and/or choosing appropriate parameters [23, 25]

3.1. THE PARRINELLO-RAHMAN-NOSÉ EXTENDED LAGRANGIAN

The starting point of our derivation of the multilevel integrator for the NPT ensemble is the Parrinello-Rahman-Nosé Lagrangian for a molecular system with N molecules or groups [8] each containing n_i atoms and subject to a potential V. In order to construct the Lagrangian we define a coordinate scaling and a velocity scaling, i.e.

$$r_{ik\alpha} = R_{i\alpha} + l_{ik\alpha} = \sum_{\beta} h_{\alpha\beta} S_{i\beta} + l_{ik\alpha}, \tag{45}$$

$$\dot{R}'_{i\alpha} = \dot{R}_{i\alpha} s \tag{46}$$

---

[7]When P is not in boldface, we imply that the stress is *isotropic*

[8]For large molecules it may be convenient to further subdivide the molecule into groups. A group, therefore encompasses a conveniently chosen subset of the atoms of the molecule

$$l''_{ik\alpha} \;=\; l_{ik\alpha} s$$

Here, the indices i and k refer to molecules and atoms, respectively, while Greek letters are used to label the Cartesian components. $r_{ik\alpha}$ is the α component of the coordinates of the k-th atom belonging to the i-th molecule; $R_{i\alpha}$ is the center of mass coordinates; $S_{i\beta}$ is the scaled coordinate of the i-th molecular center of mass. $l_{ik\alpha}$ is the coordinate of the k-th atom belonging to the i-th molecule expressed in a frame parallel at any instant to the fixed laboratory frame, but with origin on the instantaneous molecular center of mass. The set of $l_{ik\alpha}$ coordinates satisfies $3N$ constraints of the type $\sum_{k=1}^{n_i} l_{ik\alpha} = 0$.

The matrix $\mathbf{h}$ and the variable s control the pressure an temperature of the extended system, respectively. The columns of the matrix $\mathbf{h}$ are the Cartesian components of the cell edges with respect to a fixed frame. The elements of this matrix allow the simulation cell to change shape and size and are sometimes called the "barostat" coordinates. The volume of the MD cell is related to $\mathbf{h}$ through the relation

$$\Omega \;=\; \det(\mathbf{h}). \tag{47}$$

s is the coordinates of the so-called "Nosé thermostat" and is coupled to the intramolecular and center of mass velocities,

We define the "potentials" depending on the thermodynamic variables P and T

$$
\begin{aligned}
V_P &\;=\; P \det(\mathbf{h}) \\
V_T &\;=\; \frac{g}{\beta} \ln s.
\end{aligned}
\tag{48}
$$

Where P is the *external* pressure of the system, $\beta = k_B T$, and g is a constant related to total the number of degrees of freedom in the system. This constant is chosen to correctly sample the NPT distribution function.

The extended NPT Lagrangian is then defined as

$$\mathcal{L} \;=\; \frac{1}{2}\sum_{i}^{N} M_i s^2 \dot{\mathbf{S}}_i^t \mathbf{h}^t \mathbf{h} \dot{\mathbf{S}}_i + \frac{1}{2}\sum_{ik} m_{ik} s^2 \dot{\mathbf{l}}^t_{ik}\dot{\mathbf{l}}_{ik} + \frac{1}{2} W s^2 tr(\dot{\mathbf{h}}^t \dot{\mathbf{h}}) \tag{49}$$

$$+ \; \frac{1}{2} Q \dot{s}^2 - V - P_{ext}\Omega - \frac{g}{\beta} \ln s \tag{50}$$

The arbitrary parameters W and Q are the "masses" of the barostat and of the thermostats, respectively[9]. They do not affect the sampled distribution

---

[9] W has actually the dimension of a mass, while Q has the dimension of a mass time a length squared

function but only the sampling efficiency [24, 65, 66]. For a detailed discussion of the sampling properties of this Lagrangian the reader is referred to Refs. [25, 62].

3.2. THE PARRINELLO-RAHMAN-NOSÉ HAMILTONIAN AND THE EQUATIONS OF MOTION

In order to derive the multiple time step integration algorithm using the Liouville formalism described in the preceding sections we must switch to the Hamiltonian formalism. Thus, we evaluate the conjugate momenta of the coordinates $S_{i\alpha}$, $l_{ik\alpha}$, $h_{\alpha\beta}$ and s by taking the derivatives of the Lagrangian in Eq. (50) with respect to corresponding velocities, i.e.

$$\mathbf{T}_i = M_i \mathbf{G} s^2 \dot{\mathbf{S}}_i \tag{51}$$

$$\mathbf{p}_{ik} = m_{ik} s^2 \dot{\mathbf{l}}_{ik} \tag{52}$$

$$\mathbf{P}_h = s^2 W \dot{\mathbf{h}} \tag{53}$$

$$p_s = Q\dot{s}. \tag{54}$$

Where we have defined the symmetric matrix

$$\mathbf{G} = \mathbf{h}^t \mathbf{h} \tag{55}$$

The Hamiltonian of the system is obtained using the usual Legendre transformation [43]

$$H(p,q) = \sum \dot{q}p - \mathcal{L}(q,\dot{q}). \tag{56}$$

One obtains

$$H = \frac{1}{2}\sum_i^N \frac{\dot{\mathbf{T}}_i \mathbf{G}^{-1}\dot{\mathbf{T}}}{Ms^2} + \frac{1}{2}\sum_{ik} \frac{\mathbf{p}^t_{ik}\mathbf{p}_{ik}}{m_{ik}s^2} + \frac{1}{2}\frac{tr\left(\mathbf{P}^t_h\mathbf{P}_h\right)}{s^2 W} + \frac{p_s^2}{2Q}$$

$$+ V + P\Omega + \frac{g\ln s}{\beta} \tag{57}$$

In the extended systems formulation we always deal with real and *virtual* variables. The virtual variables in the Hamiltonian (57) are the scaled coordinates and momenta while the unscaled variables (e.g $\mathbf{R}_i = \mathbf{h}\mathbf{S}_i$ or $p'_{ik\alpha} = p_{ik\alpha}/s$ are the real counterpart. The variable s in the Nosé formulation plays the role of a time scaling [51, 57, 65]. The above Hamiltonian is given in terms of *virtual variables* and in term of a virtual time and is indeed a *true* Hamiltonian function and has corresponding equation of motions that can be obtained applying Eq. (3) with $\mathbf{x} \equiv S_{i\alpha}, l_{ik\alpha}, h_{\alpha\beta}, s, T_{i\alpha}, p_{ik\alpha}, \pi_{\alpha\beta}, p_s$ in a standard fashion. Nonetheless, the equations of motions in terms of

356

these virtual variable are inadequate for several reasons since for example one would deal with a fluctuating time step [57, 65]. It is therefore convenient to work in terms of real momenta and real time. The real momenta are related to the virtual counterpart through the relations

$$T_{i\alpha} \rightarrow T_{i\alpha}/s \tag{58}$$

$$p_{ik\alpha} \rightarrow p_{ik\alpha}/s \tag{59}$$

$$(\mathbf{P}_h)_{\alpha\beta} \rightarrow (\mathbf{P}_h)_{\alpha\beta}/s \tag{60}$$

$$p_s \rightarrow p_s/s \tag{61}$$

$$\tag{62}$$

It is also convenient [24] to introduce new center of mass momenta as

$$\mathbf{P}_i \equiv \mathbf{G}^{-1}\mathbf{T}_i. \tag{63}$$

such that the corresponding velocities may be obtained directly without the knowledge of the "coordinates" $\mathbf{h}$ in $\mathbf{G}^{10}$, namely

$$\dot{\mathbf{S}}_i = \frac{\mathbf{P}_i}{M}. \tag{64}$$

Finally, a real time formulation and a new dynamical variable η are adopted:

$$t \rightarrow t/s \tag{65}$$

$$\eta \equiv \ln s \tag{66}$$

The equations of motions for the newly adopted set of dynamical variables are easily obtained from the true Hamiltonian in Eq. (57) and then using Eqs. (58-66) to rewrite the resulting equations in terms of the new momenta. In so doing, we obtain:

$$\dot{\mathbf{l}}_{ik} = \frac{\mathbf{p}_{ik}}{m_{ik}}, \qquad \dot{\mathbf{S}}_i = \frac{\mathbf{P}_i}{M_i}, \qquad \dot{\mathbf{h}} = \frac{\mathbf{P}_h}{W}, \qquad \dot{\eta} = \frac{p_\eta}{Q} \tag{67}$$

$$\dot{\mathbf{p}}_{ik} = \mathbf{f}_{ik}^c - \frac{p_\eta}{Q}\mathbf{p}_{ik}, \tag{68}$$

$$\dot{\mathbf{P}}_i = \mathbf{h}^{-1}\mathbf{F}_i - \mathbf{G}^{-1}\dot{\mathbf{G}}\mathbf{P}_i - \frac{p_\eta}{Q}\mathbf{P}_i, \tag{69}$$

$$\dot{\mathbf{P}}_h = \left(\mathcal{V} + \mathcal{K} - \mathbf{h}^{-1}P_{ext}\det\mathbf{h}\right) - \frac{p_\eta}{Q}\mathbf{P}_h, \tag{70}$$

$$\dot{p}_\eta = \mathcal{F}_\eta, \tag{71}$$

---

[10]This allows to maintain Verlet-like breakup while integrating the equation of motions [24].

It can be verified that the conserved quantity $\mathcal{H}$ is associated with the above equations of motion, namely

$$\mathcal{H} = \frac{1}{2}\sum_{i=1}^{N}\frac{\mathbf{P}_i^t\mathbf{G}\mathbf{P}_i}{M_i} + \frac{1}{2}\sum_{i=1}^{N}\sum_{k=1}^{n_i}\frac{\mathbf{p}_{ik}^t\mathbf{p}_{ik}}{m_{ik}} + \frac{1}{2}\frac{tr\left(\mathbf{P}_h^t\mathbf{P}_h\right)}{W} +$$

$$+ \frac{1}{2}\frac{p_\eta p_\eta}{Q} + V + P_{ext}\det\mathbf{h} + gk_BT\eta. \tag{72}$$

The atomic force $\mathbf{f}_{ik}^c = \frac{\partial V}{\partial r_{ika}} - \frac{m_{ik}}{M_i}\mathbf{F}_i$ includes a constraint force contribution which guarantees that the center of mass in the intramolecular frame of the l_{ika} coordinates remains at the origin. V and $\mathcal{K}$ are the virial and ideal gas contribution to the internal pressure tensor $\mathbf{P}_{int} = V + \mathcal{K}$ and they are defined as[11]

$$V = \sum_{i=1}^{N}\mathbf{F}_i\mathbf{S}_i^t$$

$$\mathcal{K} = \sum_{i=1}^{N}M_i\left(\mathbf{h}\dot{\mathbf{S}}_i\right)\dot{\mathbf{S}}_i^t. \tag{73}$$

Finally $\mathcal{F}_\eta$ is the force driving the Nosé thermostat

$$\mathcal{F}_\eta = \frac{1}{2}\sum_{i=1}^{N}M_i\dot{\mathbf{S}}_i^t\mathbf{G}\dot{\mathbf{S}}_i + \frac{1}{2}\sum_{i=1}^{N}\sum_{k=1}^{n_i}\frac{\mathbf{p}_{ik}^t\mathbf{p}_{ik}}{m_{ik}} - gk_BT, \tag{74}$$

with g equal to the number of all degrees of freedom N_f including those of the barostat[12].

Eqs. (58-65) define a generalized coordinates transformation of the kind of Eq. (4). This transformation is non canonical, i.e. the Jacobian matrix of the transformation from the virtual coordinates does not obey Eq. (8). This means that $\mathcal{H}$ in terms of the new coordinates Eq. (72) is "only" a constant of motion, but is no longer a true Hamiltonian: application of Eq. (1) does not lead to Eqs. (67-71). Simulations using the real variables are not Hamiltonian in nature in the sense that the phase space of the real variables is compressible [67] and that Liouville theorem is not satisfied [62]. This "strangeness" in the dynamics of the real variables in the extended systems

---

[11] In presence of bond constraints and if the scaling is group-based instead of molecular based, these expression should contain a contribution from the constraints forces. Complications due to the constraints can be avoided altogether by defining groups so that no two groups are connected through a constrained bond [25]. In that case V does not include any constraint contribution.

[12] The thermostat degree of freedom must be included [57, 62] in the count when working in *virtual* coordinates. Indeed in Eq. (57) we have $g = N_f + 1$

358

does not of course imply that the sampling of the configurational *real* space is incorrect. To show this, it suffices to evaluate the partition function for a microcanonical distribution of the kind $\delta(\mathcal{H} - E)$, with $\mathcal{H}$ being given by Eq. (72). The Jacobian of the transformation of Eqs. (58-66) must be included in the integration with respect to the real coordinates when evaluating the partition function for the extended system. If the equations of motion in terms of the transformed coordinates are known, this Jacobian, $\mathcal{J}$, can be readily computed from the relation [44]:

$$\frac{d\mathcal{J}}{dt} = -\mathcal{J}\left(\frac{\partial}{\partial \mathbf{y}} \cdot \dot{\mathbf{y}}\right).$$

(75)

Where $\mathbf{y}$ has the usual meaning of phase space vector containing all independent coordinates and momenta of the systems. Inserting the equations of motion of Eq. (71) into Eq. (75) and integrating by separation of variables yields

$$\mathcal{J} = e^{N_f \eta}\,[\det \mathbf{h}]^{6N}.$$

(76)

Using (76) and integrating out the thermostat degrees of freedom, the partition function can be easily shown [62, 68] to be equivalent to that that of NPT ensemble, i.e.

$$\Delta_{NPT} \propto \int d\mathbf{h} e^{-\beta P_{text}\,\det(\mathbf{h})} Q(\mathbf{h})$$

(77)

with $Q(\mathbf{h})$ being the canonical distribution of a system with cell of shape and size define by the columns of $\mathbf{h}$.[13]

3.3. EQUIVALENCE OF ATOMIC AND MOLECULAR PRESSURE

The volume scaling defined in Eq. (45) is not unique. Note that only the equation of motion for the center of mass momentum, Eq. (69), has a velocity dependent term that depends on the coordinates of the barostat through the matrix $\mathbf{G}$ defined in Eq. (55). The atomic momenta, Eq. (68), on the contrary, are not coupled to the barostat. This fact is also reflected in the equations of motion for the barostat momenta, Eq. (70), which is driven by the internal pressure due only to the molecular or group center of masses. In defining the extended Lagrangian one could as well have defined an *atomic* scaling of the form

$$r_{ik\alpha} = \sum_{\beta} h_{\alpha\beta} s_{i\alpha k}.$$

(78)

---

[13]Actually in ref. [25, 62] is pointed out that the virial theorem implied by the distribution (77) is slightly different from the exact virial in the NPT ensemble. Martyna *et al.* [62] proposed an improved set of equations of motion that generates a distribution satisfying exactly the virial theorem.

Atomic scaling might be trivially implemented by eliminating the kinetic energy, which depends on the $\dot{l}_{ik\alpha}$ velocities, from the starting Lagrangian (50) and replacing the term $\frac{1}{2}\sum_i^N M_i s^2 \dot{\mathbf{S}}_i^t \mathbf{h}^t \mathbf{h}\dot{\mathbf{S}}_i$ with $\frac{1}{2}\sum_{ik} m_{ik} s^2 \dot{\mathbf{s}}_{ik}^t \mathbf{h}^t \mathbf{h}\dot{\mathbf{s}}_{ik}$. The corresponding equations of motions for atomic scaling are then

$$\dot{\mathbf{r}}_{ik} = \frac{\mathbf{p}_{ik}}{m_{ik}}, \qquad \dot{\mathbf{h}} = \frac{\mathbf{P}_h}{W}, \qquad \dot{\eta} = \frac{p_\eta}{Q} \tag{79}$$

$$\dot{\mathbf{p}}_{ik} = \mathbf{h}^{-1}\dot{\mathbf{p}}_{ik} - \mathbf{G}^{-1}\dot{\mathbf{G}}\mathbf{p}_{ik} - \frac{p_\eta}{Q}\dot{\mathbf{p}}_{ik}, \tag{80}$$

$$\dot{\mathbf{P}}_h = \left(\mathcal{V} + \mathcal{K} - \mathbf{h}^{-1} P_{ext}\det\mathbf{h}\right) - \frac{p_\eta}{Q}\mathbf{P}_h, \tag{81}$$

$$\dot{p}_\eta = \mathcal{F}_\eta \tag{82}$$

where the quantities $\mathcal{V}, \mathcal{K}, \mathcal{F}_\eta$ depend now on the *atomic* coordinates

$$\mathcal{V} = \sum_{i=1k}^{N} \mathbf{f}_{ik}\mathbf{s}_{ik}^t$$

$$\mathcal{K} = \sum_{i=1}^{N} M_i\left(\mathbf{h}\mathbf{s}_{ik}\right)\mathbf{s}_{ik}^t \tag{83}$$

$$\mathcal{F}_\eta = \frac{1}{2}\sum_{i=1}^{N}\sum_{k=1}^{n_i} \frac{\mathbf{p}_{ik}^t \mathbf{p}_{ik}}{m_{ik}} - gk_BT. \tag{84}$$

In case of atomic, Eq. (78), or molecular scaling, Eq. (45), the internal pressure entering in Eqs. (70,81) is then

$$P_{int} = \langle P_{atom}\rangle = \left\langle \frac{1}{3V}\sum_i\sum_k\left(\frac{\mathbf{p}_{ik}^2}{m_{ik}} + \mathbf{r}_{ik}\bullet\mathbf{f}_{ik}\right)\right\rangle \tag{85}$$

$$P_{int} = \langle P_{mol}\rangle = \left\langle \frac{1}{3V}\sum_i\left(\frac{\mathbf{P}_i^2}{M_i} + \mathbf{R}_i\bullet\mathbf{F}_i\right)\right\rangle \tag{86}$$

respectively. Where the molecular quantities can be written in term of the atomic counterpart according to:

$$\mathbf{R}_i = \frac{1}{M_i}\sum_k m_{ik}\mathbf{r}_{ik} \tag{87}$$

$$\mathbf{P}_i = \sum_k \mathbf{p}_{ik} \tag{88}$$

$$\mathbf{F}_i = \sum_k \mathbf{f}_{ik} \tag{89}$$

The equation of motion for the barostat in the two cases, Eqs. (81, 70), has the same form whether atomic or molecular scaling is adopted. The

360

internal pressure in the former case is given by Eq. (85) and in the latter is given by Eq. (86). The two pressures, Eqs. (85,86), differ instantaneously. Should the difference persist after averaging, then it would be obvious that the equilibrium thermodynamic state in the NPT ensemble depends on the scaling method. The two formulas (85,86) are fortunately equivalent. To prove this statement, we closely follow the route proposed by H. Berendsen and reported by Ciccotti and Ryckaert [69] and use Eqs. (87-89) to rearrange Eq. (86). We obtain

$$\sum_i \langle \mathbf{R}_i \bullet \mathbf{F}_i \rangle = \sum_i \frac{1}{M_i} \sum_{kl} \langle m_{ik} \mathbf{r}_{ik} \bullet \mathbf{f}_{il} \rangle. \tag{90}$$

Adding and subtracting $m_{ik} \mathbf{r}_{il} \bullet \mathbf{f}_{il}$, we get

$$= \sum_i \frac{1}{M_i} \sum_{kl} \langle m_{ik}(\mathbf{r}_{ik} - \mathbf{r}_{il}) \bullet \mathbf{f}_{il} + m_{ik} \mathbf{r}_{il} \bullet \mathbf{f}_{il} \rangle \tag{91}$$

which can be rearranged as

$$= \sum_i \frac{1}{M_i} \left\{ \sum_{kl} \left[\frac{1}{2} \langle (\mathbf{r}_{ik} - \mathbf{r}_{il}) \bullet (m_i \mathbf{f}_{il} - m_j \mathbf{f}_{ik}) \rangle \right] + \sum_l \langle \mathbf{r}_{il} \bullet \mathbf{f}_{il} \rangle \right\} \tag{92}$$

using the newton law $\mathbf{f}_{ik} = m_{ik} \mathbf{a}_{ik}$, where $\mathbf{a}_{ik}$ is the acceleration, we obtain

$$= \sum_i \frac{1}{M_i} \left\{ \sum_{kl} \left[\frac{1}{2} \langle m_j m_i (\mathbf{r}_{ik} - \mathbf{r}_{il}) \bullet (\mathbf{a}_{il} - \mathbf{a}_{ik}) \rangle \right] + \sum_l \langle \mathbf{r}_{il} \bullet \mathbf{f}_{il} \rangle \right\}. \tag{93}$$

The first term in the above equation can be decomposed according to:

$$(\mathbf{r}_{ik} - \mathbf{r}_{il}) \bullet (\mathbf{a}_{il} - \mathbf{a}_{ik}) = \frac{d}{dt} [(\mathbf{r}_{ik} - \mathbf{r}_{il}) \bullet (\mathbf{v}_{il} - \mathbf{v}_{ik})] + (\mathbf{v}_{il} - \mathbf{v}_{ik})^2 \tag{94}$$

The first derivative term on the right hand side is zero rigorously for rigid molecules or rigid group and is zero on average for flexible molecules or groups, assuming that the flexible molecules or groups do not dissociate. This can be readily seen in case of ergodic systems, by evaluating directly the average of this derivatives as

$$\langle \frac{d}{dt} [(\mathbf{r}_{ik} - \mathbf{r}_{il}) \bullet (\mathbf{v}_{il} - \mathbf{v}_{ik})] \rangle = \lim_{\tau \to \infty} \frac{1}{\tau} \int_0^\infty \frac{d}{dt} [(\mathbf{r}_{ik} - \mathbf{r}_{il}) \bullet (\mathbf{v}_{il} - \mathbf{v}_{ik})] \rangle \, dt \tag{95}$$

$$= \lim_{\tau \to \infty} \frac{1}{\tau} [(\mathbf{r}_{ik}(\tau) - \mathbf{r}_{il}(\tau)) \bullet (\mathbf{v}_{il}(\tau) - \mathbf{v}_{ik}(\tau)) + C] \tag{96}$$

So if the quantity $\mathbf{r}_{ikl}(\tau) \bullet \mathbf{v}_{ilk}(\tau)$ remains bounded (which is true if the potential is not dissociative, since k, l refers to the same molecule i), the average in Eq. (96) is zero[14]. Thus, we can rewrite the average of Eq. (93) as

$$\langle \sum_i \mathbf{R}_i \bullet \mathbf{F}_i \rangle \;=\; \sum_i \frac{1}{2M_i} \sum_{kl} m_{ik} m_{il} \langle (\mathbf{v}_{il} - \mathbf{v}_{ik})^2 \rangle + \sum_{ik} \langle \mathbf{r}_{ik} \bullet \mathbf{f}_{ik} \rangle. \quad (97)$$

The first term on the right hand side of the above equation can be further developed obtaining the trivial identity:

$$\sum_{kl} m_{ik} m_{il} \langle (\mathbf{v}_{il} - \mathbf{v}_{ik})^2 \rangle \;=\; \sum_{kl} m_{ik} m_{il} \langle \mathbf{v}_{ik}^2 \rangle + \sum_{kl} m_{ik} m_{il} \langle \mathbf{v}_{il}^2 \rangle -$$
$$-\; \sum_{kl} 2 m_{ik} m_{il} \langle \mathbf{v}_{il} \bullet \mathbf{v}_{ik} \rangle \quad (98)$$
$$=\; 2M_i \sum_k m_{ik} \langle \mathbf{v}_{ik}^2 \rangle - 2 \langle \mathbf{P}_i^2 \rangle \quad (99)$$

Substituting Eq. (99) in Eq. (97) we get

$$\langle \sum_i \mathbf{R}_i \bullet \mathbf{F}_i \rangle \;=\; \sum_{ik} m_{ik} \langle \mathbf{v}_{ik}^2 \rangle - \frac{1}{M_i} \langle \mathbf{P}_i^2 \rangle + \sum_{ik} \langle \mathbf{r}_{ik} \bullet \mathbf{f}_{ik} \rangle \quad (100)$$

Substituting Eq. (100) into Eq. (86) leads speedily to (85) which completes the proof. As a consequence of the above discussion, it seems likely that both the equilibrium and non equilibrium properties of the MD system are not affected by coordinate scaling. We shall see later that this is actually the case.

3.4. LIOUVILLEAN SPLIT AND MULTIPLE TIME STEP ALGORITHM FOR THE *NPT* ENSEMBLE

We have seen in section 2.2 that the knowledge of the Liouvillean allows us to straightforwardly derive a multi-step integration algorithm. Thus, for simulation in the *NPT* ensemble, the Liouvillean $iL = \dot{\mathbf{y}} \nabla_\mathbf{y}$ is readily available from the equations of motion in (67-71). For sake of simplicity, to build our *NPT* multiple time step integrator we assume that the system potential contains only a fast *intramolecular* V_0 term and a slow *intermolecular* term V_1, as discussed in Sec. 2.3. Generalization to multiple intra and inter-molecular components is straightforward. We define the following

---

[14]The statement *the molecule of group does not dissociate* is even too restrictive. It is enough to say that the quantity (96) remains bound.

362

components of the NPT Liouvillean

$$iL_x = -\sum_i \mathbf{P}_i \frac{p_\eta}{Q}\nabla_{\mathbf{P}_i} - \sum_{ik} \mathbf{p}_{ik}\frac{p_\eta}{Q}\nabla_{\mathbf{p}_{ik}} - \sum_{\alpha\beta}(\mathbf{P}_h)_{\alpha\beta}\frac{p_\eta}{Q}(\nabla_{\mathbf{P}_h})_{\alpha\beta} \tag{101}$$

$$iL_y = \mathcal{F}_\eta\nabla_{p_\eta} \tag{102}$$

$$iL_z = \sum_i \mathbf{G}^{-1}\dot{\mathbf{G}}\mathbf{P}_i\nabla_{\mathbf{P}_i} \tag{103}$$

$$iL_u = \sum_{\alpha\beta}\left(\mathcal{K} - \mathbf{h}^{-1}P_{ext}\det\mathbf{h}\right)_{\alpha\beta}(\nabla_{\mathbf{P}_h})_{\alpha\beta} \tag{104}$$

$$iL_s = \sum_i \mathbf{J}_i\nabla_{\mathbf{P}_i} + \sum_{ik}\mathbf{f}_{ik}^c\nabla_{\mathbf{p}_{ik}} + \sum_{\alpha\beta}(\mathcal{V})_{\alpha\beta}(\nabla_{\mathbf{P}_h})_{\alpha\beta} \tag{105}$$

$$iG_0 = \sum_i \frac{\mathbf{P}_i}{M_i}\nabla_{\mathbf{S}_i} + \sum_{ik}\frac{\mathbf{p}_{ik}}{m_{ik}}\nabla_{\mathbf{l}_{ik}} + \sum_{\alpha\beta}\frac{(\mathbf{P}_h)_{\alpha\beta}}{W}(\nabla_\mathbf{h})_{\alpha\beta} +$$
$$+ \frac{p_\eta}{Q}\nabla_\eta - \nabla_{\mathbf{l}_{ik}}V_0\nabla_{\mathbf{p}_{ik}}, \tag{106}$$

where in Eq. (105) the scaled forces $\mathbf{F}_i$ have been replaced by its real space counterparts, i.e. $\mathbf{J}_i = \mathbf{h}^{-1}\mathbf{F}_i$.

The atomic scaling version of this Liouvillean breakup is derived on the basis of Eqs. (82). One obtains

$$iL_x = -\sum_{ik}\mathbf{p}_{ik}\frac{p_\eta}{Q}\nabla_{\mathbf{p}_{ik}} - \sum_{\alpha\beta}(\mathbf{P}_h)_{\alpha\beta}\frac{p_\eta}{Q}(\nabla_{\mathbf{P}_h})_{\alpha\beta} \tag{107}$$

$$iL_y = \mathcal{F}_\eta\nabla_{p_\eta} \tag{108}$$

$$iL_z = \sum_{ik}\mathbf{G}^{-1}\dot{\mathbf{G}}\mathbf{p}_{ik}\nabla_{\mathbf{p}_{ik}} \tag{109}$$

$$iL_u = \sum_{\alpha\beta}\left(\mathcal{K} - \mathbf{h}^{-1}P_{ext}\det\mathbf{h}\right)_{\alpha\beta}(\nabla_{\mathbf{P}_h})_{\alpha\beta} \tag{110}$$

$$iL_s = \sum_{ik}\mathbf{j}_{ik}\nabla_{\mathbf{p}_{ik}} + \sum_{\alpha\beta}(\mathcal{V})_{\alpha\beta}(\nabla_{\mathbf{P}_h})_{\alpha\beta} \tag{111}$$

$$iG_0 = \sum_{ik}\frac{\mathbf{p}_{ik}}{m_{ik}}\nabla_{\mathbf{l}_{ik}} + \sum_{\alpha\beta}\frac{(\mathbf{P}_h)_{\alpha\beta}}{W}(\nabla_\mathbf{h})_{\alpha\beta} +$$
$$+ \frac{p_\eta}{Q}\nabla_\eta - \nabla_{\mathbf{l}_{ik}}V_0\nabla_{\mathbf{p}_{ik}}, \tag{112}$$

where $\mathbf{j}_{ik} = \mathbf{h}^{-1}\mathbf{f}_{ik}$ and $\mathcal{V}, \mathcal{K}, \mathcal{F}_\eta$ are given in Eqs. (83,84). For the time scale breakup in the NPT ensemble we have the complication of the extra degrees of freedom whose time scale dynamics can be controlled by varying the parameter Q and W. Large values of Q and W slow down the time dynamics of the barostat and thermostat coordinates. The potential V determines the time scale of the iG_0 term (the fast component) and of the iL_s contribution (the slow component). All other sub-Liouvilleans either handle the coupling of the true coordinates to the extra degrees of freedom (iL_x expresses the coupling of all momenta (including barostat momenta) to the thermostat momentum, while iL_z is a coupling term between the center of mass momenta and the barostat momentum), or drive the evolution of the extra coordinates of the barostat and thermostat (iL_y and iL_u). The time scale dynamics of these terms depends not only on the potential subdivision and on the parameters W and Q, but also on the type of scaling [25]. When the molecular scaling is adopted the dynamics of the virial term $\mathcal{V}$ contains contributions only from the intermolecular potential since the barostat is coupled only to the center of mass coordinates (see Eq. (73)). Indeed, the net force acting on the molecular center of mass is independent on the intramolecular potential, since the latter is invariant under rigid translation of the molecules. When atomic scaling or group (i.e. sub-molecular) scaling is adopted, the virial $\mathcal{V}$ (see Eq. (83)) depends also on the fast intramolecular such as stretching motions. In this case the time scale of the barostat coordinate is no longer slow, unless the parameter W is changed. For standard values of W, selected to obtain an efficient sampling of the NPT phase space [51, 70], the barostat dependent Liouvilleans, Eqs. (104,103), have time scale dynamics comparable to that of the intramolecular Liouvillean iG_0 and therefore must be associated with this term[15].

Thus, the <u>molecular</u> split of the Liouvillean is hence given by

$$
\begin{aligned}
iL_1 &= iL_x + iL_y + iL_z + iL_u + iL_s \\
iL_0 &= iG_0
\end{aligned}
\tag{113}
$$

whereas the <u>atomic</u> split is

$$
\begin{aligned}
iL_1 &= iL_x + iL_y + iL_s \\
iL_0 &= iG_0 + iL_z + iL_u
\end{aligned}
\tag{114}
$$

For both scaling, a simple Hermitian factorization of the total time propagator e^{iLt} yields the double time discrete propagator

$$
e^{iL_1 + iL_0} = e^{iL_1 \Delta t_1/2} (e^{iL_0 \Delta t_0})^n e^{iL_1 \Delta t_1/2}
\tag{115}
$$

---

[15]Similar considerations hold for the thermostat coordinate which in principle depends on the kinetic energy of all degrees of freedom, modulated hence by the fast motion also. In this case, however, the value of the thermostat inertia parameter Q can be chosen to slow down the time scale of the η coordinates without reducing considerably the sampling efficiency.

where Δt_0, the small time step, must be selected according to the intramolecular time scale whereas Δt_1, the large time step, must be selected according to the time scale of the intermolecular motions. We already know that the propagator (115) cannot generate a symplectic. The alert reader may also have noticed that in this case the symmetric form of the multiple time step propagator Eq. (115) does not imply necessarily time reversibility. Some operators appearing in the definition of L_1 (e.g. iL_z and iL_s) for the molecular scaling and in the definitions of iL_1 and iL_0 for the atomic scaling are in fact non commuting. We have seen in section 2.2 that first order approximation of non commuting propagators yields time irreversible algorithms. We can render the propagator in Eq. (115) time reversible by using second order symmetric approximant (i.e. Trotter approximation) for any two non commuting operators. For example in the case of the molecular scaling, when we propagate in Eq. (115) the slow propagator $e^{iL_1\Delta t/2}$ for half time step, we may use the following second order $O(\Delta t^3)$ split

$$e^{iL_1\frac{\Delta t_1}{2}} \simeq e^{iL_y\frac{\Delta t_1}{4}}e^{iL_z\frac{\Delta t_1}{2}}e^{iL_y\frac{\Delta t_1}{4}}e^{iL_x\frac{\Delta t_1}{4}}e^{i(L_s+L_u)\frac{\Delta t_1}{2}}e^{iL_x\frac{\Delta t_1}{4}} \quad (116)$$

An alternative simpler and equally accurate approach when dealing with non commuting operators is simply to preserve the unitarity by reversing the order of the operators in the first order factorization of the right and left operators of Eq. (115) without resorting to locally second order $O(\Delta t^3)$ approximation like in Eq. (116). Again for the molecular scaling, this is easily done by using the approximant

$$\left(e^{iL_1\frac{\Delta t_1}{2}}\right)_{\text{left}} = e^{iL_x\frac{\Delta t_1}{2}}e^{iL_y\frac{\Delta t_1}{2}}e^{iL_z\frac{\Delta t_1}{2}}e^{iL_u\frac{\Delta t_1}{2}}e^{iL_s\frac{\Delta t_1}{2}} \quad (117)$$

for the left propagator, and

$$\left(e^{iL_1\frac{\Delta t_1}{2}}\right)_{\text{right}} = e^{iL_s\frac{\Delta t_1}{2}}e^{iL_u\frac{\Delta t_1}{2}}e^{iL_z\frac{\Delta t_1}{2}}e^{iL_y\frac{\Delta t_1}{2}}e^{iL_x\frac{\Delta t_1}{2}} \quad (118)$$

for the rightmost propagator. Note that

$$\left(e^{iL_1\frac{\Delta t_1}{2}}\right)^{-1}_{\text{left}} = \left(e^{iL_1\frac{\Delta t_1}{2}}\right)_{\text{right}} \quad (119)$$

Inserting these approximations into (115) the overall integrator is found to be time reversible and second order. Time reversible integrators are in fact always even order and hence *at least* second order [39, 42]. Therefore the overall molecular and atomic (or group) discrete time propagators are given by

$$e^{iL_{\text{mol}}\Delta t_1} = e^{iL_s\frac{\Delta t_1}{2}}e^{iL_u\frac{\Delta t_1}{2}}e^{iL_z\frac{\Delta t_1}{2}}e^{iL_y\frac{\Delta t_1}{2}}e^{iL_x\frac{\Delta t_1}{2}}(e^{iG_0\Delta t_0})^n \times$$

$$\times \quad e^{iL_x \frac{\Delta t_1}{2}} e^{iL_y \frac{\Delta t_1}{2}} e^{iL_z \frac{\Delta t_1}{2}} e^{iL_u \frac{\Delta t_1}{2}} e^{iL_s \frac{\Delta t_1}{2}} \tag{120}$$

$$e^{iL_{\text{atom}}\Delta t_1} = e^{iL_s \frac{\Delta t_1}{2}} e^{iL_y \frac{\Delta t_1}{2}} e^{iL_x \frac{\Delta t_1}{2}} \times$$

$$\times \quad (e^{iL_u \Delta t_0/2} e^{iL_z \Delta t_0/2} e^{iG_0 \Delta t_0} e^{iL_z \Delta t_0/2} e^{iL_u \Delta t_0/2})^n \times$$

$$\times \quad e^{iL_x \frac{\Delta t_1}{2}} e^{iL_y \frac{\Delta t_1}{2}} e^{iL_s \frac{\Delta t_1}{2}} \tag{121}$$

The propagator $e^{iG_0 \Delta t_0}$, defined in Eq. (106), is further split according to the usual velocity Verlet breakup of Eq. (22). Note that in case of molecular scaling the "slow" coordinates ($\mathbf{S}$, $\mathbf{h}$, η) move with constant velocity during the n small times steps since there is no "fast" force acting on them in the inner integration. The explicit integration algorithm may be easily derived for the two propagators in Eqs. (120) and (121) using the rule in Eq. (23) and its generalization:

$$e^{ay\nabla_y} f(y) = f(ye^a)$$
$$e^{\mathbf{a}\mathbf{y}\nabla_{\mathbf{y}}} f(\mathbf{y}) = f(e^{\mathbf{a}}\mathbf{y}) \tag{122}$$

Where a and $\mathbf{a}$ are a scalar and a matrix, respectively. The exponential matrix $e^{\mathbf{a}}$ on the right hand side of Eq. (122) is obtained by diagonalization of $\mathbf{a}$.

As stated before the dynamics generated by Eqs. (67-71) or (79-82) in the NPT ensemble is not Hamiltonian and hence we cannot speak of symplectic integrators [71] for the t-flow's defined by Eqs. (120, 121). The symplectic condition Eq. (8) is violated at the level of the transformation (58-66) which is not canonical. However, the algorithms generated by Eqs. (120,121) are time reversible and second order like the velocity Verlet. Several recent studies have shown [23–25] that these integrators for the non microcanonical ensembles are also stable for long time trajectories, as in case of the symplectic integrators for the NVE ensemble.

3.5. GROUP SCALING AND MOLECULAR SCALING

We have seen in section 3.3 that the center of mass or *molecular* pressure is equivalent to the atomic pressure. The atomic pressure is the natural quantity that enters in the virial theorem [10] irrespectively of the form of the interaction potential among the particles. So, in principle it is safer to adopt atomic scaling in the extended system constant pressure simulation. For systems in confined regions, the equivalence between atomic or true pressure and molecular pressure (see sec. 3.3) holds for any definition of the *molecular* subsystem irrespectively of the interaction potentials. In other words we could have defined virtual molecules made up of atoms selected on different *real* molecules. We may expect that, as long as the system, no matter how its unities or *particles* are defined, contains a sufficiently large number of particles, generates a distribution function identical to that generated by using the "correct" atomic scaling.

From a computational standpoint molecular scaling is superior to atomic scaling. The fastly varying Liouvillean in Eq. (114) for the atomic scaling contains the two terms iL_z, iL_u. These terms are slowly varying when molecular scaling is adopted and are assigned to the slow part of the Liouvillean in Eq. (113). The inner part of the time propagation is therefore expected to be more expensive for the multiple time step integration with atomic scaling rather than with molecular scaling. Generally speaking, given the equivalence between the molecular and atomic pressure, molecular scaling should be the preferred choice for maximum efficiency in the multiple time step integration. [16] For large size molecules, such as proteins, molecular scaling might be inappropriate. The size of the molecule clearly restricts the number of *particles* in the MD simulation box, thereby reducing the statistics on the instantaneous calculated molecular pressure which may show unphysically large fluctuations. Group scaling [25] is particularly convenient for handling the simulation of macromolecules. A statistically significant number of groups can be selected in order to avoid all problems related to the poor statistics on molecular pressure calculation for samples containing a small number of large size particles. Notwithstanding, for *solvated* biomolecules and provided that enough solvent molecules are included, molecular scaling again yields reliable results. In Ref. [25] Marchi and Procacci showed that the scaling method in the NPT ensemble does not affect neither the equilibrium structural and dynamical properties nor the kinetic of non equilibrium MD. For group-based and molecular-based scaling methods in a system of one single molecule of BPTI embedded in a box of about a 1000 water molecules, they obtained identical results for the system volume, the Voronoi volumes of the proteins and for the mean square displacement of both solvent and protein atoms under normal and high pressure.

3.6. SWITCHING TO OTHER ENSEMBLES

The NPT extended system is the most general among all possible extended Lagrangians. All other ensemble can be in fact obtained within the

---

[16]There are also other less material reasons to prefer molecular scaling: atomic scaling and molecular scaling yield different dynamical properties because the equations of motions are different. Dynamical data computed via extended system simulations should always be taken with caution. With respect to pure Newtonian dynamics, however, the NPT dynamical evolution is slightly modified by a barostat coupled to the molecular center of mass [24] but is brutally damaged when the barostat is coupled to the fast degrees of freedom. For example in liquid flexible nitrogen at normal pressure and 100 K, atomic scaling changes the internal frequency by 20 cm^{-1} while no changes are detected when the barostat is coupled to the centers of mass.

same computational framework. We must stress [25] that the computational overhead of the extended system formulation, due to the introduction and handling of the extra degrees of freedom of the barostat and thermostat variables, is rather modest and is negligible with respect to a NVE simulation for large samples ($N_f > 2000$) [23–25]. Therefore, a practical, albeit inelegant way of switching among ensembles is simply to set the inertia of the barostat and/or thermostat to a very large number. This must be of course equivalent to decouple the barostat and/or the thermostat from the true degrees of freedom. In fact, by setting W to infinity[17] in Eqs. (67-71) we recover the NVT canonical ensemble equations of motion. Putting instead Q to infinity the NPH equations of motion are obtained. Finally, setting both W and Q to infinity the NVE equations of motion are recovered. Switching to the NPT isotropic stress ensemble is less obvious. One may define the kinetic term associated to barostat in the extended Lagrangian as

$$K = \frac{1}{2} \sum_{\alpha\beta} W_{\alpha\beta} s^2 \dot{h}_{\alpha\beta}^2 \tag{123}$$

such that a different inertia may in principle be assigned to each of 9 extra degrees of freedom of the barostat. Setting for example

$$W_{\alpha\beta} = W \quad \text{for } \alpha \le \beta \tag{124}$$
$$W_{\alpha\beta} = \infty \quad \text{for } \alpha > \beta \tag{125}$$
$$\tag{126}$$

one inhibits cell rotations [25].

This trick does not work, unfortunately, to change to isotropic stress tensor. In this case, there is only one independent barostat degrees of freedom, namely the volume of the system. In order to simulate isotropic cell fluctuations a set of five constraints on the **h** matrix are introduced which correspond to the conditions:

$$\frac{h_{\alpha\beta}}{h_{11}} - \frac{h_{\alpha\beta}^0}{h_{11}^0} = 0$$
$$\dot{h}_{\alpha\beta} - \frac{h_{\alpha\beta}^0}{h_{11}^0} \dot{h}_{11} = 0 \qquad \text{for } \alpha \le \beta \tag{127}$$

with $\mathbf{h}^0$ being some reference **h** matrix. These constraints are implemented naturally in the framework of the multi time step velocity Verlet using

---

[17]The value of W which works as "infinity" depends on the "force" that is acting on barostat coordinate expressed by the Eq. (69), i.e. on how far the system is from the thermodynamic equilibrium. For a system near the thermodynamic equilibrium with $N_f \simeq 10000$ a value of $W = 10^{20}$ a.m.u. is sufficient to prevent cell fluctuations.

368

the RATTLE algorithm which evaluates iteratively the constraints force to satisfy the constraints on both coordinates $\mathbf{h}$ and velocities $\dot{\mathbf{h}}$ [25]. In Ref. [25] it is proved that the phase space sampled by the NPT equations with the addition of the constraints Eq. (127) correspond to that given by NPT distribution function.

4. Multiple time steps algorithms for large size flexible systems with strong electrostatic interactions

In the previous sections we have described how to obtain multiple time step integrators *given* a certain potential subdivision and have provided simple examples of potential subdivision based on the inter/intra molecular separation. Here, we focus on the time scale separation of model potentials of complex molecular systems. Additionally, we provide a general potential subdivision applying to biological systems, as well as to many other interesting chemical systems including liquid crystals. This type of systems are typically characterized by high flexibility and strong Coulombic intermolecular interactions. Schematically, we can then write the potential V as due to two contributions:

$$V \;=\; V_{\mathrm{bnd}} + V_{\mathrm{nbn}}. \tag{128}$$

Here, the "bonded" or intramolecular part V_{bnd} is fast and is responsible for the flexibility of the system. The "non bonded" or intermolecular (or intergroup) term V_{nbn} is dominated by Coulombic interactions. The aim of the following sections is to describe a general protocol for the subdivision of such forms of the interaction potential and to show how to obtain reasonably efficient and transferable multiple time step integrators valid for any complex molecular system.

4.1. SUBDIVISION OF THE "BONDED" POTENTIAL

As we have seen in Sec. 2.3 the idea behind the multiple time step scheme is that of the reference system which propagates for a certain amount of time under the influence of some *unperturbed* reference Hamiltonian, and then undergoes an impulsive *correction* brought by the remainder of the potential. The exact trajectory spanned by the complete Hamiltonian is recovered by applying this impulsive correction onto the "reference" trajectory. We have also seen in the same section that, by subdividing the interaction potential, we can determine as many "nested" reference systems as we wish.

The first step in defining a general protocol for the subdivision of the bonded potential for complex molecular systems consists in identifying the various time scales and their connection to the potential. The interaction bonded potential in almost all popular force field is given as a function of the stretching, bending and torsion internal coordinates and has the general form

$$V_{\text{bnd}} = V_{\text{stretch}} + V_{\text{bend}} + V_{\text{tors}}, \tag{129}$$

where

$$V_{\text{stretch}} = \sum_{\text{Bonds}} K_r \left(r - r_0\right)^2$$

$$V_{\text{bend}} = \sum_{\text{Angles}} K_\theta \left(\theta - \theta_0\right)^2.$$

$$V_{\text{tors}} = \sum_{\text{Dihedrals}} V_\phi \left[1 + \cos\left(n\phi - \gamma\right)\right]. \tag{130}$$

Here, K_r and K_θ are the bonded force constants associated with bond stretching and angles bending respectively, while r_0 and θ_0 are their respective equilibrium values. In the torsional potential, V_{tors}, ϕ is the dihedral angle, while K_ϕ, n and γ are constants.

The characteristic time scale of a particular internal degrees of freedom can be estimated assuming that this coordinate behaves like a harmonic oscillator, uncoupled form the rest the other internal degrees of freedom. Thus, the criterion for guiding the subdivision of the potential in Eq. (129) is given by the characteristic frequency of this uncoupled oscillator. We now give, for each type of degree of freedom, practical formula to evaluate the harmonic frequency from the force field constants given in Eq. (130).

Stretching: The stretching frequencies are given by the well known expression

$$\nu_s = \frac{1}{2\pi} \left(\frac{K_r}{\mu}\right)^{(1/2)}, \tag{131}$$

where μ is reduced mass.

Bending: We shall assume for the sake of simplicity that the *uncoupled* bending frequencies depends on the masses of the atom 1 and 3 (see Fig. 3), that is mass 2 is assumed to be infinity. This turns out to be in general an excellent approximation for bending involving hydrogens and a good approximation for external bendings in large molecules involving masses of comparable magnitude.

The frequency is obtained by writing the Lagrangian in polar coordinates for the mechanical system depicted in Fig. 3. The Cartesian coordinates are expressed in terms of the polar coordinates as

$$x_1 = r_{12}\sin(\alpha/2) \quad y_1 = r_{12}\cos(\alpha/2) \tag{132}$$

$$x_3 = r_{32}\sin(\alpha/2) \quad y_3 = r_{32}\cos(\alpha/2) \tag{133}$$

370

where the distance r_{32} and r_{12} are constrained to the equilibrium values. The velocities are then

$$\dot{x}_1 \;=\; -r_{12}\cos(\alpha/2)\frac{\dot{\alpha}}{2} \qquad \dot{y}_1 = r_{12}\sin(\alpha/2)\frac{\dot{\alpha}}{2} \tag{134}$$

$$\dot{x}_3 \;=\; -r_{32}\cos(\alpha/2)\frac{\dot{\alpha}}{2} \qquad \dot{y}_3 = r_{32}\sin(\alpha/2)\frac{\dot{\alpha}}{2} \tag{135}$$

The Lagrangian for the uncoupled bending is then

$$\mathcal{L} \;=\; \frac{1}{2}\left(m_1\dot{x}_1^2 + m_1\dot{y}_1^2 + m_3\dot{x}_3^2 + m_3\dot{y}_3^2\right) + V_{bend} \tag{136}$$

$$\;=\; \frac{1}{8}\left(m_1 r_{12}^2 + m_2 r_{32}^2)\right)\dot{\alpha}^2 + \frac{1}{2}k_\theta(\alpha - \alpha_0)^2. \tag{137}$$

The equation of motion $\frac{d}{dt}\frac{\partial \mathcal{L}}{\dot{\alpha}} - \frac{\partial \mathcal{L}}{\alpha} = 0$ for the α coordinate is given by

$$\ddot{\alpha} + \frac{4K_\theta}{I_b}(\alpha - \alpha_0)^2 = 0. \tag{138}$$

Where, $I_b = m_1 r_{12}^2 + m_3 r_{32}^2$ is the moment of inertia about an axis passing by atom 3 and perpendicular to the bending plane. Finally, the *uncoupled* bending frequency is given by

$$\nu_b = \frac{1}{2\pi}\left(\frac{4K_\theta}{m_1 r_{12}^2 + m_3 r_{32}^2}\right)^{(1/2)} \tag{139}$$

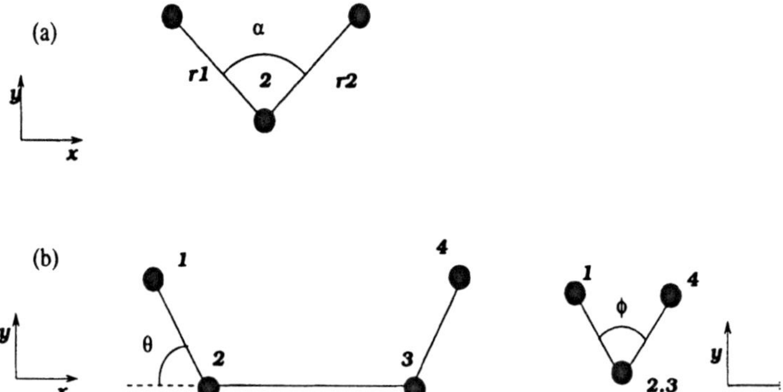

Figure 3. Bending and dihedral angles

Torsion: We limit our analysis to a purely torsional system (see Fig. 3b) where atoms 2 and 3 are held fixed, and all bond distances and the angle θ are constrained to their equilibrium values. The system has only one degree of freedom, the dihedral angle ϕ driven by the torsional potential V_ϕ. Again we rewrite the kinetic energy in terms of the bond distances, the dihedral angle and the constant bend angle θ. For the kinetic energy, the only relevant coordinates are now those of atoms 1 and 4:

$$
\begin{aligned}
x_1 &= d_{12}\cos\theta + \frac{d_{23}}{2} & x_4 &= d_{34}\cos\theta + \frac{d_{23}}{2} \\
y_1 &= d_{12}\sin\theta\cos(\phi/2) & y_4 &= d_{34}\sin\theta\cos(\phi/2) \\
z_1 &= d_{12}\sin\theta\sin(\phi/2) & z_4 &= d_{34}\sin\theta\sin(\phi/2).
\end{aligned}
\tag{140}
$$

The Lagrangian in terms of the dihedral angle coordinate is then

$$
\mathcal{L} = \frac{1}{8}I_t\dot\phi^2 - V_\phi\left[1 + \cos\left(n\phi - \gamma\right)\right],
\tag{141}
$$

where

$$
I_t = sin^2\theta\left(m_1 d_{12}^2 + m_4 d_{34}^2\right).
\tag{142}
$$

Assuming small oscillations, the potential may be approximated by a second order expansion around the corresponding equilibrium dihedral angle ϕ_0

$$
V_{tors} = \frac{1}{2}\left(\frac{\partial^2 V_{tors}}{\partial\phi^2}\right)_{\phi=\phi_0}(\phi - \phi_0)^2 = \frac{1}{2}V_\phi n^2(\phi - \phi_0)^2
\tag{143}
$$

Substituting (143) into Eq. (141) and then writing the Lagrange equation of motion for the coordinate ϕ, one obtains again a differential equation of a harmonic oscillator, namely

$$
\ddot\phi + \frac{4V_\phi n^2}{I_t}(\phi - \phi_0) = 0.
\tag{144}
$$

Thus, the *uncoupled* torsional frequency is given by

$$
\nu_t = \frac{n}{2\pi}\left(4\frac{V_\phi}{sin^2\theta\left(m_1 d_{12}^2 + m_4 d_{34}^2\right)}\right)^{(1/2)}.
\tag{145}
$$

For many all-atom force fields, improper torsions [1, 4] are modeled using a potential identical to that of the proper torsion in Eq. (130) and hence in these cases Eq. (145) applies also to the improper torsion uncoupled frequency, provided that indices 1 and 4 refer to the *lighter* atoms. In figure (4) we report the distribution of frequencies for the hydrated protein Bovine Pancreatin Trypsin Inhibitor (BPTI) using the AMBER [2] force field. The distributions might be thought as a density of the uncoupled intramolecular states of the system. As we can see in the figure there is a relevant degree of overlap for the various internal degrees of freedom. For example, "slow" degrees of freedom such as torsions may be found up to 600 wavenumber, well inside the "bending" region; these are usually improper or proper torsions involving hydrogens. It is then inappropriate to assign such "fast" torsions involving hydrogens to a slow reference system. We recall that in a multiple time simulation the integration of a supposedly slow degree of freedom with a excessively large time step is enough to undermine the entire simulation. In Fig. 4 we also notice that almost all the proper torsions fall below 350

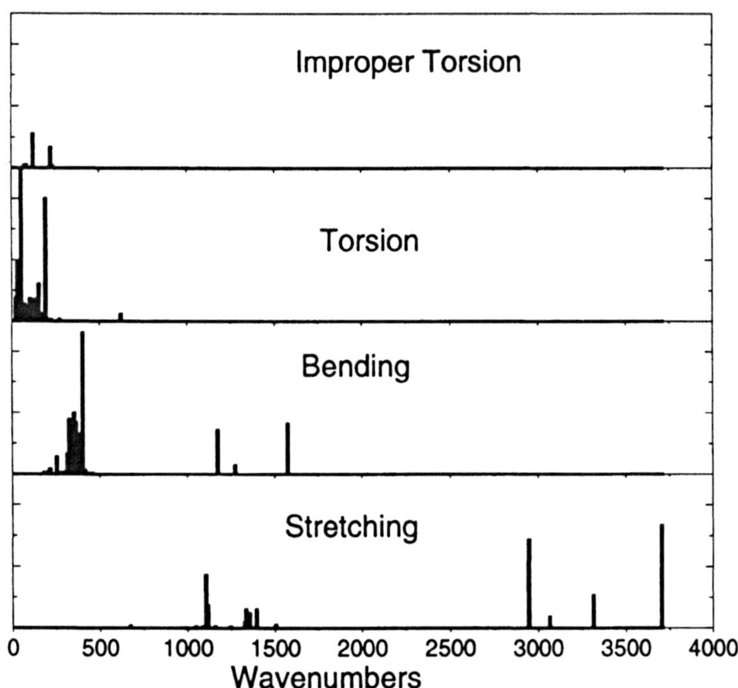

Figure 4. density of the uncoupled (see text) states for stretching, bending, proper and improper torsion obtained with the AMBER force field on the protein bovine pancreatic trypsin inhibitor (BPTI). Frequencies were calculated according to Eqs. (131,139,145)

cm^{-1}. An efficient and simple separation of the intramolecular AMBER potential [25] assigns all bendings stretching and the improper or proper torsions involving hydrogens to a "fast" reference system labeled $n0$ and all proper torsions to a slower reference system labeled $n1$. The subdivision is then

$$V_{n0} = V_{\text{stretch}} + V_{\text{bend}} + V_{\text{i-tors}} + V_{\text{p-tors}}^{(h)}$$

$$V_{\mathrm{n1}} \;=\; V_{\mathrm{p-tors}} \tag{146}$$

Where with $V_{\mathrm{p-tors}}^{(h)}$ we indicate proper torsions involving hydrogens. For the reference system V_{n0}, the hydrogen stretching frequencies are the fastest motions and the Δt_{n0} time step must be set to 0.2–0.3 fs. The computational burden of this part of the potential is very limited, since it involves mostly two or three body forces. For the reference system V_{n1}, the fastest motion is around 300 cm^{-1} and the time step Δt_{n1} should be set to 1-1.5 fs. The computational effort for the reference system potential V_{n1} is more important because of the numerous proper torsions of complex molecular systems which involve more expensive four body forces calculations. One may also notice that some of the bendings which were assigned to the n0 reference system fall in the torsion frequency region and could be therefore integrated with a time step much larger than $\Delta t_{\mathrm{n0}} \simeq 0.2\text{-}0.3$. However, in a multiple time step integration, this overlap is just inefficient, but certainly not dangerous. Indeed, no instability may derive for integrating slow degrees of freedom with exceedingly small time steps.

4.2. THE SMOOTH PARTICLE MESH EWALD METHOD

Before we discuss the non bonded multiple time step separation it is useful to describe in some details one of the most advanced techniques to handle long range forces. Indeed, this type of non bonded forces are the most cumbersome to handle and deserve closer scrutiny.

In the recent literature, a variety of techniques are available to handle the problem of long range interactions in computer simulations of charged particles at different level of approximation [10, 28, 29]. In this section, we shall focus on the Ewald summation method for the treatment of long range interactions in periodic systems [30, 31, 72]. The Ewald method gives the *exact* result for the electrostatic energy of a periodic system consisting of an infinitely replicated neutral box of charged particles. The method is the natural choice in MD simulations of complex molecular system with PBC. The Ewald potential [31] is given by

$$V'_{\mathrm{qd}} = \frac{1}{2} \sum_{ij}^{N} q_i q_j \sum_{n} \frac{1}{|\mathbf{r}_{ij} + \mathbf{r}_n|} \mathrm{erfc}\left(\alpha \left|\mathbf{r}_{ij} + \mathbf{r}_n\right|\right) \tag{147}$$

$$V_{\mathrm{qr}} = \left[\frac{1}{2\pi V} \sum_{\mathbf{m}\neq 0}^{\infty} \frac{\exp\left(-\pi^2 |\mathbf{m}|^2 / \alpha^2\right)}{|\mathbf{m}|^2} S\left(\mathbf{m}\right) S\left(-\mathbf{m}\right) - \frac{\alpha}{\pi^{1/2}} \sum_{i} q_i^2 \right] - V_{intra}. \tag{148}$$

374

with

$$S(\mathbf{m}) = \sum_i^N q_i \exp(2\pi i m \cdot \mathbf{r}_i) \qquad (149)$$

$$V_{\text{intra}} = \sum_{ij-\text{excl.}} q_i q_j \frac{\text{erf}(\alpha r_{ij})}{r_{ij}}, \qquad (150)$$

where, $\mathbf{r}_i$ is the vector position of the atomic charge q_i, $r_{ij} = \mathbf{r}_i - \mathbf{r}_j$, $\mathbf{r}_n$ is a vector of the direct lattice, $\text{erfc}(x) = \pi^{-1/2} \int_x^\infty e^{-t^2} dt$ is the complementary error function, $\text{erf}(x) = 1 - erfc(x)$, V the unit cell volume, $\mathbf{m}$ a reciprocal lattice vector and α is the Ewald convergence parameter. In the direct lattice part, Eq. (147), the prime indicates that intramolecular excluded contacts[18] are omitted. In addition, in Eq. (148) the term V_{intra} subtracts, in direct space, the intra–molecular energy between bonded pairs, which is automatically included in the right hand side of that equation. Consequently, the summation on i and j in Eq. (150) goes over all the excluded intra-molecular contacts. We must point out that in the Ewald potential given above, we have implicitly assumed the so-called "tin-foil" boundary conditions: the Ewald sphere is immersed in a perfectly conducting medium and hence the dipole term on the surface of the Ewald sphere is zero [31]. For increasingly large systems the computational cost of standard Ewald summation, which scales with N^2, becomes too large for practical applications. Alternative algorithms which scale with a smaller power of N than standard Ewald have been proposed in the past. Among the fastest algorithms designed for periodic systems is the particle mesh Ewald algorithm (PME)[32, 33], inspired by the particle mesh method of Hockney and Eastwood [34]. Here, a multidimensional piecewise interpolation approach is used to compute the reciprocal lattice energy, V_{qr}, of Eq. 148, while the direct part, V_{qd}, is computed straightforwardly. The low computational cost of the PME method allows the choice of large values of the Ewald convergence parameter α, as compared to those used in conventional Ewald. Correspondingly, shorter cutoffs in the direct space Ewald sum V_{qd} may be adopted. If $\mathbf{u}_j$ is the scaled fractional coordinate of the i-th particle, the charge weighted structure factor, $S(\mathbf{m})$ in Eq. (149), can be rewritten as:

$$S(\mathbf{m}) = \sum_{j=1}^N q_j \exp\left[2\pi i\left(\frac{m_1 u_{j1}}{K_1} + \frac{m_2 u_{2j}}{K_2} + \frac{m_3 u_{3j}}{K_3}\right)\right]. \qquad (151)$$

Where, N is the number of particles, K_1, K_2, K_3 and m_1, m_2, m_3 are integers. The α component of the scaled fractional coordinate for the i-th atom

---

[18]By excluded contacts we mean interactions between charges on atoms connected by bonds or two bonds apart.

can be written as:[19]

$$u_{i\alpha} = K_\alpha \mathbf{k}_\alpha \cdot \mathbf{r_i}, \qquad (152)$$

where $\mathbf{k}_\alpha$, $\alpha = 1, 2, 3$ are the reciprocal lattice basic vectors.

$S(\mathbf{m})$ in Eq. (151) can be looked at as a discrete Fourier transform (FT) of a set of charges placed irregularly within the unit cell. Techniques have been devised in the past to approximate $S(\mathbf{m})$ with expressions involving Fourier transforms on a regular grid of points. Such approximations of the weighted structure factor are computationally advantageous because they can be evaluated by fast Fourier transforms (FFT). All these FFT–based approaches involve, in some sense, a smearing of the charges over nearby grid points to produce a regularly gridded charge distribution. The PME method accomplishes this task by interpolation. Thus, the complex exponentials $\exp(2\pi i m_\alpha u_{i\alpha}/K_\alpha)$, computed at the position of the i-th charge in Eq. (151), are rewritten as a sum of interpolation coefficients multiplied by their values at the nearby grid points. In the smooth version of PME (SPME) [33], which uses cardinal B-splines in place of the Lagrangian coefficients adopted by PME, the sum is further multiplied by an appropriate factor, namely:

$$\exp\left(2\pi i m_\alpha u_{i\alpha}/K_\alpha\right) = b(m_\alpha) \sum_k M_n(u_{i\alpha} - k) \exp\left(2\pi i m_\alpha k/K_\alpha\right), \quad (153)$$

where n is the order of the spline interpolation, $M_n(u_{i\alpha} - k)$ defines the coefficients of the cardinal B-spline interpolation at the scaled coordinate $u_{i\alpha}$. In Eq. (153) the sum over k, representing the grid points, is only over a finite range of integers, since the functions $M_n(u)$ are zero outside the interval $0 \le u \le n$. It must be stressed that the complex coefficients $b(m_i)$ are independent of the charge coordinates $\mathbf{u_i}$ and need be computed only at the very beginning of a simulation. A detailed derivation of the $M_n(u)$ functions and of the b_α coefficients is given in Ref. [33]. By inserting Eq. (153) into Eq. (151), $S(\mathbf{m})$ can be rewritten as:

$$S(\mathbf{m}) = b_1(m_1)b_2(m_2)b_3(m_3)\mathcal{F}[Q](m_1, m_2, m_3), \qquad (154)$$

where $\mathcal{F}[Q](m_1, m_2, m_3)$ stands for the discrete FT at the grid point m_1, m_2, m_3 of the array $Q(k_1, k_2, k_3)$ with $1 \le k_i \le K_i$, $i = 1, 2, 3$. The gridded charge array, $Q(k_1, k_2, k_3)$, is defined as:

$$Q(k_1, k_2, k_3) = \sum_{i=1,N} q_i M_n(u_{i1} - k_1) M_n(u_{i2} - k_2) M_n(u_{i3} - k_3) \qquad (155)$$

---

[19]The scaled fractional coordinate is related to the scaled coordinates in Eqs (45,78) by the relation $s_{i\alpha} = 2u_{i\alpha}/K_\alpha$.

376

Inserting the approximated structure factor of Eq. (154) into Eq. (148) and using the fact that $\mathcal{F}[Q](-m_1,-m_2,-m_3) = K_1 K_2 K_3 \mathcal{F}^{-1}[Q](m_1,m_2,m_3)$, the SPME reciprocal lattice energy can be then written as

$$
\begin{aligned}
V_{\mathrm{qr}} &= \frac{1}{2} \sum_{m_1=1}^{K_1} \sum_{m_2=1}^{K_2} \sum_{m_3=1}^{K_3} B(m_1,m_2,m_3) C(m_1,m_2,m_3) \times \\
&\quad \times \ \mathcal{F}[Q](m_1,m_2,m_3) \mathcal{F}[Q](-m_1,-m_2,-m_3) \qquad (156) \\
&= \frac{1}{2} \sum_{m_1=1}^{K_1} \sum_{m_2=1}^{K_2} \sum_{m_3=1}^{K_3} \mathcal{F}^{-1}[\Theta_{rec}](m_1,m_2,m_3) \mathcal{F}[Q](m_1,m_2,m_3) \times \\
&\quad \times \ K_1 K_2 K_3 \mathcal{F}^{-1}[Q](m_1,m_2,m_3), \qquad (157)
\end{aligned}
$$

with

$$
\begin{aligned}
B(m_1,m_2,m_3) &= |b_1(m_1)|^2 |b_2(m_2)|^2 |b_3(m_3)|^2 & (158) \\
C(m_1,m_2,m_3) &= (1/\pi V) \exp(-\pi^2 \mathbf{m}^2/\alpha^2)/\mathbf{m}^2 & (159) \\
\Theta_{rec} &= \mathcal{F}[BC]. & (160)
\end{aligned}
$$

Using the convolution theorem for FFT the energy (157) can be rewritten as

$$
V_{\mathrm{qr}} = \frac{1}{2} \sum_{m_1=1}^{K_1} \sum_{m_2=1}^{K_2} \sum_{m_3=1}^{K_3} \mathcal{F}^{-1}[\Theta_{\mathrm{rec}} \star Q](m_1,m_2,m_3) \mathcal{F}[Q](m_1,m_2,m_3)
$$
$$(161)$$

We now use the identity $\sum_{\mathbf{m}} F(A)(\mathbf{m}) B(\mathbf{m}) = \sum_{\mathbf{m}} A(\mathbf{m}) F(B)(\mathbf{m})$ to arrive at

$$
V_{\mathrm{qr}} = \frac{1}{2} \sum_{m_1=1}^{K_1} \sum_{m_2=1}^{K_2} \sum_{m_3=1}^{K_3} (\Theta_{rec} \star Q)(m_1,m_2,m_3) Q(m_1,m_2,m_3)
$$
$$(162)$$

We first notice that Θ_{rec} does not depend on the charge positions and that $M_n(u_{i\alpha} - k)$ is differentiable for $n > 2$ (which is always the case in practical applications). Thus the force on each charge can be obtained by taking the derivative of Eq. (162), namely

$$
F_{i\alpha}^{(\mathrm{qr})} = -\frac{\partial V_{\mathrm{qr}}}{\partial r_{i\alpha}} = \sum_{m_1=1}^{K_1} \sum_{m_2=1}^{K_2} \sum_{m_3=1}^{K_3} \frac{\partial Q(m_1,m_2,m_3)}{\partial r_{i\alpha}} (\Theta_{rec} \star Q)(m_1,m_2,m_3).
$$
$$(163)$$

In practice, the calculation is carried out according to the following scheme: i) At each simulation step one computes the grid scaled fractional coordinates $u_{i\alpha}$ and fills an array with Q according to Eq. (155). At this stage, the derivative of the M_n functions are also computed and stored in memory. ii) The array containing Q is then overwritten by $\mathcal{F}[Q]$, i.e. Q's 3-D Fourier transform. iii) Subsequently, the electrostatic energy is computed via Eq. (157). At the same time, the array containing $\mathcal{F}[Q]$ is overwritten by the product of itself with the array containing BC (computed at the very beginning of the run). iv) The resulting array is then Fourier transformed to obtain the convolution $\Theta_{rec} \star Q$. v) Finally, the forces are computed via Eq. (163) using the previously stored derivatives of the M_n functions to recast $\partial Q / \partial r_{i\alpha}$.

The memory requirements of the SPME method are limited. $2K_1K_2K_3$ double precision reals are needed for the grid charge array Q, while the calculation of the functions $M_n(u_{i\alpha} - j)$ and their derivatives requires only $6 \times n \times N$ double precision real numbers. The K_α integers determines the fineness of the grid along the α-th lattice vector of the unit cell. The output accuracy of the energy and forces depends on the SPME parameters: The α convergence parameter, the grid spacing and the order n of the B-spline interpolation. For a typical $\alpha \simeq 0.4$ Å $^{-1}$ a relative accuracies between $10^{-4} - 10^{-5}$ for the electrostatic energy are obtained when the grid spacing is around 1 Å along each axis, and the order n of the B-spline interpolation is 4 or 5. A rigorous error analysis and a comparison with standard Ewald summation can be found in Refs. [33] and [73]. For further readings on the PME and SPME techniques we refer to the original papers [27, 32, 33, 73].

The power of the SPME algorithm, compared to the straightforward implementation of the standard Ewald method, is indeed astonishing. In Fig. (5) we report CPU timing obtained on a low end 43P/160MH IBM workstation for the evaluation of the reciprocal lattice energy and forces via SPME for cyanobiphenil as a function of the number of atoms in the system. Public domain 3-D FFT routines were used. The algorithm is practically linear and for 12,000 particles SPME takes only 2 CPU seconds to perform the calculation. A standard Ewald simulation for a box $64 \times 64 \times 64$ Å^3 (i.e. with a grid spacing in k space of $k = 2\pi/64 \simeq 0.01$ Å^{-1}) for the same sample and at the same level of accuracy would have taken several minutes.

4.3. SUBDIVISION THE NON BONDED POTENTIAL

In addition to the long range electrostatic contributions, V_{qr} and V_{qd}, given in Eqs. (147,148), more short range forces play a significant role in the total non bonded potential energy. The latter can be written as:

$$V_{nbn} = V_{vdw} + V_{qr} + V_{qd} + V_{14}. \tag{164}$$

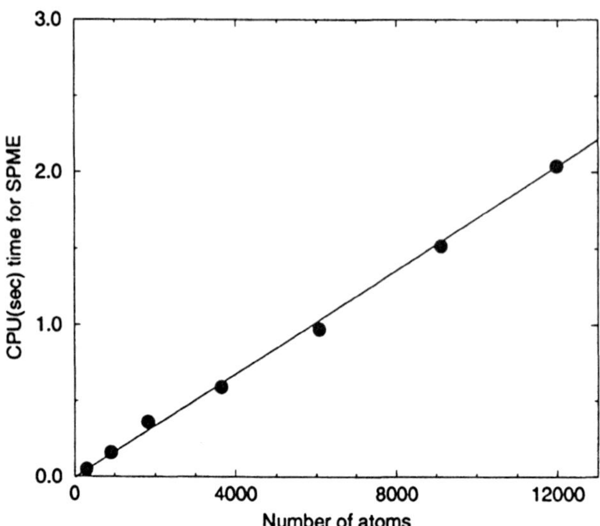

Figure 5.   CPU time versus number of particles for the SPME algorithm as measured on a 43P/160MH IBM workstation

Where, V_{vdw} is the Lennard-Jones potential, namely

$$V_{\text{vdw}} = \sum_{i<j}^{N}{}' 4\epsilon_{ij} \left[\left(\frac{\sigma_{ij}}{r_{ij}} \right)^{12} - \left(\frac{\sigma_{ij}}{r_{ij}} \right)^{6} \right]. \tag{165}$$

Here, the prime on the sum indicates that interactions between atoms separated by less than three consecutive bonds must be omitted. The term V_{14} is typical for force fields of complex molecular systems [1, 3]. While non bonded forces between atoms involved in the same covalent bond or angle bending interaction are generally excluded, the potential between atoms separated by three covalent bonds is retained and readjusted in various ways. In all cases, the V_{14} term remains in general a very stiff and, hence, a fastly varying term. The computational cost of the V_{14} contribution is very small compared to other non bonded interactions. Thus, it is safer to assigns this potential term to the slowest intramolecular reference system potential V_{n1} of Eq. (146).

The V_{qr} reciprocal lattice term, *including* the correction due to the excluded or partially excluded (i.e. the electrostatic part of V_{14}) interactions cannot be split when using SPME and must be assigned altogether to only one reference system. The time scale of the potential V_{qr} depends on the convergence parameter α. Indeed, this constant controls the relative weights of the reciprocal lattice energy V_{qr}, and of the direct lattice energy V_{qd}. By increasing α, one increases the weight of the reciprocal lattice contribution V_{qr} to the total Coulomb energy. When using SPME the cost of the reciprocal lattice sums is cut down dramatically and, therefore, the use of *large* α's becomes helpful to reduce the computational burden of the direct lattice calculation. For a value of α increased beyond a certain limit, there is no longer a computational gain, since the pair distances must always be evaluated in direct space until convergence of the Lennard-Jones energy (usually occurring at a 10 Å cutoff). Furthermore, the larger is α, the more short-ranged and fastly varying becomes the potential V_{qr}, thus requiring short time steps to integrate correctly the equations of motion. A good compromise for α, valid for cell of any shape and size, is $\alpha = 0.4\text{-}0.5$.

The direct space potential is separated [11, 25] in three contributions according to the interaction distance. The overall non bonded potential breakup is therefore

$$
\begin{aligned}
V_{nl} &= V_{14} \\
V_m &= V_{vdw}^{(1)} + V_{qd}^{(1)} \\
V_l &= V_{vdw}^{(2)} + V_{qd}^{(2)} + V_{qr} \\
V_h &= V_{vdw}^{(3)} + V_{qd}^{(3)}
\end{aligned}
\tag{166}
$$

where the superscripts m, l, h of the direct space term V_{vdw} and V_{qd} refer to the short, medium and long range non–bonded interactions, respectively. The m-th reference system includes non–bonded direct space interactions at short range, typically between 0 to 4.3-5.3 Å. V_l contains both the medium range direct space potential, with a typical range of 4.3-5.3 to 7.3-8.5 Å, and the reciprocal space term, V_{qr}. Finally, the h-th reference system, which is the most slowly varying contains, the remaining direct space interactions from 7.3-8.3 Å to cut-off distance. As the simulations proceeds the particles seen by a target particle may cross from one region to an other, while the number of two body contacts in one distance class [17] or reference system potential must be continuously updated. Instabilities caused by this flow across potential shell boundaries are generally handled by multiplying the pair potential by a group-based switching function [22][20]. Thus, at any distance r the direct space potential V can be written schematically as:

$$
V = V_1 + V_2 + V_3
\tag{167}
$$

with

$$
V_1 = V S_1
\tag{168}
$$

[20]Here, the word group has a different meaning that in Sec. 3 and stands for sub ensemble of contiguous atoms defined as having a total charge of approximately zero.

380

$$V_2 = V(S_2 - S_1) \tag{169}$$

$$V_3 = V(S_3 - S_2) \tag{170}$$

$$\tag{171}$$

where S_j is the switching function for the three shells, $j = m, l, h$ defined as:

$$S_j(R) = \left\{ \begin{array}{ll} 1 & R_{j-1} \leq R < R_j \\ S_{3p}^{(j)} & R_j \leq R < R_j + \lambda_j \\ 0 & R_j + \lambda_j < R. \end{array} \right\} \tag{172}$$

Here, R is the intergroup distance and λ_j is the healing interval for the

TABLE 2. Potential breakup and relative time steps for complex systems with interactions modeled computed using the SPME method

Component	Contributions	Spherical Shells	Time step
V_{n0}	$V_{\text{stretch}} + V_{\text{bend}} +$ $+V_{\text{i-tors}} + V_{\text{p-tors}}^{(h)}$	-	$\Delta t_{n0} = 0.33\ fs$
V_{n1}	$V_{\text{p-tors}} + V_{14}$	-	$\Delta t_{n1} = 1.0\ fs$
V_m	$V_{\text{LJ}}^{(1)} + V_{\text{qd},\alpha=0.43}^{(1)}$	$0 < r < 4.5\ \text{\AA}$	$\Delta t_m = 2.0\ fs$
V_l	$V_{\text{LJ}}^{(2)} + V_{\text{qd},\alpha=0.43}^{(2)} +$ $+V_{\text{qr},\alpha=0.43}$	$4.5 \leq r < 7.5\ \text{\AA}$	$\Delta t_l = 4.0\ fs$
V_h	$V_{\text{LJ}}^{(3)} + V_{\text{qd},\alpha=0.43}^{(3)}$	$7.5 \leq r < 10.0\ \text{\AA}$	$\Delta t_h = 12.0\ fs$

j-th shell. While R_0 is zero, $R_1 = R_m$, $R_2 = R_l$, and $R_3 = R_h$ are the short, medium, long range shell radius, respectively. The switching $S_{3p}^{(j)}(R)$ is 1 at R_j and goes monotonically to 0 at $R_j + \lambda_j$. Provided that $S_{3p}^{(j)}$ and its derivatives are continuous at R_j and $R_j + \lambda_j$, the analytical form of S_{3p} in the healing interval is arbitrary [11, 14, 18, 22]. The full breakup for an AMBER type force field along with the integration time steps, valid for any complex molecular system with strong electrostatic interactions, is summarize in table II. The corresponding five time steps integration algorithm for the NVE ensemble is given by

$$
\begin{aligned}
e^{iL\Delta t_h} \quad &= \exp\left[-\frac{\partial V_h}{\partial \mathbf{r}_i}\frac{\partial}{\partial \mathbf{p}_i}\frac{\Delta t_h}{2}\right] \left\{ \exp\left[-\frac{\partial V_l}{\partial \mathbf{r}_i}\frac{\partial}{\partial \mathbf{p}_i}\frac{\Delta t_l}{2}\right] \left\{ \exp\left[-\frac{\partial V_m}{\partial \mathbf{r}_i}\frac{\partial}{\partial \mathbf{p}_i}\frac{\Delta t_m}{2}\right] \right. \right. \\
&\left\{ \exp\left[-\frac{\partial V_{n1}}{\partial \mathbf{r}_i}\frac{\partial}{\partial \mathbf{p}_i}\frac{\Delta t_{n1}}{2}\right] \left\{ \exp\left[-\frac{\partial V_{n0}}{\partial \mathbf{r}_i}\frac{\partial}{\partial \mathbf{p}_i}\frac{\Delta t_{n0}}{2}\right] \exp\left[\dot{\mathbf{r}}_i\frac{\partial}{\partial \mathbf{r}_i}\Delta t_0\right] \right. \right. \\
&\left. \exp\left[-\frac{\partial V_{n0}}{\partial \mathbf{r}_i}\frac{\partial}{\partial \mathbf{p}_i}\frac{\Delta t_{n0}}{2}\right] \right\}^{n_{n0}} \exp\left[-\frac{\partial V_{n1}}{\partial \mathbf{r}_i}\frac{\partial}{\partial \mathbf{p}_i}\frac{\Delta t_{n1}}{2}\right] \right\}^{n_{n1}} \\
&\left. \exp\left[-\frac{\partial V_m}{\partial \mathbf{r}_i}\frac{\partial}{\partial \mathbf{p}_i}\frac{\Delta t_m}{2}\right] \right\}^{n_m} \exp\left[-\frac{\partial V_l}{\partial \mathbf{r}_i}\frac{\partial}{\partial \mathbf{p}_i}\frac{\Delta t_l}{2}\right] \right\}^{n_l} \exp\left[-\frac{\partial V_h}{\partial \mathbf{r}_i}\frac{\partial}{\partial \mathbf{p}_i}\frac{\Delta t_h}{2}\right],
\end{aligned}
\tag{173}
$$

where $n_l = \Delta t_h/\Delta t_l$, $n_m = \Delta t_l/\Delta t_m$, $n_{n1} = \Delta t_m/\Delta t_{n1}$, $n_{n0} = \Delta t_{n1}/\Delta t_{n0}$. The explicit integration algorithm can be easily derived applying the five-fold discrete time propagator (173) to the state vector $\{\mathbf{p}, \mathbf{q}\}$ at time 0 using the rule Eq. (23). The efficiency and accuracy for energy conservation of this r-RESPA symplectic and reversible integrator have been discussed extensively in Refs. [11, 74]. Extension of this subdivision to non NVE simulation is described in Ref. [25].

4.4. ELECTROSTATIC CORRECTIONS FOR THE MULTIPLE TIME STEP SIMULATION

In *flexible* molecular systems of large size, the Ewald summation presents computational problems which are crucial to constructing efficient and *stable* multiple time step integrators [74, 75]. We have seen that intra-molecular Coulomb interactions between bonded atoms or between atoms bonded to a common atom are excluded in most of the standard force fields for protein simulation. In any practical implementation of the Ewald method, the intra-molecular energy V_{intra} is automatically included in the reciprocal space summation and is subtracted in *direct space* (see Eqs. (150,148). In actual simulations the reciprocal space sum is computed with a finite accuracy whereas the intra-molecular term V_{intra}, due to the excluded Coulombic interactions, is computed exactly. This clearly prevents the cancellation of the intra-molecular forces and energies. When the stretching and bending forces are integrated explicitly, the intra-molecular term due to the excluded contacts varies rapidly with time and so does the cancellation error. Consequently, instability may be observed when integrating the reciprocal lattice forces in reference systems with large time steps. The correction due to the truncation can be evaluated by approximating the reciprocal lattice sum for the excluded contacts in Eq. (148) to an integral in the 3-dimensional k space and evaluating this integral from the cut-off $k_{cut} \equiv 2\pi|\mathbf{m}|_{max}$ to infinity in polar coordinates. The neglected reciprocal lattice intra-molecular energy is then [76]

$$V_{\text{corr}} = \frac{\alpha}{\pi^{1/2}}\text{erfc}(k_{cut}/2\alpha)\sum_i q_i^2 + \sum_{ij-\text{excl.}} q_i q_j \chi(r_{ij}, k_{\text{cut}}, \alpha) \tag{174}$$

with

$$\chi(r, k_{\text{cut}}, \alpha) = \frac{2}{\pi}\int_{k_{\text{cut}}}^{\infty} e^{-k^2/4\alpha^2}\frac{\sin(kr)}{kr}dk. \tag{175}$$

The first constant term in (174) refers to the self energy, while the second accounts for the intra-molecular excluded interactions[21]. This correction must be included in the same reference systems to which V_{qr} is assigned, e.g. V_l in our potential separation (see Table 2).

In principle the correction in Eq. (174) applies only to standard Ewald and not to the reciprocal lattice energy computed via SPME. We can still, however, use the correction Eq. (174), if a spherical cut-off k_{cut} is applied to SPME. This can be done easily by setting $\exp(-\pi^2 \mathbf{m}^2/\alpha^2)/\mathbf{m}^2 = 0$ for $2\pi\mathbf{m} > k_{cut} \equiv f_f \pi N_f/L$ where L is the side length of the cubic box and N_f is the number of grid points in each direction. The factor f_f must be chosen slightly less than unity. This simple device decreases the effective cut-off in reciprocal space while maintaining the same grid spacing, thus reducing the B-spline interpolation error (the error in the B-spline interpolation of the complex exponential is, indeed, maximum precisely at the tail of the reciprocal sums [33]). In Ref. [76] the effect of including or not such correction in electrostatic systems using multiple time step algorithms is studied and discussed thoroughly.

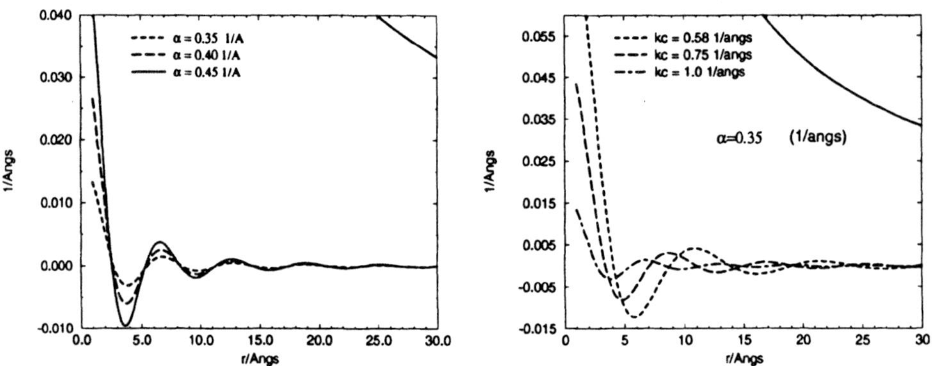

Figure 6. The correction potential $\chi(r, k_c, \alpha)$ as a function of the distance for different values of the parameters α (left) and k_c (right). The solid line on the top right corner is the bare Coulomb potential $1/r$

The potential $\chi(r, k_{cut}, \alpha)$ yields, *in direct space*, the neglected reciprocal energy due to the truncation of the reciprocal lattice sums, and must, in principle, be included for each atom pair distance in direct space. Thus, the

---

[21]Note that $\lim_{k\to 0} \chi(r, k, \alpha) = \mathrm{erf}(\alpha r)/r$.

corrected direct space potential is then

$$V'_{\text{qd}} = \frac{1}{2}\sum_{ij}^{N} q_i q_j \left[\sum_n \frac{\text{erfc}\alpha|\mathbf{r}_{ij} + \mathbf{r}_n|}{|\mathbf{r}_{ij} + \mathbf{r}_n|} + \chi(|r_{ij} + \mathbf{r}_n|, k_{\text{cut}}, \alpha) \right] \quad (176)$$

which is then split as usual in short-medium-long range according to (166). The correction is certainly more crucial for the excluded intramolecular contacts because V_{corr} is essentially a short-ranged potential which is non negligible only for intramolecular short distances. For systems with hydrogen bonds, however, the correction is also important for intermolecular interactions.

In Fig. 6 the correction potential is compared to the Coulomb potential (solid line in the top right corner) for different value of the reciprocal space cut-off k_c and of the convergence parameter α. For practical values of α and k_c, the potential is short ranged and small compared to the bare $1/r$ Coulomb interaction. In the asymptotic limit V_{corr} goes to zero as $\sin(ar)/r^2$ where a is a constant. This oscillatory long range behavior of the correction potential V_{corr} is somewhat nasty: In Fig. 7 we show the integral

$$I(r, k_c, \alpha) = \int_0^r \chi(x, k_c, \alpha) x^2 dx \quad (177)$$

as a function of the distance. If this integral converges then the $\chi(r, k)$ is absolutely convergent in 3D. We see that the period of the oscillations in

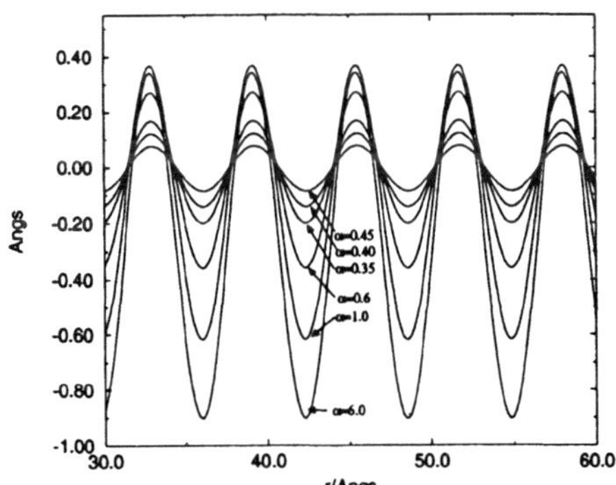
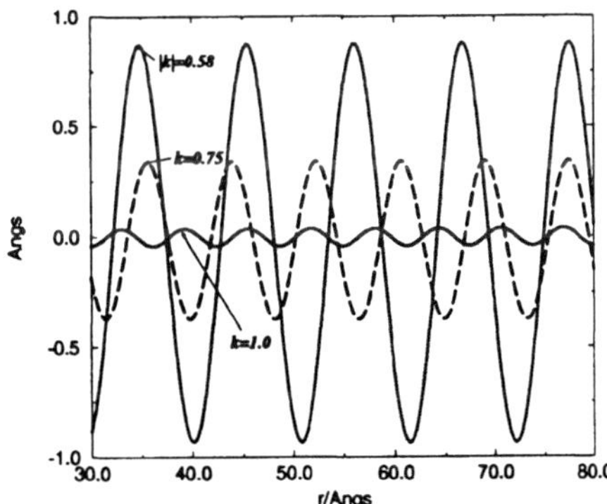

Figure 7. The integral $I(r)$ of Eq. (177) as a function of the distance for different values of the parameters α (left) and k_c (right)

$I(r)$ increases with k_c while α affects only the amplitude. The total energy

384

is hence again conditionally convergent, since the limit $lim_{r\to\infty}I(r)$ does not exist. However, unlike for the $1/r$ bare potential, the energy integral remains in this case bounded. Due to this, a cut-off on the small potential V_{corr} is certainly far less dangerous that a cutoff on the bare $1/r$ term. In order to verify this, we have calculated some properties of liquid water using the SPC model[77] from a 200 ps MD simulation in the NPT ensemble at temperature of 300 K and pressure of 0.1 MPa with i) a very accurate Ewald sum (column EWALD in Table 3), ii) with inaccurate Ewald but corrected in direct space using Eq. (176) (CORRECTED) and iii) with simple cut-off truncation of the bare Coulomb potential and no Ewald (CUTOFF). Results are reported in Table 3

	EWALD	CORRECTED	CUTOFF
Coulomb energy (KJ/mole)	-55.2 ±0.1	-55.1 ±0.1	-56.4 ±0.1
Potential energy (KJ/mole)	-46.2 ±0.1	-46.1 ±0.1	-47.3 ±0.1
Heat Capacity (KJ/mole/K)	74 ±24.5	94 ±22.0	87 ±23.2
Volume (cm^3)	18.2 ±0.1	18.3 ±0.1	18.1 ±0.1
Volume Fluctuation (Å^3)	136.9±3.5	147.0 ±3.5	138.7±3.5
R_{0-0} (Å)	2.81 ±0.01	2.81 ±0.01	2.81 ±0.01
Dielectric constant	59 ±25.8	47 ±27.3	3 ±2

TABLE 3. Properties of liquid water computed from a 200 ps simulation at 300 K and 0.1 Mpa on a sample of 343 molecules in PBC with accurate Ewald ($\alpha = 0.35$ $\AA^{-1}$; $k_c = 2.8$ $\AA^{-1}$) and no correction Eq. (174) (column EWALD), with inaccurate Ewald ($\alpha = 0.35$ $\AA^{-1}$; $k_c = 0.9$ $\AA^{-1}$) but including the correction Eq. (174) and with no Ewald and cut-off at 10.0 Å. R_{0-0} is the distance corresponding to the first peak in the Oxygen-Oxygen pair distribution function.

We notice that almost all the computed properties of water are essentially independent, within statistical error, of the truncation method. The dielectric properties, on the contrary, appear very sensitive to the method for dealing with long range tails: Accurate and inaccurate Ewald (corrected in direct space through 176) yields, within statistical error, comparable results whereas the dielectric constant predicted by the spherical cut-off method is more than order of magnitude smaller. We should remark that method ii) (CORRECTED) is almost twice as efficient as the "exact" method i).

5. Conclusion

We have discussed the application of multiple time scale methods to the simulation of complex systems at the atomistic level. We have seen how multiple time step schemes allow to account for all the flexibility of system without reducing the efficiency of the simulation. These methods can be easily generalized to extended system simulations and they can be combined effectively with powerful techniques for the evaluation of long range electrostatic interactions such as the SPME method.

This Chapter was aimed to describe in some detail the basic concept of multilevel integration of complex molecular systems and was not intended to be a review of the application of these ideas and principles in the recent literature. The most recent and accurate force fields for the simulation of complex molecular systems in the condensed phases, including protein and nucleic acids, are certainly those of MacKerrell *et al.* [7] and of Cornell *et al.* [2]. For further reading on symplectic integrators the reader is encouraged to see Refs. [39, 42]. The r-RESPA algorithm is described in many papers and textbook. A seminal paper is of course Tuckerman Martyna and Berne [18]. An excellent discussion about r-RESPA can also be found in the book of Frankel and Smit [57] and in a recent review by Berne [78]. The simulation of extended systems using r-RESPA techniques are described in detail in Refs. [18, 21, 23, 25]. Practical application of r-RESPA to the simulation of complex molecular systems with strong electrostatic interactions can be found in refs. [11, 20, 38, 74]. A clear and exhaustive discussion about long range electrostatic interaction and Ewald techniques is in refs. [31, 72]. The SPME method is described thoroughly in ref. [33].

Acknowledgements

This work supported by Italian CNR and MURST. We are in debt to G. Cardini and R. G. Della Valle for carefully proofreading the manuscript and for valuable suggestions.

References

1. Wiener, S.J., Kollmann, P.A., Nguyen, D.T. and Case, D.A. (1986) *J. Comput. Chem.*, **7**, 230.
2. Cornell, W.D., Cieplak, P., Bavly, C.I., Gould, I.R., Merz, K.M. Jr., Ferguson, D.M., Spellmeyer, D.C., Fox, T., Caldwell, J.W. and Kollmann, P. (1995) *J. Am. Chem. Soc.*, **117**, 5179.
3. Brooks, B.R., Bruccoeri, R.E., Olafson, B.D., States, D.J., Swaminanthan, S. and Karplus, M. (1983) *J. Comput. Chem.*, **4**, 187.
4. van Gunsteren, W.F. and Berendsen, H.J.C. (1987) *Groningen Molecular Simulation (GROMOS) Library Manual.* Biomos, Groningen.
5. MacKerrel, A.D., Wirkeiwicz-Kuczera, J. and Karplus, M. (1995) *J. Am. Chem. Soc*, **117**, 11946.
6. Pavelites, J.J., Gao, P.A. and MacKerrel, A.D. (1997) *Biophysical J.*, **18**, 221.
7. MacKerell Jr., A.D., Bashford, D., Bellott, M., Dunbrack, R.L., Evanseck, J.D.,

Field, M.J., Fischer, S., Gao, J., Guo, H., Ha, S., Joseph-McCarthy, D., Kuchnir, L., Kuczera, K., Lau, F.T.K., Mattos, C., Michnick S., Ngo, T., Nguyen, D.T., Prodhom, B., Reiher III, W.E., Roux, B., Schlenkrich, M., Smith, J.C., Stote, R., Straub, J., Watanabe, M., Wiorkiewicz-Kuczera, J., Yin, D. and Karplus, M. (1998) *J. Phys. Chem. B*, **102**, 3586.

8. Ryckaert, J.P., Ciccotti, G., and Berendsen, H.J.C (1977) *J. Comput. Phys.*, **23**, 327.
9. Ciccotti, G. and Ryckaert, J.P. (1986) *Comp. Phys. Report*, **4**, 345.
10. Allen, M.P. and Tildesley, D.J. (1989) *Computer Simulation of Liquids*. Oxford University Press, Walton Street, Oxford OX2 6DP.
11. Procacci, P., Darden, T., and Marchi, M. (1996) *J. Phys. Chem*, **100**, 10464.
12. Street, W.B., Tildesley, D.J. and Saville, G. (1978) *Mol. Phys.*, **35**, 639.
13. Teleman, O. and Joensonn, B. (1986) *J. Comput. Chem.*, **7**, 58.
14. Tuckerman, M.E., Martyna, G.J. and Berne, B.J. (1991) *J. Chem. Phys.*, **94**, 6811.
15. Tuckerman, M.E. and Berne, B.J. (1991) *J. Chem. Phys.*, **95**, 8362.
16. Tuckerman, M.E., Berne, B.J. and Rossi, A. (1990) *J. Chem. Phys.*, **94**, 1465.
17. Grubmuller, H., Heller, H., Winemuth, A. and Schulten, K. (1991) *Mol. Simul.*, **6**, 121.
18. Tuckerman, M.E., Berne, B.J. and Martyna, G.J. (1992) *J. Chem. Phys.*, **97**, 1990.
19. Tuckerman, M.E., Berne, B.J. and Martyna, G.J. (1993) *J. Chem. Phys.*, **99**, 2278.
20. Humphreys, D.D., Friesner, R.A. and Berne, B.J. (1994) *J. Phys. Chem.*, **98**, 6885.
21. Procacci, P. and Berne, B.J. (1994) *J. Chem. Phys.*, **101**, 2421.
22. Procacci, P. and Marchi, M. (1996) *J. Chem. Phys.*, **104**, 3003.
23. Martyna, G.J., Tuckerman, M.E., Tobias, D.J. and Klein, M.L. (1996) *Mol. Phys.*, **87**, 1117.
24. Procacci, P. and Berne, B.J. (1994) *Mol. Phys.*, **83**, 255.
25. Marchi, M. and Procacci, P. (1998) *J. Chem. Phys.*, **109**, 5194.
26. Saito, M. (1994) *J. Chem. Phys.*, **101**, 4055.
27. Lee, H., Darden, T.A. and Pedersen, L.G. (1995) *J. Chem. Phys.*, **102**, 3830.
28. Barker, J.A. and Watts, R.O. (1973) *Mol. Phys.*, **26**, 789.
29. Barker, J.A. (1980) *The Problem of Long-range Forces in the Computer Simulation of Condensed Matter*. volume 9, NRCC Workshop Proceedings, p. 45.
30. Ewald, P. (1921) *Ann. Phys.*, **64**, 253.
31. deLeeuw, S.W., Perram, J.W. and Smith, E.R. (1980) *Proc. R. Soc. London A*, **373**, 27.
32. Darden, T., York, D. and Pedersen, L. (1993) *J. Chem. Phys.*, **98**, 10089.
33. Essmann, U., Perera, L., Berkowitz, M.L., Darden, T., Lee, H. and Pedersen, L.G. (1995) *J. Chem. Phys.*, **101**, 8577.
34. Hockney, R.W. (1989) *Computer Simulation Using Particles*. McGraw-Hill, New York.
35. Petersen, H.G., Soelvanson, D. and Perramm, J.W. (1994) *J. Chem. Phys*, **101**, 8870.
36. Greengard, L. and Rokhlin, V. (1987) *J. Comput. Phys.*, **73**, 325.
37. Shimada, J., Kaneko, H. and Takada, T. (1994) *J. Comput. Chem.*, **15**, 28.
38. Zhou, R. and Berne, B.J. (1996) *J. Chem. Phys.*, **103**, 9444.
39. Sanz-Serna, J.M. (1992) *Acta Numerica*, **1**, 243.
40. Grey, S.K., Noid, D.W. and Sumpter, B.G. (1994) *J. Chem. Phys.*, **101**, 4062.
41. Biesiadecki, J.J. and Skeel, R.D. (1993) *J. Comp. Physics.*, **109**, 318.
42. Channel, P.J. and Scovel, C. (1990) *Nonlinearity*, **3**, 231.
43. Goldstein, H. (1980) *Classical Mechanics*. Addison-Wesley, Reading MA.
44. Arnold, V.I. (1989) *Mathematical Methods of Classical Mechanics*. Springer-Verlach, Berlin.
45. Trotter, H.F. (1959) *Proc. Am. Math Soc.*, **10**, 545.
46. De Raedt, H. and De Raedt, B. (1983) *Phys. Rev. A*, **28**, 3575.
47. Yoshida, H. (1990) *Phys. Letters A*, **150**, 262.

48. Toxvaerd, S.J. (1987) *J. Chem. Phys.*, **87**, 6140.
49. Andersen, H.C (1983) *J. Comput. Phys.*, **52**, 24.
50. Tuckerman, M.E. and Parrinello, M. (1994) *J. Chem. Phys.*, **101**, 1302.
51. Nose, S. and Klein, M.L. (1983) *Mol. Phys.*, **50**, 1055.
52. Herzberg, G. (1950) *Spectra of Diatomic Molecules.* Van Nostrand, New York.
53. Watanabe, M. and Karplus, M. (1995) *J. Phys. Chem.*, **99**, 5680.
54. Kjems, J.K. and Dolling, G. (1975) *Phys. Rev. B*, **11**, 16397.
55. Medina, F.D. and Daniels, W.B. (1976) *J. Chem. Phys.*, **64**, 150.
56. Cardini, G. and Schettino, V. (1990) *Chem. Phys.*, **146**, 147.
57. Frenkel, D. and Smit, B. (1996) *Understanding Molecular Simulations.* Academic Press, San Diego.
58. Rapaport, D.C. (1995) *The Art of Molecular Dynamics Simulation.* Cambridge University Press, Cambridge (UK).
59. Nosé, S. (1991) in *Computer Simulation in Materials Science.* Meyer, M. and Pontikis, V. (eds.), Kluwer Academic Publishers, p. 21.
60. Nosé, S. (1991) *Prog. Theor. Phys. Supp.*, **103**, 1.
61. Ferrario, M. (1993) in *Computer Simulation in Chemical Physics.* Allen, M.P. and Tildesley, D.J. (eds.), Kluwer Academic Publishers, p. 153.
62. Martyna, G.J., Tobias, D.J. and Klein, M.L. (1994) *J. Chem. Phys.*, **101**, 4177.
63. Andersen, H.C. (1980) *J. Chem. Phys.*, **72**, 2384.
64. Parrinello, M. and Rahman, A. (1980) *Phys. Rev. Lett.*, **45**, 1196.
65. Nosé, S. (1984) *Mol. Phys.*, **52**, 255.
66. Ferrario, M. and Ryckaert, J.-P. (1985) *Mol. Phys.*, **78**, 7368.
67. Tuckerman, M.E., Mundy, C.J. and Klein, M.L. (1997) *Phys. Rev. Lett.*, **78**, 2042.
68. Melchionna, S., Ciccotti, G. and Holian, B.L. (1993) *Mol. Phys.*, **78**, 533.
69. Berendsen, H.J.C. (1986) Lectures notes unpublished; reported by Ciccotti, G. and Ryckaert, J.P. *Comp. Physics Report*, **4** , 345.
70. Paci, E. and Marchi, M. (1996) *J. Phys. Chem.*, **104**, 3003.
71. Toxvaerd, S. (1993) *Phys. Rev. B.*, **47**, 343.
72. Hansen, J.-P. (1986) Molecular-dynamics Simulation of Coulomb Systems in Two and Three Dimensions, *Molecular Dynamics Simulation of Statistical-Mechanics Systems.* Proceedings of the International School of Physics "Enrico Fermi". North Holland Physics, Amsterdam.
73. Petersen, H.G. (1995) *J. Chem. Phys.*, **103**, 3668.
74. Procacci, P., Darden, T., Paci, E. and Marchi, M. (1997) *J. Comp. Chem.*, **18**, 1848.
75. Stuart, S.J., Zhou, R. and Berne, B.J. (1996) *J. Chem. Phys.*, **105**, 1426.
76. Procacci, P., Marchi, M. and Martyna, G.J. (1998) *J. Chem. Phys.*, **108**, 8799.
77. Rahman, A. and Stillinger, F.H. (1971) *J. Chem. Phys.*, **55**, 3336.
78. Berne, B.J. (1998) Molecular Dynamics Methods for the Enhanced Sampling of the Phase Space, *Classical and Quantum Dynamics in Condensed Phase Simulations.* Berne, B.J., Ciccotti, G. and Coker, D.F. (eds.), Proceeding of the International School 'Enrico Fermi', p. 127.

PARALLEL MOLECULAR DYNAMICS TECHNIQUES FOR THE SIMULATION OF ANISOTROPIC SYSTEMS

M.R. WILSON

Department of Chemistry, University of Durham,
South Road, Durham, DH1 3LE, United Kingdom.

1. Introduction to parallelisation

1.1. THE NEED FOR PARALLELISM

With the rapid increases in the speed of modern computers over the past few years, many people pose the question, "why bother with parallel computing?" For many researchers, the inherent problems of learning new computational techniques and of porting large computer codes, (or even re-writing them from scratch), far outweigh the benefits of parallelisation. Recently however such attitudes have begun to change. The production of parallel compilers, that can automatically parallelise many programs, have provided a painless route to parallelisation. In addition, the development of that allows purpose-written parallel programs to be ported across many different platforms has removed the problems of *lack of portability* that deterred many people from writing parallel applications.

With many of the barriers to the actual process of parallelisation removed, the real benefits of parallel computing become apparent,

- the ability to tackle really large problems, that cannot be attempted on today's single processor machines,
- the ability to split memory usage between separate processors,
- the ability to take advantage of an array of cheap mass-produced processors, instead of investing in the latest costly "superchip",
- the ability to increase the speed of computers even when size constraints (as processors become smaller and reach the mesoscopic regime) start to place limits on the speed of individual processors.

In addition to these benefits, many problems in science are inherently parallel and benefit from a parallel approach to finding a solution.

P. Pasini and C. Zannoni (eds.), Advances in the Computer Simulations of Liquid Crystals, 389–415.

1.2. TYPES OF PARALLEL COMPUTER

In the early days, most computers were based on SISD (single instruction stream-single data stream) architectures. Parallel computers however come in two main guises [1, 2], SIMD (single instruction stream-multiple data stream) and MIMD (multiple instruction stream-multiple data stream) machines. On SIMD machines, such as the AMT DAP [3], a single instruction is processed by each element in a processor array. On MIMD machines however, each processing element is able, in principle, to act independently. The MIMD category covers a huge range of modern machines from purpose built large scale systems (such as a Cray T3E), to simple clusters of workstations linked via ethernet connections. Some of these systems are able to take advantage of shared memory, whilst others must rely on the memory being associated only with individual processors.

In addition to multi-processor parallelism, many modern computers also take advantage of parallelisation *on chip*, with the processor able to carry out more than one operation per clock cycle.

1.3. MESSAGE PASSING FOR MIMD MACHINES

On MIMD machines, a method is required to allow the independent processors to exchange data when needed. In the early days of parallel computing, each manufacturer would produce a series of communication routines that could be called from within a program to carry out message passing tasks. Nowadays, the situation has dramatically improved with the implementation of standard message passing routines across a range of platforms. This allows parallel programs to be run without significant changes on everything from workstation clusters to large scale parallel supercomputers. The two most common (and useful) interfaces are PVM (parallel virtual machine [4, 5]) and MPI (message passing interface [6, 7]) both of which work with Fortran or C code.

1.4. MIMD METHODOLOGIES

1.4.1. *N-identical workers*

One of the most useful MIMD methodologies is the N-identical workers method. In this method, shown schematically in figure 1, the same program is run on a set of processors. Each processor has its own unique ID (task ID or TID) and when parallelisation is possible the TIDs are used to determine which processor does which part of the task. Message passing routines are called whenever a processor needs to know what has happened on another processor or needs data that is stored on another processor. This method works best with machines that have identical processors (e.g. Cray

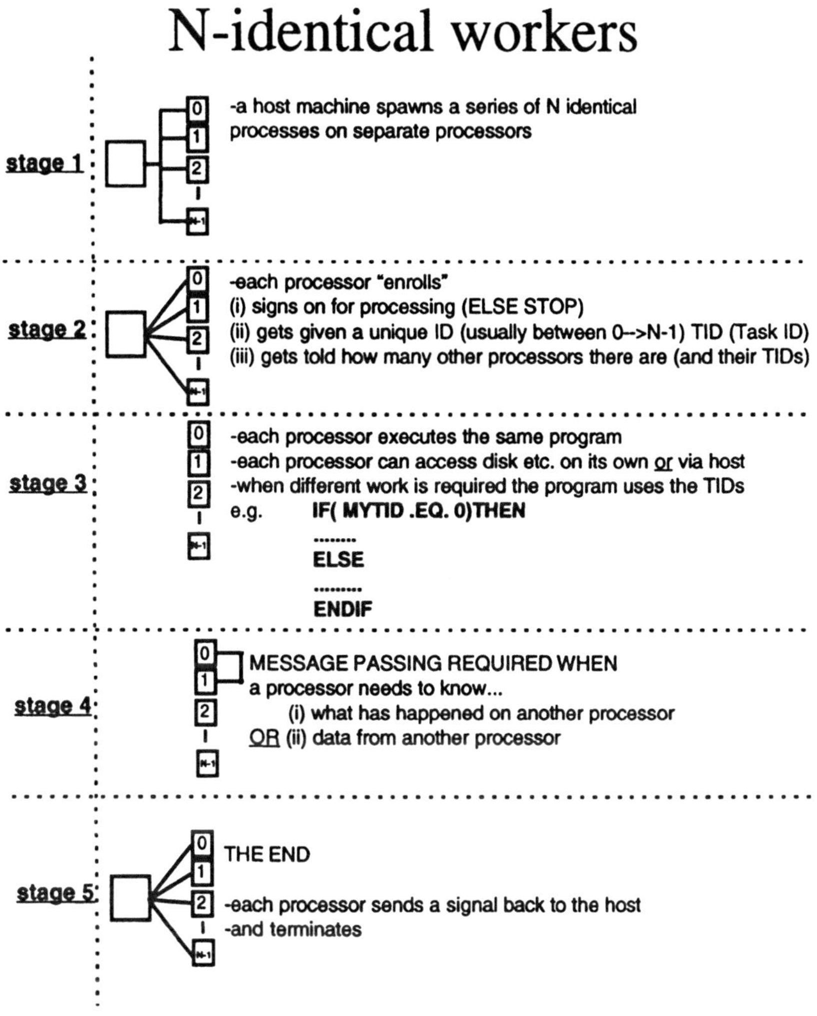

Figure 1. MIMD Methodology (i): N-identical Workers

T3E, Intel hypercube) but with careful load balancing of the work done by each CPU, can also work efficiently on parallel systems with processors of significantly different speeds (e.g. a heterogeneous workstation cluster). The N-identical worker approach is well-suited to parallel molecular dynamics simulations (see section 2).

Master/Slave

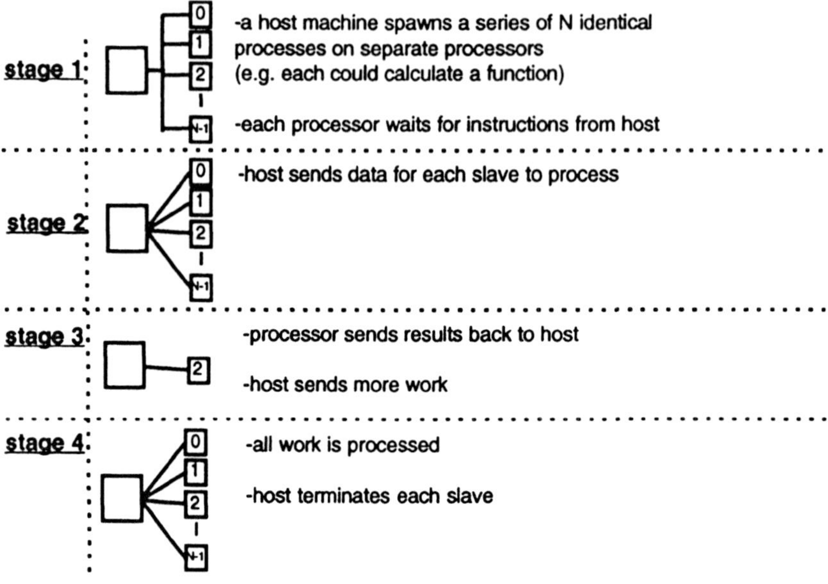

Figure 2. MIMD Methodology (ii): Master/slave scenario.

1.4.2. *The master/slave approach*

The master/slave approach is shown schematically in figure 2. It relies on a master process spawning a series of separate processes to do the same repetitive task. As soon as a slave has finished the task, results are returned to the master and new starting data is sent to each slave. This continues until no more data remains to be processed. Because each slave gets new work only when it has2 finished its existing task, this method usually has good load balancing characteristics and works well on systems of heterogeneous processors.

The master/slave approach provides an excellent method for farming out separate Monte Carlo or molecular dynamics simulations to individual processors. This is of particular use when surveying part of a phase diagram.

1.5. PARALLEL OPERATIONS

Typically message passing software can allow for a number of key parallel operations,

- synchronization of nodes,
- blocking send,
- non-blocking send,
- blocking receive,
- non-blocking receive,
- global sum operation,
- returning the node id of a processor,
- returning the total number of processors.

The sending and receiving operations can in general be blocking or non-blocking in nature. A blocking operation waits for completion before allowing the processor to move on to its next task. In contrast, in a non-blocking operation, the processor can continue as soon as the communication has been initiated. This can be advantageous in allowing parallel communications to be overlapped with computational tasks. The draw-back to this type of *asynchronous* communication is that the programmer must ensure that none of the communicated data is used before the communications process has completed.

The global sum is probably the key parallel operation in most parallel molecular simulations. It arises when the calculation of a quantity A has been slit between all the processors on the machine. Each processor i has calculated part of the answer A_i, but each must know the final answer $A = \sum_i A_i$. Often A is a single thermodynamic quantity, such as the total potential energy of the system. However, in some instances, A can also be an array of data points, such as the force array in a molecular dynamics simulation. For an n-particle system, a global sum would then be required for each element of the array A. For large n this provides a major communication load, which can dramatically slow down a simulation program.

There are a number of ways of carrying out a global sum operation, three of these are illustrated in figure 3. The key to an efficient global sum operation involves minimising the amount of message passing required. Accordingly, the hypercube method (figure 3c), which works for 2^N processors, is one of the most efficient global sum methods requiring just N overlapped communication steps.

1.6. EFFICIENT PARALLEL PROGRAMMING

For parallel molecular simulation, there are two key tasks that must always be addressed in producing efficient programs,

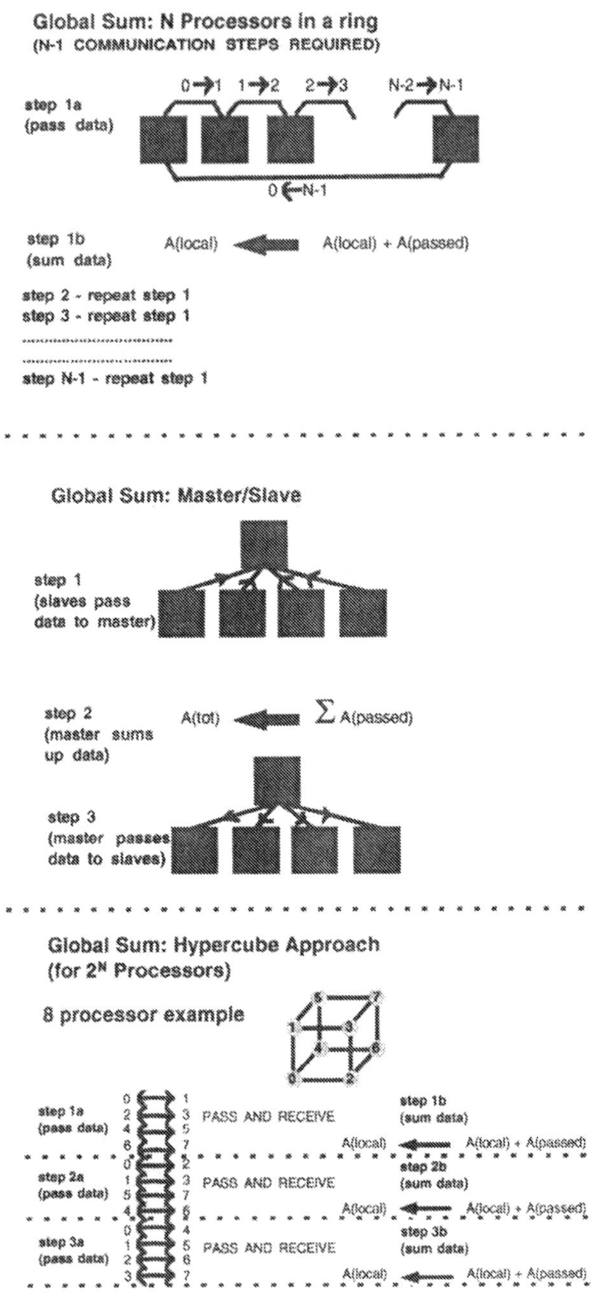

Figure 3. Three approaches to the global sum operation.

- (i) load-balancing,
- (ii) minimisation of non-overlapped communications relative to computation.

Efficient load-balancing simply means ensuring that each processor gets an approximately equal share of the work, and that no processors are left idle for any significant length of time. Minimising communications is significantly harder. On machines with fast communications this is less of a problem than (say) workstation clusters. However, even on machines such as the Cray T3E with cache-to-cache transfers, communication times can quickly limit the scaling of codes to large numbers of processors.

Traditional types of Monte Carlo simulations, (where one molecule is moved and the change in energy is calculated), rarely translate efficiently into parallel code. This is because the amount of computational time required for this operation is small relative to the communication costs of carrying out a global sum on the energy. Also, for cheap computational tasks, load-balancing is usually poor. In contrast, in molecular dynamics simulations the cost of one MD step for the whole system is usually quite large. As a consequence MD algorithms tend to translate well to parallel machines.

The main part of this article will concentrate on describing some of the parallel molecular dynamics techniques that have been developed for the simulation of anisotropic systems.

2. Replicated data methods for parallel molecular dynamics

2.1. INTRODUCTION

The replicated data method is the most popular technique for parallel molecular dynamics [8–10]. This is due to the inherent simplicity of the method; often only minor changes are required to convert a serial program into a parallel form. The essence of the method is that each compute node runs the same MD program, undertaking the same operations up to the point where a task can be carried out in parallel. At this point each node takes part of the parallel task, and at the end of this task any data required by all nodes must be passed and received in a communication step.

The draw back of the replicated data method is that it is relative expensive in terms of memory, all compute nodes must have a copy of most of the data (e.g. coordinates, forces etc.). In addition, it is relatively expensive in terms of communication costs.

Loops (e.g. Do-loop in Fortran) are the obvious place for parallelisation. Providing the executable statements within the loop are independent of the loop counter, each pass through the loop can be executed in parallel on separate processors. This data structure can be recognised by parallel compilers, and readily parallelised automatically on some parallel machines without intervention by the user.

2.2. SIMULATION OF ISOTROPIC PARTICLES

The first stage of any parallelisation task involves the design of the parallel algorithm. Many serial MD programs use variants of the leap-frog algorithm (or algorithms that are closely related to it e.g. Verlet, velocity-Verlet algorithms) [11];

- forces $\mathbf{f}_i$ are calculated from the coordinates,

$$\mathbf{f}_i = -\sum_j \nabla U_{ij}(r_{ij}); \tag{1}$$

- the forces are used to update the velocities $\mathbf{v}_i$ over a time step dt,

$$\mathbf{v}_i(t + dt/2) = \mathbf{v}_i(t - dt/2) + dt\ \mathbf{f}_i/m_i \tag{2}$$

for particles of mass m_i;
- the velocities are used to update the coordinates,

$$\mathbf{r}_i(t + dt) = \mathbf{r}_i(t) + dt\ \mathbf{v}_i(t + dt/2). \tag{3}$$

Timings of the leap-frog algorithm for typical systems of atoms indicate that at least 80% of the simulation time is used solely in the evaluation of forces (and this percentage increases as the number of particles in the system increases), and approximately 10% of the time is used in solving the equations of motion (equations 2-3). The rest of the time used is spent on a variety of tasks including initalisation of variables, reading and writing of data etc. Consequently, within the replicated data method it is only necessary to concentrate on the paralleisation of the force loop (and possibly the integration steps) to provide excellent parallel speed-ups.

A simple example of the main force calculation is shown schematically for pseudo-Fortran code in figure 4. The serial code contains a nested do-loop to evaluate all the pair forces for the system. The parallel code simply divides the work up so that each compute node computes the forces for its own set of atoms. So for eight processors, node 0 gets atoms 1,9,17..., node 1 gets atoms 2,10,18..., etc. At the end of the parallel force loop a global sum is required to distribute the correctly summed forces to each processor.

2.3. EFFICIENT LOAD BALANCING

The major problem with the parallel code in figure 4 is that load balancing will be extremely poor. This is because the number of pair-forces evaluated

```
c     loop over all pairs of interactions
      DO I = 1, N-1

        DO J = I+1, N

c         calculate pair distance
c         evaluate forces f_ij
c         sum the forces
            f_i^tot ← f_i^tot + f_ij

            f_j^tot ← f_j^tot - f_ij
        ENDDO
      ENDDO
```

scalar code

```
c     loop over all pairs of interactions
      DO  I = IDNODE+1, N-1, NODES

        DO  J = I+1, N

c         calculate pair distance
c         evaluate forces f_ij
c         sum the forces
            f_i^tot ← f_i^tot + f_ij

            f_j^tot ← f_j^tot - f_ij
        ENDDO
      ENDDO
```

parallel code

Figure 4. Parallelisation of the force loop. (NODES = number of nodes in the system, IDNODE = node id for the compute node executing the code.)

continually decreases with increases in the loop-counter I. This results, for instance, in less work being done by node 1 than node 0. It is therefore desirable for the inner loop in figure 4 to be of similar length for each element I. This is easily achieved by making a Verlet neighbour list, and using Brode-Ahlrichs decomposition [12] to achieve a similar number of neighbours for each atom in the list. This is shown schematically in figure 5 for 9 and 10 atom systems, where the rows represent the neighbour list elements for each atom in the system. Pseudo-code to generate a parallel neighbour list using the Brode-Ahlrichs method is shown in figure 6. Using this code, each compute node stores its own neighbour list. This has the advantage of distributing the memory required to store the neighbour list

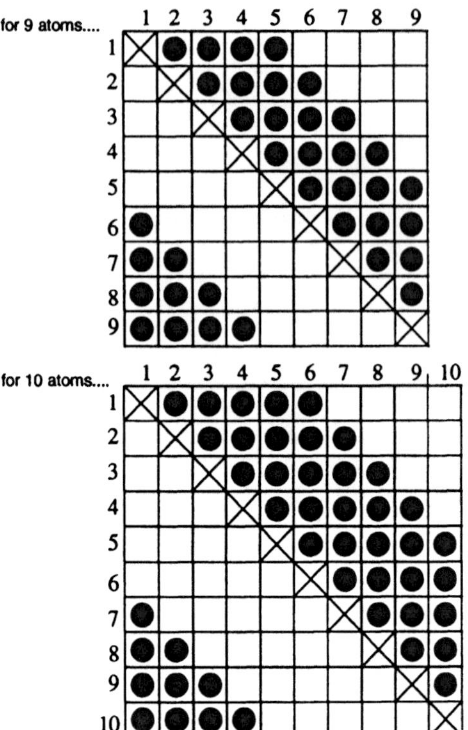

Figure 5. Brode-Ahlrichs decomposition of a Verlet neighbour list for systems with odd (9) and even (10) numbers of particles. Rows represent the neighbour list for individual atoms. Filled-circles indicate which atoms are members of each atomic neighbour list. Each atomic neighbour list has exactly the same length for odd numbers of particles.

over all the nodes in the system.

2.4. SIMULATION OF ANISOTROPIC PARTICLES

The replicated data method outlined above can easily be adapted to axially symmetric potentials (e.g. the Gay-Berne potential). Forces on the centre of mass are handled in the same way as forces on atoms; and the motion about the centre of mass can be handled by an extension to the leap-frog algorithm [11]

$$dt\ \lambda_i(t) = -2\mathbf{u}_i(t - dt/2) \cdot \mathbf{e}_i(t), \tag{4}$$

$$\mathbf{u}_i(t + dt/2) = \mathbf{u}_i(t - dt/2) + dt\frac{\mathbf{g}_i^{\perp}}{I} + dt\ \lambda_i(t)\mathbf{e}_i(t) \tag{5}$$

$$\mathbf{e}_i(t + dt) = \mathbf{e}_i(t) + dt\ \mathbf{u}_i(t + dt/2) \tag{6}$$

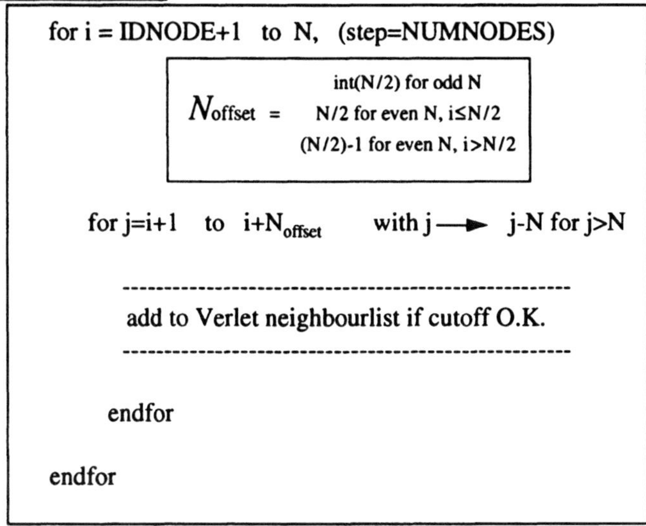

Figure 6.   Pseudo-code to make a Verlet neighbour list with Brode-Ahlrichs decomposition. (NUMNODES = number of nodes in the system, IDNODE = node id for the compute node executing the code.)

where the axis vector of molecule i is given by $\mathbf{e}_i$, $\mathbf{u}_i = \dot{\mathbf{e}}_i$, I is the moment of inertia and

$$\mathbf{g}_i^{\perp} = \mathbf{g}_i - (\mathbf{g}_i \cdot \mathbf{e}_i)\,\mathbf{e}_i \tag{7}$$

where

$$\mathbf{g}_i = \sum_j \mathbf{g}_{ij} = -\sum_j \nabla_{\mathbf{e}_i} U_{ij} = -\sum_j \left(\frac{\partial U_{ij}}{\partial e_{x_i}}, \frac{\partial U_{ij}}{\partial e_{y_i}}, \frac{\partial U_{ij}}{\partial e_{z_i}} \right). \tag{8}$$

The same parallel Brode-Ahlrichs form of the Verlet neighbour list as used above (in section 2.3) is implemented to evaluate the forces $\mathbf{f}_{ij}$ and the so-called gorques $\mathbf{g}_{ij}$ [10]. With a global sum required for $\mathbf{f}_i$ and $\mathbf{g}_i^{\perp}$ (6N quantities).

For 256 Gay-Berne particles approximately 96% of CPU time is spent in evaluating $\mathbf{f}_{ij}$ and $\mathbf{g}_{ij}$ for the serial code. So large parallel speed ups are possible. Parallel performance however is limited by the speed of the global sum operations which become expensive for many processors.

For biaxial particles a quaternion form of the leap-frog algorithm [11] can be used in place of equations 4-6. However, the strategy to parallelise

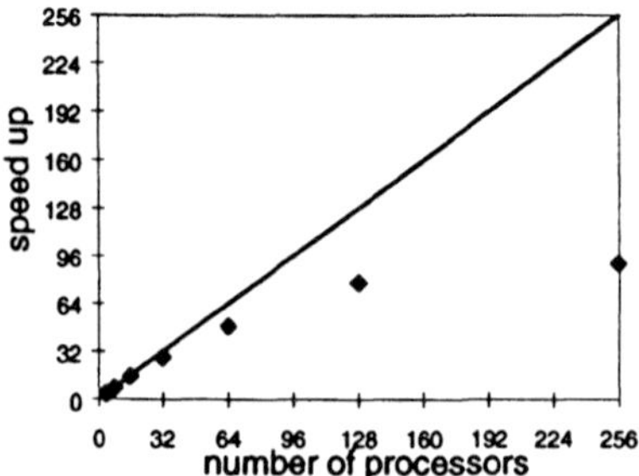

Figure 7. The performance of the replicated data algorithm for 1024 Gay-Berne particles on a Cray T3D.

the calculation of forces and (in this cases) torques remains exactly the same [10].

2.5. PERFORMANCE OF THE REPLICATED DATA ALGORITHM

The parallel performance of the replicated data algorithm is shown in figure 7 for 1024 Gay-Berne particles. The performance curve shows the point at which the global sum operation starts to severely limit the speed of the algorithm (at 128 processors). Eventually, for a large enough number of processors, this leads to the code slowing down as the number of compute nodes is increased still further.

Increasing the ratio of computational time to communication time dramatically improves the performance of the algorithm. Consequently, adding point dipoles to the Gay-Berne particles and carrying out a computationally expensive Ewald sum, leads to a much improved parallel performance over figure 7. Simply increasing the number of particles in the system has exactly the same effect.

2.6. PARALLELISATION OF THE INTEGRATION LOOP

The replicated data approach can easily be used to divide up the integration steps (equations 2–3 and 4–6). However, a global sum of the coordinates/orientations (and the velocities/orientational derivatives if these are to be stored) is required if this part of the algorithm is to be parallelised. On many machines the speed of the global sum is such that it is more efficient to carry out a scalar integration of these equations.

2.7. EXTENSION TO MULTI-SITE MOLECULES

2.7.1. *Molecular potential*
In general multi-site molecular systems require both intramolecular and intermolecular terms in the potential energy and these pose new problems for parallelisation. A typical multi-site potential is the GBMOL potential [13, 14] given by

$$E = \sum_{\text{bonds}} E_{\text{bond}} + \sum_{\text{angles}} E_{\text{angle}} + \sum_{\text{dihedrals}} E_{\text{dihedral}} \qquad (9)$$
$$+ \sum_{\langle ij \rangle} E_{\text{LJ}} + E_{\text{el}} + \sum_{\langle ij \rangle} E_{\text{LJ-GB}} + \sum_{\langle ij \rangle} E_{\text{GB-GB}};$$

where E_{bond} is a bond distortion energy, E_{angle} is an angle distortion energy, E_{dihedral} is the dihedral angle energy, E_{LJ} is the Lennard-Jones pair potential, E_{el} is an electrostatic pair potential, $E_{\text{GB-GB}}$ is the Gay-Berne potential, $E_{\text{LJ-GB}}$ is the pair potential for the interaction between a Lennard-Jones site and a Gay-Berne site, and $\langle ij \rangle$ represents a sum over all pairs of sites.

The intramolecular terms in the potential are usually trivial to parallelise via the replicated data method (see section 2.7.2). However, some care must be taken in order to parallelise the expensive non-bonded interactions efficiently (see section 2.7.4).

2.7.2. *Intramolecular multi-site potentials*
Scalar code to handle multi-site potentials, such as a bond angles, usually involves a simple loop, with each iteration of the loop calculating the forces for an individual bond angle. An example of this is shown schematically in figure 8. In such cases the replicated data method is trivial to implement and works extremely well. The do-loop in figure 8 has been parallelised in the usual way (as in section 2.2); and after completion of the loop a global sum is carried out for the total angle contribution to the potential energy. The expensive global sum for the forces is carried out once only in the simulation program after all the intramolecular and intermolecular forces have been evaluated. Consequently, the parallelisation of the intramolecular interactions adds no substantial communications overheads to the program.

2.7.3. *SHAKE constraints*
The SHAKE procedure is an iterative scheme to constrain bond lengths to specified lengths. This avoids the need for bond stretching potentials and in practice allows the use of longer time-steps in molecular dynamics simulations (typically 2 fs instead of 0.5 fs without SHAKE). However, the

```
            DO NUM = IDNODE+1, NANGLES, NUMNODES

     C           ATOMS NUMBERS
                 I = ANG1(NUM)
                 J = ANG2(NUM)
                 K = ANG3(NUM)

     C    WORK OUT ENERGY AND FORCES FOR I,J,K
          ------------------------------------------

          ENDDO
```

Figure 8. Fortran code for the parallelisation of bond angle interactions (NUMNODES = number of nodes in the system, IDNODE = node id for the compute node executing the code.)

iterative nature of SHAKE, and the fact that bond constraints cannot be applied individually without influencing neighbouring bonds, means that a rather complicated algorithm is required to carry out an efficient replicated data form of SHAKE.

Such a procedure has been implemented in the program DL_POLY [15] and involves each node calculating incremental bond correction vectors for its own list of bonds, and then sending this information to all neighbouring nodes that share the same atoms. Naturally, this procedure works best for small molecules where the amount of communication required is small.

2.7.4. *Non-bonded interactions*

In evaluating non-bonded interactions it is usual to compile an *excluded atom list*. This list contains all pairs of atoms that are not involved in non-bonded interactions (usually atoms that are bonded together, or involved in 3-site angle interactions etc.). In order to parallelise the non-bonded interactions efficiently, the excluded atom list must be broken down into separate lists for each node. This decomposition is done in the usual way, with each node making an excluded atom list for the atoms it usually handles when processing the Verlet neighbour list. The excluded atom lists on individual nodes must also conform to the usual Brode-Ahlrichs atom order, this means that they can be processed linearly when each node compiles its own Verlet neighbour list. For simulations involving different types of non-bonded interactions (as in equation 9), it is usually most efficient to compile separate Verlet neighbour lists for each interaction type. This is particularly important when different non-bonded terms have different cutoff ranges.

3. Systolic loop methods for parallel molecular dynamics

3.1. OUTLINE OF THE SYSTOLIC LOOP METHOD

A schematic diagram of the various stages in the systolic loop algorithm (for anisotropic particles) is shown in figure 9. For an n-processor system, the coordinates/orientations are divided into (2n-1) groups, and each group is allocated a *home* processor. Processor 0 takes just one group, whilst all the other processors have two groups as shown in stage 1 of figure 9. In stage 2 of the algorithm, node 0 calculates the *intra*-group forces and gorques for all its resident coordinates. At the same time, all the other processors in the system calculate the *inter*-group forces and gorques for their two groups of resident data. When all these calculations are completed a *systolic data pulse* occurs. Data is passed around the system as indicated in stage 3 of figure 9, where each of the data packets consist of coordinates and orientations plus force and gorque accumulators for the relevant atoms. The accumulation of forces and gorques, and the systolic data pulses, continue until 2n-1 cycles have been completed. At this point each data packet has returned to its *home* processor, and all the forces and gorques have been summed. To complete a MD step, each processor is then able to integrate the equations of motion for its own group(s) of coordinates.

3.2. ASSESSMENT OF THE SYSTOLIC LOOP ALGORITHM

The algorithm works quite well in practice [8, 16]. It is a purpose-designed parallel algorithm, so unlike the the replicated data approach cannot be run efficiently on scalar machines. It is also considerably more complicated to implement than the replicated data approach. The load-balancing characteristics of the algorithm are usually quite good. Node 0 does less work in computing forces than the other nodes. However, this is not usually a problem, particularly if the number of nodes is large and node 0 is also kept busy with various other house-keeping tasks. Memory usage is quite good because coordinates, orientations, velocities, orientational derivatives, forces and gorques can all be distributed across each node in the system.

In practice, the systolic loop algorithm tends to scale better than the replicated data method for large numbers of particles N, but is poorer than the replicated data method for small N [8]. It also tends to work best for systems with long-range potentials, where all pair-forces are required (e.g. for electrostatics).

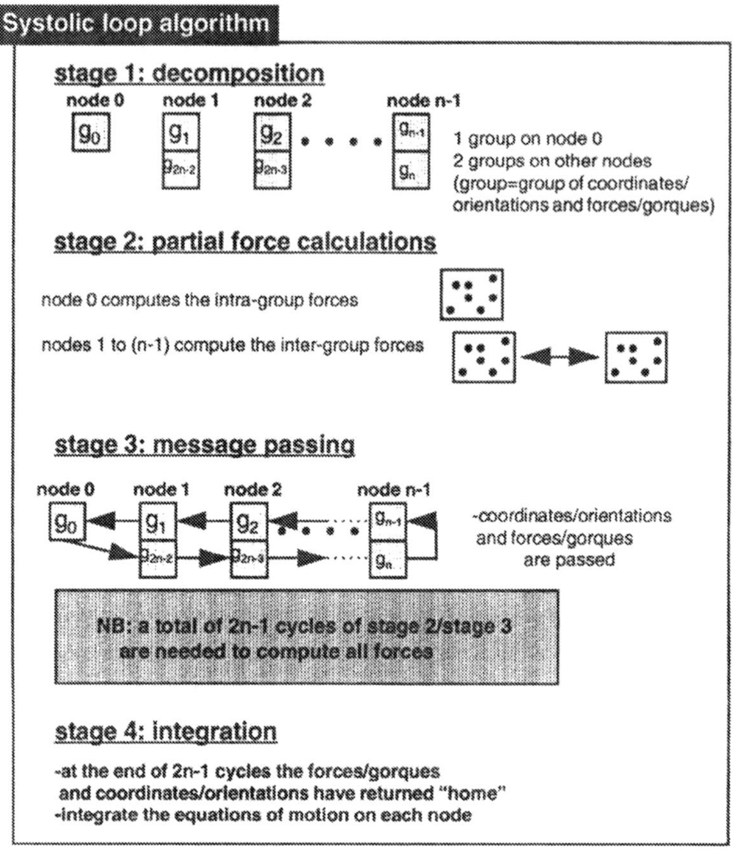

Figure 9. Schematic diagram illustrating the four stages of the systolic loop algorithm.

4. Domain decomposition methods for molecular dynamics

4.1. INTRODUCTION

The domain decomposition algorithm represents an extremely powerful approach to parallel molecular dynamics simulations that use short-ranged potentials [8,10,17–23]. The algorithm uses a spatial decomposition of the simulation box as indicated in figure 10, where we can define a hierarchy consisting of a domain (whole system), n-regions (one for each processor) and individual cells within the regions. The cuboidal cells of figure 10 are defined so that the smallest length of a cell is always larger than the cutoff on the non-bonded potential. This means that an atomic (or molecular site if anisotropic particles are being used) can only interact with another site

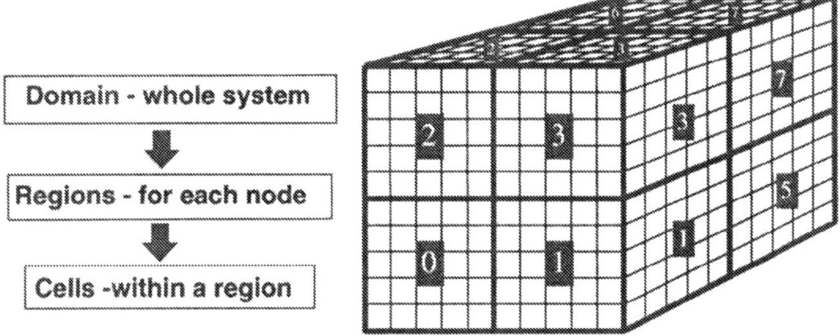

Figure 10.   Diagram showing the division of the simulation box (domain) into regions and cells using the domain decomposition approach.

if that site is present in the same cell, or present in a neighbouring cell (i.e. one that shares a face, edge or vertex with the original cell).

The advantage of this approach stems from the fact that each processor is able to act independently in calculating forces for most sites in its own region. The only exception arises from sites that appear in the cells on the boundary of a region. For these sites, message passing must take place to communicate vital data from neighbouring processors. However, the quantity of data that must be communicated is small compared to that required in the large global sums of the force array in the replicated data method. This means that the domain decomposition algorithm can work extremely efficiently with minimal communications overheads.

The algorithm scales well with the number of particles in a system [8,10]. As each region gets larger, the relative proportion of particles in cells on the boundary of a region becomes smaller, and so the relative contribution of communications to the overall simulation time diminishes.

4.2. CALCULATION OF FORCES USING DOMAIN DECOMPOSITION

There are two possible schemes for evaluating forces and gorques using the domain decomposition algorithm. These involve two different approaches to the transfer of data from cells on the boundary of a region, and are illustrated schematically for a 2-dimensional example (with just four-processors) in figure 11.

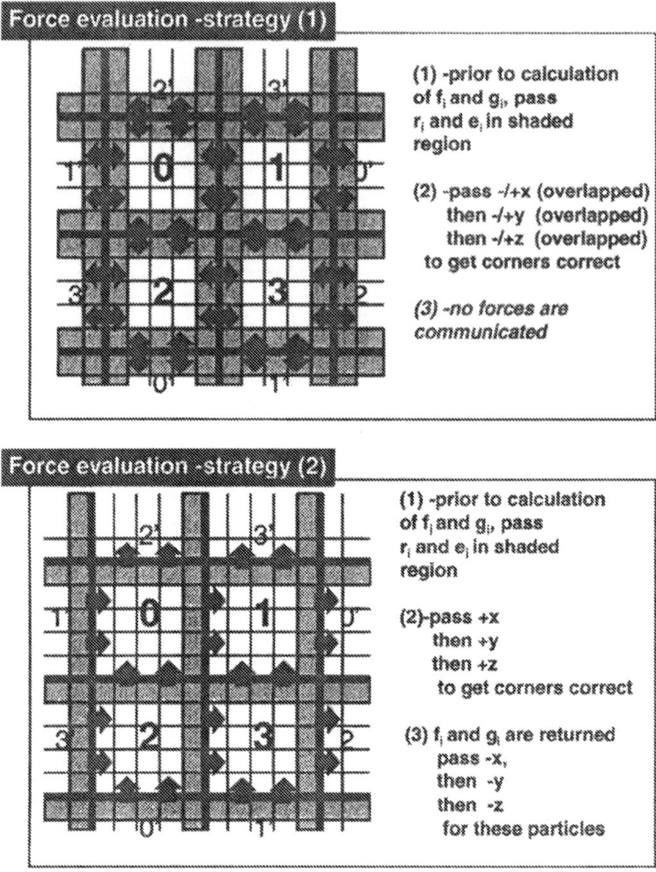

Figure 11. Two alternative strategies for the evaluation of forces and gorques using the domain decomposition approach to molecular dynamics for a 2-dimensional example.

4.2.1. Force evaluation strategy 1

In strategy 1, sites in a region are initially allocated to individual cells prior to the force evaluation. Neighbouring nodes then exchange copies of coordinate/orientation data for sites occupying boundary cells as indicated in the first part of figure 11, with coordinate data being passed and received simultaneously in two directions. A linked-list (neighbour list) can then be made in the usual way [24]. The linked-list is used to calculate pair forces only for those sites that are resident on a particular processor. There is no need to communicate any force data back to neighbouring processors.

It should be noted that in passing the coordinate/orientation data it is necessary to carry out the communications in three separate stages [8]

- pass and receive data in the plus and minus x direction (North/South),
- pass and receive data in the plus and minus y direction (East/West),
- pass and receive data in the plus and minus z direction (Up/down).

After the North/South and East/West data passes, it is necessary to do some additional sorting of data so that a node will receive all necessary data from cells which connect to it via a vertex or edge.

4.2.2. *Force evaluation strategy 2*

In strategy 2, (illustrated in the second part of figure 11), neighbouring sites exchange coordinate/orientation data in only one set of directions (e.g. North, East and Up). These data are then used to compile a linked-list as in strategy 1. However, in this alternative strategy the forces/gorques on sites in the cells on the South, West and Down boundary directions are evaluated and sent back in the reverse direction to the coordinates/orientations (South, West and Down) to be added to the forces in the appropriate (North, East and Up) boundary cells of the neighbouring regions.

4.2.3. *Comparison of force evaluation strategies*

Strategy 1, incorporates some inefficiency in the force/gorque evaluation, due to the duplication of some of the effort in calculating forces/gorques for boundary region sites. Strategy 2, avoids this problem but requires two separate sets of communications steps (coordinates/orientations, forces-/gorques) that (unlike strategy 1) cannot be overlapped. In practice there seems to be little to choose between these two strategies with regard to the performance of the algorithm.

4.3. INTEGRATION OF EQUATIONS OF MOTION

The integration of the equations of motion is straightforward and can occur in parallel on each node without the need for communications. However, at the end of the integration it is necessary to check to see which sites have left the region of space controlled by their host processor. For these atoms it is necessary to transfer the coordinates/orientations and the velocities/orientational derivatives to the appropriate neighbouring processor to be ready for the next force evaluation. A flow diagram summarising the various steps required in the domain decomposition algorithm is shown in figure 12.

4.4. PERFORMANCE OF THE DOMAIN DECOMPOSITION ALGORITHM

The parallel performance of the domain decomposition algorithm is excellent. Figure 13 shows a linear scaling of the algorithm as the number of processors increases on a Cray T3D. A similar linear speed up is achieved

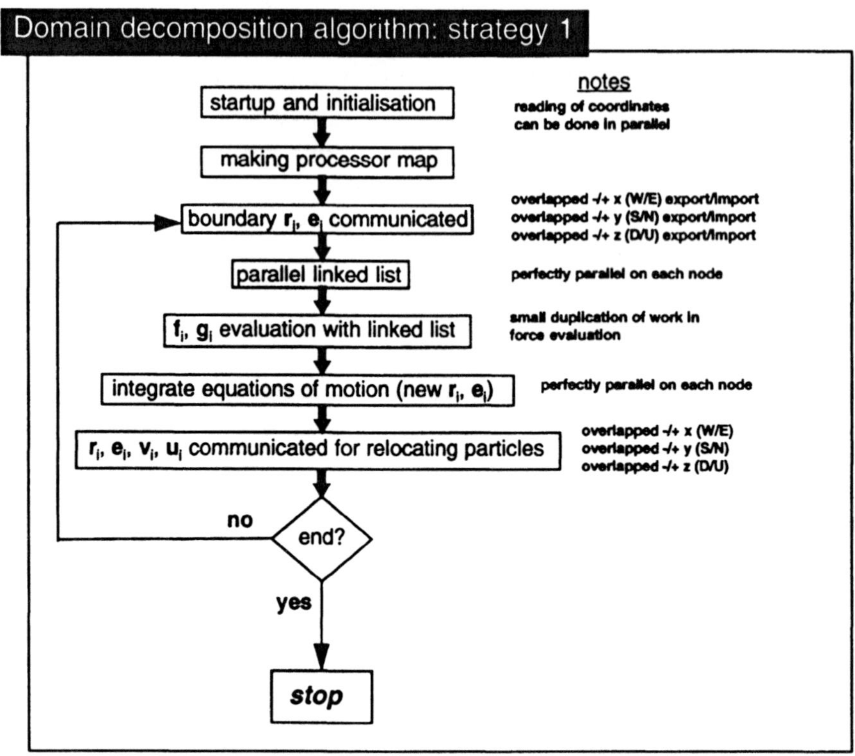

Figure 12. Flow chart illustrating the sequence of step required in the domain decomposition algorithm using strategy 1 for the evaluation of the forces and gorques.

for system sizes of between 1024 and 65536 Gay-Berne particles. For large numbers of particles, the algorithm shows substantially lower simulation times than a competing replicated data algorithm [10] ($\approx 13\times$ faster for 65536 molecules on 256 nodes).

4.5. DOMAIN DECOMPOSITION ALGORITHM WITH TWISTED PERIODIC BOUNDARY CONDITIONS

Recently a nice technique has been devised to facilitate the simulation of twisted (pseudo-chiral) structures. This involves the use of twisted periodic boundary conditions, as illustrated in figure 14a. The two shaded arrows at each end of the main simulation box represent two molecules. These molecules interact through the usual minimum image convention and so the closest interaction distance is given by considering the periodic image of the black arrow (which is illustrated in the image box). However, instead of using the usual boundary conditions to compute the periodic image, the

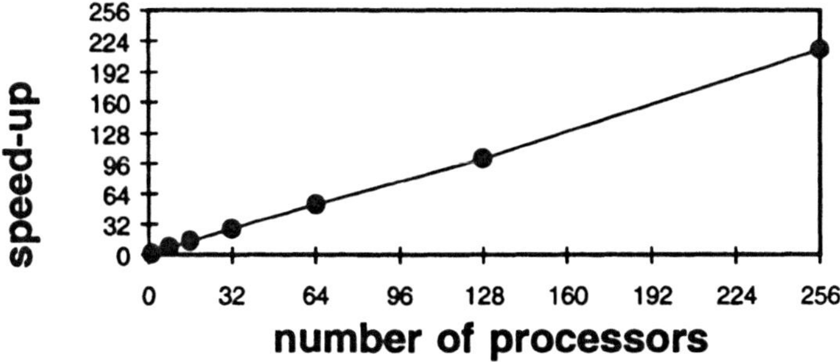

Figure 13. The performance of the domain decomposition algorithm for 65536 Gay-Berne particles on a Cray T3D.

position and orientation of the image of the black arrow are twisted through 90 degrees. The same ($+90°$ or $-90°$) operation is applied whenever two molecules interact through the ends of the simulation box along the z-axis, and also when a molecule itself moves across the z-boundaries of the simulation box. Normal periodic boundary conditions are enforced in the x and y directions.

In order to implement twisted periodic boundary conditions within the domain decomposition frame-work, it is first necessary to implement a *twisted map* to define processor connectivity, (as illustrated in figure 14b for an eight-processor example). The system is divided up into regions in the usual way, using the normal spatial decomposition to allocate processors to regions. However, a 90 degree twist is then applied to the processor mapping across the end boundary of the simulation box. So processor 0 (in figure 14b) passes data to processor 4 at the processor boundary within the simulation box, and to processor 5 at the processor boundary that lies at the end of the simulation box.

During the simulation, the twisted periodic boundary conditions are implemented as illustrated in figure 14c. Molecule 1 is due to move in the z-direction across the end of the simulation box. It first undergoes a rotation about the centre of the region handled by its own processor (taking it from position 1 to position 1'). Then, in moving across the end of the simulation box, the *twisted map* is used to take it to the x,y coordinates given by position 1".

Recently this scheme has been used on the Cray T3D in Edinburgh to facilitate the simulation of twisted nematic structures, and an analogue of the twist-grain-boundary phase for system sizes of 84,000 particles [25].

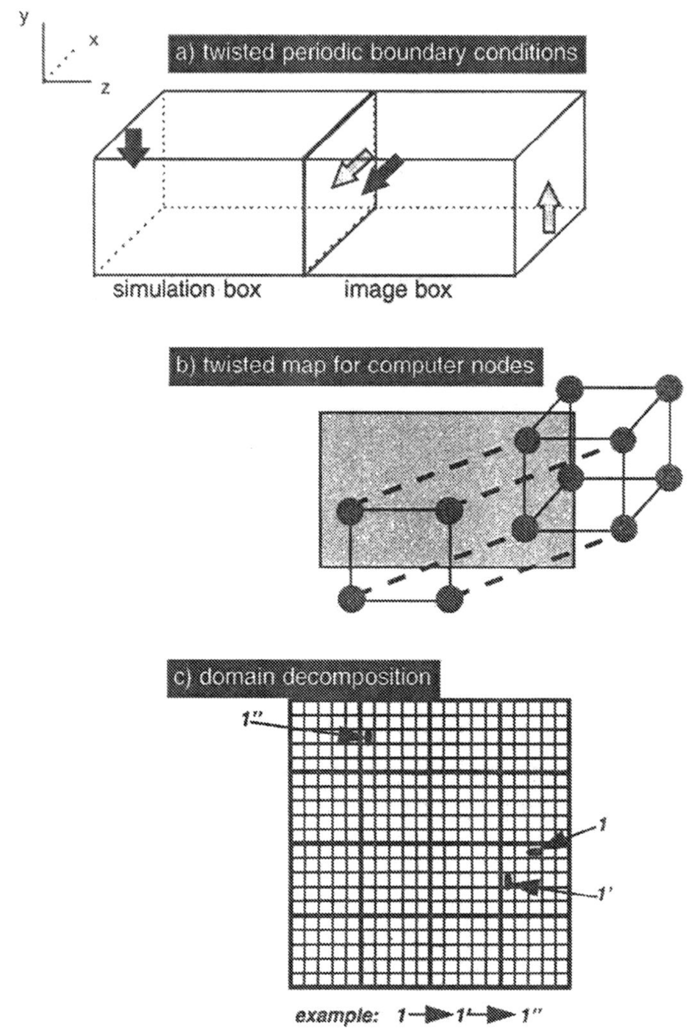

Figure 14. Twisted periodic boundary conditions. a) Illustration of molecules in a simulation box with a 90 degree twist. b) Processor map showing the connectivity of processors for a domain decomposition algorithm with twisted periodic boundaries. c) Rotations of a molecule within the domain decomposition algorithm. Molecule 1 is first rotated about the centre of its region, then undergoes a translational shift as it is transferred between processors using the twisted map of part b).

4.6. EXTENSION TO MULTI-SITE MOLECULES

A number of research groups have considered the extension of the domain decomposition algorithm to multi-site molecules [18, 21, 22]. Such systems present no great problems for replicated data methods (see section 2.7), but the presence of three-site and four-site potentials cause major problems for the domain decomposition algorithm. Most of these problems can be overcome (see below), but they greatly add to the complexity of the algorithm.

4.6.1. *Multi-site potentials*

The major problem associated with multi-particle potentials is that the particles involved can be distributed across different cells and across different regions (and therefore across different processors). In practice this problem is helped by the fact that in all molecular systems two atoms that are involved in a bond, or an angle, or a dihedral angle, will never be separated by a distance greater than the nonbonded cutoff. This means that the normal practice of only communicating information from cells on the boundary of a region should be quite sufficient to gather the necessary data for the calculation of bond, bond angle and dihedral interactions. However, the problem remains of making sure that all multi-particle interactions are calculated by exactly one processor. One solution to this problem has been provided by Esselink and Hilbers by their *efficient tuple scan* (ETS) method [18] and an alternative *triplet and quadruplet search* method has been devised by Brown and co-workers [21]. Finally, a rather different approach involving a *bonded data list* (BDL) has recently been described by Jabbarzadeth et al. [22] and used to simulate systems n-hexadecane.

4.6.2. *Parallel SHAKE*

The SHAKE method is intrinsically difficult to implement within a domain decomposition strategy. The SHAKE constraints must be applied to each molecule in the system and large molecules may span several processor regions. In principle this can lead to large communication costs for each iteration in the SHAKE procedure. Brown et al. have considered two approaches to a parallel SHAKE algorithm and implemented these in the simulation of polymer chains [21]. However, these approaches are only moderately successful. On current computers inter-processor communication times are still sufficently long as to significantly slow down parallel approaches to SHAKE constraints.

4.6.3. *Handling long-range forces*

By its very nature, the domain decomposition algorithm is not designed to handle long range forces. However, the decomposition of coordinates/orientations between individual nodes opens up possibilities for fitting together a combination of the domain decomposition method and the systolic loop method. The latter would be used to evaluate the long-range forces, whilst

412

the domain decomposition algorithm would be used solely for the short-ranged forces. This would work most efficiently within a multi-time step algorithm where the long-range forces were calculated less frequently than the short-range ones.

5. Other parallel approaches to molecular dynamics

Is is perfectly possible to design MD algorithms that are hybrids of the three main methods; replicated data, systolic loop and domain decomposition. One recent novel scheme (the Comfort program) has been designed by Tildesley and co-workers and is shown in outline in the flow diagram of figure 15. This algorithm works well for the situation where the majority of sites in the system are involved in short-range repulsion-dispersion interactions and a smaller number of sites are involved in long-range electrostatic interactions.

The coordinates for figure 15 are distributed via molecules across the various nodes in the system. This means that each node can compute the intramolecular interactions for its own molecules without need for communication of coordinates. Additionally, when the equations of motion are integrated, (in the final step of the algorithm), and SHAKE constraints are applied, there is also no need for communications between processors.

The calculation of the Ewald sum for charged sites and the repulsion-dispersion interaction for uncharged sites are carried out in separate procedures using, in both instances, a replicated data type of strategy. This necessitates a global communication of the coordinates prior to both calculations, so that each node has access to all relevant data. In figure 15 it can be seen that the algorithm removes the cost of these expensive communications by starting them prior to the computation of the intramolecular interactions. This means that the smaller number of charged coordinates have completed their communication step before the Ewald sum starts, and the larger number of uncharged coordinates have completed their communication step before the repulsion-dispersion calculation starts. The one communications bottle-neck involves the gathering of forces to each home processor, which must occur immediately prior to the force calculation. However, the overlapping of coordinate communications and computations does succeed in speeding up the algorithm significantly.

6. Conclusions

This article has provided a snapshot of the current progress in the use of parallel computational techniques in molecular simulation, with particular emphasis on parallel methods for molecular dynamics.

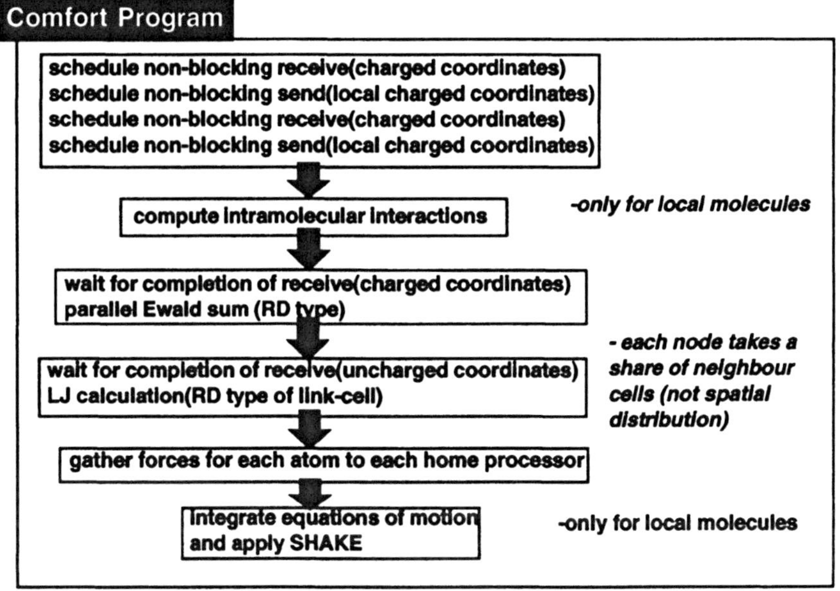

Figure 15. Flow chart illustrating the sequence of steps in the Comfort program.

The replicated data method described in section 2 is the easiest form of parallelisation to use and works quite well in situations where the amount of inter-processor communication is small relative to the computation done by each processor. However, for short range potentials, the global communication sums that are required in this algorithm can limit parallel efficiency. The algorithm requires more memory than either systolic loop or domain decomposition methods.

At the current time, domain decomposition techniques provide the best parallel performances for single site isotropic and anisotropic potentials. However, the extension of these methods to the simulation of molecular systems, composed of flexible molecules, is both complicated and difficult. Unlike the situation for replicated data simulation, a good *off-the-shelf* domain decomposition program, that can tackle generalised molecular systems, does not currently exist.

It is clear that there is still some work to be done in producing optimised parallel code for the simulation of molecular systems. Such code could use some of the latest developments in multi-timestep algorithms. These allow fast motion such as bond stretching and bond bending to be separated from nonbonded interactions, thus removing the need for SHAKE constraints which can slow down parallel applications. Additionally, multi-timestep algorithms allow short-range repulsion-dispersion interactions to be separated from more slowly varying long range forces (such as electrostatics). This would allow the short range forces to be handled by a domain decomposition strategy and the long-range forces to be treated by another mechanism (e.g. the systolic loop method).

Finally, with the continued development of cheap and fast mass produced processors, and concurrent developments in faster inter-processor communications, it seems likely that parallel computing will play a key role in the future of molecular simulation.

References

1. Fountain, T.J. (1994) *Parallel Computing Principles and Practice*. Chapter 2. Cambridge University Press, Cambridge.
2. Hwang, K. and Briggs, F.A. (1985) *Computer Architecture and Parallel Processing*. McGraw-Hill Book Company, New York.
3. Fountain, T.J. (1994) *Parallel Computing principles and practice*. Chapter 7F. Cambridge University Press, Cambridge.
4. Excellent on-line references for PVM can be found at http://www.netlib.org/pvm3/ and http://www.epm.ornl.gov/pvm/. (PVM source code is also available free from the latter.).
5. Geist, A., Beguelin, A., Dongarra, J., Jiang, W., Manchek, R. and Sunderam, V. *PVM, Parallel Virtual Machine a Users' Guide and Tutorial for Networked Parallel Computing*. This on-line book for PVM can be found at http://www.netlib.org/pvm3/book/.
6. Free implementations of MPI and excellent on-line MPI references can be found at http://www.erc.msstate.edu/mpi/ or by following links from this page.
7. Geist, G.A., Kohl, J.A. and Papadopoulos, P.M. *PVM and MPI, a Comparison of Features*. This on-line article discusses the differences between PVM and MPI, and can be found at http:/www.epm.ornl.gov/pvm/.
8. Smith, W. (1991) *Comp. Phys. Comm.*, **62**, 229.
9. Smith, W. (1992) *Comp. Phys. Comm.*, **67**, 392.
10. Wilson, M.R., Allen, M.P., Warren, M.A., Sauron, A. and Smith, W. (1997) *J. Comput. Chem.*, **18**, 478.
11. Allen, M.P. and Tildesley, D.J. (1987) *Computer Simulation of Liquids*. Chapter 3. Oxford University Press, Oxford.
12. Brode, S. and Ahlrichs, R. (1986) *Comp. Phys. Comm.*, **42**, 51.
13. Wilson, M.R. (1997) *J. Chem. Phys.*, **107**, 8654.
14. Wilson, M.R. (1996) GBMOL, *A replicated data molecular dynamics program to simulate combinations of Gay-Berne and Lennard-Jones sites*. University of Durham.
15. DL_POLY is a package of molecular simulation routines written by Smith, W. and Forester, T. R. copyright The Council for the Central Laboratory of the Research Councils, Daresbury Laboratory at Daresbury, Nr. Warrington (1996). The replicated data form of SHAKE is called RD-SHAKE and is described in section 2.6.8 of the DL_POLY_2.0 reference manual.
16. Raine, A.R.C., Fincham, D. and Smith, W. (1989) *Comp. Phys. Comm.*, **55**, 13.
17. Esselink, K., Smit, B. and Hilbers, P.A.J. (1993) *J. Comput. Phys.*, **106**, 101.
18. Esselink, K. and Hilbers, P.A.J. (1993) *J. Comput. Phys.*, **106**, 108.
19. Rapaport, D.C. (1988) *Comp. Phys. Rep.*, **9**, 1.
20. Brown, D., Clarke, J.H.R., Okuda, M. and Yamazaki, T. (1993) *Comp. Phys. Comm.*, **74**, 67.

21. Brown, D., Clarke, J.H.R., Okuda, M. and Yamazaki, T. (1994) *Comp. Phys. Comm.*, **83**, 1.
22. Jabbarzadeth, A., Atkinson, J.D. and Tanner, R.I. (1997) *Comp. Phys. Comm.*, **107**, 123.
23. Hilbers, P. and Esselink, K. (1992) Parallel Computing and Molecular Dynamics Simulations, in *Computer Simulations in Chemical Physics*. Allen, M.P. and Tildesley, D.J. (eds.), Kluwer, The Netherlands, p. 473.
24. Allen, M.P. and Tildesley, D.J. (1987) *Computer Simulation of Liquids*. Chapter 5. Oxford University Press, Oxford.
25. Allen, M.P., Warren, M.A. and Wilson, M.R. (1998) *Phys. Rev. E*, **57**, 5585.

INDEX

9 780792 360988